Helpful Geometric Formulas for Solving Applied Problems

1. Square of side s:	area $= s^2$ perimeter $= 4s$
2. Rectangle of length l and width w:	area $= lw$ perimeter $= 2l + 2w$
3. Circle of radius r:	area $= \pi r^2$ circumference $= 2\pi r$
4. Triangle of base a and height h:	area $= \dfrac{1}{2}ah$
5. Cube of side s:	volume $= s^3$ surface area $= 6s^2$
6. Rectangle of length l, width w, and height h:	volume $= lwh$ surface area $= 2lw + 2lh + 2wh$
7. Right circular cylinder of radius r and height h:	volume $= \pi r^2 h$ lateral surface area $= 2\pi rh$ total surface area $= 2\pi rh + 2\pi r^2$
8. Sphere of radius r:	volume $= \dfrac{4}{3}\pi r^3$ surface area $= 4\pi r^2$

College Algebra
and Its Applications

College Algebra and Its Applications

Larry Joel Goldstein

Homewood, IL 60430
Boston, MA 02116

On the cover...

Chambered Nautilus by Stewart Dickson ©1992 All Rights Reserved

Created using the Wolfram Research, Inc. *Mathematica*® system for doing
Mathematics by computer. The *Mathematica*® routine MakePolygons
simply creates a polygon mesh on a rectangular list of three-dimensional
coordinates, in the same manner as ParametricPlot3D:

The shell is a three-dimensional parametric surface:

$$X = (-1+e^{\frac{u}{6\pi}})(1+\cos v)\cos\left(u - \frac{\sin\left(2v + \frac{\pi}{2}\right)}{2}\right); \quad Y = (-1+e^{\frac{u}{6\pi}})(1+\cos v)\sin\left(u - \frac{\sin\left(2v + \frac{\pi}{2}\right)}{2}\right);$$

$$Z = (-1 + e^{\frac{u}{6\pi}})\sin v;$$

shell = MakePolygons$\left[\text{Table}\left[\{X, Y, Z\}, \{u, 0, 6\ \text{Pi}, \text{Pi}/12\}, \{v, -\text{Pi}, 0, \text{Pi}/12\}\right]\right]$;

The chambers are modified cross-sections derived from the shell equations;

$$x = u(-1 + e^{\frac{w}{6\pi}})(1 + \cos v)\cos\left(w - \frac{\sin\left(2v + \frac{\pi}{2}\right)}{2} - (1 - u)\right);$$

$$y = u(-1 + e^{\frac{w}{6\pi}})(1 + \cos v)\sin\left(w - \frac{\sin\left(2v + \frac{\pi}{2}\right)}{2} - (1 - u)\right);$$

$$z = u(-1 + e^{\frac{w}{6\pi}})\sin v;$$

chambers = Flatten$\Big[\text{Table}\big[\text{MakePolygons}[\text{Table}[\text{N}[\{x, y, z\}], \{u, 1/12, 1, 1/12\},$
$\{v, -\text{Pi}, 0, \text{Pi}/12\}]], \{w, 22\ \text{Pi}/6, 17\ \text{Pi}/3, \text{Pi}/6\}\big], 1\Big]$;

© RICHARD D. IRWIN, INC., 1993

Sponsoring editor: *Tom Tucker*
Developmental editor: *Jim Minatel*
Marketing manager: *Robb Linsky*
Project editor: *Jean Lou Hess*
Production manager: *Bob Lange*
Designer: *Heidi J. Baughman*
Art manager: *Kim Meriwether*
Photo research coordinator: *Patricia A. Seefelt*
Compositor: *Publication Services, Inc.*
Typeface: *10/12 Times Roman*
Printer: *Von Hoffmann Press*

Library of Congress Cataloging-in-Publication Data

Goldstein, Larry Joel.
   College algebra and its applications/ Larry Joel Goldstein.
      p.   cm.
    ISBN 0-256-10267-8   ISBN 0-256-12930-4 (Instructor's Edition)
    1. Algebra.   I. Title.
   QA154.2.G644   1993
   512.9—dc20   92–25404

Printed in the United States of America
1 2 3 4 5 6 7 8 9 0 VH 9 8 7 6 5 4 3 2

For Melissa,
Whose talent and courage are an inspiration to us all.

THE GOLDSTEIN SERIES

Algebra and Trigonometry and Their Applications
College Algebra and Its Applications
Trigonometry and Its Applications

A Student Solutions Manual written by Larry Joel Goldstein is available for use with this text. It contains detailed solutions to selected odd-numbered exercises from the text. It can be purchased at your college bookstore.

About the Author

Larry Joel Goldstein is a well-known mathematician, computer scientist, and educator. He received his B.A. (summa cum laude with distinction in mathematics) from the University of Pennsylvania in 1965 and his M.A. and Ph.D. from Princeton University in 1967. He was the Josiah Willard Gibbs Professor of Mathematics at Yale University and later an associate professor of mathematics at the University of Maryland. In 1972 he was promoted to full professor, the youngest full professor ever appointed at Maryland.

Dr. Goldstein's research earned him the Allan C. Davis Medal of the Maryland Academy of Sciences as 1975's Outstanding Maryland Scientist of the Year. He has earned many research grants and directed a number of Ph.D. students. Also known for his work in mathematics education, Dr. Goldstein's college-level textbooks have set national trends. His prominence in education and research was recognized in 1977 when he was appointed Distinguished Teacher/Scholar by the University of Maryland.

Dr. Goldstein currently resides in Maryland and teaches developmental math and other courses at both the community college and university level. He continues to spend the majority of his time writing in the areas of mathematics and computer systems and science.

IRWIN'S ACCURACY ASSURANCE PROGRAM

Linda Miller

Paula Norris

Jane Kulesza

Mitchel Levi

Larry Stephens

Phyllis Hammer

At the very beginning of this textbook project, the publisher and I decided that the most important thing we could do was ensure that these texts are accurate. As a longtime classroom teacher and user of textbooks I am well aware of the frustrations caused by errors in text or in the answer section. Therefore, throughout the writing and production of this textbook we have single-mindedly pursued a process designed to eliminate errors. While one cannot guarantee that a textbook will be completely error free, I would like to describe the steps we have taken to make this text as clean as possible.

Exhaustive proofreading took place along two fronts—within the chapters themselves and the answer sections. This check for accuracy was maintained all through the textbook process, from final manuscript—to the typesetting stage—and to final pages.

Linda Miller and Paula Norris at Delta State University read the final manuscript checking the narrative, examples, and graphs. Jane Kulesza solved all the exercises and compared the solutions to the answers appendix. Corrected manuscript and answers were entered into my computer disk.

Typesetting was done from my computer disks, minimizing the chance of errors being introduced at this stage. The typeset manuscript was again exhaustively checked; the writing, examples, and figures by myself and Jane Kulesza. The answer appendix was checked by Mitchel Levi of Broward Community College.

This same process was repeated once final pages were produced. Larry Stephens and John Konbalina of the University of Nebraska at Omaha checked the exercise answers and Phyllis Hammer of Sampson Community College read the chapters. As a result of this process, every word, example, figure, graph, and answer has been independently checked by different people at all three crucial stages of the textbook process: final manuscript, after typesetting, and after final pages. As the author I feel completely confident that this textbook is as error free as possible, and that it will serve you well in this regard.

THE ACCURACY
ASSURANCE PROCESS

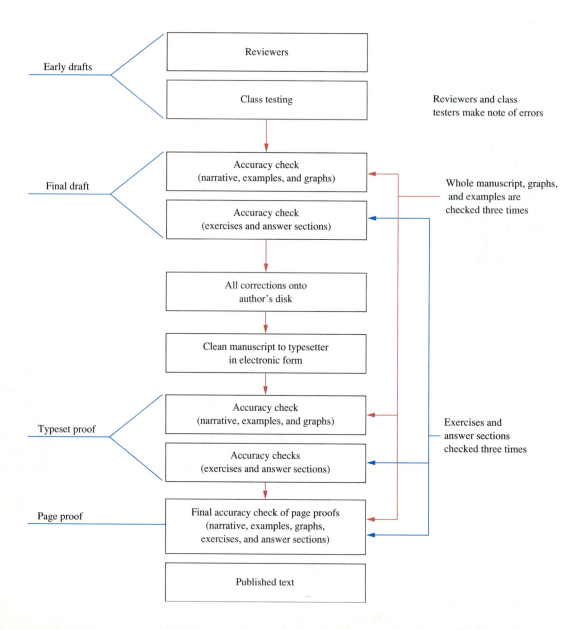

Early drafts

Reviewers

Class testing

Reviewers and class
testers make note of errors

Final draft

Accuracy check
(narrative, examples, and graphs)

Accuracy check
(exercises and answer sections)

Whole manuscript, graphs,
and examples are
checked three times

All corrections onto
author's disk

Clean manuscript to typesetter
in electronic form

Typeset proof

Accuracy check
(narrative, examples, and graphs)

Accuracy checks
(exercises and answer sections)

Exercises and
answer sections
checked three times

Page proof

Final accuracy check of page proofs
(narrative, examples, graphs,
exercises, and answer sections)

Published text

PREFACE FOR THE STUDENT

The *College Algebra* course plays a very significant role in the undergraduate mathematics curriculum. On the one hand, it provides preparation and background for calculus and other higher-level mathematical courses. And on the other, it serves as a quantitative literacy requirement in many core curricula.

In recent years, the course has confronted many problems and changes, some unique to the subject itself and others reflecting the general turmoil and uncertainty in the undergraduate mathematics curriculum. The students, always a heterogeneous group in this course, today exhibit a wider range of mathematical preparation and maturity. This brings into question the proper level for the course and the content which should be included. New technology has been developed and cries out to be integrated into the traditional curriculum. New teaching methodology is being discussed and applied. New supplements are required to enhance student learning and to ease the burdens of a teaching staff which has more students and fewer resources to handle the load.

It is with these problems and changes in the background that I have written the present text. The text presents a modern approach to a traditional curriculum. It is expected that students will have two years of high school algebra, although students with slightly more or less preparation will also be well served by this text.

FEATURES OF THE TEXT

Mathematical Precision   Mathematics is a precise language. Like any other educational product involving language, a text must be precise to be of value. While I have omitted much of the forbidding language and formalism of many earlier books, I recognize the importance of mathematical precision for you. I have incorporated a high degree of mathematical precision into the discussion, particularly in definitions, theorems, and key results. This will be of the greatest advantage to those who continue on to other mathematics courses, but it will help you see the value of mathematically precise thinking.

Motivation   Throughout the text, I have tried in every way to motivate the material. Formal definitions and proofs are introduced only after they have been intuitively explained or motivated by examples, many of them applied. In fact, applications are used extensively in the examples and problem sets so you can appreciate how mathematics occurs in the world around you. In addition, I have added a few excursions into mathematical history.

I have attempted to avoid the "It can be shown ..." trap and attempt to provide motivation or show justification wherever possible, given the level of mathematics being presented. I believe it is important that the underlying concept be explained. For an example of this style see the discussion of scientific notation in section 1.2 or the introduction to functions in section 3.4.

Emphasis on Graphical Analysis

In applying algebra to other disciplines and to higher mathematics, visualization and graphical analysis is one of the most important skills required. Throughout the book I emphasize understanding graphs and applying them to solve problems. You will be given much practice in recognizing key graphs and understanding their relations to the functions they represent. The sections on transforming the graphs of basic functions (3.6) and the geometric interpretation of the slope of a straight line (3.3) are good examples of this emphasis. There are over 660 figures and graphs in the text, many of them in the exercises. It is my belief that this geometric approach greatly helps you to understand the algebraic theory and to solve problems.

Problem Solving

Developing an intuition for problem solving is one of the most elusive skills that students need to gain from an algebra course. One of the goals of this text is to aid you in developing this skill. To help with this, I have included strategies for solving many types of problems and included illustrations. For example, see section 2.2 on solving word problems. I have also included cautions about common errors. These should serve as your guides as you encounter new types of problems in other courses and in the world outside the classroom.

Examples

Worked examples form the body of any algebra text. This one is no exception. I have included almost 460 examples. The examples illustrate algebraic techniques, problem-solving strategies, and applications. I have included verbal descriptions of the processes involved so that the examples "talk" to you like a live instructor might. Care has been taken to make the solutions thorough and to include all necessary steps so that you are not left guessing.

Exercise Sets

The exercise sets are probably the most important part of any mathematics text. They provide the practice needed to master the concepts and skills. Therefore, I have gone to great effort to provide a tremendous number and broad range of exercise sets.

There are over 4200 exercises in this book. They begin with drill problems that allow you to practice technical "skills" and build into more involved exercises and applications that allow you to develop your problem solving and critical thinking. In addition, there are optional "Technology" exercises that allow you to use graphing calculators or computers to explore concepts and solve problems. Lastly, "In Your Own Words" problems require you to develop your problem-solving skills by explaining mathematical ideas and procedures in writing.

Applications

Mathematics does not exist in a vacuum. Mathematical ideas are often present in physical problems and mathematics can be used to model real-world phenomena. In this text, I present over 500 applications, many using real data, to show how algebra is used in the real world. These applications emphasize the connection of algebra to many fields, such as astronomy, management, and zoology.

Technology   In addition to the technology exercises described above, the text contains many optional discussions of technology use in algebra. Included are discussions of graphing calculators, scientific calculators, spreadsheets, graphing software, and symbolic manipulation packages that are first introduced in section 1.2. These optional subsections show how these tools can be used to perform many tasks discussed in the book and to explore concepts beyond traditional pencil and paper methods. These optional technology sections help further emphasize the role of graphical analysis as most are related to graphing. You can find these optional subsections highlighted in red in the table of contents. Special thanks to Howard Wilson of Oregon State University for spending a day with me and the editors providing suggestions on how to incorporate the use of technology into the text.

Chapter Review   Each chapter ends with a review section. This section summarizes key results and formulas covered in the chapter in the form of a table, including references to the first introduction of the topic in the chapter. These will be very useful in preparing for exams. The chapter reviews conclude with an extensive set of review exercises to reinforce the skills and concepts from the chapter.

Mathematical History   I have indulged my interest in mathematical history by including inserts throughout the text that describe how algebra came to be and some of the interesting personalities who developed the subject. I hope that this will stir your interest in mathematical history as well.

Design   The look and visual appeal of a text impacts on how you will learn from it. A well laid-out page will hold interest longer and be easier to draw important information from. I have worked closely with the designers and editors to ensure a quality design that incorporates the functional use of color. Multiple colors highlight examples, definitions, and key results. Multiple colors are also especially useful to clarify graphs and figures. Color photographs enhance the visual appeal of the text.

Supplements

A modern college textbook requires an extensive set of supplements. I was fortunate to be closely involved in writing and overseeing many of these supplements.

Student Solutions Manual   This contains detailed solutions to one-third of the text exercises. All of the solutions are from the odd-numbered exercises.

Graphing Utility Guide   This contains numerous short laboratory-type graphing calculator explorations and exercises. Many topics in algebra are covered. Keystroke instructions are provided in separate appendices for the TI, HP, Casio, and Sharp calculators. Keystrokes are kept to a minimum in the utility guide, so this manual can be used with any graphing utility. The pages are perforated, so you can turn in your assignments.

Tutorial Software   This software is used to provide computer-based interactive learning and testing in your school's computer lab.

The Plotter   This software is for graphing and analyzing functions. A manual is included that describes operations and includes student exercises. The software is menu driven and has an easy-to-use window-type interface. The high-quality graphics can also be used for classroom presentation and demonstration. If you go on to calculus classes you will want to keep the software for future use as powerful calculus functions are included.

ACKNOWLEDGMENTS

Reviewers   IRWIN reviewed four drafts of the manuscript with over 40 different professors and instructors. Their suggestions and comments helped shape the text as well as point out errors or inconsistencies. I would like to thank the following individuals who reviewed all or part of the manuscript.

Norma Agras
Miami Dade Community College

Lee Armstrong
University of Central Florida

Judy Barclay
Cuesta College

Ray Boersema
Front Range Community College

Elaine Brooks
Hudson Valley Community College

Dick Clark
Portland Community College

Ken Dodaro
Florida State University

Betty Fein
Oregon State University

August Garver
University of Missouri, Rolla

John Gosselin
University of Georgia

Donald Gray
Iowa Western Community College

William Grimes
Central Missouri State University

Shirley Huffman
Southwest Missouri State University

Paula Kemp
Southwest Missouri State University

Vuryl Klassen
California State University, Fullerton

Charles Klein
De Anza College

John Krajewski
University of Wisconsin, Eau Claire

Giles Maloof
Boise State University

James Darwin Mann
Morehead State University

Jill McClain
Eastern Michigan University

Marilyn McCollum
North Carolina State University

Barbara McLachlan
San Diego Mesa College

Jack Motta
Sante Fe Community College

Jeannette O'Rourke
Middlesex County College

James Peake
Iowa State University

Elise Price
Tarrant County Junior College

Donald Redmond
University of Southern Illinois, Carbondale

Gina Reed
Gainesville College

Richard Rockwell
Pacific Union College

Stephen Rodi
Austin Community College

Thomas Roe
South Dakota State University

Jean E. Rubin
Purdue University

Dan Scanlon
Orange Coast College

Samuel Self
El Paso Community College

Frank A. Smith
Kent State University

George Szoke
University of Akron

Marvel Townsend
University of Florida

Arnold Villone
San Diego State University

Beverly Weatherwax
Southwest Missouri State University

Howard Wilson
Oregon State University

Gordon Woodward
University of Nebraska

Jan Wynn
Brigham Young University

Jean Sanders

Class Testing In addition to thorough reviewing, this manuscript was class tested in manuscript form by Linda Miller at Delta State University and Jean Sanders of the University of Wisconsin at Platteville. Many thanks for their detailed lists of corrections and their valuable suggestions.

Focus Group I had the unique opportunity to participate in a focus group at a recent MAA/AMS meeting. A group of nine professors shared their thoughts on the course and their teaching. Their insight shaped and guided much of the development of the text. I wish to thank the participants.

C. P. Barton
Stephen F. Austin State University

R. G. Dean
Stephen F. Austin State University

William Durand
Henderson State University

Joseph Fiedler
California State University, Bakersfield

John Gregory
Southern Illinois University

Ed Huffman
Southwest Missouri State University

Shirley Huffman
Southwest Missouri State University

Thomas Kelley
Metro State College of Denver

Anita McDonald
University of Arizona

If I haven't left anyone off the lists, 54 individual professors and instructors made significant contributions to the text. Thank you everyone for your help and unfailing conscientiousness.

Larry Joel Goldstein
Brookeville, Maryland

CONTENTS

INDEX OF APPLICATIONS

PHYSICAL SCIENCES

Astronomical distances, 53
Atomic mass, 53
Atomic number, 331
Boiling point of aluminum, 53
Boiling point of water, 87
Boyle's law, 215
Bullet cap explosion time, 26
Chemical mixture, 74, 123
Communications satellite orbit, 455
Distance as a function of time, 272
Distance measurement, 112, 128, 130–31
Distance-time-rate calculation, 68, 73, 107, 123, 124, 156–57, 173, 195, 216
Earth's orbit, 454–55
Earthquake magnitude, 312
Electrical attraction/repulsion, 210–11
Electrical resistance, 48, 98, 206–7, 227, 295
Electronic particle wavelength, 210
Electron volts, 26
Escape velocity, 26
Expansion of a material, 48
Gas pressure, 210
Gas temperature, 209
Gas volume/pressure, 163
Gravitational attraction, 22–23, 25, 207, 210
Hooke's law, 205
Illumination, 209
Kelvin temperature, 216
Kepler's law, 210
Law of motion, 424, 440
Light filtration, 288
Magnitude of a star, 314
Moon orbit, 26
Motion of a particle, 464
Newton's law of cooling, 315
Newton's law of gravitation, 207, 210
Optics, 272
Parabola reflection property, 448
Pendulum swing problem, 123
Period of a pendulum, 209
pH balance, 312
Radiation energy, 209
Radioactive decay, 284–85, 287–88, 303–4, 305
Radioisotope power supply, 320
Relative motion, 94–95, 97, 98
Skydiving terminal velocity, 292–93, 294
Sound loudness, 215, 272, 307–8, 312
Speed of falling objects, 26, 166, 173, 227, 314–15
Speed of light, 25, 26
Speed of moving objects, 85, 86–87, 108, 124, 167, 172, 215, 223–24
Speed of sound, 24, 27, 98, 123, 462
Stopping distance, 209
Temperature conversion, 74, 123, 215, 331
Volume maximization, 233
Volume of a box, 260
Weight above the earth, 272
Wind speed, 335, 346

ENGINEERING

Architectural design, 83, 172, 224–25, 227, 255–56
Automobile glass thickness, 288
Bullet cap explosion time, 26
Communications engineering, 428, 430
Construction costs, 157
Deflection of a beam, 233, 260
Distance calculation, 12–13
Electrical resistance, 48
Expansion of materials, 48
Heating costs, 357
Oil storage tank size, 51
Power line cost, 173
Road width calculation, 73
Safe load calculation, 215
Semiconductor voltage, 184
Suspension bridge cables, 448
Swimming pool size, 123
Technology, 23–24
Telescope mirror, 448
Ventilating system size, 97–98
Water cistern size, 330
Water seepage, 53
Windmill power, 210

GENERAL INTEREST

Accident rate, 163
Air travel time, 332
Ancestors, 424
Assignment of draftees, 437
Baseball diamond size, 26, 195
Baseball player combinations, 430
Basketball floor size, 353
Batted ball force, 215
Batting averages, 48, 53, 400
Beam length safety, 210
Book arrangements, 432
Box volume, 36
Bricklaying time, 335
Bridge hand combinations, 431
Building blocks, 439
Carnival game, 437
Club membership, 431
Coin flipping, 435–36
Coin combinations, 67
Committee membership, 431, 432, 440, 441
Computer language conversion, 49, 53
Concert selections, 431
Dice roll, 53, 434, 435, 437
Door prize winner, 437
Earned run average, 107
Elevator capacity, 353
Game of Tong, 441
Gardening, 36
Government spending, 424
Grade point averages, 37, 48
Guest arrival order, 437

CHAPTER 1

REVIEW OF ELEMENTARY ALGEBRA

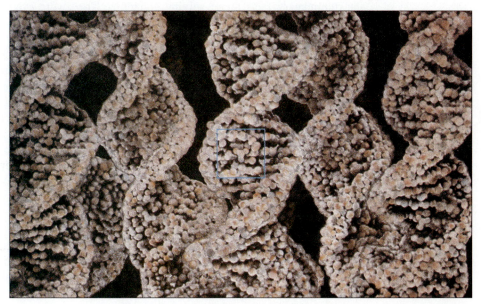

A close-up of the genetic material DNA. The molecules shown here are 10^{-8} meters in diameter.

Algebraic ideas were conceived in ancient times by the Greek geometers. A quite sophisticated use of symbols to represent numbers was evident in the works of Euclid and Pythagoras several hundred years before Christ. They interpreted numbers geometrically, as lengths, areas, and volumes. For this reason, they did not develop the notion of the number 0 or the notion of a negative number. These concepts were developed by the Hindus in the early centuries after Christ.

The arithmetic of the Hindus was refined into symbolic form about 700 A.D. by the Arabs, who gave birth to the subject we now call **algebra**. The name *algebra* was derived from the title of a text written by the Arab mathematician al-Khwarizmi (780–850 A.D.).

The Arabs knew much of what we now call elementary algebra, including the rudimentary facts about solving equations. They were even able to solve quadratic equations using the familiar quadratic formula, which we will discuss later in the book.

During the thousand years since the Arabs created algebra, the subject has grown to encompass a huge body of knowledge. Included within algebra are methods for solving many types of equations, including those of the third, fourth, and higher degrees. The subject includes methods for manipulating algebraic expressions that arise in applied problems in fields ranging from biology to astrophysics.

Algebra has evolved into a major field that is a prerequisite to learning and applying many branches of mathematics, including calculus, statistics, and numerical analysis, to mention but a few. The branches of mathematics that rely on algebra are, in turn, the foundation for solving problems in almost all fields in which quantitative problems arise: business, biology, chemistry, physics, law, sociology, psychology, medicine, and computer science.

In this book, we will provide a thorough look at algebra as a prerequisite to its application in other mathematics courses and in other fields. In order to enhance your perspective on the subject, we will attempt to keep both the history and the applications of the subject in view.

1.1 THE REAL NUMBER SYSTEM

The real number system is the foundation upon which elementary algebra rests. In this section, we summarize the most elementary facts about the real numbers as a basis for our subsequent development of algebra.

Real Numbers

Arithmetic and geometric problems involve numbers such as the following:

$$5, \ -18, \ 1.782, \ \frac{1}{3}, \ \frac{7}{5}, \ \pi, \ \sqrt{2}, \ 97\%$$

These numbers are examples of **real numbers**. Such numbers are not new to you. You have been using them since your earliest days of grade school arithmetic. Indeed, the elementary properties of real numbers are usually covered in courses that precede this one. So rather than discuss them in detail, we provide in this section only a brief review, excluding proofs and detailed logical development.

A real number can be written using decimal notation. A real number whose decimal expression requires only finitely many digits is called a **terminating decimal**. Here are some examples of terminating decimals:

$$-58.354, \ 12, \ 0.0000031$$

A decimal expression that contains infinitely many nonzero digits is called a **nonterminating decimal**. Here are some examples of nonterminating decimals:

$$\frac{1}{3} = 0.3333333\ldots, \qquad \sqrt{2} = 1.4142135\ldots$$

The three dots ($\ldots$) indicate that the decimal expression continues indefinitely. A decimal expression that has a repeating pattern, such as $12.247247247\ldots$, is called a **repeating decimal**. A decimal expression such as that of $\sqrt{2}$ given above, which has no repeating pattern, is called a **nonrepeating decimal**.

We will typically denote real numbers by lower-case letters, such as $a, b, c, \ldots, w, x, y, z$.

Sets

A **set** is a collection of objects. The collection of all real numbers is an example of a set. Sets are customarily denoted using capital letters, such as A, B, C. We will reserve the notation **R** to mean the set of real numbers.

An object in a set is called an **element** of the set and is said to **belong** to the set. Thus, $1/3$ and -3.75 are elements of the set **R**. If every element of the set A is also an element of the set B, then we say that A is a **subset** of B. For example, let A denote the set consisting of the numbers 0 and 1, and B the set consisting of all real numbers between 0 and 1 inclusive. Then A is a subset of B. We write this symbolically as $A \subseteq B$.

If A and B have the same elements, then we say that A and B are **equal** and write $A = B$. If A and B are not equal (that is, they do not consist of exactly the same elements), then we write

$$A \neq B$$

Throughout this book, unless we explicitly say otherwise, all sets will be subsets of the real numbers.

One way of describing a set is to list its elements within braces. For instance, the set A consisting of the integers 1, 2, 3, 4, 5 can be written as

$$A = \{1, 2, 3, 4, 5\}$$

Another method for describing sets is to give one or more conditions that specify the elements. This method is called **set-builder notation**. Using this notation, the set of nonzero real numbers x may be written as

$$\{x \mid x \neq 0\} \qquad \text{or} \qquad \{x : x \neq 0\}$$

This notation is read: "The set of all x such that x is not equal to 0."

Subsets of the Real Numbers

Let's now explore in some detail two important subsets of the real numbers, the **integers** and the **rational numbers**. We shall meet these sets throughout the book.

Integers

The set of **integers** is the subset of real numbers consisting of the counting numbers, their negatives, and 0. Every integer is indeed a real number since it is a decimal expression (which terminates just to the left of the implied decimal point). Here are some examples of integers:

$$3, \ 3{,}254, \ 0, \ -17$$

The set of all integers is denoted by **I**. Since every integer is also a real number, the set **I** of integers is a subset of the set **R** of real numbers.

The integers $1, 2, 3, \ldots$ are called the **counting numbers**.

The arithmetic operations for real numbers—addition, subtraction, multiplication, and division—may be applied to integers. If a and b are integers, then $a + b$, $a - b$, and ab are also integers. Note, however, that the quotient of one integer divided by another may or may not be an integer. For instance, the quotient 4/2 yields the integer 2, whereas the quotient 6/4 yields the noninteger 1.5.

Suppose that a and b are integers, $b \neq 0$. We say that b is a **factor** of a (or that a **is divisible by** b) provided that $a = bc$ for some integer c. Equivalently, b is a factor of a provided that the quotient a/b is an integer. For example, 3 is a factor of 12 since $12 = 3 \cdot 4$. However, 5 is not a factor of -12 since $-12/5$ is not an integer.

The factors of a small integer are not difficult to determine using trial and error. The factors of 7 are $1, -1, 7,$ and -7. The factors of 12 are $1, -1, 2, -2, 3, -3, 4, -4, 6, -6, 12,$ and -12.

A **prime number** is an integer greater than 1 whose only positive factors are 1 and itself. For example, 2, 3, 5, 7, and 31 are all primes. An integer greater than 1 that has positive factors other than 1 and itself is called **composite**. For example, 12 is a composite since it has the factors 2 and 3 (among others).

FUNDAMENTAL THEOREM OF ARITHMETIC

An integer a greater than 1 can be written as a product of primes (some possibly repeated):

$$a = p_1 \cdot p_2 \cdot \cdots \cdot p_n$$

Moreover, this factorization can be accomplished in only one way (without regard to order).

Here are some factorizations of integers into products of primes:

$$24 = 2 \cdot 2 \cdot 2 \cdot 3$$
$$15 = 5 \cdot 3$$
$$7 = 7$$

We should make several important observations about these factorizations. Note that in the first example, the factorization

$$24 = 2 \cdot 2 \cdot 2 \cdot 3$$

is not considered different from

$$24 = 2 \cdot 3 \cdot 2 \cdot 2$$

since one is a rearrangement of the other. In the last example, 7 is a prime itself. In this case, the factorization consists of a single prime factor.

The process of factoring an integer into prime factors can be done by testing for factors, starting with 2, as shown in the following example.

➤ **EXAMPLE 1**
Factoring Integers

Factor 2244 into a product of primes.

Solution

We begin by testing for the factor 2. Division shows that $2244/2 = 1122$, so that we have the factorization

$$2244 = 2 \cdot 1122$$

We test again for the factor 2 by dividing 1122 by 2. The result is the integer 561, indicating another factor of 2. We now have the factorization

$$2244 = 2 \cdot 2 \cdot 561$$

Since 561 is odd, there are no other factors of 2. We search through the odd prime numbers for a factor, beginning with 3. We find that $561/3 = 187$ and $187/11 = 17$. Therefore, we have the factorization

$$2244 = 2 \cdot 2 \cdot 3 \cdot 11 \cdot 17$$

Since all the factors are prime, the factorization into a product of primes is complete.

Historical Note: Primes and Factoring

Euclid

Primes have fascinated amateur and professional mathematicians for thousands of years. Their irregular pattern of distribution suggests many questions, some of which have not yet been answered. For example, a prime pair is a pair of primes that differ by 2, such as 17 and 19. On examining a table of primes, it appears that prime pairs occur infinitely often. This guess was first stated by the mathematician Goldbach in 1742 and is currently unproven.

Euclid proved that there are infinitely many primes. Mathematicians have delighted in exhibiting large primes. Using computers, teams of researchers have uncovered huge primes, having tens of thousands of digits. Every year or so, a larger prime is discovered.

Factorization is simple to do, in principle. Just test all possible factors using trial and error. For large numbers, however, this method becomes unwieldy. Even using a large computer, factorization of numbers involving hundreds of digits can take tens of billons of years! This fact has been used to construct so-called unbreakable codes. The key to the code is the product of two primes. To decode a message requires knowledge of the primes. With large primes, factorization of their product is an essentially impossible problem, so that decoding is impossible. Such codes are called **public key cryptography systems** and were developed in the early 1970s by researchers at Stanford University and MIT.

Rational Numbers   A **rational number** is a real number that can be written as the quotient of two integers a/b, where b is nonzero. Here are some examples of rational numbers:

$$\frac{5}{3}, \quad -\frac{117}{89}, \quad -\frac{45}{7}, \quad \frac{5}{1} = 5$$

An integer a is a rational number, since it can be written as the quotient $a/1$, so the integers form a subset of the rational numbers.

A given rational number can be written in an infinite number of ways as a quotient of integers. Indeed, if a/b is one such representation, then ak/bk is another, where k is a nonzero integer. Thus, for example, here are a number of distinct quotients, all of which represent the real number 1/3:

$$\frac{1}{3} = \frac{2}{6} = \frac{12}{36} = \frac{-3}{-9} = \cdots$$

Let a/b be a rational number. Then a is called the **numerator** and b the **denominator**. If the numerator and denominator have a common nonzero factor k, then the numerator and denominator may be divided by k without altering the value of the rational number. The process of dividing numerator and denominator by a common factor is called **simplifying** or **reducing.** If the numerator and denominator of a rational number have no factors in common other than 1 or −1, then the rational number is said to be **in lowest terms**. For example, the rational number 15/45 has a numerator and denominator both divisible by 15. This rational number may be written in lowest terms as 1/3.

One way of simplifying a fraction into lowest terms is to factor its numerator and denominator into prime factors and divide both numerator and denominator by any common factors. This procedure is illustrated in the next example.

> ➤ **EXAMPLE 2**
> Reducing Rationals
> to Lowest Terms

Reduce 280/462 to lowest terms.

Solution

Factoring the numerator and denominator into products of primes, we have

$$\frac{280}{462} = \frac{2 \cdot 2 \cdot 2 \cdot 5 \cdot 7}{2 \cdot 3 \cdot 7 \cdot 11}$$

Dividing both numerator and denominator by $2 \cdot 7$, we may reduce the fraction to

$$\frac{2 \cdot 2 \cdot 5}{3 \cdot 11} = \frac{20}{33}$$

Note that some rational numbers, such as 2/5, have terminating decimal expressions (0.4), whereas others, such as 1/3, have decimal expressions that are nonterminating *and repeating* (0.33333 . . .). It can be proved that the decimal expression of *any* rational number either terminates or repeats.

The repeating pattern of a rational number may consist of any number of digits. For example, by performing division, we find that

$$\frac{1}{7} = 0.142857142857\overline{142857}$$

That is, the decimal expansion of 1/7 ends in an infinite sequence of 142857s. It is customary to express 1/7 in a decimal notation by writing a bar over the part that repeats, as in the above display.

Conversely, every decimal that ends in a repeating pattern represents a rational number. Thus, we can state the following criterion for detecting rational numbers.

REPEATING DECIMAL THEOREM

A real number is rational if and only if its decimal representation either terminates or is repeating.

Irrational Numbers   A real number that is not a rational number is said to be **irrational.** Examples of common irrational numbers are π and $\sqrt{2}$. An irrational number has a nonterminating, nonrepeating decimal expression. Although it is easy to recognize rational numbers, it is often very difficult to prove that a particular real number is irrational. (After all, how can you be sure just by examining the start of a decimal expression that it doesn't repeat after, say, 100,000 digits?) As a result of the efforts of many mathematicians, however, we know that particular numbers are irrational. For example, the fact that $\sqrt{2}$ is irrational is a famous theorem due to the Greek geometer Euclid, whereas the irrationality of π was first proved only in the 19th century by the German mathematician Lindemann. We will have more to say about irrational numbers, particularly square roots, later in the chapter.

The Various Subsets of the Real Numbers

Every integer is a rational number, and every rational number is a real number. In terms of sets, we can describe these facts by saying that the integers are a subset of the rational numbers and the rational numbers are a subset of the real numbers. The interrelationships between the integers, rational numbers, and real numbers is illustrated in Figure 1.

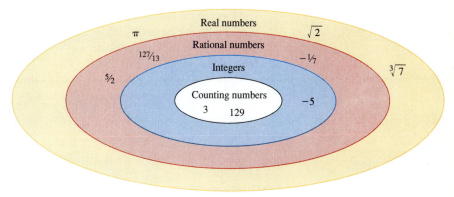

Figure 1

Arithmetic Operations on Real Numbers

The following seven properties form the basis for performing arithmetic among real numbers.

ADDITION AND MULTIPLICATION OF REAL NUMBERS

Let a, b, and c be any real numbers.

1. **Closure:** $a + b$ and $a \cdot b$ are real numbers.
2. **Commutative laws:** $a + b = b + a$ and $ab = ba$.
3. **Identities:** The real number 0 is called the **additive identity** since it has the property $a + 0 = 0 + a = a$. The number 1 is called the **multiplicative identity** since it has the property $1 \cdot a = a \cdot 1 = a$ for every real number a.
4. **Inverses:** A real number a has an **additive inverse**, denoted $-a$, with the property $a + (-a) = (-a) + a = 0$. A nonzero real number a has a **multiplicative inverse** (or **reciprocal**), denoted a^{-1}, with the property $a \cdot a^{-1} = a^{-1} \cdot a = 1$.
5. **Associative laws:** $a + (b + c) = (a + b) + c$ and $a(bc) = (ab)c$.
6. **Distributive laws:** $a(b + c) = ab + ac$ and $(a + b)c = ac + bc$.
7. **No divisors of zero:** We have $a \cdot 0 = 0$ for any real number a. If ab equals 0, then either a equals 0 or b equals 0.

Properties 1–7 of addition and multiplication provide the basis for deducing many of the facts that you routinely use in working with real numbers. For instance, the associative law for addition states that the values of $a + (b + c)$ and $(a + b) + c$ are the same. It is customary to denote their common value simply as $a + b + c$. In a similar fashion, we write $a + b + c + d$ to mean any of the sums gotten by grouping the terms, such as $a + [(b + c) + d]$. (Repeated application of the associative law shows that all such groupings lead to the same sum.)

The associative law for multiplication states that the products $a(bc)$ and $(ab)c$ are equal. Their common value can be denoted simply as abc. In a similar fashion, the associative law for multiplication says that the product of any number of real numbers is independent of the parentheses used and hence can be written without using any parentheses.

The operation of **subtraction** is defined by the relationship

$$a - b = a + (-b)$$

That is, the **difference** $a - b$ of a and b is obtained by adding $-b$ to a.

From its definition, it is clear that the operation of subtraction is closely connected with taking additive inverses. Here are some useful properties of subtraction that involve that connection. We will use these properties throughout the book, without further comment.

PROPERTIES OF NEGATIVES

$-a = (-1)a$	$0 - a = -a$
$-(-a) = a$	$(-a)(-b) = ab$
$-(a + b) = -a - b$	$(-a)b = -ab = a(-b)$
$a - 0 = a$	

Let a and b be real numbers with b nonzero. Then the **quotient** of a divided by b, denoted

$$a/b, \ \frac{a}{b}, \ \text{or} \ a \div b$$

is defined by

$$\frac{a}{b} = a \cdot \frac{1}{b} = a \cdot b^{-1}, \quad b \neq 0$$

Note that property 4 of addition and multiplication asserts that each nonzero real number a has a multiplicative inverse. However, **the number 0 does not have a multiplicative inverse**. Indeed, if b were a multiplicative inverse of 0, then

$$1 = 0 \cdot b \quad \textit{(definition of multiplicative inverse)}$$
$$1 = 0 \quad \textit{(b $\cdot$ 0 = 0, by Property 7)}$$

That is, 1 is equal to 0, which is nonsense. Thus, 0 does not have a multiplicative inverse.

Since the definition of the quotient a/b involves the multiplicative inverse of b, b must be nonzero. To put it another way, **division by 0 is undefined**.

The status of division by 0 is misunderstood by many beginning students. They think that division by 0 is undefined because mathematicians didn't want to define it. There is no way to define division by 0 so that it is consistent with the other properties of the real number system.

Quotients occur regularly in algebraic calculations, so it is important to be able to perform arithmetic operations using them. The principal arithmetic properties of quotients are stated in the following formulas.

PROPERTIES OF QUOTIENTS

1. $a^{-1} = \dfrac{1}{a}, \quad a \neq 0$

2. $\dfrac{a}{b} = a \cdot \dfrac{1}{b}, \quad b \neq 0$

3. $\dfrac{a}{b} + \dfrac{c}{d} = \dfrac{(ad + bc)}{(bd)}, \qquad \dfrac{a}{b} - \dfrac{c}{d} = \dfrac{(ad - bc)}{(bd)}, \quad b \neq 0, d \neq 0$

4. $\dfrac{a}{b} \cdot \dfrac{c}{d} = \dfrac{ac}{bd}, \quad b \neq 0, d \neq 0$

5. $\left(\dfrac{a}{b}\right)^{-1} = \dfrac{b}{a} = b \cdot \dfrac{1}{a}, \quad a \neq 0, b \neq 0$

6. $\dfrac{\dfrac{a}{b}}{\dfrac{c}{d}} = \dfrac{a}{b} \cdot \dfrac{d}{c} = \dfrac{ad}{bc}, \quad b \neq 0, c \neq 0, d \neq 0$

7. $-\dfrac{a}{b} = \dfrac{-a}{b} = \dfrac{a}{-b}, \quad b \neq 0$

The Real Number Line

It is possible to represent the real number system geometrically as the points of a line. Start out with a straight line and arbitrarily choose a point 0 on it. This point is called the **origin**. It divides the line into two half lines. Customarily, the half line to the right of 0 represents the positive direction. Next, choose a line segment whose length represents one unit and lay out unit distances starting with 0 and proceeding along both half lines. (See Figure 2.)

The line with the choices of origin, positive direction, and unit segments is called a **number line**. The choice of origin and unit line segments is called a **coordinate system**. Each point on the **real** number line corresponds to a real number. Points to the right of the origin correspond to positive real numbers, and points to the left of the origin correspond to negative real numbers. The origin corresponds to 0. Conversely, each real number corresponds to a point on the number line. Figure 3 shows a point called a that corresponds to a real number A between 3 and 4. Since each point of the line corresponds to one and only one real number, we say that there is a **one-to-one correspondence** between the real

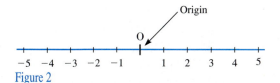

Figure 2

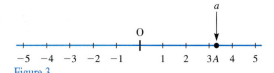

Figure 3

number system and the number line. If a is a point on the number line and A is the corresponding real number, then we say that A is the **coordinate corresponding to** a.

Inequalities

Real numbers have a natural ordering that allows any two numbers to be compared with one another. Let a and b be real numbers. We say that a is **less than** b, denoted $a < b$, provided that $b - a$ is positive. If a is less than b, we also say that b is **greater than** a, denoted $b > a$. The symbols $<, >$ are called **inequality symbols**, and statements involving them are called **inequalities**. In interpreting inequalities, remember that the inequality symbol always points to the smaller number. The following inequalities are true statements:

$$1 < 5, \qquad -1 < -0.5, \qquad 1/3 > 0.3$$

since $5 - 1 = 4$ is positive, $-0.5 - (-1) = 0.5$ is positive, and $1/3 - 0.3 = 0.0333\ldots$ is positive.

If either a is less than b or a is equal to b, then we write $a \leq b$ (read "a is less than or equal to b"). Similarly, if either a is greater than b or a is equal to b, then we write $a \geq b$ (read "a is greater than or equal to b"). Thus, for example, the following inequalities are true:

$$3 \leq 3, \qquad -5 \leq 1, \qquad 4 \geq 4$$

The number line is useful for illustrating inequalities geometrically. If a and b are real numbers, then $a < b$ is true if and only if the point with coordinate a lies to the left of the point with coordinate b. (See Figure 4.)

We say that a number x is **positive** provided that $x > 0$; we say that x is **negative** provided that $x < 0$; we say that x is **nonnegative** provided that $x \geq 0$.

The fundamental rules for manipulating inequalities are summarized in the following result.

$a < b$

Figure 4

PROPERTIES OF INEQUALITIES

Let a, b, and c be real numbers. Then the following properties of inequalities hold.

1. If $a < b$ and $b < c$, then $a < c$.
2. If $a < b$ and c is any real number, then $a + c < b + c$.
3. If $a < b$ and c is any real number, then $a - c < b - c$.
4. If $a < b$ and c is positive, then $ac < bc$.
5. If $a < b$ and c is negative, then $ac > bc$. That is, in multiplying an inequality by a negative number, it is necessary to reverse the inequality sign.

Proof   Let's prove 1 and 4. The proofs of 2, 3, and 5 are similar.

1. If $a < b$ and $b < c$, then $b - a > 0$ and $c - b > 0$. Since the sum of two positive numbers is positive, the number $(c - b) + (b - a)$ is positive. This number may be rewritten as $c - a$, which is positive. That is, $a < c$.
2. If $a < b$, then $b - a > 0$. If c is positive, then the product $(b - a)c$ is also positive, being the product of two positive numbers. But $(b - a)c = bc - ac$, so $bc - ac > 0$. Thus, $ac < bc$. ◆

The above properties state that inequalities may be treated very much like equations. You may add or subtract the same quantity from both sides, and you may multiply or divide both sides by a positive number. However, if you multiply or divide an inequality by a *negative* number, then you must reverse the direction of the inequality.

The above laws of inequalities were stated for the inequality symbol $<$. However, throughout their statement you may replace $<$ with any of the other inequality symbols $>, \leq, \geq$. (In Property 5, where two opposite symbols are used, you may replace the first symbol with any of the inequality symbols and the other with its opposite. That is, if the first symbol is replaced with $\leq$, then the second is replaced with $\geq$.)

Absolute Value

In many applications, it is necessary to consider the magnitude but not the sign of a real number. For dealing with such applications, it is convenient to define the notion of absolute value.

| **Definition 1** | Let a be a real number. The **absolute value** of a, denoted $|a|$, is the real number |
| **Absolute Value** | defined by the formula |

$$|a| = \begin{cases} a & \text{if } a \geq 0 \\ -a & \text{if } a < 0 \end{cases}$$

Absolute values occur in many applications, so it is important to be able to calculate with them. The next example gives some practice.

➤ **EXAMPLE 3**
Calculating
Absolute Values

Calculate the following:

1. $\left| -\dfrac{3}{8} \right|$ 2. $|2|$ 3. $\left| \dfrac{33}{12} - \pi \right|$

Solution

1. Since $-3/8$ is negative, we have

$$\left| -\frac{3}{8} \right| = -\left(-\frac{3}{8} \right) = \frac{3}{8}$$

2. Since 2 is positive, we have

$$|2| = 2$$

3. We have

$$\frac{33}{12} - \pi = 2.75 - 3.14159\ldots = -0.39159\ldots$$

which is less than 0. Therefore, we have

$$\left| \frac{33}{12} - \pi \right| = -\left(\frac{33}{12} - \pi \right)$$

$$= \pi - \frac{33}{12}$$

$$\doteq 0.39159$$

The following is a summary of some of the most useful properties of absolute values.

PROPERTIES OF ABSOLUTE VALUE

Let a and b be real numbers.

1. The absolute value of the additive inverse of a number is the same as the absolute value of the number. That is:
$$|-a| = |a|$$

2. The absolute value of a number is nonnegative. That is:
$$|a| \geq 0$$

3. The absolute value of a product equals the product of the absolute values. That is:
$$|ab| = |a||b|$$

4. The absolute value of a quotient equals the quotient of the absolute values. That is:
$$\left|\frac{a}{b}\right| = \frac{|a|}{|b|} \quad (b \neq 0)$$

5. $|a^2| = a^2$

Proof

1. If $a > 0$, then $-a < 0$, so that
$$|-a| = -(-a) = a = |a|$$

If $a = 0$, then
$$|-0| = 0 = |0|$$

If $a < 0$, then $-a > 0$ and
$$|-a| = -a = |a|$$

2. The proofs of other properties are left for the exercises.   ◆

Distances on the Number Line

Using absolute values, we can define the distance between points on the number line as follows:

Definition 2
Distance on the Number Line

Suppose that A and B are points on the number line with corresponding coordinates a and b. The distance between A and B, denoted $d(A, B)$, is defined by the formula
$$d(A, B) = |a - b|$$

➤ **EXAMPLE 4**
Calculating Distances

Find the distance $d(A, B)$ between the points A and B of the number line, whose respective coordinates a and b are given as follows:

1. $a = 1, b = 9$
2. $a = -1, b = -5$

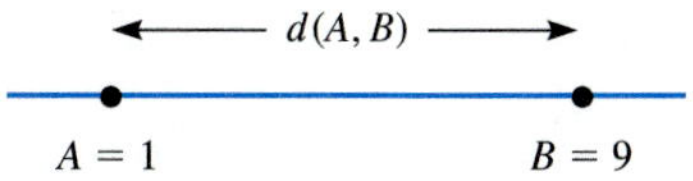

Solution

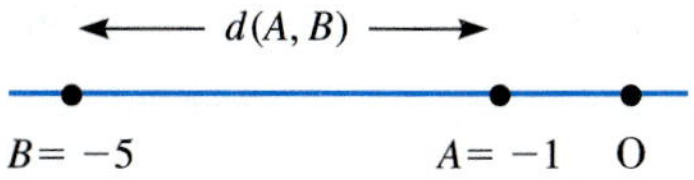

In each case, the value of $d(A, B)$ is derived using a simple absolute value calculation. Note in each case that the value of the distance corresponds with our intuitive notion of distance as illustrated in the accompanying figures (Figures 5 and 6).

A = 1   B = 9

Figure 5

B = −5   A = −1   O

Figure 6

1. $d(A, B) = |a - b| = |1 - 9| = |-8| = 8$
2. $d(A, B) = |a - b| = |-1 - (-5)| = |4| = 4$

Suppose that A is any point on the number line. We can compute the distance of A to the origin in terms of the coordinate a of A. The coordinate of the origin is 0, so that we have the formula

$$d(A, 0) = |a - 0| = |a|$$

The following is a summary of the most important properties of distance on the number line.

PROPERTIES OF DISTANCE

Let A, B, and C be points on the number line.

1. **Nonnegativity:** $d(A, B) \geq 0$
2. **Symmetry:** $d(A, B) = d(B, A)$
3. **Triangle inequality:** $d(A, B) \leq d(A, C) + d(C, B)$

Each of these properties states a fact familiar from geometry. The first asserts that the distance between two points is nonnegative. The second asserts that the distance A to B is the same as the distance from B to A. The third asserts that to get from A to B, the distance is at most the distance of going from A to C and then from C to B. Although these facts are geometrically obvious, our definition of distance in terms of absolute value allows us to give simple proofs of these facts that don't depend on our often faulty geometric intuitions. You will be asked to prove properties 1–3 of distance in the exercises.

Exercises 1.1

Consider the following numbers:

$$24, \ \frac{2}{3}, \ -1.8, \ 2\pi, \ 0, \ 93\%, \ \sqrt[5]{32}, \ \sqrt{5},$$

$$\sqrt{16}, \ \sqrt{\frac{1}{4}}, \ -234, \ -\frac{16}{7}, \ 0.0035$$

1. List the numbers that are integers.
2. List the numbers that are rational.
3. List the numbers that are irrational.
4. List the numbers that are real.

Find decimal notation for each of the following.

5. $\dfrac{2}{7}$ 6. $\dfrac{45}{11}$ 7. $\dfrac{23}{12}$ 8. $\dfrac{34}{9}$

Reduce the following rational numbers to lowest terms.

9. $\dfrac{35}{40}$ 10. $\dfrac{24}{21}$ 11. $\dfrac{801}{702}$ 12. $\dfrac{425}{525}$

13. $\dfrac{150}{25}$ 14. $\dfrac{240}{8}$ 15. $\dfrac{864}{64}$ 16. $\dfrac{219}{438}$

For each of the following, find (a) the multiplicative inverse and (b) the additive inverse.

17. 45

18. −23

19. −3.2

20. 0.0625

21. $-\dfrac{2}{3}$

22. $\dfrac{5}{-4}$

23. 1.6%

24. 160%

Identify each number as either rational or irrational.

25. $-16\dfrac{2}{3}$

26. 74.3%

27. $-\sqrt{4}$

28. $-\sqrt{5}$

29. $5.2\overline{367}$

30. -23.897654102

What law is exemplified by each of the following sentences?

31. $3 + x = x + 3$

32. $9 \cdot 9^{-1} = 1$

33. $-2(x + y) = -2x - 2y$

34. $x + (7 + y) = (x + 7) + y$

35. $78 \cdot 1 = 78$

36. $9(pq) = (9p)q$

37. $-t + t = 0$

38. $18(a + b) = 18a + 18b$

39. $a + (7 + b) = (a + 7) + b$

40. $ab = ba$

41. $0 = t + (-t)$

42. $3x - 3y = 3(x - y)$

43. $-\dfrac{4}{5}\left(-\dfrac{5}{4}\right) = 1$

44. $\dfrac{2}{3} + \left(-\dfrac{2}{3}\right) = 0$

45. $-\dfrac{1}{x} + 0 = -\dfrac{1}{x}$

46. $-2 + 4.5$ is a real number

Fill in the blank with the correct symbol: $<, >,$ or $=$.

47. 2.3 ____ 6.8

48. -6 ____ -12

49. 0 ____ -1

50. 56.7% ____ 0.567

51. $-\sqrt{81}$ ____ -9

52. -3.216 ____ -3.220

53. $\dfrac{3}{4}$ ____ $\dfrac{5}{9}$

54. $(5 - 3)$ ____ $(3 - 5)$

55. $(6.7 - 2.8)$ ____ $(9.3 - 4.2)$

56. $\sqrt{5}$ ____ 2.36

57. -1.732 ____ $-\sqrt{3}$

58. $\sqrt{0}$ ____ 0

Evaluate the following.

59. $(-5 \cdot 8) + 4(-3) - 9(-7)$

60. $6 + \left[8 - 6\left(\dfrac{7}{6}\right)\right](4 - 2)$

61. $(3 - 8)(4 - 7)$

62. $-[4 - 2(8 + 6)]$

63. $(15 + 6)(10 - 9)$

64. $\left(\dfrac{45}{3}\right) + \left(-\dfrac{45}{9}\right) - \left(-\dfrac{54}{6}\right)$

65. $\dfrac{-100}{-5} - \dfrac{-63}{-7} + \dfrac{-125}{25}$

66. $\left(\dfrac{2}{3}\right)\left(\dfrac{9}{5}\right) - \left(\dfrac{4}{5}\right)\left(\dfrac{10}{8}\right)$

67. $\left|\dfrac{1}{4}\right|$

68. $|5 - 3|$

69. $|4 + 7|$

70. $|-5 + 0|$

71. $|-7 - 3|$

72. $|-8 + 2|$

73. $|3.28 - 3.28|$

74. $\left|-\dfrac{2}{3}\right|$

75. $\left|\sqrt{10}\right|$

76. $\left|\sqrt{10} - \pi\right|$

77. $\left|2 - \sqrt{5}\right|$

78. $\left|\pi - \sqrt{10}\right|$

Fill in the blank with the correct symbol: $<, >,$ or $=$.

79. $|4| + |-6|$ ____ $|4 + (-6)|$

80. $|7| - |9|$ ____ $|7 - 9|$

81. $|-8| - |17|$ ____ $|-8 - 17|$

82. $|-24| + |-18|$ ____ $|-24 + (-18)|$

Write without parentheses.

83. $(a - b)c$

84. $d(x + y)$

85. $(a + b)(-c)$

86. $3(x - y)$

Write as a reduced fraction.

87. $\dfrac{a - b}{c} \cdot \dfrac{p}{q}$

88. $\dfrac{\dfrac{a}{b}}{\dfrac{1}{c/d}}$

89. $\dfrac{a}{3} - \dfrac{3}{a}$

90. $\dfrac{4}{b} + \dfrac{b}{4}$

Find the distances between the points A and B of the number line, where the coordinates a and b are given.

91. $a = 7, b = 10$

92. $a = -13, b = 6$

93. $a = -\dfrac{2}{5}, b = \dfrac{3}{4}$

94. $a = 0, b = -89$

95. $a = -6.7, b = -34.2$

96. $a = \sqrt{2}, b = \pi$

Find the prime factorization.

97. 99

98. 112

99. 142

100. 300

101. 175

102. 288

103. 2700

104. 7800

105. Write set builder notation for the set of all negative real numbers.

106. Write set builder notation for the set of all real numbers that are not equal to 2.

107. Write set builder notation for the set of all real numbers that are not less than -3.

108. Write set builder notation for the set of all positive real numbers greater or equal to -7.

Let a and b be real numbers. Prove each of the following properties of absolute values.

109. $|-a| = |a|$ (*Hint:* Consider two cases: $a \geq 0$ and $a < 0$.)

110. $|ab| = |a||b|$

111. Prove that $\left|\dfrac{a}{b}\right| = \dfrac{|a|}{|b|}$

Let A, B, and C be points on the number line with coordinates a, b, and c, respectively. Prove each of the following. Use the absolute value properties where possible.

112. $d(A, B) \geq 0$

113. $d(A, B) = d(B, A)$

114. $d(A, B) \leq d(A, C) + d(C, B)$

115. For what values of x is $|x| - x = 0$?

116. Give an example to disprove $a|b + c| = |ab + ac|$

117. Solve for x:

  a. $|x| = 3$

  b. $|x| = 0$

  c. $|x| = -5$

118. Prove that
$$\frac{a + b}{c} = \frac{a}{c} + \frac{b}{c}$$

119. Show by example that
$$\frac{a}{b + c} \neq \frac{a}{b} + \frac{a}{c}$$

That is, show that division is not distributive over addition.

120. At what decimal place does 22/7 differ from π?

121. Which is a closer approximation to π, 22/7 or 3927/1250 ?

Find an equivalent expression that does not use parentheses.

122. $-(3x)$

123. $-(x + 3)$

124. $-(a + b)$

125. $-(x^2 - x + 4)$

✍ In Your Own Words

126. Consider an operation $\oplus$ defined by $a \oplus b = a + 2b$. Determine whether this operation is commutative. Is multiplication distributive over $\oplus$? That is, does the following identity hold: $a(b \oplus c) = ab \oplus ac$?

127. Suppose you are given an integer n to factor into primes. Describe in words how you would go about doing this.

128. Describe in words how you would go about reducing a rational number to lowest terms.

▌ 1.2 EXPONENTS AND RADICALS

In this section, we review the basic facts about integer and rational exponents.

Positive Exponents

Many applications involve products in which a number is repeatedly multiplied by itself. Mathematicians have introduced a shorthand notation for such products. If n is a positive integer, then we write a^n to mean a product of n factors of a. That is, we have

$$a^1 = a$$
$$a^2 = a \cdot a$$
$$a^3 = a \cdot a \cdot a$$
$$a^4 = a \cdot a \cdot a \cdot a$$

Here are some numerical examples of a^n for various values of a and n.

$$2^3 = 2 \cdot 2 \cdot 2 = 8$$
$$(-5)^3 = (-5) \cdot (-5) \cdot (-5) = -125$$
$$\left(\frac{1}{2}\right)^2 = \frac{1}{2} \cdot \frac{1}{2} = \frac{1}{4}$$

The expression a^n is read "a to the nth power" or simply "a to the nth." Computing a^n is called "raising a to the nth power." In an expression a^n, a is called the **base**, and n is called the **exponent**.

Square Roots and Other Radicals

As preparation for our discussion of rational exponents, let's review the basic facts about square roots and nth roots. We begin with the definition of square root.

Definition 1 **Square Root**	Suppose that a is a nonnegative real number. A **square root** of a is a real number b whose square is a. That is, b is a square root of a provided that $b^2 = a$.

For example, the number $a = 4$ has square roots 2 and -2 since $(2)^2 = 4$ and $(-2)^2 = 4$. The number $a = 0$ has the square root 0, since $0^2 = 0$. On the other hand, the number -3 does not have a square root **in the real number system** since the square of any real number b is nonnegative and hence cannot equal -3. For similar reasons, if a is any negative number, then a does not have a square root in the real number system.

Zero is the only real number with only one square root, namely, itself. However, each *positive* real number a has two square roots, one positive and the other negative. We define the symbol $\sqrt{a}$ to mean the positive square root. We call this square root the **principal square root** of a. We also set $\sqrt{0} = 0$. However, if a is negative, the symbol $\sqrt{a}$ is undefined in the real number system. Thus, for example, we have

$$\sqrt{4} = 2, \qquad \sqrt{\frac{1}{16}} = \frac{1}{4}, \qquad \sqrt{0} = 0$$

$\sqrt{-1}$ is undefined in the real number system

In the first three examples, the values of $\sqrt{a}$ are rational numbers. However, for many rational numbers a, the value of $\sqrt{a}$ is irrational. For example, it can be proved that $\sqrt{2}$ is irrational. The decimal expression of this real number is nonterminating and nonrepeating and begins

$$\sqrt{2} = 1.414213\ldots$$

Definition 2 *n*th **Root**	Let n be a positive integer and a be a real number. An ***n*th root** of a is a real number b whose nth power equals a. That is, $$b^n = a$$

For example, a third root of 8 is 2, since $2^3 = 8$. Unlike the case for square roots, a negative real number can have a **real** third root. The third root of -8 is -2 since $(-2)^3 = -8$.

Just as a negative number does not have a square root, a negative number does not have an nth root when n is even. Thus, for example, -2 does not have a fourth root or a sixth root. If a is positive and n is even, then a has two nth roots, which are negatives of each other. The positive value is denoted $\sqrt[n]{a}$ and is called the **principal *n*th root** of a. Thus, for example, the number 2 has two fourth roots, $\sqrt[4]{2}$ and $-\sqrt[4]{2}$. The numerical value of $\sqrt[4]{2}$ is approximately 1.189. The number 0 has just one nth root, namely, 0. In case n is odd, any real number a has a single nth root, which is denoted $\sqrt[n]{a}$.

We have

$$\sqrt[3]{8} = 2, \qquad \sqrt[4]{\frac{1}{16}} = \frac{1}{2}, \qquad \sqrt[3]{-\frac{8}{27}} = -\frac{2}{3}$$

The symbols $\sqrt{a}$ and $\sqrt[n]{a}$ are called **radicals**. Manipulations involving radicals can be handled, for the most part, using the laws of exponents applied to rational exponents, which leads us to our next topic.

Rational Exponents

Let a be a real number. To define a^r, where r is a rational number, we consider the special case where r is a number of the form $1/n$ for a positive integer n. In this case, we define $a^{1/n}$ by the formula

$$a^{1/n} = \sqrt[n]{a}$$

assuming that the radical on the right side is defined. If the radical on the right side is not defined, then the power $a^{1/n}$ is undefined. Thus, for example, we have

$$2^{1/2} = \sqrt{2}$$
$$8^{1/3} = \sqrt[3]{8} = 2$$
$$(81)^{1/4} = \sqrt[4]{81} = 3$$

When r is the positive rational number

$$r = \frac{m}{n}$$

for m and n positive integers and m/n in lowest terms, we define a^r as

$$a^r = a^{m/n} = (a^{1/n})^m = (\sqrt[n]{a})^m = \sqrt[n]{a^m}$$

This formula makes sense only if the expression on the right is defined. Here are some examples of numbers raised to rational exponents:

$$4^{3/2} = (\sqrt{4})^3 = 2^3 = 8$$
$$(-8)^{5/3} = (\sqrt[3]{-8})^5 = -32$$
$$(-2)^{3/2} = (\sqrt{-2})^3 = \text{undefined}$$

Negative and Zero Exponents

Let a be a nonzero number. We define the **zero power** of a by the formula

$$a^0 = 1, a \neq 0$$

Thus, $3^0 = 1$, $(-7)^0 = 1$, $(2/3)^0 = 1$, and so forth. Note, however, that the definition excludes the case where a is equal to 0. That is, 0^0 is undefined.

Suppose that a is nonzero and r is a rational number for which a^r is defined and $r > 0$. Then we define a^{-r} by the formula

$$a^{-r} = \frac{1}{a^r}$$

Thus, for example, we have

$$2^{-3} = \frac{1}{2^3} = \frac{1}{8}$$
$$\left(\frac{1}{7}\right)^{-2} = \frac{1}{(1/7)^2} = \frac{1}{1/49} = 49$$
$$4^{-3/2} = \frac{1}{4^{3/2}} = \frac{1}{(\sqrt{4})^3} = \frac{1}{2^3} = \frac{1}{8}$$

Rational exponents occur often in applications, as the following examples show.

➤ **EXAMPLE 1**
Calcium Injection

In order to measure the rate at which calcium is absorbed by the body, a subject is injected with a radioactive isotope of calcium. At time t days after the injection, the amount of calcium remaining is given by an expression of the form

$$t^{-3/2} \quad (t > 0.5)$$

This expression assumes that the calcium is measured in appropriate units. How many units of calcium remain after nine days?

Solution

The amount of calcium remaining after nine days is obtained by evaluation of the formula for $t = 9$:

$$9^{-3/2} = \frac{1}{9^{3/2}}$$

$$= \frac{1}{(\sqrt{9})^3}$$

$$= \frac{1}{3^3}$$

$$= \frac{1}{27}$$

That is, after nine days there remains 1/27 unit of calcium.

The Laws of Exponents

In order to manipulate algebraic expressions involving exponents, the following laws are of critical importance.

LAWS OF EXPONENTS

Let a and b be real numbers and r and s rational numbers. Then the following laws of exponents hold, provided that all of the expressions appearing in a particular equation are defined.

1. $a^r \cdot a^s = a^{r+s}$
2. $(a^r)^s = a^{rs}$
3. $(ab)^r = a^r b^r$
4. $\dfrac{a^r}{a^s} = a^{r-s}$
5. $\left(\dfrac{a}{b}\right)^r = \dfrac{a^r}{b^r}$
6. $a^{-r} = \dfrac{1}{a^r}$
7. $\left(\dfrac{a}{b}\right)^{-r} = \left(\dfrac{b}{a}\right)^r$

We will assume that these laws of exponents hold without providing any proofs.

➤ **EXAMPLE 2**
Using Laws of Exponents

Use the laws of exponents to obtain equivalent expressions that involve no parentheses.

1. $(s^5)^2$
2. $(ab)^4$
3. $(x^2)^3 x^5$

Solution

1. By Law 2, we have
$$(s^5)^2 = s^{5 \cdot 2} = s^{10}$$

2. By Law 3, we have
$$(ab)^4 = a^4 b^4$$

3. Combining Laws 3 and 1, we have
$$(x^2)^3 x^5 = x^{2 \cdot 3} x^5 = x^{6+5} = x^{11}$$

➤ **EXAMPLE 3**
Eliminating Negative
and Zero Exponents

Write the following expressions in a form that involves no negative or zero exponents.

1. $(x^3 y^{-2})^2$
2. $\left(\dfrac{x^{-5}}{x^{-3}} \right)^{-3}$

Solution

1. We have
$$
\begin{aligned}
(x^3 y^{-2})^2 &= x^{3 \cdot 2} y^{-2 \cdot 2} \quad &\textit{(Law 2)} \\
&= x^6 y^{-4} \\
&= x^6 \cdot \frac{1}{y^4} \quad &\textit{(Law 6)} \\
&= \frac{x^6}{y^4}
\end{aligned}
$$

2. We have
$$
\begin{aligned}
\left(\frac{x^{-5}}{x^{-3}} \right)^{-3} &= \frac{x^{(-5) \cdot (-3)}}{x^{(-3) \cdot (-3)}} \quad &\textit{(Laws 2 and 5)} \\
&= \frac{x^{15}}{x^9} \\
&= x^{15-9} \quad &\textit{(Law 4)} \\
&= x^6
\end{aligned}
$$

➤ **EXAMPLE 4**
Eliminating Negative
Exponents and Radicals

Write the following expressions in a form that involves no negative exponents or radicals.

1. $(x^{1/2})^{-4}$
2. $\sqrt{\dfrac{x}{x^{-3/2}}}$

Solution

1. By Law 2 and Law 6, we have

$$(x^{1/2})^{-4} = x^{1/2 \cdot (-4)}$$
$$= x^{-2}$$
$$= \frac{1}{x^2}$$

2. By the definition of a rational exponent, we have

$$\sqrt{\frac{x}{x^{-3/2}}} = \left(\frac{x}{x^{-3/2}}\right)^{1/2}$$
$$= \frac{x^{1/2}}{(x^{-3/2})^{1/2}}$$
$$= \frac{x^{1/2}}{x^{-3/4}}$$
$$= x^{(1/2)-(-3/4)}$$
$$= x^{5/4}$$

As a consequence of the laws of exponents, we deduce the following useful properties of radicals.

PROPERTIES OF RADICALS

For a and b real and not both negative:

1. $\sqrt[n]{a}\,\sqrt[n]{b} = \sqrt[n]{ab}$
2. $\sqrt[n]{\dfrac{a}{b}} = \dfrac{\sqrt[n]{a}}{\sqrt[n]{b}}\quad b \neq 0$
3. $\sqrt[n]{a^n} = |a|\quad n$ even
4. $\sqrt[n]{a^n} = a,\, n$ odd

Note that these formulas hold only for values of a, b, and n for which all the radicals appearing are defined.

The next example illustrates the use of these formulas in manipulating radicals.

➤ **EXAMPLE 5**
Simplifying Radical Expressions

Simplify the following expressions by removing radicals where possible, and by using as simple as possible an expression under the radical otherwise.

1. $\sqrt[3]{64y^3z^9}$ 2. $\sqrt[3]{18x^4} \cdot \sqrt[3]{12x^2}$ 3. $\sqrt[3]{54x^3}/\sqrt[3]{16x^2}$

Solution

1. Applying Property 1 of radicals yields

$$\sqrt[3]{64} \cdot \sqrt[3]{y^3} \cdot \sqrt[3]{z^9} = 4yz^3$$

2. Applying Property 1 of radicals and Law 1 of exponents yields

$$\sqrt[3]{18x^4 \cdot 12x^2} = \sqrt[3]{3^2 \cdot 2 \cdot 3 \cdot 2^2 x^6}$$
$$= \sqrt[3]{2^3 3^3 x^6}$$
$$= 2 \cdot 3x^2$$
$$= 6x^2$$

3. From Property 2, we have

$$\sqrt[3]{\frac{54x^3}{16x^2}} = \sqrt[3]{\frac{27x}{8}} = \frac{3}{2}\sqrt[3]{x}$$

In our discussion above, we have defined powers of a real number with rational exponents. It is possible to also define powers corresponding to irrational exponents, for example:

$$2^{\sqrt{2}}, 5^\pi$$

We will provide only some of the details of the definition in this book. However, you should note that once the definition is made, all of the laws of exponents stated above hold for those powers as well.

Precedence of Powers

In performing calculations, raising to a power has a higher precedence than addition, subtraction, multiplication, or division. That is, all raising to powers occurs before the other operations are performed. In the expression

$$5/10 - 3 \cdot 2^3$$

we first perform all raising to powers, proceeding from left to right to get the expression

$$5/10 - 3 \cdot 8$$

Next we perform multiplications and divisions, proceeding from left to right, to obtain

$$0.5 - 24$$

Finally we perform additions and subtractions, proceeding from left to right, to obtain the result -23.5.

Scientific Notation

Powers of 10 are commonly used in scientific work, especially in describing numbers that are very large or very small. For example, the distance light travels in a year is approximately 10^{16} meters. The diameter of a DNA molecule is approximately 10^{-8} meters. To get a feel for these powers, review the following table containing the first few positive and negative powers of 10.

n	10^n	n	10^n
0	1	-1	0.1
1	10	-2	0.01
2	100	-3	0.001
3	1000	-4	0.0001
4	10,000	-5	0.00001

Multiplication by 10^n is equivalent to shifting the decimal point n places. If n is positive, the shift is to the right, whereas if n is negative, the shift is to the left. Here are some examples:

$$15.354 \times 10^3 = 15{,}354 \qquad \textit{(shift right 3 places)}$$
$$15.354 \times 10^{-3} = 0.015354 \qquad \textit{(shift left 3 places)}$$

By shifting the decimal point an appropriate number of places, we can write any number as a power of 10 times a decimal with exactly one nonzero digit to the left of the decimal point. For example

$$15.354 = 1.5354 \times 10^1$$
$$0.0058723 = 5.8723 \times 10^{-3}$$
$$1487 = 1.487 \times 10^3$$

This form of writing a decimal is called **scientific notation** and is commonly used in science. Its advantage is that very large and very small numbers can be expressed in a very concise format. For instance, 1 followed by 100 zeros can be written in scientific notation as

$$1.0 \times 10^{100}$$

whereas a decimal point followed by 99 zeros and a 1 can be written in scientific notation as

$$1.0 \times 10^{-100}$$

In scientific notation, the initial decimal is called the **mantissa**, and the power of 10 is called the **exponent**. According to the laws of exponents, to multiply numbers written in scientific notation we multiply mantissas and add the exponents. For example

$$(2.0 \times 10^5) \cdot (3.5 \times 10^{-2}) = \left[(2.0) \cdot (3.5) \cdot (10^5 \cdot 10^{-2})\right] = 7.0 \times 10^3$$

➤ **EXAMPLE 6**

Gravitation

Newton's Law of Universal Gravitation asserts that the force F of gravitational attraction between two bodies of masses m_1 and m_2 that are at a distance r from one another is given by the formula

$$F = G\frac{m_1 m_2}{r^2}$$

where G is the universal gravitational constant, which in the metric system is equal to

$$G = 6.670 \times 10^{-11}$$

In the metric system, F is measured in newtons, r in meters, and m_1 and m_2 in kilograms. The mass of the earth is approximately equal to 5.96×10^{24} kg. Determine the gravitational force exerted by the earth on an asteroid of mass 10^{18} kg at a distance 10^{10} meters from the center of the earth.

Solution

From the given formula, we have

$$F = G\frac{m_1 m_2}{r^2}$$

$$= (6.670 \times 10^{-11})\frac{(5.96 \times 10^{24}) \cdot (10^{18})}{(10^{10})^2}$$

Applying the laws of exponents to multiply the powers of 10, we have

$$F = (6.670) \cdot (5.96) \cdot \frac{10^{-11+24+18}}{10^{20}}$$

$$= 39.75 \times \frac{10^{31}}{10^{20}}$$

$$= 39.75 \times 10^{31-20}$$

$$= 3.975 \times 10^{12} \quad \text{newtons}$$

In computer or calculator work, the multiplication sign and the 10 are replaced by the letter E, which stands for *exponent*. For instance, the number

$$1.275 \times 10^{-3}$$

is written for computer or calculator usage as

$$1.275E-3$$

Many real numbers are nonterminating decimals. To deal with them requires that we use approximations obtained by rounding the decimal expression of the number to a given number of places. The digits used in the approximation are called **significant digits**. For example, to approximate the number π to five significant digits, we round the decimal expression $3.1415926535\ldots$ to the fifth digit to obtain 3.1416. In this text, we have used approximations with four significant figures to obtain numerical results.

Algebra and Technology

We live in an age distinguished by incredible advances in technology. Many technological achievements affect the way in which algebra and trigonometry are taught and applied. In this book, we will attempt to give some insight into the roles that technology plays.

The four most important applications of technology to algebra and trigonometry are:

1. *Scientific calculators.* These calculators provide numerical calculating capability, which includes evaluation of scientific functions such as powers, trigonometric functions, exponentials, and logarithms. We will provide many examples of how scientific calculators may be applied to perform the numerical calculations required in this text.

2. *Graphing calculators.* In recent years, graphing calculators have provided the capability to input and graph user-defined functions. These calculators eliminate much of the tedium of plotting individual points in order to graph functions. Throughout this text, we will point out how graphing calculators may be applied to graph the functions that arise and to provide insight into the mathematical ideas we develop.

3. *Symbolic mathematics programs.* A number of programs perform manipulation of algebraic expressions. They can, for example, multiply polynomials and add algebraic fractions. Such programs perform what is called **symbolic calculation**, to distinguish them from calculators, which provide only numerical answers. Some popular symbolic mathematics programs are *Mathematica, Maple,* and *Derive.* Most symbolic packages require a computer to run. However, some calculators have symbolic programs built-in.

4. *"Scratchpad" programs.* There are many programs that provide "scratch-pads" for entering calculations in much the same manner as you would on a sheet of paper. Scratchpads perform only numerical calculations but often have capability for graphing data. The most common form of scratchpad is provided by a spreadsheet program, such as *1–2–3* or *Excel*. A popular nonspreadsheet scratchpad program is *MathCad*. We will illustrate how scratchpad programs may be used to tabulate and analyze data.

This categorization of mathematical technology is rather arbitrary in that it reflects what has been developed so far. Technology is being developed very rapidly, and new categories of mathematical technology are certain to arise in the years ahead. In this book, we will describe how to use the technology that currently exists. However, you should approach technology in a very flexible way. In order to solve a particular type of mathematical problem (e.g., approximating the solution to a polynomial equation), you must choose the most effective technology. The technology you choose will depend on the tools that have been developed and are available to you.

The following problem illustrates the sort of numerical calculations that are best handled using a scientific calculator.

➤ **EXAMPLE 7**

Speed of Sound

The speed of sound varies with the air temperature. At temperature T (given in degrees Celsius), the speed of sound V (given in meters per second) is given by the formula

$$V = 331.5 \left(\frac{T + 273}{273} \right)^{1/2}$$

1. If the air temperature is 20°C, what is the speed of sound?
2. If the air temperature is 100°C, what is the speed of sound? Give your answer to four significant figures.

Solution

1. The speed of sound is obtained by replacing T with 20 in the formula

$$V = 331.5 \left(\frac{20 + 273}{273} \right)^{1/2} = 331.5 \sqrt{\frac{20 + 273}{273}}$$

The best way of obtaining a numerical value for this expression is to use a scientific calculator. Each calculator is a little different, so to determine the precise keystrokes to perform this calculation, you will need to consult your calculator reference manual. To four significant digits, the value obtained is 343.4 m/sec.

2. In this case, replace T in the formula with 100 to obtain

$$V = 331.5 \left(\frac{100 + 273}{273} \right)^{1/2} = 331.5 \sqrt{\frac{100 + 273}{273}} = 387.5 \text{ m/sec.}$$

The following exercise set contains material for practicing arithmetic operations using your calculator.

Exercises 1.2

Write the following expressions in a form that does not involve any negative or zero exponents or parentheses and that
contains as few multiplications as possible.

1. $5^3 \cdot 5^{-3}$ 2. $2^{-8} \cdot 2^{-9}$ 3. $a^{-5} \cdot a^2$ 4. $t^{-7} \cdot t^7$

5. $q^3 \cdot q^{-6} \cdot q^7$ 6. $m^{-3} \cdot m^{-5} \cdot m^{20}$

7. $(3x^5)(-4x^6)$ 8. $(6y^{-4})(8y^{-10})$

9. $(4x^3y^3)(9x^{-3}y^{-5})$ 10. $(-12p^{-6}t^{11})(-5p^8t^{-7})$

11. $(3ab^4)^2$ 12. $(-5x^{-2}y^4)^{-3}$ 13. $(2y)^4(3x)^4$

14. $(2ab)^3(3ab)^2$ 15. $\dfrac{t^{-12}}{t^{-7}}$ 16. $\dfrac{a^{10}}{a^8}$

17. $\dfrac{a^3b^{-3}}{a^{-2}b}$ 18. $\dfrac{10x^4y^{-4}}{-2x^{-1}y^3}$ 19. $(-3x^2b^{-4})^{-2}$

20. $(-5p^{-3}q^4r)^{-3}$ 21. $\left(\dfrac{2a^3b^{-2}}{3a^4b^{-3}}\right)^3$

22. $\left(\dfrac{-4p^5y^{-2}}{5p^{-1}q^6}\right)^{-4}$ 23. $\dfrac{(4x^2y^{-1}z^{-3})^{-2}}{(8xy^{-1}z^2)^{-1}}$

24. $\dfrac{(10^{-1}a^{-4}c^5)^{-2}}{(5^{-2}a^3c^6)^{-3}}$ 25. $(-2)^{-3} \cdot 5^2$

26. $-6^2 - 3^2$ 27. $\dfrac{(-3)^2}{3^3}$ 28. $(2^2 + 3^2)^2$

29. $(5^2 - 4^2)^2$ 30. $(2^{-2} - 5^{-2})^{-2}$

31. $\dfrac{(-2)^{-2} + (-2)^{-3}}{(-2)^5}$ 32. $\dfrac{(-1)^5 + (-2)^{-3}}{(-2)^2}$

Simplify the following expressions so that the result involves
no radicals or fractional exponents.

33. $\sqrt[3]{-125}$ 34. $\sqrt[5]{32}$ 35. $8^{5/3}$ 36. $81^{3/4}$

37. $27^{2/3}$ 38. $4^{5/2}$ 39. $9^{3/2}$ 40. $16^{5/2}$

41. $64^{2/3}$ 42. $8^{2/3}$

Write the following expressions in a form that involves no
negative exponents or radicals.

43. $\sqrt[3]{y^2}$ 44. $\sqrt[5]{t^3}$ 45. $\sqrt[6]{y^{-17}}$

46. $\sqrt{(x^2 - 5)^{-1}}$ 47. $(a^{4/5})^{-20}$ 48. $(x^{-1/3})^{-12}$

49. $\sqrt{\dfrac{x}{x^{4/5}}}$ 50. $\sqrt[3]{\dfrac{a}{a^{-2/3}}}$

51. $\sqrt[3]{a^{-18}}$ 52. $\sqrt{p^4q^8}$

Simplify the following expressions. Assume that the variables represent positive numbers.

53. $\sqrt{75}$ 54. $\sqrt{72}$ 55. $\sqrt[3]{16}$ 56. $\sqrt[4]{162}$

57. $\sqrt{49x^2}$ 58. $\sqrt{36t^2}$ 59. $\sqrt{32(t + 1)^2}$

60. $\sqrt{49x^3}$ 61. $\sqrt{3x} \cdot \sqrt{6x}$ 62. $\sqrt{8y} \cdot \sqrt{2y}$

63. $\sqrt{4x^2y^4}$ 64. $\sqrt[3]{64z^6}$ 65. $\sqrt{98a^2b^{-6}}$

66. $\sqrt[5]{-1 \cdot x^{10}y^{-15}}$ 67. $\sqrt[3]{9x^2y} \cdot \sqrt[3]{6xy^3}$

68. $\sqrt[4]{90ab^3} \cdot \sqrt[4]{16ab}$ 69. $\dfrac{\sqrt[3]{54t}}{\sqrt[3]{2t}}$

70. $\dfrac{\sqrt{80a}}{\sqrt{5a}}$ 71. $\dfrac{\sqrt{3(2x^2)^3}}{\sqrt{6x}}$ 72. $\dfrac{\sqrt[3]{9xy}}{\sqrt[3]{3x^{-2}}}$

73. $\dfrac{\sqrt[3]{625a^4b^7}}{\sqrt[3]{5ab^2}}$ 74. $\dfrac{\sqrt[3]{40x^5y^2}}{\sqrt[3]{2xy}}$

75. $\sqrt{2(x + 1)^2} \cdot \sqrt{4(x + 1)}$

76. $\sqrt[3]{2(y - 1)^2} \cdot \sqrt[3]{4(y - 1)^6}$

77. $\sqrt[4]{\dfrac{243a^6b^{-13}c^{15}}{3a^2b^{-9}c^7}}$ 78. $\sqrt[5]{\dfrac{160x^9y^{12}}{5xy^2}}$

Convert to scientific notation.

79. 2,000,000,000 lbs. (the amount of coffee consumed annually in the United States)

80. 0.00000000006672 (the universal gravitational constant, in N·m^2/kg^2)

81. 0.0000000000000000016 (1 electron volt, in ft.-lb.)

82. 5,878,000,000,000 (the distance, in miles, that light
   travels in one year. Also known as 1 light-year)

83. $3.56E12$ 84. $1.89E-13$

Convert to decimal notation.

85. $\$1.2 \times 10^{10}$ (the amount of money spent on lotteries in
   1989)

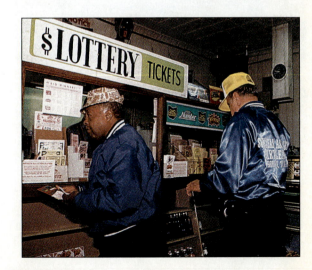

86. $\$5.8 \times 10^9$

87. 3.83×10^{-23} (the number of kilocalories in 1 electron volt)

88. 10^{-6} sec. (the amount of time it takes a bullet cap to explode)

89. $5.78E-18$

90. $7.02E\,16$

♦ Applications

Solve the following problems and give the answers in scientific notation.

91. It takes light 2,200,000 years to travel to earth from the constellation Andromeda. How far is it from earth to Andromeda? (See Exercise 82.)

92. The sun is 93,000,000 miles from the earth. How long does it take for light to travel to earth from the sun?

93. The moon travels about 1,500,078 miles in one orbit about the earth. About how far is the moon from the center of the earth? (Assume that the orbit of the moon is circular.)

94. The United States imported 6 million barrels of oil each day in a recent year. How many barrels of oil did it import in the entire year?

Solve the following problems.

95. A baseball diamond is also a square whose sides have length 90 ft. It is 60.5 ft. from home plate to the pitcher's rubber. (See figure.) About how far is it from the pitcher's rubber to second base? (Use the Pythagorean equation for right triangles, $a^2 + b^2 = c^2$ or $c = \sqrt{a^2 + b^2}$, where c is the length of the hypotenuse and a and b are the lengths of the legs.)

96. A circle with area 113.04 m² has a square inscribed as shown. (See figure.) Find the area of the square.

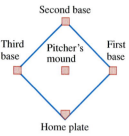

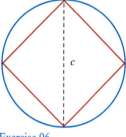

Exercise 95   Exercise 96

97. The time t, in seconds, it takes an object to fall s feet is given by $t = \sqrt{s}/4$. TV station KTHI in Fargo, ND, has a transmitting antenna 2063 ft. high. A tool is dropped from the top by a repairman. (See figure.) How long will it take to fall to the ground?

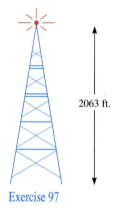

2063 ft.

Exercise 97

The *escape velocity* V_0 of a projectile is the initial velocity needed in order for it to escape the gravitational pull of the planet. Escape velocity (in m/sec) is given by

$$V_0 = \sqrt{\frac{2GM}{R}}$$

where G is the universal gravitational constant, 6.672×10^{-11} N · m²/kg², M is the mass of the planet, in kilograms, and R is its radius, in meters.

98. The mass of the earth is 5.97×10^{24} kg, and its radius is 6.37×10^6 m. What velocity is necessary for a rocket to escape the pull of the earth?

99. The mass of Mars is 6.57×10^{23} kg, and its radius is 3.45×10^6 m. What is the escape velocity for a rocket leaving Mars?

100. Heron's formula for the area A of a triangle (see figure) is given by

$$A = \sqrt{s(s-a)(s-b)(s-c)}$$

where $s = (a + b + c)/2$ and a, b, and c are the lengths of the sides of the triangle. What is the area of a triangle with sides of lengths 5.5 ft., 7.2 ft., and 2.1 ft.?

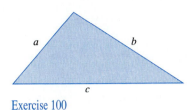

Exercise 100

101. Suppose an amount of money P is invested in a savings account for time t, in years, at an interest rate of i, compounded annually, and grows to an amount A. The interest rate is given by

$$i = \left(\frac{A}{P}\right)^{1/t} - 1$$

Suppose $2000 grows to an amount of $3500 in 10 years. What is the interest rate?

102. On a warm day at the beach the temperature is 25°C. What is the speed of sound on such a day? (See Example 7.)

103. On a cold winter day the temperature is −3°C. What is the speed of sound on such a day? (See Example 7.)

104. Show that the area of a square is one-half the product of the lengths of the diagonals.

105. Find numbers a and b to show that $\sqrt{a + b} \neq \sqrt{a} + \sqrt{b}$.

Simplify the following expressions.

106. $\sqrt{\sqrt{\sqrt{\sqrt{65{,}536}}}}$

107. $\dfrac{(-1)^{n+3}}{(-1)^{n+1}}$

108. $\left(\sqrt[12]{\sqrt[6]{a^{80}}}\right)^{9}\left(\sqrt[12]{\sqrt[6]{a^{80}}}\right)^{9}$

109. $\left[\dfrac{(5x^{a}y^{b})^{4}}{(-5x^{a}y^{b})^{3}}\right]^{5}$

Simplify. Use your calculator for approximations to six decimal places where appropriate.

110. $\left[\left(\dfrac{3^3}{3} - 6 \cdot 3^2 + 36 \cdot 3\right) - \left(\dfrac{0^3}{3} - 6 \cdot 0^2 + 36 \cdot 0\right)\right] - 27$

111. $\dfrac{1}{2}\left[\left(\dfrac{3}{2} \cdot 6^2 + 6\right) - \left(\dfrac{3}{2} \cdot 2^2 + 2\right)\right]$

112. $\dfrac{1}{4}\left[\left(4 \cdot 2 - \dfrac{2^3}{3}\right) - \left(4 \cdot (-2) - \dfrac{(-2)^3}{3}\right)\right]$

113. $\left[\dfrac{5^2}{8} + 5 \cdot 5 - \dfrac{5^2}{12}\right] - \left[\dfrac{(-4)^2}{8} + 5 \cdot (-4) - \dfrac{(-4)^3}{12}\right]$

114. $\left[\dfrac{2^2}{8} + \dfrac{2}{2} - \dfrac{2^3}{12}\right] - \left[\dfrac{(-1)^2}{8} + \left(-\dfrac{1}{2}\right) - \dfrac{(-1)^3}{12}\right]$

115. $\left(\dfrac{11 - 1}{2 \cdot 5}\right)\pi[3^2 + 2(5^2) + 2(6^2) + 2(6^2) + 2(7^2) + 10^2]$

116. $\sqrt{(8\sqrt{17 - \pi})^2 - (-\sqrt{3} + \sqrt{\pi})^2}$

117. $\left[\dfrac{4\pi^2\left(1.069 \times 10^5 \times \dfrac{10{,}784}{2}\right)}{(6.673 \times 10^{-8})(5.976 \times 10^{27})}\right]^{1/2}$

✐ In Your Own Words

118. Discuss the merits of scientific notation. When would you find scientific notation more convenient than standard decimal notation? Less convenient?

119. Consult your calculator manual to determine how to enter a number in scientific notation. Describe the procedure in words.

120. Consult your calculator manual to determine the upper and lower limits on a number in scientific notation. Test these limits by attempting to enter a number outside them. What happens?

1.3 POLYNOMIALS

An **algebraic expression** is obtained by combining arithmetic operations and forming radicals using real numbers and variables. Here are some examples of algebraic expressions involving the single variable x:

$$3x + 5$$

$$\dfrac{x^2 - 3}{\sqrt{x}}$$

$$\sqrt[3]{3x^2 - 3x + \dfrac{2}{\sqrt{x}}}$$

Here are some algebraic expressions involving several variables:

$$xy + y^2$$
$$-13xzy$$
$$\pi r^2 h$$

Polynomial Expressions in a Single Variable

A **monomial** in the variable x is an algebraic expression of the form

$$ax^n$$

where a is a constant and n is a nonnegative integer. The number a is called the **coefficient** of the monomial, and n is called its **degree**. Some examples of monomials are:

$$5x^2, \frac{1}{3}x^9, 5x^0 \text{ or } 5$$

To multiply two monomials in the same variable x, we multiply the coefficients and add the corresponding degrees:

$$ax^n \cdot bx^m = abx^{m+n}$$

Thus, for example,

$$3x^2 \cdot 4x^5 = (3 \cdot 4)x^{2+5} = 12x^7$$

Definition 1 **Polynomial Expression**	A **polynomial expression in x** (**polynomial** for short) is a sum of monomials in x. That is, a polynomial in x is an expression of the form $$a_n x^n + a_{n-1} x^{n-1} + \cdots + a_1 x + a_0$$ where n is a nonnegative integer. The real numbers $a_n, a_{n-1}, \ldots, a_1, a_0$ are called the **coefficients** of the polynomial.

Here are some examples of polynomials in the variable x:

$$4x^3 - 2x^2 + x - 5, \qquad -5x + 3, \qquad x^2 + 1, \qquad -1$$

It is customary to omit any terms that have coefficient 0. For example, we write

$$3x^2 - 1$$

rather than

$$3x^2 + 0x - 1$$

Note that all of the exponents appearing in a polynomial must be nonnegative integers. Thus, the following are *not* polynomials:

$$2x^{-2} + 1, \qquad 5x^{1/2} + x$$

The monomials that form a polynomial are called its **terms**. The nonzero term of highest degree is called the **leading term** of the polynomial, its coefficient the **leading coefficient** of the polynomial, and its degree the **degree of the polynomial**. The term corresponding to the zero power of the variable is called the **constant term**. For example, the polynomial $-3x^2 + 2x + 1$ is of degree 2, has leading coefficient -3, leading term $-3x^2$, and constant term 1. The various terms are illustrated in Figure 1.

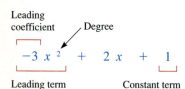

Leading coefficient / Degree

$$-3 x^2 + 2 x + 1$$

Leading term   Constant term

Figure 1

Two polynomials P and Q are said to be **equal** provided that the polynomials are of the same degree and that their corresponding coefficients are equal. The polynomial with all coefficients 0 is called the **zero polynomial** and is denoted **0**. Mathematicians agree not to assign this polynomial a degree.

Special terms are used to describe polynomials of low degrees. A polynomial of degree 0 has the form a, where a is a nonzero constant, and is called a **constant polynomial**. A polynomial of degree 1 has the form $ax + b$, where a and b are constants ($a \neq 0$). Such a polynomial is called **linear**. A polynomial of degree 2 has the form $ax^2 + bx + c$, where a, b, and c are constants. Such a polynomial is called **quadratic**. Polynomials of degrees 3 and 4 are called **cubic** and **quartic**, respectively.

Polynomials often arise in business applications. For instance, let x be the number of units produced by a factory, $R(x)$ be the revenue derived by selling x units, and $C(x)$ be the cost of producing x units. Then $R(x)$ and $C(x)$ are often given by polynomial expressions. Furthermore, the difference $R(x) - C(x)$ gives the profit earned by producing and selling x units.

Polynomials in Several Variables

The discussion above has been concerned exclusively with monomials and polynomials in a single variable. As a generalization, we may consider monomials and polynomials in several variables. A monomial in several variables is a product of a constant and nonnegative integer powers of the variables. Here are examples of monomials in variables x, y, and z:

$$-3xyz, \qquad 14x^2y^3, \qquad 0.5xy^3z^9$$

The degree of a monomial in several variables is the sum of the exponents of the variables that appear in the monomial. For instance, the degree of xyz is $1 + 1 + 1 = 3$, and the degree of the monomial x^2yz^3 is $2 + 1 + 3 = 6$.

A polynomial in several variables is a sum of monomials in the variables. Here are some examples of polynomials in the variables x, y, and z:

$$-3xyz + 14x^2y^3 - 5xy^3z^9, \qquad x + y + z, \qquad xy + yz + xz^8$$

The degree of a polynomial in several variables is the highest degree of the monomials that appear. For instance, the degrees of the three polynomials above are 13, 1, and 9, respectively.

Polynomials with different coefficients are regarded as different. For instance, $x^3 + 1$ is not the same polynomial as x^4, and we write

$$x^3 + 1 \neq x^4$$

Addition and Subtraction of Polynomials

Two terms of a polynomial are said to be **similar** if they have the same variables raised to the same powers. For example, the terms $-2xy^2$ and $4xy^2$ are similar, but the terms $4xy^2$ and $4x^2y$ are not similar.

Polynomials may be added and subtracted by combining similar terms. For example, if $P = 2x^2 + 3x - 1$ and $Q = -3x^2 + 4x - 4$, then we have

$$P + Q = [2 + (-3)]x^2 + (3 + 4)x + [(-1) + (-4)]$$
$$= -x^2 + 7x - 5$$
$$P - Q = [2 - (-3)]x^2 + (3 - 4)x + [(-1) - (-4)]$$
$$= 5x^2 - x + 3$$

Multiplying Polynomials and Other Algebraic Expressions

To multiply two polynomials, we repeatedly apply the distributive law and the laws of exponents, as illustrated in the following examples.

➤ **EXAMPLE 1**
Multiplication of Polynomials

Calculate the product

$$3x^2(4x^3 - 3x^2 - 1)$$

Solution

By the distributive law, the desired product can be obtained by multiplying the monomial by each term of the polynomial and adding the resulting terms.

$$3x^2(4x^3 - 3x^2 - 1) = (3x^2 \cdot 4x^3) + \left[3x^2 \cdot (-3x^2)\right] + \left[3x^2 \cdot (-1)\right]$$
$$= 12x^5 - 9x^4 - 3x^2$$

➤ **EXAMPLE 2**
More Multiplication of Polynomials

Determine the following product:

$$(x^2 - 1)(2x^2 - x - 3)$$

Solution

$$(x^2 - 1)(2x^2 - x - 3) = x^2(2x^2 - x - 3) - 1(2x^2 - x - 3) \quad \text{\textit{Distributive law}}$$
$$= 2x^4 - x^3 - 3x^2 - 2x^2 + x + 3 \quad \text{\textit{Laws of exponents}}$$
$$= 2x^4 - x^3 - 5x^2 + x + 3 \quad \text{\textit{Combining like terms}}$$

Product Identities

Many products occur so often that it is useful to memorize them.

PRODUCT IDENTITIES

Square of a binomial sum: $(A + B)^2 = A^2 + 2AB + B^2$
Square of a binomial difference: $(A - B)^2 = A^2 - 2AB + B^2$
Product of sum and difference: $(A + B)(A - B) = A^2 - B^2$

In these formulas, the variables have been denoted by A and B to emphasize that the formulas can be used with A and B replaced by any algebraic expression. For instance, suppose that we wish to calculate the product

$$(3x^2 - 4y)(3x^2 + 4y)$$

We immediately recognize this as the product of a sum and a difference, so we apply the product of sum and difference formula with A replaced by $3x^2$ and B replaced by $4y$. We obtain the result:

$$A^2 - B^2 = (3x^2)^2 - (4y)^2 = 9x^4 - 16y^2$$

We can also apply the various product identities with A and B replaced by expressions involving radicals, as the following example shows.

➤ **EXAMPLE 3**
Applying Product Identities

Determine the following products.

$$1. \ (2\sqrt{x} - 3x)\left(\frac{2}{x} + x^4\right) \quad 2. \ \left(\sqrt{x} + \frac{1}{\sqrt{x}}\right)^2 \quad 3. \ (\sqrt{x-1} - 2)(\sqrt{x-1} + 2)$$

Solution

1. We apply the distributive law. The products that arise are then calculated using the laws of exponents. The result is

$$(2\sqrt{x} - 3x)\left(\frac{2}{x} + x^4\right) = 4\frac{\sqrt{x}}{x} + 2x^4 \cdot \sqrt{x} - 6\frac{x}{x} - 3x^5$$

$$= \frac{4x^{1/2}}{x} + 2x^{1/2} \cdot x^4 - 6 - 3x^5$$

$$= \frac{4}{x^{1/2}} + 2x^{9/2} - 6 - 3x^5, \quad x > 0$$

2. By the Square of a Binomial Sum formula, with A replaced by $\sqrt{x}$ and B replaced by $1/\sqrt{x}$, we have

$$\left(\sqrt{x} + \frac{1}{\sqrt{x}}\right)^2 = (\sqrt{x})^2 + 2 \cdot (\sqrt{x}) \cdot \left(\frac{1}{\sqrt{x}}\right) + \left(\frac{1}{\sqrt{x}}\right)^2$$

$$= x + \frac{2\sqrt{x}}{\sqrt{x}} + \frac{1}{x} \qquad \textit{(since } (\sqrt{x})^2 = x)$$

$$= x + 2 + \frac{1}{x}, \quad x > 0$$

3. Apply the Product of Sum and Difference formula, with A replaced by $\sqrt{x} - 1$ and B replaced by 2. We then derive the product

$$(\sqrt{x-1} - 2)(\sqrt{x-1} + 2) = (\sqrt{x-1})^2 - 2^2$$

$$= (x - 1) - 4$$

$$= x - 5$$

➤ **EXAMPLE 4**
Square of a Binomial

Show that $x^2 + 1$ is not equal to $(x + 1)^2$.

Solution

We are asked to show that

$$x^2 + 1 \neq (x + 1)^2$$

Multiplying out the right side, this amounts to showing that

$$x^2 + 1 \neq x^2 + 2x + 1$$

This is clear. Both sides are polynomials, but they are not the same. It is important to note that even when two polynomials are unequal, they may be equal when a particular value is substituted for x. For instance, in this example, if we substitute 0 for x, then both sides have the value 1. This, however, does not mean that the *polynomials* are equal. In order for two polynomials to be equal, they must be equal for all values of x.

Factoring

Factoring is the process of writing a polynomial as a product of polynomials. For example, the polynomial $x^2 - x$ can be written as the product $x(x - 1)$. Factoring is an essential ingredient in the solution of many algebraic problems, such as finding the solutions to polynomial equations.

The simplest factors of a polynomial are monomial factors. In factoring a polynomial, we start by finding common monomial factors. This process is called **removing the greatest common factor**. It is illustrated in the following example.

➤ **EXAMPLE 5**
Removing Monomial Factors

Find common monomial factors and write each polynomial as a product of factors.

1. $3x^3 - 24x^2 - 18x$ 2. $\frac{1}{2}x^2 - 4x$

Solution

1. Each term of the polynomial contains as a factor $3x$, so we may remove the common factor $3x$ and write the polynomial in the form

$$3x^3 - 24x^2 - 18x = 3x \cdot x^2 - 3x \cdot 8x - 3x \cdot 6$$
$$= 3x(x^2 - 8x - 6)$$

2. In factoring, it is always a good strategy to factor out fractional coefficients as part of a monomial. In this example, factoring out the fraction $1/2$ will result in all integer coefficients in the remaining factor. In addition, all terms have a common factor x. So we have

$$\frac{1}{2}x^2 - 4x = \frac{1}{2}x \cdot x - \frac{1}{2}x \cdot 8$$
$$= \frac{1}{2}x(x - 8)$$

When factoring, always check your results by multiplication.

FACTORIZATION IDENTITIES

Square of binomial sum: $A^2 + 2AB + B^2 = (A + B)^2$
Square of binomial difference: $A^2 - 2AB + B^2 = (A - B)^2$
Difference of squares: $A^2 - B^2 = (A + B)(A - B)$
Sum of cubes: $A^3 + B^3 = (A + B)(A^2 - AB + B^2)$
Difference of cubes: $A^3 - B^3 = (A - B)(A^2 + AB + B^2)$

The following example illustrates how these factorization identities may be used to factor a variety of polynomial expressions.

➤ **EXAMPLE 6**
Applying Factorization
Identities

Factor the following polynomials.

1. $4x^2 - 12x + 9$
2. $16x^6 - 25y^4$
3. $8x^4 - 8x$
4. $27x^6 + y^3$

Solution

1. The expression can be written

$$4x^2 - 12x + 9$$
$$(2x)^2 - 2 \cdot 2x \cdot 3 + 3^2$$
$$A^2 - 2 \cdot A \cdot B + B^2$$

where

$$A = 2x$$
$$B = 3$$

So the Square of a Binomial Difference formula yields

$$4x^2 - 12x + 9 = (2x - 3)^2$$

2. This expression is a difference of two perfect squares, namely,

$$16x^6 - 25y^4$$
$$(4x^3)^2 - (5y^2)^2$$

Therefore, by the Difference of Squares formula, with A replaced by $4x^3$ and B replaced by $5y^2$, we have

$$16x^6 - 25y^4 = (4x^3 + 5y^2)(4x^3 - 5y^2)$$

3. Each term has a factor of $8x$. Before we attempt to apply any factorization formula, we factor out this monomial factor:

$$8x^4 - 8x$$
$$8x(x^3 - 1)$$

We recognize the second factor as being given by the Difference of Cubes formula, which then yields the factorization

$$8x^4 - 8x = 8x(x - 1)(x^2 + x + 1)$$

4. By the Sum of Cubes formula, with $A = 3x^2$ and $B = y$, we have

$$27x^6 + y^3 = (3x^2)^3 + y^3$$
$$= (3x^2 + y)\left[(3x^2)^2 - 3x^2y + y^2\right]$$
$$= (3x^2 + y)(9x^4 - 3x^2y + y^2)$$

➤ **EXAMPLE 7**

Factoring Polynomials

Factor the following polynomials into linear factors with integer coefficients, if possible.

1. $x^2 + 9x + 14$ 2. $15x^2 + 2x - 1$ 3. $x^2 + x + 1$

Solution

1. Suppose that we have a factorization of the form

$$(ax + b)(cx + d) = acx^2 + (ad + bc)x + bd$$

Matching coefficients, we must have $ac = 1$, $ad + bc = 9$, and $bd = 14$. Use the first and last equations to narrow the choices. A quick check of the various possibilities shows that

$$a = c = 1, \quad b = 7, \quad d = 2$$

So the factorization is

$$(x + 7)(x + 2)$$

2. We are looking for factors of the form $(ax + b)(cx + d)$, where a, b, c, and d are integers. The product ac must equal 15, and the product bd must equal -1. Furthermore, $ad + bc$ must equal 2. Testing all possibilities shows that the only one that works is the factorization

$$(5x - 1)(3x + 1)$$

3. In trying out all possibilities for a, b, c, and d in the Product of Binomials formula, we find no combination of linear factors that works. This quadratic polynomial cannot be factored into linear factors with integer coefficients.

Another technique of factoring is to group the terms so that you can apply the distributive law in reverse. The next example illustrates this technique.

➤ **EXAMPLE 8**

Factoring by Grouping

Factor the following polynomials.

1. $6x^3 - 9x^2 + 4x - 6$
2. $t^3 + 2t^2 - t - 2$
3. $x^3 - xy^2 + 5x^2 - 5y^2$

Solution

1. Use parentheses to group the terms:

$$(6x^3 - 9x^2) + (4x - 6) = 3x^2(2x - 3) + 2(2x - 3)$$
$$= (2x - 3)(3x^2 + 2)$$

2. In this example, after factoring by grouping, one of the factors may be factored as a difference of squares:

$$t^3 + 2t^2 - t - 2 = (t^3 + 2t^2) + 1 \cdot (-t - 2)$$
$$= t^2(t + 2) - 1 \cdot (t + 2)$$
$$= (t^2 - 1)(t + 2)$$
$$= (t + 1)(t - 1)(t + 2)$$

3. In this example, we apply factoring by grouping to a polynomial in two variables:

$$(x^3 - xy^2) + (5x^2 - 5y^2) = x(x^2 - y^2) + 5(x^2 - y^2)$$
$$= (x + 5)(x^2 - y^2)$$
$$= (x + 5)(x + y)(x - y)$$

Exercises 1.3

Determine whether each of the following expressions is a polynomial. If not, explain why. If it is, state the leading coefficient and the degree of the polynomial.

1. $34x^5 - 12x^4 + x^2 - 3x + 4.5$

2. $-3x + \frac{5}{4}y^2 + 10$

3. $-7x^{-2} + 3x^{-1} + 5x^2$

4. $\frac{2}{3}x^3 + 5x^2 + x - 3 + x^{-1}$

5. $3.4y^3 - 5y^2 + x - 3 + x^{-1}$

6. $x^2 + z^3 + \frac{2}{z}$

7. $m^4 + \sqrt{m} - m^3 + 10$

8. $t^{1/2} + 4t^{2/3} - t^3 - 10t^4 + 4t^5$

Perform the indicated operations.

9. $(45x^3 - 23x^2 - 7x + 12)$
$+ (-28x^3 + 16x^2 + 10x - 13)$

10. $(-29x^3 - 10x^2 + 4x - 5) + (12x^3 + 18x^2 - 7x + 3)$

11. $(45x^3 - 23x^2 + 4x - 5) - (12x^3 + 18x^2 - 7x + 3)$

12. $(-29x^3 - 10x^2 + 4x - 5) - (12x^3 + 18x^2 - 7x + 3)$

13. $\left(\frac{1}{4}x^3 + \frac{1}{3}x^2 - \frac{1}{4}x + \frac{1}{5}\right) - \left(\frac{1}{4}x^4 + 2x^2 + \frac{1}{2}x - 1\right)$

14. $\left(\frac{1}{3}x^3 + \frac{1}{2}x^2 - \frac{1}{4}x + \frac{1}{4}\right) - \left(\frac{1}{4}x^4 + 2x^2 + \frac{1}{2}x - 1\right)$

15. $\left(\frac{2}{5}x^4 - \frac{3}{2}x^3 - x + \frac{2}{3}\right) - \left(\frac{2}{3}x^3 + \frac{3}{4}x^2 - \frac{7}{8}x + \frac{1}{3}\right)$

16. $\left(\frac{5}{6}x^4 - \frac{2}{3}x^3 - \frac{1}{4}x + \frac{3}{2}\right) + \left(\frac{1}{6}x^3 + \frac{3}{4}x^2 - \frac{1}{2}x - \frac{1}{4}\right)$

17. $(4x^2 - 2x^3y - 5x^2y^2 + 2xy^3) + (4x^3y - 6xy^3 + y^4)$

18. $(4a^2 - 3a + 9) - (a^2 - 3a + 9)$

19. $(3p^4 - 9p^3q^2 - 2q^4) - (-p^4 - 2p^2q^3 + q^4)$

20. $(6m^2n - 3mn^2 + 4n^3) + (2m^3 - 5m^2n - 8mn^2)$

21. $(0.6x^4 - 0.5xy + 1.2y^2) + (3.2xy + 1.8x^2 - 0.7y^2)$

22. $(1.5ab - 0.02a^2 + 0.8b^2) + (0.5a^2 + 0.3ab - 1.7b^2)$

23. $(1.6x^2 - 0.5xy + 1.2y^2 + 1.2)$
$- (0.5x^2 + 0.3xy - 0.7y^2)$

24. $(0.3xy - 1.2x^2 + 1.4y^2) - (0.5x^2 + 0.3xy - 2.7y^2)$

25. $(3x\sqrt[3]{x} - 4x\sqrt{x} + 5x\sqrt[4]{x})$
$- (2x\sqrt{x} - 5x\sqrt[3]{x} - 4x\sqrt[4]{x})$

26. $(8x\sqrt[3]{x} - 4x\sqrt{x} + 5x\sqrt[4]{x})$
$+ (4\sqrt[3]{x^2} + 7\sqrt{x} - 10\sqrt[3]{x})$

27. $(4x\sqrt{y} - 3y\sqrt{x} + 2\sqrt{xy})$
$+ (4\sqrt{xy} - 7x\sqrt{y} - 2y\sqrt{x})$

28. $\left(\frac{2}{3}x^{-1} - \frac{1}{2}y\sqrt{x} - y\right) - \left(\frac{1}{2}x^{-1} - \frac{3}{4}y\sqrt{x} + 2y\right)$

In Exercises 29 and 30, an expression is given for the total revenue $R(x)$ from the sale of x units of a product and the total cost $C(x)$ of producing x units. Find an expression for the total profit $P(x)$, where

$$P(x) = R(x) - C(x)$$

29. $R(x) = 9x - 2x^2, C(x) = x^3 - 3x^2 + 4x + 1$

30. $R(x) = 100x - x^2,$

$$C(x) = \frac{1}{3}x^3 - 6x^2 + 89x + 100$$

Perform the indicated operations.

31. $(4x + 3)(2x + 5)$ 32. $(a - 5)(9a - 2)$

33. $(9a + 2)(9a - 2)$ 34. $(a - 2b)(a + 2b)$

35. $(a - 2)(a^2 + 2a + 4)$

36. $(m + n)(m^2 - mn + n^2)$

37. $(2a^2b^2 - 5ab)(3ab^2 + 2a^2b)$

38. $(2pq^2 + 10pq)(3p^2q - 5pq)$

39. $(5x - 3y)^2$ 40. $(a + 7t)^2$

41. $(2ab + 3bc)^2$

42. $(a - b)(a + b)(a^2 + b^2)$

43. $(4x^2 - 7xy)^2$

44. $(3x + t)(3x - t)(9x^2 + t^2)$

45. $(a + b)^2(a - b)^2$ 46. $(x + 2)^2(x - 2)^2$

47. $(a + b - y^3)(a + b + y^3)$

48. $(2x - y^2 + 5)(2x - y^2 - 5)$

49. $(\sqrt{2}xy - 3x)^2$ 50. $(\sqrt{5}x + 2x^2)^2$

51. $(\sqrt{3}a + \sqrt{2}b)(\sqrt{3}a - \sqrt{2}b)$

52. $(\sqrt{7}t - \sqrt{5})(\sqrt{7}t + \sqrt{5})$

53. $\left(\frac{4}{5}x - \frac{3}{7}y\right)\left(\frac{4}{5}x + \frac{3}{7}y\right)$ 54. $\left(\frac{4}{5}x + \frac{5}{6}y\right)^2$

55. $\left(\frac{2}{3}t^2 - \frac{3}{5}y^3\right)^2$ 56. $(1.1x - 2y)^2$

57. $\left(\frac{1}{4}x^2 - y^2\right)\left(\frac{1}{2}x^2 - 4xy + \frac{1}{2}y^2\right)$

58. $(3x - 2y)^2(3x + 2y)$

59. $(3x^2 + 4x - 5 + 3x + 2 - x^2)$
$\times (5x + 13 - 2x^2 - 10 - 6x + x^2)$

60. $[2ab + 4a^2 - 3(b^2 + a^2) + 2(b^2 - ab)]$
$\times [a^2 - 6ab + 4b^2 - 3(b^2 - 2ab)]$

61. Find a formula for $(A + B)^4$.

62. Find a formula for $(A - B)^4$.

Use the formulas found in Exercises 61 and 62 to do the following multiplications.

63. $(x + 8)^4$ 64. $(x - 1)^4$

65. $(6x^2 + 1)^4$ 66. $(3 - 4y^3)^4$

Factor the following expressions.

67. $-6x - 3x^2y + 9x$ 68. $8a^2 - 2a + 12ab$

69. $x^2 - 4x - 21$ 70. $x^2 - 11x + 30$

71. $(5x - 2)(x + 2) + (3x - 8)(5x - 2)$

72. $x(\sqrt{5} + \sqrt{7}) - y(\sqrt{5} + \sqrt{7})$

73. $14a^2b - 12ab^2$

74. $21b^2 + 18xb$

75. $xy - zy - xw + zw$

76. $ab + ac - mb - mc$

77. $2x^3 + x^2 - 8x - 4$

78. $x^3 - 2x^4 + 8y^3 - 16xy^3$

79. $529 - 324x^2$

80. $49 - 9y^2$

81. $8ax - 12x^2 + 4a^2x - 20x$

82. $6x - 3x - 10x^2$

83. $x^3 - 27$

84. $2z^3 - 16$

85. $8ax^2 + 24ax + 18a$

86. $2x^5 + 12x^3 + 18x$

87. $6x^2 + 29x + 35$

88. $9a^2 + 3a - 42$

89. $0.25x^2 - 0.49y^2$

90. $\frac{1}{4}a^2b^2 - \frac{1}{9}c^2$

91. $-2t^2 + 11t - 12$

92. $-6m^2 - m + 1$

93. $2a^3 - 16$

94. $16a^4b + 0.002ab^4$

95. $0.3x^2 - 0.2y^2$

96. $\frac{5}{3}a^2b^2 - \frac{8}{3}c^2d^2$

97. $27x^6 - 8y^3$

98. $2(x - 1)^2 + 3(x - 1) + 1$

99. $x^4 - 13x^2 + 36$

100. $(2x - 1)^2 - (x - 3)^2$

101. $(a + 2)^3 + 8$

102. $2xy + 2y - x^2 - x$

103. $x^2 + 2xy + y^2 - 25$

104. $a^2 - b^2 + 6b - 9$

105. $3x^6 - 24y^6$

106. $x^6 - y^6$

107. $\frac{1}{6}x^2 - \frac{1}{72}x - \frac{1}{12}$

108. $0.06x^2 - 0.07x - 0.2$

♦ **Applications**

109. Two rooms are being painted and carpeted. Room A is 5 m by 4 m, and room B is 6 m by 4 m. Both rooms are 2.5 m high. The cost of paint is p dollars to cover a square meter, and the cost of carpet is c dollars per square meter. Find a polynomial expression for the cost of finishing (a) room A, (b) room B, and (c) both rooms.

110. A gardener is preparing two areas for seeding. The first is a rectangle 6 m by 7 m. The second is a circle 8 m in diameter. It costs Z dollars per square meter for fertilizer and C dollars per meter for fencing. Find a polynomial expression for the cost of preparing (a) the first area, (b) the second area, and (c) both areas.

111. To prepare a swimming pool for summer, the surface area must be painted and it must be filled with water. Two circular pools are being prepared: one is 4 ft. deep and 10 ft. in diameter, and the other is 7 ft. deep and 20 ft. in diameter. It costs P dollars to paint a square foot and W dollars to fill one cubic foot. Find a polynomial expression for the cost of preparing (a) the smaller pool, (b) the larger pool, and (c) both pools.

112. One side of a rectangle is 3 m longer than the other side. Find a polynomial expression for the area of the rectangle in terms of the shorter side. Find an expression for the area in terms of the longer side.

113. A square rug lies in the center of a rectangular room. There are 2 ft. uncovered on two sides of the rug and 3 ft. on the other two sides. Find a polynomial expression for the area of the room in terms of a side of the rug.

114. A box is 10 ft. longer than it is high and 3 ft. wider than it is long. Find an expression for the volume of the box in terms of its height. Find an expression for the volume in terms of its length.

Simplify the following expressions. Assume that the variables in the exponents represent rational numbers.

115. $\left[\dfrac{3x^2 + (2x)^2}{(3x^2)^2 + 5x^4}\right]^{-2}$

116. $\left[\dfrac{2y^2 - (2y)^3}{5y^3 + (2y^2)^2}\right]^{-1}$

117. $(3x^a - 2y^c)^2$

118. $(x^a)^{a+b}(x^{-b})^{a-b}$

119. $(5t^n + 7)(4t^n - 5)$

120. $(x^n - y^n)(x^n + y^n)$

121. $(x^{m+n})^{m-n}(x^{m+n})^{m+n}$

122. $(x - 2)(x^2 + 2x + 4)(x + 2)(x^2 - 2x + 4)$

Factor the following expressions.

123. $a^{2n} - b^{2n}$

124. $10x^{2y} + x^y - 3$

125. $36t^{2n} - 60t^n + 25$

126. $a^{6n} - b^{3n}$

127. $1 + \dfrac{x^{12}}{1000}$

128. $a^{2x} + 10a^x + 25 - 36y^{2x}$

Verify each of the following statements.

129. $(3x + 5)^2 \neq 9x^2 + 25$

130. $(x - 2)^2 \neq x^2 - 4$

131. $(3x - 5y)(3x - 5y) \neq 9x^2 - 25y^2$

132. $x^3 - 1 \neq (x - 1)(x^2 - 2x + 1)$

1.4 RATIONAL EXPRESSIONS

A **rational expression in the variable** x is an algebraic expression of the form

$$\frac{P(x)}{Q(x)}$$

where $P(x)$ and $Q(x)$ are polynomials in x. We call $P(x)$ the **numerator** and $Q(x)$ the **denominator**.

Suppose that $R(x) = P(x)/Q(x)$ is a rational expression in x. We may evaluate $P(x)$ and $Q(x)$ for all real numbers x. This allows us to evaluate $R(x)$ for all values of x for which the denominator $Q(x)$ is nonzero. (Such values of x would lead to division by 0.)

For example, consider the rational expression

$$R(x) = \frac{x + 1}{(x - 2)(x - 1)}$$

Then for x equal to 0, the value of $R(x)$ is equal to

$$\frac{0 + 1}{(0 - 2)(0 - 1)} = \frac{1}{2}$$

However, $R(x)$ is undefined for x equal to either 1 or 2, since for these values of x, the denominator has the value 0.

A rational expression in several variables is a quotient of polynomials in the variables. Here are some examples of rational expressions in the variables x, y, and z:

$$\frac{x + y}{x}, \qquad \frac{x^2 + y^2 + 3xyz}{2x - 3y}$$

Rational expressions in several variables often occur in applications as the following example illustrates.

➤ **EXAMPLE 1**

Grade Point Averages

In many colleges and universities, a student earns 4 quality points for each credit hour of an A, 3 quality points for a B, 2 quality points for a C, 1 quality point for a D, and 0 quality points for an F. The total number of quality points earned is given by the polynomial

$$4a + 3b + 2c + d$$

where the student has earned a credit hours of A, b credit hours of B, c credit hours of C and d credit hours of D. The student's grade point average, or GPA, is given by the rational expression

Course	Grade	Credit Hours
Chemistry	B	3
Trigonometry	A	4
Business	C	4
History	C	3
English	B	3

$$\text{GPA} = \frac{4a + 3b + 2c + d}{a + b + c + d + f}$$

where f is the number of credit hours of F. One semester, a student receives the grades shown in the accompanying table. What is the student's GPA?

Solution

The number of credits of A is 4, the number of credits of B is $3 + 3 = 6$, and the number of credits of C is $4 + 3 = 7$. Substituting a equals 4, b equals 6, c equals 7, and d and f both equal 0 in the formula for GPA, we find that

$$\begin{aligned}
\text{GPA} &= \frac{4a + 3b + 2c + d}{a + b + c + d + f} \\
&= \frac{4 \cdot 4 + 3 \cdot 6 + 2 \cdot 7 + 0}{4 + 6 + 7 + 0 + 0} \\
&= \frac{48}{17} \\
&= 2.8
\end{aligned}$$

Reducing Rational Expressions

Two algebraic expressions are said to be equivalent provided that one can be changed into the other using a sequence of permissible algebraic operations. For example, the two expressions

$$(x + 1)^2, \qquad x^2 + 2x + 1$$

are equivalent to one another, since the second can be derived from the first using the associative, commutative, and distributive laws, as expressed in the rules for multiplication of polynomials.

Equivalent expressions may be used interchangeably in solving problems. For solving a particular problem, one equivalent expression may be easier to work with than another.

The procedure of **reducing a rational expression to lowest terms** is a technique of replacing a rational expression with an equivalent one, usually to obtain a simpler rational expression. Consider a quotient a/b of real numbers a and b, with $b \neq 0$, and suppose that c is a nonzero number. Then, from the properties of quotients in Section 1.1, we have the equalities

$$\frac{ac}{bc} = \frac{a}{b} \cdot \frac{c}{c} = \frac{a}{b} \cdot 1 = \frac{a}{b} \quad \text{if } b, c \neq 0$$

This sequence of equalities shows that the expression ac/bc is equivalent to the expression a/b. This equivalence comes about by removing the common nonzero factor c from both the numerator and denominator. We have used this procedure to reduce rational numbers to lowest terms. However, the procedure applies as well when $a, b,$ and c are values of algebraic expressions. Thus, we have the following general principle.

REDUCING RATIONAL EXPRESSIONS TO LOWEST TERMS

If $A(x)$, $B(x)$, and $C(x)$ are algebraic expressions, then we have the equivalence of algebraic expressions

$$\frac{A(x)C(x)}{B(x)C(x)} = \frac{A(x)}{B(x)}$$

which is valid for all values of x for which $B(x)$ and $C(x)$ are nonzero.

This principle asserts, roughly, that any common factors of the numerator and denominator of a rational expression may be eliminated. The process of removing common factors from the numerator and denominator of a rational function is called **reduction**. If the numerator and denominator of a rational expression have no nonconstant polynomials in common, then we say that the rational expression is in **lowest terms**. This terminology is analogous to that used in the case of rational numbers. In the case of rational numbers, we showed how the process of reduction to lowest terms could be accomplished using factorization. The same is true in the case of rational expressions, as the following example illustrates.

➤ **EXAMPLE 2**
Reducing a Rational Expression to Lowest Terms

Reduce the following rational expression to lowest terms.

$$\frac{x^5 - x^3}{x^2 + 3x + 2}$$

Solution

Factor the numerator and denominator to obtain the equivalent expression

$$\frac{x^3(x^2 - 1)}{(x + 1)(x + 2)} = \frac{x^3(x - 1)(x + 1)}{(x + 2)(x + 1)}$$

The denominator is 0 if x has either of the values -2 or -1, so to avoid division by 0, we must exclude these values. We see that the numerator and denominator have the common factor $x + 1$. So, the expression is equivalent to

$$\frac{x^3(x - 1)}{x + 2}, \quad x \neq -1, -2$$

It is clear that the numerator and denominator of this rational expression have no factors in common, so that it is in lowest terms. It is acceptable to leave the rational expression in factored form. Or, alternatively, we may multiply out the numerator and denominator and write it in the equivalent form. It is important to note that the final expression is perfectly well-behaved for $x = -1$. By removing the common factor $x + 1$ from the numerator and denominator, the problem with division by 0 is eliminated. However, this does not entitle you to say that the final expression equals the original expression when $x = 1$, since the original expression is undefined for this value.

Warning: When simplifying an expression by removing common factors of the numerator and denominator, the results are valid only for values of the variable(s) for which the common factor is nonzero. This is easy to overlook but may lead to strange results if you later substitute a value that is forbidden by the simplification.

➤ **EXAMPLE 3**
Another Reduction

Reduce the following rational expression to lowest terms:

$$\frac{3x^2 + 3x}{x^6 - 1}$$

Solution

We factor the numerator and denominator

$$\frac{3x^2 + 3x}{x^6 - 1} = \frac{3x(x + 1)}{(x^3 - 1)(x^3 + 1)}$$

$$= \frac{3x(x + 1)}{(x - 1)(x^2 + x + 1)(x + 1)(x^2 - x + 1)}$$

$$= \frac{3x}{(x - 1)(x^2 + x + 1)(x^2 - x + 1)}$$

Since the final form of the numerator and denominator have no common factors, the last expression is in lowest terms.

Addition and Subtraction of Rational Expressions

Let $A(x)/Q(x)$ and $B(x)/Q(x)$ be rational expressions with the same denominator. These expressions may be added or subtracted according to the formula

$$\frac{A(x)}{Q(x)} \pm \frac{B(x)}{Q(x)} = \frac{A(x) \pm B(x)}{Q(x)}$$

To add or subtract rational expressions having *different* denominators, it is necessary to write them in equivalent forms having the same denominators. To see how this may be done, consider the rational expressions

$$\frac{x}{x-1}, \qquad \frac{x+1}{x+2}$$

The denominator of the first expression is $x - 1$ and that of the second is $x + 2$. We may replace each expression by one in which numerator and denominator are multiplied by a common factor. Let's multiply the numerator and denominator of the first expression by $x + 2$ and the numerator and denominator of the second by $x - 1$. The two expressions are then replaced, respectively, by the equivalent expressions

$$\frac{x(x+2)}{(x-1)(x+2)}, \qquad \frac{(x+1)(x-1)}{(x+2)(x-1)}$$

Note that the two expressions now have the *same* denominator and so can be added. The denominator $(x + 2)(x - 1)$ is a common multiple of the denominators of the original expressions. In fact, it is the *least* common multiple, in the sense that any multiple of the two denominators must have as factors at least $x - 1$ and $x + 2$.

In general, to determine the least common multiple of two polynomials, it is simplest to first factor them and form the polynomial that consists of the product of the highest power of each factor that appears. For instance, consider the three polynomials

$$(x-1)^3(x-2), \qquad (x-1)(x-2)^4, \qquad (x+1)^2$$

To form the least common multiple, we take the product of the highest power of each factor that appears, namely

$$(x-1)^3(x-2)^4(x+1)^2$$

The following example illustrates how rational functions may be added and subtracted by replacing them with rational expressions having as a common denominator the least common multiple of the given denominators.

➤ **EXAMPLE 4**
Adding Rational Expressions

Perform the indicated additions of rational expressions.

1. $\dfrac{1}{x} + \dfrac{1}{x-1}$ 2. $\dfrac{2x}{x-1} + \dfrac{2}{x+1}$ 3. $\dfrac{1}{x^2+x+1} + \dfrac{1}{x-1}$

Solution

1. The least common multiple of the given denominators is $x(x - 1)$. We multiply numerator and denominator of the first expression by $x - 1$ and those of the second expression by x. This results in expressions with denominator equal to $x(x - 1)$:

$$\frac{1}{x} + \frac{1}{x-1} = \frac{1}{x} \cdot \frac{x-1}{x-1} + \frac{1}{x-1} \cdot \frac{x}{x}$$

$$= \frac{(x-1)+x}{x(x-1)}$$

$$= \frac{2x-1}{x^2-x}, \qquad x \neq 0, 1$$

2. The least common multiple of the denominators is $(x - 1)(x + 1)$.

$$\frac{2x}{x - 1} + \frac{2}{x + 1} = \frac{2x}{(x - 1)} \cdot \frac{x + 1}{x + 1} + \frac{2}{(x + 1)} \cdot \frac{x - 1}{x - 1}$$

We may now add the rational expressions

$$\frac{2x(x + 1) + 2(x - 1)}{(x - 1)(x + 1)} = \frac{2x^2 + 2x + 2x - 2}{(x + 1)(x - 1)}$$

$$= \frac{2x^2 + 4x - 2}{x^2 - 1} \quad (x \neq 1, -1)$$

(applying the Sum and Difference product formula)

3. In this example, the least common multiple of the denominators is $(x - 1)(x^2 + x + 1)$. Therefore, we have

$$\frac{1}{x^2 + x + 1} + \frac{1}{x - 1} = \frac{1}{x^2 + x + 1} \cdot \frac{x - 1}{x - 1} + \frac{1}{x - 1} \cdot \frac{x^2 + x + 1}{x^2 + x + 1}$$

$$= \frac{(x - 1) + (x^2 + x + 1)}{(x^2 + x + 1)(x - 1)}$$

$$= \frac{x^2 + 2x}{x^3 - 1} \quad (x \neq 1) \qquad \textit{(Difference of Cubes)}$$

➤ **EXAMPLE 5**

Adding Three Rational Expressions

Express the following as a rational expression in lowest terms (or reduced form).

$$\frac{8y}{y^2 - 1} - \frac{2}{1 - y} + \frac{4}{y + 1}$$

Solution

We must first determine the least common multiple of the three denominators. The first denominator may be written in the form $(y + 1)(y - 1)$. Moreover, the second denominator can be written in the form $-(y - 1)$. Therefore, the expression can be written in the form

$$\frac{8y}{(y + 1)(y - 1)} - \frac{2}{-(y - 1)} + \frac{4}{y + 1} = \frac{8y}{(y + 1)(y - 1)} + \frac{2}{y - 1} + \frac{4}{y + 1}$$

From the last expression, we see that the least common multiple of the denominators is $(y + 1)(y - 1)$. So we write the three rational expressions with $(y + 1)(y - 1)$ as denominator:

$$\frac{8y}{(y - 1)(y + 1)} + \frac{2(y + 1)}{(y - 1)(y + 1)} + \frac{4(y - 1)}{(y + 1)(y - 1)}$$

$$= \frac{8y + 2y + 2 + 4y - 4}{(y - 1)(y + 1)}$$

$$= \frac{14y - 2}{(y - 1)(y + 1)}$$

$$= \frac{2(7y - 1)}{(y + 1)(y - 1)}, \quad (y \neq -1, 1)$$

Multiplication and Division of Rational Expressions

Multiplication of rational expressions is carried out using the following formula:

$$\frac{a}{b} \cdot \frac{c}{d} = \frac{ac}{bd}$$

The next example illustrates how this formula may be applied.

➤ **EXAMPLE 6**
Multiplying Rational Expressions

Perform the indicated multiplications of rational expressions.

1. $\dfrac{x}{x-1} \cdot \dfrac{2}{x+1}$

2. $\dfrac{x^2}{3x+1} \cdot \dfrac{x}{5x+1}$

Solution

1. $\dfrac{x}{x-1} \cdot \dfrac{2}{x+1} = \dfrac{2x}{(x+1)(x-1)}$

$$= \frac{2x}{x^2-1}, \quad x \neq 1, -1 \qquad \textit{(by the Sum and Difference product formula)}$$

2. $\dfrac{x^2}{3x+1} \cdot \dfrac{x}{5x+1} = \dfrac{x^2 \cdot x}{(3x+1)(5x+1)}$

$$= \frac{x^3}{15x^2+8x+1}, \quad x \neq -\frac{1}{3}, -\frac{1}{5}$$

(Law 1 of exponents, distributive law)

Division of rational expressions may be carried out using the formula:

$$\frac{a/b}{c/d} = \frac{a}{b} \cdot \frac{d}{c} = \frac{ad}{bc}$$

The next example shows how to apply this formula.

➤ **EXAMPLE 7**
Dividing Rational Expressions

Calculate the following quotients.

1. $\dfrac{x^2-1}{x} \div \dfrac{1}{x-1}$

2. $\dfrac{a-b}{a+b} \div \dfrac{a+2b}{a-b}$

Solution

1. $\dfrac{\dfrac{x^2-1}{x}}{\dfrac{1}{x-1}} = \dfrac{x^2-1}{x} \cdot \dfrac{x-1}{1}$

$$= \frac{(x^2-1)(x-1)}{x}$$

$$= \frac{x^3-x^2-x+1}{x}, \quad x \neq 1, 0$$

2. $\dfrac{\dfrac{a-b}{a+b}}{\dfrac{a+2b}{a-b}} = \dfrac{a-b}{a+b} \cdot \dfrac{a-b}{a+2b}$

$$= \frac{(a-b)^2}{(a+b)(a+2b)}, \quad a \neq b, -b, -2b$$

The next example combines all of the operations on rational expressions that we have discussed to simplify complex rational expressions.

➤ **EXAMPLE 8**

Reducing a Complex Rational Expression

Write the following as a rational expression in lowest terms.

$$\frac{a^{-1} + b^{-1}}{a^{-3} + b^{-3}}$$

Solution

The given expression may be written in the form

$$\frac{\dfrac{1}{a} + \dfrac{1}{b}}{\dfrac{1}{a^3} + \dfrac{1}{b^3}}$$

Multiply both the numerator and denominator by $a^3 b^3$, to obtain the equivalent expression

$$\frac{\left(\dfrac{1}{a} + \dfrac{1}{b}\right) \cdot (a^3 b^3)}{\left(\dfrac{1}{a^3} + \dfrac{1}{b^3}\right) \cdot (a^3 b^3)} = \frac{\dfrac{a^3 b^3}{a} + \dfrac{a^3 b^3}{b}}{\dfrac{a^3 b^3}{a^3} + \dfrac{a^3 b^3}{b^3}}$$

$$= \frac{a^2 b^3 + a^3 b^2}{b^3 + a^3}$$

$$= \frac{a^2 b^2 (b + a)}{b^3 + a^3}$$

And, by the Sum of Cubes factorization formula, we can rewrite the expression in the form

$$\frac{(a^2 b^2)(a + b)}{(a + b)(a^2 - ab + b^2)}$$

Simplifying this last expression to lowest terms gives us the result

$$\frac{a^2 b^2}{a^2 - ab + b^2}, \quad a, b \neq 0, a \neq -b$$

Second Method   Write the expression in the form

$$\frac{\dfrac{1}{a} + \dfrac{1}{b}}{\dfrac{1}{a^3} + \dfrac{1}{b^3}} = \frac{\dfrac{1}{a} + \dfrac{1}{b}}{\left(\dfrac{1}{a}\right)^3 + \left(\dfrac{1}{b}\right)^3}$$

Introduce new variables x and y defined by $x = 1/a$, $y = 1/b$. In terms of these variables the given expression may be written

$$\frac{x + y}{x^3 + y^3} = \frac{x + y}{(x + y)(x^2 - xy + y^2)} \qquad \textit{(Sum of Cubes formula)}$$

Simplifying the last expression to lowest terms and converting back to the original variables gives

$$\frac{1}{x^2 - xy + y^2} = \frac{1}{\left(\frac{1}{a}\right)^2 - \left(\frac{1}{a}\right)\left(\frac{1}{b}\right) + \left(\frac{1}{b}\right)^2}$$

Adding the rational expressions in the denominator gives us

$$\frac{1}{\dfrac{b^2}{a^2b^2} - \dfrac{ab}{a^2b^2} + \dfrac{a^2}{a^2b^2}} = \frac{1}{\dfrac{b^2 - ba + a^2}{a^2b^2}}$$

$$= \frac{a^2b^2}{a^2 - ab + b^2}, \quad a, b \neq 0, a \neq -b$$

Rationalizing Denominators and Numerators

Expressions often involve radicals in the denominator. By multiplying numerator and denominator by a suitable common expression, you can often arrive at a form in which the radicals appear only in the numerator. For example, consider the expression

$$\frac{1}{\sqrt{2}}$$

Here a radical appears in the denominator. To remove the radical from the denominator, multiply both numerator and denominator by $\sqrt{2}$ to obtain

$$\frac{1}{\sqrt{2}} = \frac{1 \cdot (\sqrt{2})}{(\sqrt{2}) \cdot (\sqrt{2})} = \frac{\sqrt{2}}{2}$$

The process of removing a radical from a denominator is called **rationalizing the denominator**. This may be accomplished in most situations by multiplying both numerator and denominator by a suitable expression. Choosing the right expression to multiply by takes a bit of practice. However, as a guide, use the various product identities we have discussed in order to arrive at a product in the denominator that contains no radicals. For instance, suppose that the denominator is the expression

$$1 + \sqrt{x}$$

The Sum and Difference of Squares formula suggests that this denominator may be rationalized by multiplying by the expression

$$1 - \sqrt{x}$$

Indeed, the product of the two expressions is equal to

$$(1 + \sqrt{x})(1 - \sqrt{x}) = 1^2 - (\sqrt{x})^2 = 1 - x$$

As another example, suppose that the denominator of an expression is equal to

$$\sqrt[3]{x}$$

We may rationalize the denominator by multiplying both numerator and denominator by $x^{2/3}$. Indeed, by the laws of exponents, the new denominator is equal to

$$\sqrt[3]{x} \cdot x^{2/3} = x^{1/3} x^{2/3} = x^{1/3+2/3} = x$$

The following example provides some practice in rationalizing denominators.

➤ **EXAMPLE 9**
Rationalizing Denominators

Rationalize the denominator of the following expressions.

1. $\dfrac{1}{x + \sqrt{x}}$

2. $\dfrac{\sqrt[3]{x}}{\sqrt[3]{x} - 2}$

Solution

1. The Difference of Squares product formula suggests that we multiply the numerator and denominator by the expression $x - \sqrt{x}$. Doing so gives us the following equivalent expression:

$$\frac{1}{x + \sqrt{x}} = \frac{1}{x + \sqrt{x}} \cdot \frac{x - \sqrt{x}}{x - \sqrt{x}}$$

$$= \frac{x - \sqrt{x}}{x^2 - (\sqrt{x})^2}$$

$$= \frac{x - \sqrt{x}}{x^2 - x}$$

2. The Difference of Cubes product formula, with $A = \sqrt[3]{x}$ and $B = 2$, suggests that we multiply the numerator and denominator by $A^2 + AB + B^2$. The denominator will then be equal to

$$A^3 - B^3 = (\sqrt[3]{x})^3 - 2^3 = x - 8$$

Carrying out this calculation, we are able to rationalize the denominator:

$$\frac{\sqrt[3]{x}}{\sqrt[3]{x} - 2} = \frac{\sqrt[3]{x}}{\sqrt[3]{x} - 2} \cdot \frac{\sqrt[3]{x^2} + 2\sqrt[3]{x} + 2^2}{\sqrt[3]{x^2} + 2\sqrt[3]{x} + 2^2}$$

$$= \frac{x + 2\sqrt[3]{x^2} + 4\sqrt[3]{x}}{x - 8}$$

In calculus, it is often necessary to write expressions so that they have no radicals in the *numerator*. This process is called **rationalizing the numerator** and is illustrated in the following example.

➤ **EXAMPLE 10**
Rationalizing a Numerator

Write the expression

$$\frac{\sqrt{x + h} - \sqrt{x}}{h}$$

in the form that involves no radicals in the numerator.

Solution

The Difference of Squares formula suggests that we multiply both numerator and denominator by the expression

$$\sqrt{x+h} + \sqrt{x}$$

This gives us the equivalent expression

$$\frac{(\sqrt{x+h} - \sqrt{x}) \cdot (\sqrt{x+h} + \sqrt{x})}{h \cdot (\sqrt{x+h} + \sqrt{x})}$$

Applying the Sum and Difference formula to the numerator, we have

$$\frac{(\sqrt{x+h})^2 - (\sqrt{x})^2}{h(\sqrt{x+h} + \sqrt{x})}$$

We can simplify the numerator to read

$$\frac{(x+h) - x}{h(\sqrt{x+h} + \sqrt{x})} = \frac{h}{h(\sqrt{x+h} + \sqrt{x})}$$

$$= \frac{1}{\sqrt{x+h} + \sqrt{x}}$$

Thus,

$$\frac{\sqrt{x+h} - \sqrt{x}}{h} = \frac{1}{\sqrt{x+h} + \sqrt{x}} \qquad (h \neq 0)$$

➤ **EXAMPLE 11**
Another Numerator
to Rationalize

Rationalize the numerator of the expression

$$\frac{\sqrt[3]{x+2} - \sqrt[3]{x}}{\sqrt[3]{x}}$$

Solution

We must come up with an expression whose product with the numerator has no radicals present. When looking for such an expression, you should always keep in mind the various product/factoring formulas. In this case, the cube roots suggest the Difference of Cubes formula:

$$A^3 - B^3 = (A - B)(A^2 + AB + B^2)$$

If we set $A = \sqrt[3]{x+2}$ and $B = \sqrt[3]{x}$, then this formula reads

$$2 = (\sqrt[3]{x+2})^3 - (\sqrt[3]{x})^3 = (\sqrt[3]{x+2} - \sqrt[3]{x})(\sqrt[3]{(x+2)^2} + \sqrt[3]{x(x+2)} + \sqrt[3]{x^2})$$

Using this formula, we may rationalize the desired numerator:

$$\frac{\sqrt[3]{x+2} - \sqrt[3]{x}}{\sqrt[3]{x}} = \frac{(\sqrt[3]{x+2} - \sqrt[3]{x})}{\sqrt[3]{x}} \cdot \frac{(\sqrt[3]{(x+2)^2} + \sqrt[3]{x(x+2)} + \sqrt[3]{x^2})}{(\sqrt[3]{(x+2)^2} + \sqrt[3]{x(x+2)} + \sqrt[3]{x^2})}$$

$$= \frac{2}{\sqrt[3]{x}(\sqrt[3]{(x+2)^2} + \sqrt[3]{x(x+2)} + \sqrt[3]{x^2})}$$

Exercises 1.4

Write the following rational expressions in lowest terms.

1. $\dfrac{x^2 - 1}{x^3 - 1}$

2. $\dfrac{a^3 + b^3}{a^2 - b^2}$

3. $\dfrac{4x^2 - 4}{8x^2 + 28x + 12}$

4. $\dfrac{x^2 - 9}{x^2 + 2x + 1}$

5. $\dfrac{a^4 + 2a^3 + a^2}{a^2 + 2a + 1}$

6. $\dfrac{t(t^2 + 7t + 12)}{t^2 + t - 12}$

7. $\dfrac{2r + 2s}{2(r^2 + s^2 + 2rs)}$

8. $\dfrac{(xy)^3 - 1}{xy - 1}$

9. $\dfrac{x^3 - 1}{(x^2 + x + 1)^2}$

10. $\dfrac{2(x - 1)^2 - (2x - 1)(2x - 2)}{(x - 4)^4}$

11. $\dfrac{(x^2 - 2x + 4)(1) - (x + 1)(x + 2)}{(x^2 - 2x + 4)^2}$

12. $\dfrac{(x^2 + y^2)(2x) - (x^2 - y^2)(2x)}{(x^2 + y^2)^2}$

Perform the indicated operations and simplify.

13. $\dfrac{x^2 + 5x + 6}{x^2 + 8x + 15} \cdot \dfrac{x^2 + 9x + 20}{x^2 + 6x + 8}$

14. $\dfrac{5x^2 + 30x + 45}{6x^2 - 24} \cdot \dfrac{3x^2 + 12x + 12}{10x^2 - 90}$

15. $\dfrac{t^3 - 1}{t^2 - 1} \div \dfrac{t^2 + t + 1}{t - 1}$

16. $\dfrac{a^3 + c^3}{4a - 4c} \div \dfrac{a^3 + a^2ac^2}{2a^2 - 2c^2}$

17. $\dfrac{x^3 + 2x^2 + x + 2}{2x^4 + 6x^2 + 4} + \dfrac{2x^3 + 4x^2 + 6x + 12}{3x^4 - 3x^2 - 18}$

18. $\dfrac{x^2 + 2xy + y^2 + x + y}{x^2 + 6x^2 + 4} + \dfrac{2x^2 + 3xy + y^2 + 2x + y}{x^2 + 3xy + 2y^2y^2}$

19. $\dfrac{3x - 5}{2x + 1} - \dfrac{5 - 3x}{2x + 1}$

20. $\dfrac{7a + 3}{2a + 7} - \dfrac{4a + 7}{2a + 7}$

21. $\dfrac{3x - 7}{x - 3} + \dfrac{2x + 3y}{3 - x}$

22. $\dfrac{3a}{a - b} + \dfrac{2b + 4}{b - a} + \dfrac{6a - b}{a - b}$

23. $\dfrac{2t^2 - 3t + 5}{t^2 + 3t - 5} - \dfrac{t^2 - t + 2}{5 - 3t - t^2}$

24. $\dfrac{x^2 + 6x + 12}{x^2 - 3x - 8} + \dfrac{2x^2 - 6x + 4}{8 + 3x - x^2}$

25. $\dfrac{x - (3x + 2)}{x^2 - 3x - 8} + \dfrac{2x^2 - 6x + 4}{8 + 3x - x^2}$

26. $\dfrac{a - 4 - (5 - a)}{a^3 + 2a^2 + a} - \dfrac{2a - 4 - (a - 3)}{a^2 - 1}$

27. $\dfrac{t}{t^2 + 5t + 6} - \dfrac{2}{t^2 + 3t + 2}$

28. $\dfrac{t}{t^2 + 11t + 30} - \dfrac{5}{t^2 + 9t + 20}$

29. $\dfrac{6}{x + 5} - \dfrac{2}{x - 3} + \dfrac{4x - 1}{x^2 + 2x - 15}$

30. $\dfrac{4a}{a - b} + \dfrac{3b}{a + b} - \dfrac{2ab}{a^2 - b^2}$

31. $\dfrac{1}{x - 1} - \dfrac{2}{x^2 - 1} - \dfrac{3}{x^3 - 1}$

32. $\dfrac{1}{a + b} + \dfrac{1}{a^2 + 2ab + b^2} + \dfrac{1}{a^3 + 3a^2b + 3ab^2 + b^3}$

33. $\dfrac{n(n + 1)(n + 2)}{2 \cdot 3} + \dfrac{(n + 1)(n + 2)}{2}$

34. $\dfrac{n(n + 1)(n + 1)(n + 3)}{2 \cdot 3 \cdot 4} + \dfrac{(n + 1)(n + 2)(n + 3)}{2 \cdot 3}$

Simplify.

35. $\dfrac{ab^{-1} + ba^{-1}}{a^{-1} + b^{-1}}$

36. $\dfrac{2a^{-1} + 2y^{-1}}{(x + y)y^{-1}}$

37. $\dfrac{\dfrac{1}{x} + \dfrac{1}{y} + \dfrac{1}{z}}{\dfrac{1}{xy} + \dfrac{1}{yz}}$

38. $\dfrac{\dfrac{x}{y} + \dfrac{y}{x} + 2}{\dfrac{y + 1}{x} + \dfrac{y + 1}{y}}$

39. $\dfrac{\dfrac{1}{x + h} - \dfrac{1}{x}}{h}$

40. $\dfrac{\dfrac{1}{(x + h)^2} - \dfrac{1}{x^2}}{h}$

41. $\dfrac{\dfrac{1}{(x + h)^3} - \dfrac{1}{x^3}}{h}$

42. $\dfrac{a + 1 - 2a^{-1}}{a + 4 + 4a^{-1}}$

43. $\dfrac{\left[(x + x^{-1})^{-1}\right]^{-1}}{x + x^{-1}}$

44. $\dfrac{(a + h)^3 - a^3}{h}$

45. $\dfrac{(a + h)^2 - a^2}{h}$

46. $\dfrac{(x + h)^2 - 5(x + h) - (x^2 - 5x)}{h}$

47. $\dfrac{(x + h)^2 + 3(x + h) + 2 - (x^2 + 3x + 2)}{h}$

48. $\dfrac{\dfrac{1}{2(x + h) + 1} - \dfrac{1}{2x + 1}}{h}$

49. $\dfrac{\dfrac{x + h}{x + h + 1} - \dfrac{x}{x + 1}}{h}$

◆ Applications

50. Use the formula for grade point average given in Example 1. Suppose a student gets the following grades for one term:

Course	Grade	Number of Hours in Course
Accounting	B	4
College Algebra	A	5
English	A	5
French	C	3
Phys. Ed.	F	1

Find the student's GPA.

51. A baseball hitter's slugging percentage, P, is given by the rational expression

$$P = \frac{s + 2d + 3t + 4h}{B}$$

where s = number of singles, d = number of doubles, t = number of triples, h = number of home runs, and B = number of at-bats. In his career, Hank Aaron had 2294 singles, 624 doubles, 98 triples, and 755 home runs in 12,364 at-bats. What was his slugging percentage?

52. In the theory of electricity, the total resistance R for two resistors in parallel between points a and b (see figure) is given by the rational expression

$$R = \frac{1}{\dfrac{1}{R_1} + \dfrac{1}{R_2}}$$

where R_1 and R_2 are the values of the separate resistors.

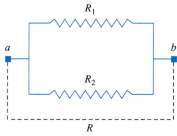

Exercise 52

a. Simplify this rational expression.

b. What is the total resistance of a circuit with a 10-ohm and a 5-ohm resistor in parallel?

53. The *coefficient of linear expansion* α of a material (see figure) describes how it expands as it is heated and is given by the rational expression

$$\alpha = \frac{L_2 - L_1}{L_1 (t_2 - t_1)}$$

where L_1 is the length at the beginning temperature t_1 and L_2 is the length at temperature t_2. Find the coefficient of linear expansion of aluminum if the length of a bar is 10 m at 10°C and is 10.02 m at 100°C. How long will the bar be at 250°C?

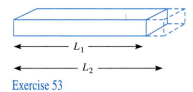

Exercise 53

54. Refer to the preceding problem. Find the coefficient of linear expansion of steel if the length of a bar is 6 m at 20°C and is 6.012 m at 200°C.

Rationalize the denominators of the following expressions.

55. $\dfrac{\sqrt{3x}}{\sqrt{4y}}$

56. $\dfrac{\sqrt[3]{2}}{\sqrt[3]{12}}$

57. $\dfrac{4}{\sqrt{5} + \sqrt{6}}$

58. $\dfrac{6}{\sqrt{3} - \sqrt{8}}$

59. $\dfrac{1}{\sqrt{a} - \sqrt{b}}$

60. $\dfrac{3}{\sqrt{x} + \sqrt{h}}$

61. $\dfrac{\sqrt{a} + \sqrt{2}}{\sqrt{a} - \sqrt{b}}$

62. $\dfrac{2\sqrt{x} + 3\sqrt{3}}{\sqrt{x} - \sqrt{3}}$

Rationalize the numerator. Many of these exercises are like those encountered in calculus.

63. $\dfrac{\sqrt{a + h} - \sqrt{a}}{h}$

64. $\dfrac{2\sqrt{x} + 3\sqrt{y}}{x}$

65. $\dfrac{\sqrt{a} + \sqrt{2}}{\sqrt{a} - \sqrt{2}}$

66. $\dfrac{2\sqrt{x} + 3\sqrt{3}}{\sqrt{x} - \sqrt{3}}$

67. $\dfrac{\sqrt{n + 1} - \sqrt{n}}{\sqrt{n}}$

68. $\sqrt{n^2 + 1} - n$

69. $\dfrac{1 - \sqrt{x}}{1 - x}$

70. $\dfrac{\sqrt{4 + h} - 2}{h}$

71. $\dfrac{\sqrt{x + h} - (x + h) - (\sqrt{x} - x)}{h}$

72. $\dfrac{\sqrt{2(x + h) - 1} - \sqrt{2x - 1}}{h}$

Simplify. Many of these exercises are like those frequently seen in calculus.

73. $\sqrt{x^2 - 4} + \dfrac{x^2 - 4}{\sqrt{x^2 - 4}}$

74. $\dfrac{\dfrac{1}{\sqrt{x + h}} - \dfrac{1}{\sqrt{x}}}{h}$

75. $\dfrac{(x - 2)^{n+2}}{(n + 1)^2} \div \dfrac{(x - 2)^n}{n}$

76. $\dfrac{(-1)^n (x - 1)^{n+1}}{n + 1} \div \dfrac{(-1)^{n-1}(x - 1)^n}{n}$

77. $60\left[(x^3 + 15x)^{19} 2x \right.$
$\left. + 19(x^3 + 15x)^{18}(3x^2 + 15)(x^2 + 5) \right]$

78. $\dfrac{3(x - 1)^2 + (x - 1)^3}{(x - 1)^2}$

79. $\dfrac{1 - \dfrac{x}{\sqrt{x^2 + 1}}}{x - \sqrt{x^2 + 1}}$

80. $\dfrac{(x^2 + y^2)^{1/2}(1) - \left(\dfrac{1}{2}\right)(x^2 + y^2)^{-1/2}(2x)}{\left[\sqrt{x^2 + y^2} \right]^2}$

81. $\dfrac{(x^2 - 4)^{1/3}(5) - \left(\dfrac{1}{3}\right)(x^2 - 4)^{-2/3}(2x)(5x)}{\left[(x^2 - 4)^{1/3} \right]^2}$

82. Show that

$$\left(\sqrt{8 - \sqrt{48}} \right)^2 = (\sqrt{2} - \sqrt{6})^2$$

To program a computer in a language known as BASIC, one has to be able to convert between BASIC and algebraic symbolism. The following chart shows the BASIC symbols and their corresponding algebraic symbols.

BASIC Notation	Algebraic Notation
A+B	$a + b$
A−B	$a - b$
A*B	ab
A/B	$\dfrac{a}{b}$, or $a \div b$
A^B	a^b
( )	( )

The following examples illustrate common conversions.

BASIC Notation	Algebraic Notation
(86*43/2^5)	$\left(\dfrac{86 \cdot 43}{2^5}\right)$
1/2*x	$\dfrac{1}{2}x$
1/(2*x)	$\dfrac{1}{2x}$
B^2−4*A*C	$b^2 - 4ac$
(A^3−B^3)/(A+B)	$\dfrac{a^3 - b^3}{a + b}$
A−(C^3/D)+B	$a - \dfrac{c^3}{d} + b$

Convert each of the following to BASIC notation.

83. $a^2 - 2ab + b^2$

84. $\dfrac{7}{x - y}$

85. $\left(\dfrac{2}{3}\right)x$

86. $\dfrac{a}{b} - \dfrac{c}{d}$

87. $\dfrac{a^2}{b} - 3d + \left(\dfrac{a}{b}\right)^3$

88. $\dfrac{a + b}{a^3 + b^3}$

Convert each of the following to algebraic notation.

89. 3*A^4−5*D

90. (C−4)/(4*C)

91. A−2/B

92. (A−B)^4

93. 4*X^3−3*X^2+2*X−17

94. (A^3/B^4)^(1/2)

95. Calculate (2/3)*X and 2/3*X for X=−2, X=5, and X=−8.

✎ In Your Own Words

96. What are the benefits of BASIC notation over standard mathematical notation? The drawbacks?

97. Describe in words the procedure for multiplying two polynomials.

98. In what ways are polynomial expressions and rational expressions similar? In what ways are they different?

99. Explain how a polynomial expression can be thought of as a special type of rational expression.

1.5 CHAPTER REVIEW

Important Concepts, Properties, and Formulas—Chapter 1

Commutative laws	$a + b = b + a$ $ab = ba$	p. 7
Associative laws	$a + (b + c) = (a + b) + c$ $a(bc) = (ab)c$	p. 7
Inverses	$a + (-a) = (-a) + a = 0$ $a \cdot a^{-1} = 1, \qquad a \neq 0$ $a \cdot \dfrac{1}{a} = 1, \qquad a \neq 0$	p. 7
Distributive laws	$a(b + c) = ab + ac$ $(a + b)c = ac + bc$	p. 7
Integers	$\{\ldots, -3, -2, -1, 0, 1, 2, 3, \ldots\}$	p. 3
Rational number	Ratio of two integers with nonzero denominator, infinitely repeating decimal notation.	p. 5
Irrational number	Real number that is not a ratio of two integers, nonrepeating decimal notation.	p. 6
Real numbers	Rational numbers and irrational numbers.	p. 2
Absolute value	$\|a\| = a$ if $a \geq 0$ $\|a\| = -a$ if $a < 0$ $\|-a\| = \|a\|, \qquad \|a\| \geq 0$ $\|ab\| = \|a\| \cdot \|b\|$ $\|a + b\| \leq \|a\| + \|b\|$ $\left\|\dfrac{a}{b}\right\| = \dfrac{\|a\|}{\|b\|} \quad (b \neq 0)$	p. 12
Distance between numbers	$d(A, B) = \|a - b\|$	p. 12

Laws of exponents	$a^r \cdot a^s = a^{r+s}, \qquad \dfrac{a^r}{a^s} = a^{r-s}$ $(a^r)^s = a^{rs}$ $(ab)^r = a^r b^r$ $\left(\dfrac{a}{b}\right)^r = \dfrac{a^r}{b^r}$ $a^{-r} = \dfrac{1}{a^r}$ $\left(\dfrac{a}{b}\right)^{-r} = \left(\dfrac{b}{a}\right)^r$	p. 18
Properties of radicals	$\sqrt[n]{ab} = \sqrt[n]{a}\,\sqrt[n]{b}$ $\sqrt[n]{\dfrac{a}{b}} = \dfrac{\sqrt[n]{a}}{\sqrt[n]{b}}, \qquad b \neq 0$ $\sqrt[n]{a^n} = \begin{cases} a & \text{if } n \text{ is odd} \\ |a| & \text{if } n \text{ is even} \end{cases}$	p. 20
Formulas for multiplication and factoring	Product of Sum and Difference $\quad (A + B)(A - B) = A^2 - B^2$	p. 32
	Square of a Binomial Sum $\quad (A + B)^2 = A^2 + 2AB + B^2$	
	Square of a Binomial Difference $\quad (A - B)^2 = A^2 - 2AB + B^2$	
	Product of Two Binomials $\quad (aA + b)(cA + d) = acA^2 + (ad + bc)A + bd$	
	Sum of Cubes $\quad A^3 + B^3 = (A + B)(A^2 - AB + B^2)$	
	Difference of Cubes $\quad A^3 - B^3 = (A - B)(A^2 + AB + B^2)$	

Cumulative Review Exercises—Chapter 1

Translate each of the following to an algebraic expression.

1. The square root of the sum of two numbers.

2. The square root of the sum of the squares of two numbers.

3. A number minus 37% of itself.

4. The sum of two numbers divided by the difference of the same two numbers.

5. The total surface area S of a right circular cylinder with base of radius r and height h is given by

$$S = 2\pi rh + 2\pi r^2$$

An oil storage tank has a base with a radius of 10 ft. and a height of 25 ft. Find its total surface area.

Consider the following numbers.

$$-54.7, \ \sqrt[3]{-25}, \ 4^{5/2}, \ 17, \ -4, \ -\frac{1}{7}, \ 6^0,$$
$$-3^2, \ 9^{3/2}, \ 67.2\%, \ 0.892, \ 2.\overline{345}, \ \sqrt{3}$$

6. List the numbers that are positive.

7. List the numbers that are real numbers.

8. List the numbers that are irrational.

9. List the numbers that are rational.

10. List the numbers that are integers.

What law is exemplified by each of the following exercises?

11. $3(ab) = (3a)b$

12. $-9 + 0 = -9$

13. $x + (-x) = 0$

14. $a \cdot a^{-1} = 1$ for $a \neq 0$

15. $3x + 3y = 3(x + y)$

Fill in the blank with the correct symbol: $<$, $>$, or $=$.

16. 0.019 ____ 0.02 17. 78% ____ 0.77

18. $\sqrt{35}$ ____ 5.917 19. $\sqrt[3]{16}$ ____ $2\sqrt[3]{2}$

20. Prove the following for any real number a:
$$\left| a^2 \right| = a^2$$

Perform the indicated operations and simplify.

21. $(2x - 3)(3x^2 + 4x + 1)$

22. $3(a^2 - 2)(a + 2) - (a^2 - 7a)$

23. $\dfrac{2}{t} - \dfrac{t}{3}$ 24. $\dfrac{5}{y^2} + \dfrac{4}{y} - \dfrac{3}{y^3}$

25. $\dfrac{2a}{1 - 2a} - \dfrac{3}{4a^2 - 1} + \dfrac{3a}{2a + 1}$

26. $(\sqrt{7t} + \sqrt{5})(\sqrt{7t} - \sqrt{5})$

27. $\sqrt[3]{8x^2} + \sqrt[3]{x^5} + \sqrt{4x^3}$

28. $\dfrac{x + 4}{x^2 - x - 2} \cdot \dfrac{x - 2}{3x^2 - 4x + 1}$

29. $\dfrac{x}{x^2 + 17x + 72} - \dfrac{8}{x^2 + 15x + 56}$

30. $\dfrac{1}{(x + 1)^2} - \dfrac{1}{(x + 1)^3}$

31. $(x - 9)^3 - 6(x - 9)^2 - 135(x - 9)$

32. $(x - 2)(x - 11) + (x - 2)(x - 13) + (x - 11)(x + 13)$

33. $\dfrac{a - b}{ab} + \dfrac{b - c}{bc} - \dfrac{a - c}{ac}$

34. $(|a| + |b|)^2$

35. $\left[\left(\dfrac{x^2 + 8x + 16}{5x^3 - 5} \cdot \dfrac{6x^2 + 6x + 6}{x^2 - 16} \right) \div \dfrac{2x + 8}{5x - 20} \right]$

Simplify the following expressions using the laws of exponents. Write the answers in a form that does not involve any negative or zero exponents.

36. $(9x^{-2})(-3x^{18})^{-4}$ 37. $(8a^3b^5)(-4a^{-5}b^2)$

38. $\left(\dfrac{x^{-5}}{x^{-3}} \right)^{-2}$ 39. $(10x^{-2}y^3)(25x^{-3}y^2)^{-1}$

40. $\left(\dfrac{27a^6b^{-5}c^2}{9a^{-4}b^3c^{-4}} \right)^{-3}$ 41. $\dfrac{(2x^{-2}y^4z^{-5})^{-4}}{(3x^5y^4z)^{-2}}$

Write the following expressions in a form that involves no negative exponents or radicals.

42. $\sqrt{y^2}$ 43. $\sqrt[3]{z^4}$ 44. $\sqrt[4]{t^2}$

45. $\sqrt{(a^2 + b^2)^3}$ 46. $\sqrt{\dfrac{x^{2/3}}{x^9}}$ 47. $\sqrt[4]{a^{-8}b^{12}}$

Simplify the following expressions. Where possible, eliminate fractional or negative exponents, and use as few terms as possible.

48. $-8^{2/3}$ 49. $16^{3/4}$ 50. $25^{5/2}$

51. $27^{2/3} - \left(\dfrac{4}{9} \right)^{-3/2}$ 52. $\dfrac{x^{-3} - y^{-3}}{x^{-2} - y^{-2}}$

53. $\dfrac{\dfrac{1}{a} + \dfrac{1}{b}}{a^2 + 2ab + b^2}$ 54. $\dfrac{1}{x} + \dfrac{\dfrac{1}{x^3}}{1 - \dfrac{4}{x^4}}$

55. $\dfrac{ab - 1}{a^3b^3 - 1}$ 56. $\sqrt{12x^2yz^3}$ 57. $\sqrt[3]{\dfrac{27a^3}{64}}$

58. $\dfrac{\sqrt[3]{625a^4b^7}}{\sqrt[3]{5ab^2}}$ 59. $\sqrt{(x + y)^5}\sqrt{(x + y)^8}$

60. $\sqrt[4]{(81a^4b^8)^3}$ 61. $4\sqrt[2]{\dfrac{160x^9y^{12}}{2\sqrt[5]{5xy^2}}}$

62. $\left(\dfrac{2^2}{8} + \dfrac{2}{2} - \dfrac{2^3}{12} \right)$
$- \left(\dfrac{(-1)^2}{8} + \dfrac{-1}{2} - \dfrac{(-1)^3}{12} \right)$

63. $\dfrac{\dfrac{2^{n+1}x^{n+1}}{(n + 1)^5}}{\dfrac{2^n x^n}{n^5}}$ 64. $\dfrac{\dfrac{\sqrt{(x - 1)^{n/2}}}{(x - 1)^{2n}}}{(x - 1)^{-n/4}}$

Factor if possible.

65. 456 66. 2025

67. $8x^2 - 6 - 8x$ 68. $49x^2 + 25y^2$

69. $1 + 125a^3$

70. $6(a + 1)^2 + 7(a + 1) - 5$

71. $a^3 - 3a^2b + 3ab^2 - b^3$ 72. $6x^{2a} - 5x^a - 1$

73. $x^{3a} + 27$ 74. $7x^2y - 28y$

75. $4x^3 + 30x^2 + 14x$ 76. $4a^6 + 108b^3$

77. $55 - 6x - x^2$ 78. $a^3 - \dfrac{1}{8}$

79. $ax^2 + 2axy + ay^2 + bx^2 + 2bxy + by^2$

80. $x^2 - 10xy + 25y^2$

81. $36t^2 - 16m^2$ 82. $a^{2x+4} - b^{6x+10}$

83. $\dfrac{1}{16}a^2 + \dfrac{1}{4}ab + \dfrac{1}{4}b^2 - \dfrac{1}{9}a^2 b^2$

84. $x^2 - 7$ *Hint:* Let $7 = (\sqrt{7})^2$

85. $x^3 - 7$ *Hint:* Let $7 = (\sqrt[3]{7})^3$

Rationalize the denominators of the following expressions.

86. $\dfrac{5}{\sqrt{6}}$

87. $\dfrac{1 - \sqrt{a}}{1 + \sqrt{a}}$

88. $\dfrac{4}{\sqrt{3} - \sqrt{2}}$

89. $\dfrac{\sqrt[3]{18}}{\sqrt[3]{5}}$

Rationalize the numerators of the following expressions.

90. $\dfrac{\sqrt{2}}{2}$

91. $\dfrac{\sqrt[4]{t}}{5t}$

92. $\dfrac{\sqrt{3 + h} - \sqrt{3}}{h}$

93. $\sqrt{n^2 - n} - n$

94. $\dfrac{\sqrt{x + 3} - \sqrt{x - 3}}{\sqrt{x + 3} - \sqrt{x + 3}}$

Convert to scientific notation.

95. 2467 (the boiling point of aluminum, in degrees Celsius)

96. 0.00003 (the probability that Joe DiMaggio would get at least one hit in each of 56 consecutive games)

Convert to decimal notation.

97. 6.48×10^{-2} (the number of grams in one grain)

98. 2.4×10^{13} (the distance of the star *Alpha Centauri* from earth)

99. One mole of any element contains 6.02×10^{23} atoms. Given the mass of one mole, as follows, find the mass of one atom of:

   a. Hydrogen, 1.0079 g/mole

   b. Aluminum, 26.98154 g/mole

   c. Lead, 207.2 g/mole

♦ **Applications**

100. A tank contains water H m deep. A hole is pierced in a wall of the tank h m below the water surface. (See figure.) The stream of water coming from that hole will hit the floor x meters from the wall, where x is given by

$$x = \sqrt{h(H - h)}$$

An aboveground swimming pool has water in it that is 3 m deep. A hole is pierced in the pool 2.5 m above the ground. How far from the swimming pool will the water hit the ground?

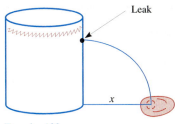

Exercise 100

101. The hypotenuse of a triangle is twice as long as the shortest leg of the triangle. Find a formula for the length of the other leg.

102. A circle is inscribed in a square. The area of the circle is 200 cm^2. (See figure.) Find the length of a side of the square. (Use 3.14 for π.)

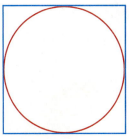

Exercise 102

103. A 45°, 45°, 90° triangle has legs of length a. Find the length of the hypotenuse.

104. A die has six faces, each with one of the numbers 1 through 6. The probability that a 6 is obtained on a roll is 1/6. A die is rolled 50 times. The probability that at most four rolls will be 6s is

$$\left(\frac{1}{6}\right)^{50} + 50\left(\frac{1}{6}\right)^{49}\left(\frac{5}{6}\right) + \frac{50 \cdot 49}{1 \cdot 2}\left(\frac{1}{6}\right)^{48}\left(\frac{5}{6}\right)^{2}$$

$$+ \frac{50 \cdot 49 \cdot 48}{1 \cdot 2 \cdot 3}\left(\frac{1}{6}\right)^{47}\left(\frac{5}{6}\right)^{3} + \frac{50 \cdot 49 \cdot 48 \cdot 47}{1 \cdot 2 \cdot 3 \cdot 4}\left(\frac{1}{6}\right)^{46}\left(\frac{5}{6}\right)^{4}$$

Use your calculator to approximate the probability.

Convert to BASIC notation.

105. $\dfrac{x - y}{a + b}$

106. $(a - b)(a^2 + ab + b^2)$

Convert to algebraic notation.

107. `45*63^2+2*A−B^2`

108. `−B−(B^2−4*A)*C^(1/2)`

Perform the indicated operations.

109. $(\sqrt{xy} + 2\sqrt{x})(3\sqrt{xy} + 2\sqrt{y})$

110. $\sqrt{1 - a^2} + a^2(1 - a^2)^{1/2}$

111. $(x^{\sqrt{3}} + x^{\sqrt{2}})^2$

112. $\dfrac{\sqrt{1 + x}}{\sqrt{1 - x}} - \dfrac{2(x - 1)}{\sqrt{1 - x^2}} - \dfrac{\sqrt{4 - 4x}}{\sqrt{1 + x}}$

113. Rationalize the denominator.

$$\frac{1}{\sqrt{7} - \sqrt{8} + \sqrt{12}}$$

114. Rationalize the numerator.

$$\frac{\sqrt{2(x + h) - 3} - \sqrt{2x - 3}}{h}$$

115. Simplify

$$\left[(\sqrt{32})^{(6/5 + \sqrt{6/5})} \right]^{(6/5 - \sqrt{6/5})}$$

EQUATIONS AND INEQUALITIES

V. Kantorovic invented the method of linear programming, a method for solving systems of linear inequalities. This technique, invented only decades ago, is now an important mathematical tool in economic planning. For his invention, Kantorovic was awarded the Nobel prize in Economics.

P roblems in many disciplines, such as biology, chemistry, engineering, and business, require variables used to represent physical quantities: temperatures, prices, blood pressures, forces. Such problems often impose restrictions on the variables; some of these restrictions are specified as equations, and others are specified as inequalities. Equations and inequalities, in addition to their importance in other disciplines, are the cornerstone on which algebra is built.

In this chapter, we study both equations and inequalities, the methods available for solving the most elementary types of each, and some of the applications to which they may be put. This chapter also lays the foundation for using equations and for problem solving that we will carry through the rest of the book.

2.1 EQUATIONS AND IDENTITIES

An **equation** is a statement that two algebraic expressions are equal. Here are some examples of equations:

$$2 = 1 + 1$$
$$(x + 1)^2 = x^2 + 2x + 1$$
$$2x - 3 = 4$$
$$3x + 4y = 8$$

There are three types of equations: **identities**, **conditional equations**, and **inconsistent equations**. An **identity** is an equation that is true for all variable values. The second equation above is an example of an identity since it states an algebraic fact that remains true for all values of x.

A **conditional equation** is an equation that is true for only some values of the variables. In this chapter, most of our attention will be directed to conditional equations. The third equation above is a conditional equation because it is true when 7/2 is substituted for x:

$$2\left(\frac{7}{2}\right) - 3 = 7 - 3 = 4$$

but it is not valid when any other value is substituted for x. For example, if $x = -1$, then

$$2(-1) - 3 = -5 \neq 4$$

So the equation is not valid if $x = -1$.

An **inconsistent equation** is an equation which states an impossibility, such as $x + 2 = x + 3$, which is true for no value of x.

Determining the set of values of the variables for which a conditional equation is true is called **solving the equation**. The values of the variable for which an equation holds are called **solutions** (or **roots**) of the equation. The set of all of these values is called the **solution set**.

Most equations will include restrictions on the solutions. Some equations (most of the ones considered in this book) are to be solved only for solutions that are real numbers, some equations arising from physical problems require that you restrict the solutions to positive real numbers, the solutions to yet other equations may be restricted to integers, and so forth.

Methods for solving equations constitute one of the main topics of algebra. In this chapter, we will be concerned with solving two of the simplest types of equations, linear and quadratic equations. In later chapters, we will learn to solve many other types of equations.

Equivalent Equations

There are many methods for solving equations. Some methods apply only to equations of a particular sort, say, linear or quadratic equations. However, certain operations can be employed as part of the solution of any equation. These operations transform an equation into an **equivalent equation**, namely, one with the same solutions. For example, adding or subtracting the same number from both sides of an equation produces an equivalent equation, as does multiplication and division of both sides by the same nonzero number.

The following table summarizes the operations that lead to equivalent equations:

OPERATIONS THAT GENERATE EQUIVALENT EQUATIONS

1. Algebraic simplification, such as removing parentheses in applying associative and distributive rules, combining like terms, performing indicated algebraic operations, and applying the laws of exponents.
2. Interchange the two sides of the equation.
3. Add or subtract the same expression from both sides of the equation.
4. Multiply or divide both sides of the equation by the same nonzero number.

Consider the equation

$$5x + 3 = 2x + 1$$

We may subtract $2x$ from both sides of the equation to obtain the equivalent equation

$$(5x + 3) - 2x = (2x + 1) - 2x$$
$$3x + 3 = 1$$

Similarly, we may divide both sides of the equation

$$9x = 11$$

by 9 to obtain the equivalent equation

$$\frac{9x}{9} = \frac{11}{9}$$
$$x = \frac{11}{9}$$

Linear Equations

Let us begin by considering the simplest equations in one variable.

Definition 1
Linear Equation

A **linear equation** in one variable is equivalent to an equation of the form
$$ax + b = 0$$
where a and b are constants and $a \neq 0$.

Here are some examples of linear equations:

$$-5x + 1 = 0$$
$$x + 3(2x - 4) = 1$$
$$5x + 2 = 3x - 1$$

By replacing an equation with an equivalent one, we may solve linear equations, as illustrated in the following example.

➤ EXAMPLE 1
Solving a Linear Equation

Solve the linear equation

$$-2(3x + 7) + 4 = 3x - 2(x + 4)$$

Solution

The first step in the solution of this equation is to simplify the expressions on both sides by removing the parentheses:

$$-6x - 14 + 4 = 3x - 2x - 8 \qquad \textit{Distributive law}$$
$$-6x - 10 = x - 8 \qquad \textit{Combine like terms}$$

Next, we use allowable operations to arrive at an equivalent equation that has all terms involving x on one side of the equation and all terms not involving x on the other:

$$-6x - x - 10 = x - 8 - x \qquad \textit{Subtract } x \textit{ from both sides}$$
$$-7x - 10 + 10 = -8 + 10 \qquad \textit{Add 10 to both sides}$$
$$-7x = 2$$
$$-\frac{1}{7} \cdot (-7x) = -\frac{1}{7} \cdot 2 \qquad \textit{Multiply both sides by } -\frac{1}{7}$$
$$x = -\frac{2}{7} \qquad \textit{Inverse property, Associative law}$$

Thus, the solution is $-2/7$.

We now check that the solution works by substituting the value $-2/7$ for x in the equation:

$$-2(3x + 7) + 4 \qquad\qquad 3x - 2(x + 4)$$

$$= -2\left[3\left(-\frac{2}{7}\right) + 7\right] + 4 \qquad\qquad = 3\left(-\frac{2}{7}\right) - 2\left(-\frac{2}{7} + 4\right)$$

$$= -2\left[-\frac{6}{7} + 7\right] + 4 \qquad\qquad = -\frac{6}{7} - 2\left[-\frac{2}{7} + \frac{28}{7}\right]$$

$$= -2\left[-\frac{6}{7} + \frac{49}{7}\right] + 4 \qquad\qquad = -\frac{6}{7} - 2\left(\frac{26}{7}\right)$$

$$= -2\left(\frac{43}{7}\right) + 4 \qquad\qquad = -\frac{58}{7}$$

$$= -\frac{58}{7}$$

➤ EXAMPLE 2
Linear Equation with Fractional Coefficients

Solve the linear equation

$$\frac{1}{3}x + \frac{3}{7} = \frac{1}{2}x + 4$$

Solution

It is usually helpful to eliminate fractions in an equation by first multiplying both sides by the least common denominator of all rational numbers that appear, in this case 42. This gives

$$42\left(\frac{1}{3}x + \frac{3}{7}\right) = 42\left(\frac{1}{2}x + 4\right)$$ *Multiply by least common denominator*

$$14x + 18 = 21x + 168$$

$$-150 = 7x$$ *Constant terms to one side and x terms to the other*

$$-\frac{150}{7} = x$$ *Multiply by $\frac{1}{7}$*

Check:

$$\frac{1}{3}x + \frac{3}{7} = \frac{1}{3}\cdot\left(-\frac{150}{7}\right) + \frac{3}{7} \qquad\qquad \frac{1}{2}x + 4 = \frac{1}{2}\cdot\left(-\frac{150}{7}\right) + 4$$

$$= -\frac{50}{7} + \frac{3}{7} \qquad\qquad\qquad\qquad = -\frac{75}{7} + 4$$

$$= -\frac{47}{7} \qquad\qquad\qquad\qquad\qquad = -\frac{47}{7}$$

In the preceding example, we eliminated fractions by multiplying by the least common denominator. This process is called **clearing the denominators**. In a similar fashion, in an equation involving decimals, it is usually helpful to clear the decimals by multiplying both sides of the equation by an appropriate power of 10. For example, to clear the decimals from the equation

$$0.1x + 0.003 = 5.75$$

we would multiply both sides by 10^3 or 1000, to obtain

$$100x + 3 = 5750$$

Equations That Reduce to Linear Equations

Many equations, after appropriate algebraic simplification, are equivalent to linear equations. The following examples provide illustrations of a number of typical cases.

➤ **EXAMPLE 3**
Reducing to a Linear Equation

Solve the following equation:

$$(3x + 1)(2x - 1) = (6x - 1)(x + 5)$$

Solution

The first step in the solution is to remove the parentheses by calculating the products on both sides of the equation:

$$6x^2 - x - 1 = 6x^2 + 29x - 5$$

At first glance, it looks as if this equation involves second powers of x and thus is not a linear equation. However, by subtracting $6x^2$ from both sides, we arrive at a linear equation:

$$-x - 1 = 29x - 5$$

$$-30x = -4$$

$$x = \frac{2}{15}$$

The solution is 2/15.

Check:

$$(3x + 1)(2x - 1)$$
$$= \left[3\left(\frac{2}{15}\right) + 1\right]\left[2\left(\frac{2}{15}\right) - 1\right]$$
$$= \left(\frac{21}{15}\right)\left(-\frac{11}{15}\right)$$
$$= -\frac{231}{225}$$

$$(6x - 1)(x + 5)$$
$$= \left[6\left(\frac{2}{15}\right) - 1\right]\left(\frac{2}{15} + 5\right)$$
$$= \left(-\frac{3}{15}\right)\left(\frac{77}{15}\right)$$
$$= -\frac{231}{225}$$

➢ **EXAMPLE 4**
A Rational Equation

Determine all solutions of the following equation:

$$\frac{1}{2x + 1} = \frac{2x}{4x^2 - 3}$$

Solution

The first step in the solution is to multiply both sides of the equation by the least common denominator, $(2x + 1)(4x^2 - 3)$:

$$(2x + 1)(4x^2 - 3)\left(\frac{1}{2x + 1}\right) = (2x + 1)(4x^2 - 3)(2x)\left(\frac{1}{4x^2 - 3}\right)$$

Simplifying the expressions on both sides gives

$$1 \cdot (4x^2 - 3) = 2x \cdot (2x + 1)$$
$$4x^2 - 3 = 4x^2 + 2x$$
$$-3 = 2x$$
$$x = -\frac{3}{2}$$

Check:

$$\frac{1}{2x + 1} = \frac{1}{2\left(-\frac{3}{2}\right) + 1}$$
$$= \frac{1}{-2}$$
$$= -\frac{1}{2}$$

$$\frac{2x}{4x^2 - 3} = \frac{2\left(-\frac{3}{2}\right)}{4\left(-\frac{3}{2}\right)^2 - 3}$$
$$= \frac{-3}{9 - 3}$$
$$= -\frac{1}{2}$$

Operations That Don't Lead to Equivalent Equations

Multiplying and dividing an equation by an expression does not necessarily lead to an equivalent equation. When multiplying or dividing by an algebraic expression, you must separately examine the values for the variable that make the expression equal to 0 or lead to division by 0. These values of the variable may or may not be solutions to the equation. The method for handling such values is illustrated in the next two examples.

➤ **EXAMPLE 5**

Dividing by an Expression
That May Equal 0

Determine all solutions to the following equation:

$$(x - 1)(x - 2) = (x - 1)(2x + 3)$$

Solution

We begin by dividing both sides by $x - 1$ and performing allowable operations:

$$\frac{(x - 1)(x - 2)}{x - 1} = \frac{(x - 1)(2x + 3)}{x - 1}$$

$$x - 2 = 2x + 3$$

$$x = -5$$

This value is a solution of the original equation since

$$[(-5) - 1][(-5) - 2] \quad \big| \quad [(-5) - 1][2(-5) + 3]$$
$$= 42 \qquad\qquad \big| \qquad\qquad = 42$$

Dividing both sides by the factor $x - 1$ is allowed provided that $x - 1 \neq 0$, that is, provided that $x \neq 1$. The excluded value, namely, $x = 1$, must be tested separately. Substituting $x = 1$ in the original equation yields

$$(x - 1)(x - 2) = (1 - 1)(1 - 2) \quad \big| \quad (x - 1)(2x + 3) = (1 - 1)(2(1) + 3)$$
$$= 0 \qquad\qquad \big| \qquad\qquad = 0$$

Because we arrive at a true statement, $x = 1$ is a solution of the original equation. So the solution set of the given equation is $\{-5, 1\}$.

➤ **EXAMPLE 6**

Extraneous Solutions

Solve the equation

$$\frac{1}{(x - 1)^2} = \frac{x}{(x - 1)^2}$$

Solution

Multiplying both sides of the equation by $(x - 1)^2$,

$$\frac{1}{(x - 1)^2} \cdot (x - 1)^2 = \frac{x}{(x - 1)^2} \cdot (x - 1)^2$$

$$1 = x$$

Thus, the only possible solution of the equation is $x = 1$. However, substitution of 1 for x in the equation leads to division by 0 on both sides of the equation, so $x = 1$ is not a solution. The given equation therefore has no solutions.

Note how multiplication by $x - 1$ in the preceding example introduced a false solution to the equation. Such a solution is called an **extraneous solution.**

Remember: Whenever you multiply both sides of an equation by an algebraic expression, the values of the variable that make the expression undefined or equal to 0 must be tested individually to determine whether or not they are solutions to the equation.

Another algebraic operation that may not always lead to equivalent equations is taking the square roots of both sides of an equation. Recall that the square root of a real number has two values, one positive (principal square root) and one negative. To obtain an equivalent equation, you must take into account both the positive and the negative values of the square root, by inserting a $\pm$ in the equation.

➤ **EXAMPLE 7**
Square Root of an Equation

Determine all solutions to the following equation:
$$(5x + 1)^2 = x^2$$

Solution

Both sides of the equation are perfect squares, so it seems natural to take the square root of both sides. We can do so provided that we remember to take into account both values of the square roots:

$$5x + 1 = \pm x$$

(Note that in taking square roots, we could have added $\pm$ preceding the left side of the equation also. However, each choice of sign on the left could be reduced to one of the equations already given.) So the solutions to the original equation are given by the solutions to the two equations

$$5x + 1 = x, \qquad 5x + 1 = -x$$

That is, x must have one of the values

$$x = -\frac{1}{4}, \qquad x = -\frac{1}{6}$$

The solutions to these equations, and hence to the original equation, are $-1/4$ and $-1/6$.

➤ **EXAMPLE 8**
An Equation with No Solutions

Determine the solutions of the equation
$$5(x + 1)(x + 2) = 5x^2 + 15x - 12$$

Solution

Multiply out the left side of the equation:

$$5(x^2 + 3x + 2) = 5x^2 + 15x - 12$$
$$5x^2 + 15x + 10 = 5x^2 + 15x - 12$$
$$10 = -12 \qquad \textit{Subtracting } 5x^2 + 15x \textit{ from each side}$$

This last statement is a contradiction, so the given equation is equivalent to an equation that has no solutions. The given equation is inconsistent.

In our discussion above, we showed how taking the square root of both sides of an equation does not necessarily lead to an equivalent equation and, in fact, might result in the loss of certain solutions. Squaring both sides of an equation also does not necessarily lead to an equivalent equation. For example, consider the equation

$$x = 1$$

If we square both sides, we obtain the equation

$$x^2 = 1$$

This equation has 1 and -1 as a solution, since $(-1)^2 = 1$. However, -1 is not a solution of the original equation! The operation of squaring both sides of the equation results in an equation that has all of the original solutions plus possibly some extraneous solutions.

Exercises 2.1

Solve.

1. $-5x + 72 = 7$

2. $3t - 13 = -39$

3. $1 - 0.2y = 0.13$

4. $0.34t + 2 = 1.8$

5. $-\dfrac{2}{3}x - \dfrac{4}{5} = \dfrac{7}{12}$

6. $\dfrac{5}{6} - \dfrac{1}{3}t = \dfrac{1}{2}$

7. $64 - 32t = 0$

8. $160 - 32t = 0$

9. $70 - 2x = 0$

10. $20 - 2x = 1$

11. $-0.02x - 0.5 = 0$

12. $-1.5x + 150 = 0$

13. $97\%x = 4850$

14. $5,611.5 = 64.5\%y$

15. $y + 11\%y = 721.50$

16. $x + 8\%x = 216$

17. $5x - 17 - 2x = 6x - 1 - x$

18. $3x + 4 - x = 2(x - 2)$

19. $6 - \dfrac{3}{4}x = \dfrac{1}{2}x - 4$

20. $5x + \dfrac{3}{4} = \dfrac{1}{2}x - 3$

21. $5(2x - 3) = 12x - (x + 8)$

22. $2y - (3y + 2) = 5 - 4(3 - y)$

23. $(3x - 1) - (5x + 2) = 5 - [5x - (x + 2)]$

24. $6(x - 4) + 3x = 3 - [2x + 3(x - 5)]$

25. $\dfrac{1}{2}(5x + 3x - 2x) = \dfrac{3}{2}(x + 1)$

26. $\dfrac{1}{2}(6x + 2) = \dfrac{1}{3}(6 + 9x)$

27. $-3(2x - 5) - [3 + 5(x - 1) - 3x]$
$= 25 - [5x + 2 - 3(x + 2)]$

28. $-4(3y + 1) - [4 - 2(2y - 5) - 3(y + 2)]$
$= 45 - [-2y + 4 - 4(y + 7)]$

29. $2.6t - 4.5t + 40 = -7.8t + 1.3t - 80$

30. $\dfrac{2}{3}y - \dfrac{1}{6}y - 2 = -\dfrac{3}{2}y + \dfrac{3}{4}y - 10$

31. $(x - 2)(x + 3) = (x + 5)(x - 7)$

32. $(y + 1)(y + 2) = (y - 2)(y + 6)$

33. $(2y + 1)(3y + 4) = (6y - 2)(y + 5)$

34. $(4t - 1)(2t - 5) = (t - 3)(8t + 3)$

35. $(x - 2)(3x + 5) = (x - 2)(2x - 7)$

36. $(x + 1)(2x + 7) = (x + 1)(3x - 4)$

37. $(x - 11)(2x + 7) = (x - 11)(3x - 4)$

38. $(x + 12)(3x + 5) = (x + 12)(2x - 7)$

39. $\dfrac{x}{15 - x} = \dfrac{2}{3}$

40. $\dfrac{y}{23 + y} = \dfrac{4}{5}$

41. $\dfrac{1}{3x + 1} = \dfrac{3x}{9x^2 - 1}$

42. $\dfrac{x - 4}{x + 5} = \dfrac{x}{x + 1}$

43. $\dfrac{y}{y - 2} = \dfrac{y - 3}{y - 4}$

44. $\dfrac{5}{5t - 3} = \dfrac{t}{t^2 - 1}$

45. $(3x + 1)^2 = x^2$

46. $(4t - 1)^2 = t^2$

47. $(4t - 1)^2 = 9$

48. $(3x + 1)^2 = 64$

49. $x + 2 = 3 + x$

50. $3x + 4 = 7(5x - 1)$

51. $t - 9 = 11 + t$

52. $x + 7 = 7 + x$

53. $\dfrac{x - 4}{x - 5} = \dfrac{1}{x - 5}$

54. $\dfrac{4}{y - 7} = \dfrac{y - 3}{y - 7}$

55. $2t - 6(4 - 3t) + 4(5 + 2t) = 0$

56. $2(6 + t) - 6(-2 - 3t) + 2t = 0$

57. $2(3 + t) + 2(4 - 2t)(-2) = 0$

58. $2(6 - t)(-1) + 2(-1 - 4t)(-4) = 0$

Determine whether each equation is an identity.

59. $x + 7 = 7 + x$

60. $5 - t = t + 5$

61. $x + 7 = 9 + x$

62. $5 - t = -3t$

63. $(x + 3)^2 = x^2 + 9$

64. $(x - 3)(x + 3) = x^2 + 9$

65. $(t + 1)(t^2 - t + 1) = t^3 + 1$

66. $(x - 2)(x + 3) = x^2 + x - 6$

67. $\sqrt{x^2 + 25} = x + 5$

68. $\dfrac{x + 4}{x + 4} = 1$

69. $(x + 1)^3 = x^3 + 3x^2 + 3x + 1$

✍ In Your Own Words

70. Find the error in the following "proof" that $2 = 1$.

$$a = 1$$
$$b = 1$$
$$a = b$$
$$a^2 = ab$$
$$a^2 - b^2 = ab - b^2$$
$$(a - b)(a + b) = b(a - b)$$
$$a + b = b$$
$$1 + 1 = 1$$
$$2 = 1$$

71. Why does cubing both sides of an equation always lead to an equivalent equation but squaring does not?

72. In general, what powers of an equation always lead to equivalent equations? Which ones do not?

2.2 APPLICATIONS OF LINEAR EQUATIONS

Linear equations arise in solving problems in many fields, including physics, chemistry, biology, engineering, business, psychology, political science, and computer science, to name only a few. In this text, in addition to developing the mathematical tools for solving linear equations, we will endeavor to acquaint you with some of the wide range of applications in which the mathematical tools can be used. We begin in this section with a survey of a few of the many applications in which linear equations are useful. We can barely scratch the surface within the body of the text. In the exercises, you will find many additional interesting problems to work out.

The applications of mathematics typically involve **word problems**, problems stated in sentence form. One of the most important things you must learn from this course is to translate the words of applied problems into algebraic language.

Many applied problems involve formulas from geometry. Here is a summary of some you are most likely to need.

HELPFUL GEOMETRIC FORMULAS FOR SOLVING APPLIED PROBLEMS

1. Square of side s:
$$\text{area} = s^2$$
$$\text{perimeter} = 4s$$

2. Rectangle of length l and width w:
$$\text{area} = lw$$
$$\text{perimeter} = 2l + 2w$$

3. Circle of radius r:
$$\text{area} = \pi r^2$$
$$\text{circumference} = 2\pi r$$

4. Triangle of base a and height h:
$$\text{area} = \frac{1}{2}ah$$

5. Cube of side s:
$$\text{volume} = s^3$$
$$\text{surface area} = 6s^2$$

6. Rectangle of length l, width w, and height h:
$$\text{volume} = lwh$$
$$\text{surface area} = 2lw + 2lh + 2wh$$

7. Right circular cylinder of radius r and height h:
$$\text{volume} = \pi r^2 h$$
$$\text{lateral surface area} = 2\pi r h$$
$$\text{total surface area} = 2\pi r h + 2\pi r^2$$

8. Sphere of radius r:
$$\text{volume} = \frac{4}{3}\pi r^3$$
$$\text{surface area} = 4\pi r^2$$

Solving Applied Problems

Here is a concrete example of an applied problem and its solution. Let's work through it and then analyze the components of the solution as a way of developing a general approach to problem solving.

➤ **EXAMPLE 1**

Farming

A rectangular field is twice as long as it is wide. What are the dimensions of the field if it requires 900 feet of fencing to go around its exterior?

Solution

After reading the problem carefully, we analyze it to determine what information is given and what information is unknown. We are given that the field is twice as long as it is wide and that the amount of fencing required to go around the perimeter is 900 ft. It is a good idea to draw a picture describing the given information. (See Figure 1.)

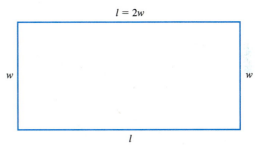

Figure 1

The next step in the problem solution is to translate the problem into algebraic language. As a first substep, we define variables to represent the unknown quantities. There are two such quantities: the length and the width, and we denote them with the variables l and w. From the given information we know that the length is twice the width, so we can express the length in terms of the width algebraically as follows:

$$l = 2w$$

From the given information we also know that perimeter $P = 900$.

From geometry we have the following formula for the perimeter of a rectangle (see Figure 1).

$$P = 2l + 2w$$

The perimeter formula has three unknowns, P, l, and w. But we substitute $2w$ for l and 900 for P in the formula to obtain

$$900 = 2(2w) + 2w$$
$$900 = 6w$$

This is a linear equation in one variable, which we can solve by dividing both sides of the equation by 6:

$$\frac{900}{6} = \frac{6w}{6}$$

$$w = 150$$

The solution of the problem does not end here. We have determined the value of the variable w, but we still have not solved the original problem, finding the dimensions of the field. We have determined only the width of the field. We can determine the length of the field by substituting the value for the width into the equation for length:

$$l = 2w = 2(150) = 300$$

The next step in solving a problem is to restate the answer in the terms used in the question asked, using units appropriate for the problem: The field is 300 feet long and 150 feet wide.

We should also check that the problem solution does indeed work by verifying the conditions set forth in the original problem. In this example, we check that the length is twice the width and that the perimeter of the field equals 900 feet: If $l = 300$ and $w = 150$, then the length is twice the width. The perimeter is:

$$2l + 2w = 2(300) + 2(150)$$
$$= 600 + 300$$
$$= 900$$

as required.

Strategy for Solving Applied Problems

From our analysis of the solution to Example 1, we can arrive at the following general strategy for problem solving.

A STRATEGY FOR SOLVING APPLIED PROBLEMS

1. Read the problem carefully, several times if necessary.
   a. Determine what information is given.
   b. Determine what information is unknown, the information that you must find.
   c. Draw a sketch illustrating the given and unknown information.
2. Translate the problem into algebraic language.
   a. Represent the unknown quantities in the problem by variables. Choose one variable and express the other unknowns in terms of that variable.
   b. Express the given information in algebraic form. This may consist of given values or relationships between the variables in the form of equations or inequalities.
   c. State any relationships that are derived from mathematical formulas, such as geometric rules or physical laws.
3. Apply algebraic techniques to solve the equation.
4. Use the solution(s) of the equation and the relationships between the variables to determine the values of the other variables, if any.
5. Use the values of the variables to obtain a solution to the applied problem originally posed.
6. Check the solution using the original words of the problem.

Practice in Solving Applied Problems

➤ **EXAMPLE 2**
Number Properties

The sum of three integers is 66. The second is 2 more than the first, and the third is 4 more than twice the first. What are the integers?

Solution

Let us first state the problem in the form of a word equation:

First integer + Second integer + Third integer $= 66$

Now we introduce symbols to represent the unknowns. We designate the first integer by x. The second would then be $x + 2$, and the third would be $2x + 4$. The condition that the sum of three numbers is 66 can then be written in the form of the equation

$$x + (x + 2) + (2x + 4) = 66$$
$$4x + 6 = 66$$
$$4x = 60$$
$$x = 15$$

Thus, the first of the numbers is 15, the second is $x + 2$, or 17, and the third is $2x + 4$, or 34. A quick check shows that these numbers do indeed add up to 66.

➤ **EXAMPLE 3**
Coin Problem

John has 20 coins in his pocket, consisting of dimes and pennies. The number of dimes exceeds twice the number of pennies by 5. How many dimes and how many pennies are there?

Solution

Let d denote the number of dimes and p the number of pennies. The total number of coins is 20, so we have the equation

$$d + p = 20$$

We can also express a relationship between the number of dimes and the number of pennies:

Number of dimes $= 2 \cdot$ Number of pennies $+ 5$
$$d = 2p + 5$$

We can now use $2p + 5$ for d in the first equation:

$$d + p = 20$$
$$(2p + 5) + p = 20$$
$$3p + 5 = 20$$
$$p = 5$$

The number of pennies p is 5. The number of dimes d is

$$d = 2p + 5 = 2(5) + 5 = 15$$

Let's check the solution just obtained. The number of coins is $15 + 5 = 20$. Moreover, the number of dimes, 15, exceeds twice the number of pennies, $2(5) = 10$, by 5. So the solution satisfies the specified conditions of the problem.

The distance moved by an object going at a constant rate for a given time is related by the equation

$$\text{Distance} = \text{Rate} \times \text{Time}$$

The following example shows how this relationship can lead to a linear equation.

➤ **EXAMPLE 4**
Distance-Rate-Time Problem

The Brown family is traveling to visit relatives in a nearby city. Mrs. Brown and the children leave from home at noon traveling 40 miles per hour. Mr. Brown leaves at 1 P.M. and travels at 55 miles per hour. How long is it before Mr. Brown overtakes Mrs. Brown?

Solution

Let t denote the number of hours after noon at which Mr. Brown overtakes Mrs. Brown. At this time, Mrs. Brown has been traveling for t hours and has gone a distance

$$\text{Rate} \times \text{Time} = 40t$$

On the other hand, Mr. Brown has been traveling $t - 1$ hours and has gone a distance of

$$\text{Rate} \times \text{Time} = 55(t - 1)$$

Since Mr. Brown overtakes Mrs. Brown, the two distances are equal. That is, we have the linear equation

$$40t = 55(t - 1)$$
$$40t = 55t - 55$$
$$-15t = -55$$
$$t = \frac{55}{15}$$
$$t = 3\frac{2}{3} \text{ hrs.}$$

That is, Mr. Brown overtakes Mrs. Brown at 3:40 P.M.

➤ **EXAMPLE 5**
Mixture Problem

A cosmetic manufacturer requires 10,000 gallons of a mixture containing 30% isopropyl alcohol and 70% water. It has on hand two mixtures. Mixture A consists of 15% isopropyl alcohol and 85% water, and mixture B consists of 40% isopropyl alcohol and 60% water. How many gallons of each should be mixed to obtain the desired mixture? (See Figure 2.)

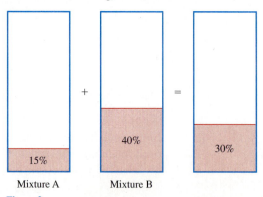

Mixture A   Mixture B

Figure 2

Solution

Let

$$x = \text{Number of gallons of mixture A}$$

Then

$$10{,}000 - x = \text{Number of gallons of mixture B}$$
$$\text{Isopropyl alcohol in mixture A} = 0.15 \cdot x$$
$$\text{Isopropyl alcohol in mixture B} = 0.4 \cdot (10{,}000 - x)$$

When the two mixtures are added together, they must contain 30% of 10,000, or 3000, gallons of isopropyl alcohol.

$$\text{Alcohol in mixture A} + \text{Alcohol in mixture } B = \text{Alcohol in final mixture}$$
$$\text{when added together}$$
$$0.15x + 0.4(10{,}000 - x) = 3000$$
$$0.15x + 4000 - 0.4x = 3000$$
$$-0.25x = -1000$$
$$x = 4000$$

The company should use 4000 gallons of mixture A and $10{,}000 - 4000 = 6000$ gallons of mixture B.

Let's check this solution. The solution contains $4000 + 6000 = 10{,}000$ gallons. The amount of isopropyl alcohol in the mixture is

$$0.15(4000) + 0.4(6000) = 600 + 2400$$
$$= 3000 \text{ gallons}$$

which is 30% of the mixture. The amount of water in the mixture is $10{,}000 - 3000 = 7000$ gallons, which is 70% of the mixture.

Suppose that P dollars is invested. We say that the investment pays **simple interest** at an annual rate R if each year the investment pays interest of PR dollars. Over a period of T years, such an investment pays total interest given by the formula

$$I = PRT$$

The following example illustrates how simple interest may be used.

➤ **EXAMPLE 6**
Financial Planning

The manager of the Amalgamated Industries Pension Fund wishes to make a one-year investment of $1,000,000 in a combination of stocks and bonds to yield simple interest income of $100,000 per year. The stocks yield 7% per year simple interest, and the bonds yield 12% per year. How much should the manager invest in stocks and how much in bonds to achieve the desired total annual interest income of $100,000?

Solution

Let x denote the amount invested in stocks. Since the numbers given all have a large number of 0s, we may simplify writing by representing the number in the problem in $100,000 units. Since a total of $1,000,000 is to be invested, the amount to be invested in bonds is $10 - x$. From the statement of the problem, the total amount of investment income must be $100,000. This is a linear equation that we can easily solve:

$$\text{Interest from stocks} + \text{Interest from bonds} = 1$$
$$0.07x + 0.12(10 - x) = 1$$
$$0.07x + 1.2 - 0.12x = 1$$
$$-0.05x = -0.2$$
$$x = 4$$

That is, $400,000 should be invested in stocks. The amount to be invested in bonds is $1,000,000 - $400,000 = $600,000. We must now check that these investments yield the desired total income:

$$\text{Interest from stocks} + \text{Interest from bonds} = 100{,}000$$
$$0.07(400{,}000) + 0.12(600{,}000) = 100{,}000$$
$$28{,}000 + 72{,}000 = 100{,}000$$

These investments in stocks and bonds do indeed produce the desired amount of income.

Solving for One Variable in Terms of Another

Often an equation contains several variables, and we must express one variable in terms of another to arrive at an equation in a single variable. We used this approach in Examples 1 and 3. Solving for one variable is especially important when using geometric formulas or formulas from physics or chemistry. The next few examples give some practice in this important skill.

➤ EXAMPLE 7
Solving in Terms of a Variable

Solve the following equation for y in terms of x:

$$xy + y - 2x + 1 = 0$$

Solution

First, we derive an equivalent equation with all terms involving y on one side of the equation and all terms that don't involve y on the other:

$$xy + y = 2x - 1$$

We can factor y from both terms of the left side:

$$y(x + 1) = 2x - 1$$

Dividing both sides by $x + 1$ now yields

$$y = \frac{2x - 1}{x + 1}, \quad x \neq -1$$

Note that we must exclude the value $x = -1$, since this value would lead to a zero denominator.

If we have a problem that gives rise to an equation containing x and y, we could now substitute $(2x - 1)/(x + 1)$ for y to arrive at an equivalent equation that involves the single variable x.

➤ EXAMPLE 8
Geometry

The volume V of a cylinder is given by the formula

$$V = \pi r^2 h$$

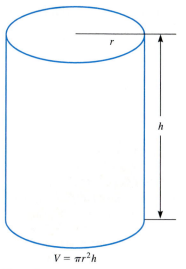

$V = \pi r^2 h$

Figure 3

where r is the radius of the cylinder and h is its height. Express the radius in terms of the volume and the height. (See Figure 3.)

Solution

We divide both sides of the equation by πh to obtain

$$r^2 = \frac{V}{\pi h}$$

We may solve for r by taking the square root of both sides. Remember that both a positive and a negative square root result:

$$r = \pm \sqrt{\frac{V}{\pi h}}$$

Since r denotes the radius of a cylinder, its value must be positive, so we can ignore the negative value. The solution then is

$$r = \sqrt{\frac{V}{\pi h}}$$

➤ **EXAMPLE 9**
Compound Interest

Most calculations of interest in financial transactions, whether savings accounts, bonds, or loans, are computed in terms of compound interest. Suppose that a passbook savings account pays compound interest at an annual rate I. If an amount A is deposited for n years, then the balance B in the account is given by the formula

$$B = A(1 + I)^n$$

Express the interest rate I in terms of A, B, and n.

Solution

$$(1 + I)^n = \frac{B}{A} \qquad \textit{Divide both sides by A}$$

$$[(1 + I)^n]^{1/n} = \left(\frac{B}{A}\right)^{1/n} \qquad \textit{Raise to the power 1/n}$$

The left side may be rewritten in the form

$$(1 + I)^{n \cdot (1/n)} = (1 + I)^1 = 1 + I \qquad \textit{Law of exponents 2}$$

Substituting $1 + I$ for the left side of the equation, we have

$$1 + I = \left(\frac{B}{A}\right)^{1/n}$$

$$I = \left(\frac{B}{A}\right)^{1/n} - 1$$

Exercises 2.2

1. What is 59% of 463?

2. 38% of 2.5 is what?

3. 28% of what number is 0.476?

4. 124 is 0.5% of what number?

5. $10.92 is what percent of $84?

6. What percent of $60 is $69?

7. A coat is marked down 25% to $69. What was the original price?

8. A 7% raise in rates by the public utility company resulted in a $2.50 monthly increase. How much was the bill before the monthly increase? After?

9. An investment grows to $1635 after one year at an interest rate of 9%, compounded annually. What amount was originally invested?

10. An investment grows to $60,760 after one year at an interest rate of 8.5%, compounded annually. What amount was originally invested?

11. The perimeter of a rectangle is 82 m. The length is 5 m longer than the width. Find the length, width, and the area.

12. The length of a rectangle is 3 ft. less than twice the width. The perimeter is 42 ft. Find the dimensions and the area.

13. The area of the shaded triangle ABC is 10 cm². Find the area of the trapezoid. (See figure.)

Formula: Area ABQP = ½ (b₁ + b₂) h

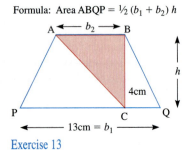

Exercise 13

14. The area of the circle is 12π yd². Find the area of the triangle. (See figure.)

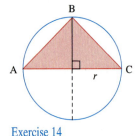

Exercise 14

15. A student has 35 math and science books on a shelf. There are 3 more math books than science books. How many of each type of book are there?

16. A lab technician has 78 vials in a case. Each vial contains either an acid or a base. There are 16 fewer vials with an acid in them than with a base. How many vials contain a base, and how many contain an acid?

17. The sum of three consecutive integers is 273. Find the integers.

18. The sum of three consecutive integers is 1803. Find the integers.

19. The sum of three consecutive even integers is 390. Find the integers.

20. The sum of three consecutive odd integers is 657. Find the integers.

21. A 97-ft. rope is cut into three pieces. The second piece is twice as long as the first, and the third is 7 ft. shorter than the first. Find the length of each piece.

22. There are four employees in a small business: the president, the janitor, and two salespersons. The president gets 10 times what the janitor gets. Each salesperson gets one and one-half times what the janitor gets. The total payroll is $161,000. What is the salary of each employee?

23. Jose leaves Sigourney's house to walk home. After 10 minutes Sigourney realizes that Jose has forgotten his keys and takes off after him. Jose is walking at a speed of 4 mph, and Sigourney is running at a speed of 5 mph. How long will it take her to catch him? *Hint*: 10 min. = 1/6 hr.

24. At 2:00 P.M. a plane leaves an airport flying due north. At 3:30 P.M. a second plane leaves the same airport flying due south. The speed of the first plane is 500 mph, and the speed of the second plane is 300 mph. When will the planes be 2750 miles apart?

25. The Indianapolis 500-mile race begins with 11 rows of three cars each. Car #92 starts in the first row and travels at an average speed of 186 mph, and car #54 is in the fifth row and travels at an average speed of 194 mph. The distance between the two cars at the start of the race is 5/16 mile. How long will it take car #54 to catch car #92?

26. A boat sailed upstream and returned in a total of 2 hours. The speed of the river was 4 km/hr., and the speed of the boat was 12 km/hr. How far did the boat travel?

27. The Newlands and Brians are going on vacation together. They must drive 268 km to get to their campground. The Brians must delay their departure for 2 hours after the Newlands due to car trouble. The Brians travel at 65 km/hr., and the Newlands travel at 58 km/hr. What is the difference, in minutes, between the times it takes each family to make the trip?

28. A rectangular lot is at the right-angle intersection of two roads. The lot has dimensions 200 ft. by 80 ft. Both roads are to be widened by the same amount in such a way that the perimeter of the lot will be decreased to 512 ft. How much width was added to the roads? (See figure.)

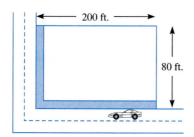

Exercise 28

Solve each formula for the given letter.

29. $F = pqV$: for V; for p.

30. $A = \frac{1}{2}bh$: for h; for b.

31. $2x + 2y + 5 = 0$: for y; for x.

32. $\pi x + 2x + 2y = 24$: for y; for x.

33. $k(36 - 2x - \pi x) + k\left(\frac{\pi}{4}\right)x = 0$: for x.

34. $-6[2x - (100 + c)] = 0$: for x; for c.

35. $A = \pi r^2$: for r. Assume $r > 0$.

36. $P = ab^2$: for b. Assume $b > 0$.

37. $P = ab^3$: for b.

38. $V = \frac{4}{3}\pi r^3$: for r. This is the formula for the volume of a sphere.

39. $p + hdq + \frac{1}{2}dv^2 = k$: for d; for q.

40. $A = P + Prt$: for P; for t.

41. $S = \dfrac{a_1 - a_1 r^n}{1 - r}$: for a_1.

42. $Q = \dfrac{k(t_2 - t_1)aT}{d}$: for t_2; for T.

43. $m_1 u_1 + m_2 u_2 = m_1 v_1 + m_2 v_2$: for m_2; for v_1.

44. $S = 2\pi rh + 2\pi r^2$: for h.

45. $x^2 + y^2 = 4\lambda^2(x^2 + y^2)$: for λ.

46. $\dfrac{2x}{16} = \dfrac{2(L - x)}{12\sqrt{3}}$: for x; for L.

♦ **Applications**

47. A bank account pays compound interest n times a year at an annual rate I. If an amount P is deposited for t years, then the balance A in the account is given by the *compound interest formula*

$$A = P\left(1 + \frac{I}{n}\right)^{nt}$$

   a. An investment of $5000 is made at 11.5% compounded semiannually (two times per year). How much will be in the account after 5 years?

   b. Solve the formula for I.

   c. An investment of $10,000 is made for 5 years at an interest rate which is compounded quarterly (four times per year). It grows to $14,859.47. What was the interest rate?

Use the compound interest formula in Exercise 47 for Exercises 48–52.

48. An investment of $2500 is made at 8.75% compounded quarterly. How much will be in the account after 16 years?

49. An investment of $12,500 is made at 9.75% compounded quarterly. How much will be in the account after 10 years?

50. Suppose $1200 is invested at 10.5%. After 4 years, how much is in the account if interest is compounded (a) annually, (b) quarterly, (c) monthly, (d) weekly (use 52 weeks per year), (e) daily (use 360 days per year)?

51. Suppose $1100 is invested at 9.625%. After 6 years, how much is in the account if interest is compounded (a) annually, (b) quarterly, (c) monthly, (d) weekly (use 52 weeks per year), (e) daily (use 360 days per year)?

52. $2000 will grow to $3338.68 in 5 years at an interest rate that is compounded daily. What is the interest rate?

53. A cosmetic manufacturer requires 20,000 gallons of a mixture that is 35% isopropyl alcohol and 65% water. It has on hand two mixtures: mixture A, which is 15% isopropyl alcohol and 85% water, and mixture B, which is 50% isopropyl alcohol and 50% water. How many gallons of each should be mixed to obtain the desired mixture?

54. A chemist has one solution of acid and water that is 65% acid, and a second that is 20%. How many gallons of each should be mixed together to get 120 gallons of a solution that is 50% acid?

55. Two investments are made totaling $12,000. For a certain year these investments yield $885 in simple interest. Part of the $12,000 is invested at 6% and the rest at 9%. How much is invested at each rate?

56. The manager of a pension fund wants to invest $5,000,000 in a combination of stocks and bonds to yield income of $496,000 per year. The stocks yield 8% per year, and the bonds yield 11% per year. How much should the manager invest in stocks and how much in bonds?

57. In order to get a major league hockey team, a city's minor league team must average 6000 per game in attendance. This year, the city's minor league team has played 9 of its 10 games and has averaged 5800. How many people must attend the 10th game in order for the city to qualify for the major league team?

58. A machine shop must average a weekly scrap rate of no more than 2%. Its scrap rates have been 1.5%, 2.4%, 2.1%, and 2.2%. If it produces the same number of parts per day, what must its scrap rate be on the fifth day to average 2% for the week?

59. *Straight-line depreciation* is a way a business computes the value of an item for tax purposes. A company buys an office machine for $9700 on January 1 of a given year. The machine is expected to last for 6 years, at the end of which time its *trade-in* or *salvage value* will be $1300. (See figure.) If the company figures the decline in value to be the same each year, then the *book value V* after t years, $0 \le t \le 6$, is given by

$$V = C - t\left(\frac{C - S}{N}\right)$$

where C = the original cost of the item ($9700), N = the years of expected life (6), and S = the salvage value ($1300).

a. Find the formula for V, for the straight-line depreciation of the office machine.

b. Determine whether this is a linear equation.

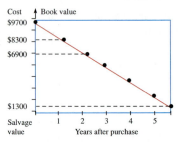

Exercise 59

c. Find the salvage value after 0 years, 1 year, 2 years, 3 years, 5 years, and 6 years. The graph describes the situation. Note that the graph is indeed a *straight line*.

60. It has been found in a study that the total length L (in mm) and the tail length t (in mm) of females of the snake species *Lampropeltis polyzona* are related by the linear equation

$$t = 0.143L - 1.18$$

a. Such a snake is found with a total length of 455 mm. Use the equation to predict its tail length.

b. Solve the equation for L. Determine whether the resulting equation is linear.

c. Use the equation in (b) to find the total length of a snake whose tail is 140 mm long.

61. The formulas for conversion between Celsius (C) temperature and Fahrenheit (F) temperature are given by the linear equations

$$F = \frac{9}{5}C + 32 \quad \text{and} \quad C = \frac{5}{9}(F - 32)$$

a. Room temperature of 20°C corresponds to what Fahrenheit temperature?

b. Water boils at 212°F. To what Celsius temperature does this correspond?

c. For what temperature are the Fahrenheit and the Celsius temperatures the same?

62. The sum of three consecutive even integers is 888. Find the integers.

Solve each equation for the indicated letter.

63. $k(4r^2 - 3x^2) = 0$, for x. Assume $r > 0$.

64. $\dfrac{m - 3w^2}{4} = 0$, for w.

65. $\dfrac{1}{3}\pi(48 - h^3) = 0$, for h.

2.3 QUADRATIC EQUATIONS

In the previous two sections, we studied linear equations, their solutions, and their applications. Let's now turn to the next important class of equations, polynomial equations of the second degree with real coefficients. In this section we will learn three methods for solving such equations and discuss a number of their applications.

Definition 1 **Quadratic Equation**	A **quadratic equation** is an equation equivalent to an equation of the form $$ax^2 + bx + c = 0$$ where a, b, and c are real numbers and $a \neq 0$.

Here are some examples of quadratic equations:

$$2x^2 + 3x + 1 = 0$$
$$x^2 - 3 = 0$$
$$x^2 - 5x = 0$$

One method for solving such equations involves factoring the left side of the equation into a product of binomials. For instance, we may write the first equation in the form

$$(2x + 1)(x + 1) = 0$$

If a product of real numbers is 0, then one of the factors must be 0. Therefore, one of the following equations must hold:

$$2x + 1 = 0 \quad \text{or} \quad x + 1 = 0$$

Solving each of these linear equations, we see that $x = -1/2$ or $x = -1$. Substituting these values of x into the original equation, we see that they are indeed solutions.

The method just described applies to any quadratic equation whose left side may be factored into a product of binomials and is called the **method of factoring**.

➤ EXAMPLE 1
Factoring

Use the method of factoring to solve the quadratic equation

$$6x^2 + 13x - 5 = 0$$

Solution

We factor the left side:

$$(3x - 1)(2x + 5) = 0$$

To determine the solutions of the quadratic equation, we set each of the linear factors equal to 0:

$$3x - 1 = 0 \quad \text{or} \quad 2x + 5 = 0$$

We solve each equation separately to obtain the solutions

$$x = \frac{1}{3} \quad \text{or} \quad x = -\frac{5}{2}$$

To check these solutions, we substitute their values into the given equation:

$$6 \cdot \left(\frac{1}{3}\right)^2 + 13 \cdot \left(\frac{1}{3}\right) - 5$$

$$= \frac{2}{3} + \frac{13}{3} - \frac{15}{3}$$

$$= 0 \qquad\qquad 0$$

$$6 \cdot \left(-\frac{5}{2}\right)^2 + 13 \cdot \left(-\frac{5}{2}\right) - 5$$

$$= \frac{75}{2} - \frac{65}{2} - \frac{10}{2}$$

$$= 0 \qquad\qquad 0$$

➤ **EXAMPLE 2**
More Factoring

Solve the following quadratic equation using the method of factoring.

$$2x^2 - 6x = x - 3$$

Solution

Subtract $x - 3$ from both sides to produce an equivalent equation:

$$2x^2 - 7x + 3 = 0$$

We factor the quadratic polynomial on the left to obtain

$$(2x - 1)(x - 3) = 0$$

and equate each linear factor to 0:

$$2x - 1 = 0 \qquad x - 3 = 0$$

$$x = \frac{1}{2} \qquad x = 3$$

The two solutions to the equation are $x = 1/2$ and 3. We then check that the two solutions work:

$$2 \cdot \left(\frac{1}{2}\right)^2 - 6 \cdot \left(\frac{1}{2}\right) \qquad \frac{1}{2} - 3$$

$$= \frac{1}{2} - \frac{6}{2} \qquad = -\frac{5}{2}$$

$$= -\frac{5}{2} \qquad = -\frac{5}{2}$$

$$2 \cdot (3)^2 - 6 \cdot 3 \qquad 3 - 3$$

$$= 18 - 18 \qquad = 0$$

$$= 0 \qquad = 0$$

Completing the Square

If one side of a quadratic equation is a perfect square and the other is a constant, then the equation may be solved by taking the square roots of both sides. For instance, consider the equation:

$$x^2 = 4$$

Here, x^2 is a perfect square, so we may solve the equation by taking square roots:

$$\sqrt{x^2} = \sqrt{4}$$
$$x = \pm 2$$

Similarly, consider the equation

$$(x - 3)^2 = 10$$

Taking square roots on both sides gives

$$x - 3 = \pm\sqrt{10}$$
$$x = 3 \pm \sqrt{10}$$

We may extend this technique into a general method for solving quadratic equations by transforming one side of the equation into a perfect square with the other side a constant. This method is called **completing the square.**

COMPLETING THE SQUARE

To complete the square of the expression $x^2 + bx$:

1. Divide the coefficient b of x by 2 to get $b/2$.
2. Square the result of 1. to obtain $(b/2)^2$.
3. Add the result of 2. to obtain the perfect square:

$$x^2 + bx + \left(\frac{b}{2}\right)^2 = \left(x + \frac{b}{2}\right)^2$$

➤ **EXAMPLE 3**
Completing the Square

Use the method of completing the square to solve the quadratic equation $x^2 + 6x + 5 = 0$.

Solution

First isolate all terms involving x on the left side of the equation:

$$x^2 + 6x = -5$$

We wish to turn the left side into a perfect square. This may be done by adding 9 to both sides:

$$x^2 + 6x + 9 = -5 + 9$$
$$(x + 3)^2 = 4$$
$$x + 3 = \pm 2$$
$$x = -3 \pm 2 = -5, -1$$

So the given equation has the two solutions −5 and −1.

➤ **EXAMPLE 4**
More Completing the Square

Use the method of completing the square to solve the following quadratic equation:

$$4x^2 + 12x + 5 = 0$$

Solution

First we subtract the constant term from both sides of the equation:

$$4x^2 + 12x = -5$$

Now we divide all terms by the coefficient of x^2.

$$x^2 + 3x = -\frac{5}{4}$$

Now we complete the square by adding to each side the square of half the coefficient of x, namely, $(3/2)^2 = 9/4$, and then factoring:

$$x^2 + 3x + \frac{9}{4} = -\frac{5}{4} + \frac{9}{4}$$

$$x^2 + 3x + \frac{9}{4} = 1$$

$$\left(x + \frac{3}{2}\right)^2 = 1 \qquad \textit{Perfect squares on both sides}$$

Let's now take the square roots of both sides.

$$x + \frac{3}{2} = \pm 1$$

$$x = \pm 1 - \frac{3}{2}$$

$$x = 1 - \frac{3}{2} = -\frac{1}{2} \qquad \text{or} \qquad x = -1 - \frac{3}{2} = -\frac{5}{2}$$

Here is a check that these two values are solutions to the given equation:

$$4 \cdot \left(-\frac{1}{2}\right)^2 + 12 \cdot \left(-\frac{1}{2}\right) + 5$$
$$= 1 - 6 + 5$$
$$= 0 \qquad \bigg| \qquad 0$$

$$4 \cdot \left(-\frac{5}{2}\right)^2 + 12 \cdot \left(-\frac{5}{2}\right) + 5$$
$$= 25 - 30 + 5$$
$$= 0 \qquad \bigg| \qquad 0$$

The Quadratic Formula

A third method for solving quadratic equations is based on the **quadratic formula**, which allows us to calculate the solutions of any quadratic equation in terms of the coefficients.

QUADRATIC FORMULA

The solutions of the quadratic equation

$$ax^2 + bx + c = 0, \qquad a \neq 0$$

are given by the formula

$$x = \frac{-b \pm \sqrt{b^2 - 4ac}}{2a}$$

The use of the $\pm$ sign in the quadratic formula means that the formula gives one solution corresponding to the positive sign and one to the negative sign.

Proof We use the method of completing the square to prove the quadratic formula. We start with the standard form of the quadratic equation and rewrite the equation with the constant term on the right:

$$ax^2 + bx = -c$$

We divide both sides by the coefficient of x^2, which is a:

$$x^2 + \frac{bx}{a} = -\frac{c}{a}$$

We next complete the square on the left by adding the square of half the coefficient of x, namely, $(b/2a)^2$, to both sides of the equation:

$$x^2 + \frac{bx}{a} + \left(\frac{b}{2a}\right)^2 = -\frac{c}{a} + \left(\frac{b}{2a}\right)^2$$

$$\left(x + \frac{b}{2a}\right)^2 = -\frac{c}{a} + \frac{b^2}{4a^2}$$

$$= \frac{-4ac + b^2}{4a^2}$$

Taking the square root of both sides, we have

$$x + \frac{b}{2a} = \frac{\pm\sqrt{b^2 - 4ac}}{2a}$$

$$x = \frac{-b \pm \sqrt{b^2 - 4ac}}{2a}$$

♦

The expression $b^2 - 4ac$ is called the **discriminant** of the quadratic equation. The sign of the discriminant determines the nature of the solutions to the equation, as follows:

$$b^2 - 4ac > 0 : \quad \text{two different real solutions}$$
$$= 0 : \quad \text{one real solution (two identical solutions)}$$
$$< 0 : \quad \text{no real solutions}$$

More specifically, from the quadratic formula, we deduce:

THE SOLUTIONS OF A QUADRATIC EQUATION

Consider the quadratic equation:

$$ax^2 + bx + c = 0$$

1. If $b^2 - 4ac > 0$, then the equation has the two different, real solutions:

$$x = \frac{-b + \sqrt{b^2 - 4ac}}{2a}, \quad x = \frac{-b - \sqrt{b^2 - 4ac}}{2a}$$

2. If $b^2 - 4ac = 0$, the equation has the single real solution:

$$x = -\frac{b}{2a}$$

3. If $b^2 - 4ac < 0$, then the equation has no real solutions.

➤ **EXAMPLE 5**
Quadratic Formula

Solve the following equations for real solutions using the quadratic formula.

1. $2x^2 - 3x - 4 = 0$
2. $9x^2 - 12x + 4 = 0$
3. $x^2 + x + 1 = 0$

Solution

1. From the equation,

$$a = 2, b = -3, c = -4$$

So the discriminant has the value

$$b^2 - 4ac = (-3)^2 - 4(2)(-4) = 41$$

Since the discriminant is positive, the equation has two real solutions. They are given by the quadratic formula as

$$x = \frac{-b \pm \sqrt{b^2 - 4ac}}{2a}$$

$$= \frac{-(-3) \pm \sqrt{41}}{2(2)}$$

$$= \frac{3 \pm \sqrt{41}}{4} = 2.351, -0.851$$

2. In this equation, $a = 9$, $b = -12$, and $c = 4$, so the discriminant has the value

$$b^2 - 4ac = (-12)^2 - 4(9)(4) = 0$$

So the equation has one real solution, that is, two identical real solutions, given by the quadratic formula as

$$x = \frac{-b \pm \sqrt{b^2 - 4ac}}{2a}$$

$$= \frac{-(-12) \pm \sqrt{0}}{2(9)}$$

$$= \frac{12}{18}$$

$$= \frac{2}{3}$$

3. In this case, the discriminant has the value

$$b^2 - 4ac = 1^2 - 4(1)(1) = -3$$

Since the discriminant is negative, the equation has no real solutions.

Equations That Reduce to Quadratic Equations

Many equations, although not quadratic themselves, can be solved most easily if we reduce them to quadratic equations through algebraic manipulations. The next two examples give illustrations of such equations.

➤ **EXAMPLE 6**

A Rational Equation

Determine all solutions of the following rational equation:

$$\frac{x}{x - 1} = \frac{6x - 5}{x + 15}$$

Solution

When confronted with rational expressions in an equation, it is usually a good idea to rewrite the equation in such a way as to remove the rational expressions. In this example, we may remove the rational expressions by multiplying both sides of the equation by the least common multiple of the denominators, namely, by $(x - 1)(x + 15)$:

$$x(x + 15) = (6x - 5)(x - 1)$$

$$x^2 + 15x = 6x^2 - 11x + 5$$

$$5x^2 - 26x + 5 = 0$$

We now have a quadratic equation that we can solve using the quadratic formula:

$$x = \frac{-(-26) \pm \sqrt{(-26)^2 - 4(5)(5)}}{2(5)}$$

$$= \frac{26 \pm \sqrt{576}}{10}$$

$$= \frac{26 \pm 24}{10}$$

$$= 5, \frac{1}{5}$$

Neither solution is 1 or -15, the values of x for which the rational expressions in the equation are undefined. Therefore, the values $x = 5$ and $x = 1/5$ are solutions to the equation. (If 1 or -15 had been a solution to the quadratic equation, we would have had to disqualify that value because it would have made one of the expressions in the equation undefined.)

➤ **EXAMPLE 7**

Another Rational Equation

Determine the solutions of the equation

$$\frac{3}{x - 5} + \frac{5}{x + 5} = 4$$

Solution

The first step in the solution is to clear the fractions by multiplying both sides by the least common multiple of the denominators, namely $(x - 5)(x + 5)$. This gives the equation

$$\frac{3}{x - 5}(x + 5)(x - 5) + \frac{5}{x + 5}(x + 5)(x - 5) = 4(x + 5)(x - 5)$$

$$3(x + 5) + 5(x - 5) = 4(x^2 - 25)$$

$$8x - 10 = 4x^2 - 100$$

$$4x^2 - 8x - 90 = 0$$

We can divide both sides of this last equation by 2, to obtain

$$2x^2 - 4x - 45 = 0$$

Let's use the quadratic formula. The discriminant is

$$b^2 - 4ac = (-4)^2 - 4(2)(-45) = 376$$

Therefore, the quadratic formula yields

$$x = \frac{-(-4) \pm \sqrt{376}}{2(2)}$$

$$= \frac{4 \pm 2\sqrt{94}}{4}$$

$$= \frac{2 \pm \sqrt{94}}{2} = 5.848, -3.848$$

The equation we ultimately solved was obtained from the given equation by multiplying by the expression $(x - 5)(x + 5)$. This expression is neither 0 nor undefined for x equal to one of the solutions found. Therefore, these solutions are indeed solutions to the original equation.

➤ **EXAMPLE 8**
A Radical Equation

Determine all solutions of the following radical equation:

$$3x + 2\sqrt{x} = 1$$

Solution

When confronted with an equation containing a radical, you should attempt to transform the equation so that the radical disappears. In this case, it is simplest to isolate the radical on one side of the equation and square both sides:

$$2\sqrt{x} = 1 - 3x$$

$$\left(2\sqrt{x}\right)^2 = (1 - 3x)^2 \qquad \textit{Squaring both sides}$$

$$4x = 1 - 6x + 9x^2$$

$$9x^2 - 10x + 1 = 0$$

We now have a quadratic equation that we can solve with the aid of the quadratic formula:

$$x = \frac{-(-10) \pm \sqrt{(-10)^2 - 4(9)(1)}}{2(9)}$$

$$= \frac{10 \pm \sqrt{64}}{18}$$

$$= \frac{10 \pm 8}{18}$$

$$= 1 \text{ or } \frac{1}{9}$$

The possible solutions of the equation are thus $x = 1$ and $x = 1/9$. However, note that we squared both sides of the equation. Recall our discussion earlier in the chapter that squaring an equation may introduce extraneous solutions. In the light of that discussion, it is necessary to check both of the possible solutions to determine whether they are in fact solutions. Substituting $x = 1/9$ into the given equation yields

$$3 \cdot \frac{1}{9} + 2\sqrt{\frac{1}{9}}$$
$$= \frac{1}{3} + 2 \cdot \frac{1}{3}$$
$$= 1 \qquad \qquad 1$$

So 1/9 is a solution. However, if we substitute $x = 1$ into the given equation, we find that

$$3 + 2\sqrt{x} = 3 \cdot 1 + 2\sqrt{1}$$
$$= 3 + 2$$
$$= 5 \qquad \qquad 1$$

The two sides are unequal. So $x = 1$ is an extraneous solution that arose from squaring the original equation.

Applications of Quadratic Equations

Quadratic equations occur in many applied problems. In the next examples, we will provide illustrations of how they are used in architecture, economics, and physics.

> **EXAMPLE 9**
Architecture

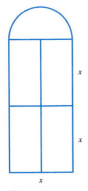

Figure 1

An architect is designing a house featuring Palladian windows, as shown in Figure 1. To use passive solar heating, each window, consisting of four rectangular panes and a semicircular pane, must have 2500 square inches of glass. Determine the dimensions of each window.

Solution

The condition we seek to satisfy is

Area of semicircular region + Area of rectangular region = 2500

The semicircular region has diameter x and radius $x/2$. From the formula for the area of a circle, we have the following formula for a semicircle:

$$\frac{1}{2}\pi r^2 = \frac{1}{2}\pi\left(\frac{x}{2}\right)^2 = \frac{\pi}{8}x^2$$

The area of the entire rectangular region is

$$2 \cdot x \cdot x = 2x^2$$

Therefore, x must satisfy the equation

$$\frac{\pi}{8}x^2 + 2x^2 = 2500$$

$$\left(\frac{\pi}{8} + 2\right)x^2 = 2500 \qquad \textit{Factoring out } x^2$$

$$x^2 = \frac{2500}{(\pi/8) + 2}$$

$$x = \pm\sqrt{\frac{2500}{(\pi/8) + 2}} \qquad \textit{Taking square roots}$$

Since x represents a physical length, we can ignore the negative value. So the width of the window is

$$x = \sqrt{\frac{2500}{(\pi/8) + 2}}$$

$$= \sqrt{\frac{20,000}{\pi + 16}} \text{ in.}$$

$$= 32.3241 \text{ in. (calculator)}$$

The height of the rectangular portion of the window is twice this amount, or 64.6482 in. The circumference of the semicircle is equal to

$$\pi \cdot \text{radius} = \pi \frac{x}{2}$$

$$= (3.14159) \cdot \frac{32.3241}{2}$$

$$= 50.7746 \text{ in.}$$

➤ **EXAMPLE 10**

Economics

The manufacturer of a very popular model of compact disc stereo system notes that the demand for the product fluctuates with the price. If the price is p dollars, then the market research department reports that monthly sales s may be predicted by the formula

$$s = 5000 - 10p$$

Suppose that the manufacturer wishes to achieve revenues of $600,000 in a given month. What should be the price of the compact disc players?

Solution

The revenues for the month are given by the product

$$\text{Revenue} = \text{Sales} \cdot \text{Price}$$

Using the given expression for sales in terms of price, we have

$$\text{Revenue} = (5000 - 10p)\, p$$

To achieve a revenue of $600,000, the following condition must be fulfilled:

$$600,000 = (5000 - 10p)\, p$$

$$-10p^2 + 5000p - 600,000 = 0$$

$$p^2 - 500p + 60,000 = 0 \qquad \textit{Dividing by } -10$$

We can solve this equation by factoring, but it will take quite a while to work out. However, we can always solve a quadratic equation using the quadratic formula. Let's proceed directly to using the quadratic formula:

$$p = \frac{500 \pm \sqrt{(-500)^2 - 4(1)(60,000)}}{2(1)}$$

$$= \frac{500 \pm \sqrt{10,000}}{2}$$

$$= \frac{500 \pm 100}{2}$$

$$= 300, \ 200$$

In other words, the company can achieve the desired revenues by choosing one of the two prices, $300 or $200, for its compact disc player. Since both prices yield the desired revenue, the company can choose the price on the basis of some other criterion than revenue. The lower price would probably be chosen to generate consumer goodwill.

> **EXAMPLE 11**

Physics

If an object is thrown vertically with an initial velocity v_0 from an initial height h_0, its height h (in feet) t seconds later is given by the following formula from the laws of physics:

$$h = -16t^2 + v_0 t + h_0$$

Suppose that a ball is thrown upward from the top of a 960-foot skyscraper with a velocity of 96 feet per second. According to the formula, the height h of the ball after t seconds is given by

$$h = -16t^2 + 96t + 960$$

1. How long does it take the ball to reach the ground?
2. After how many seconds is the ball 480 feet off the ground?

Solution

1. At the time t when the ball hits the ground, its height equals 0, so that

$$-16t^2 + 96t + 960 = 0$$
$$t^2 - 6t - 60 = 0$$
$$t = \frac{6 \pm \sqrt{(-6)^2 - 4(1)(-60)}}{2(1)}$$
$$= \frac{6 \pm \sqrt{276}}{2}$$
$$= \frac{6 \pm 2\sqrt{69}}{2}$$
$$= 3 + \sqrt{69}, \ 3 - \sqrt{69}$$
$$= 11.3066, \ -5.3066$$

Since t represents the time since the ball was thrown, the negative value must be discarded on physical grounds. A single value of t remains. That is, the ball hits the ground after 11.3066 seconds.

2. If the height of the ball is 480 feet, then t must satisfy the equation

$$-16t^2 + 96t + 960 = 480$$
$$-16t^2 + 96t + 480 = 0$$
$$t^2 - 6t - 30 = 0$$
$$t = \frac{-(-6) \pm \sqrt{(-6)^2 - 4(1)(-30)}}{2}$$
$$= \frac{6 \pm \sqrt{156}}{2}$$
$$= 3 \pm \sqrt{39}$$
$$= 9.2450 \text{ or } -3.2450$$

That is, the ball is at height 480 feet at 9.2450 seconds after it is thrown.

Exercises 2.3

Determine the real solutions of the following quadratic equations.

1. $x^2 - 3 = 0$

2. $5x^2 = 0$

3. $x^2 + 4 = 0$

4. $4x^2 + 3 = 0$

5. $3t^2 = -15$

6. $2a^2 = -11$

7. $(x - 2)^2 = 1$

8. $(x + 5)^2 = 20$

9. $(3x + 1)^2 = 9$

10. $(2x - 3)^2 = 5$

Solve the following equations, completing the square to determine all real solutions.

11. $x^2 + 5x - 10 = 0$

12. $x^2 + 8x + 7 = 0$

13. $x^2 + 6 = 4x$

14. $x^2 = 3x - 5$

15. $3x^2 + 4x = 1$

16. $8x^2 = 12x + 8$

17. $7x + 10 = 3x^2$

18. $3y^2 - 10y - 1 = 0$

Use the quadratic formula to solve the following quadratic equations for real roots.

19. $3x^2 + 4x = 0$

20. $5x^2 = 8x$

21. $2x^2 - 4x + 5 = 0$

22. $7x^2 + 10x + 12 = 0$

23. $2x^2 + x = 4$

24. $2x^2 = 5x + 14$

Determine all solutions of the following equations in real numbers. Use any method you wish.

25. $18x^2 = 10x$

26. $13t^2 + 6t = 0$

27. $6x^2 + 5x = 21$

28. $10x^2 - 3x - 1 = 0$

29. $x^2 - 5x + 10 = 0$

30. $x^2 - 3x + 5 = 0$

31. $2 + \dfrac{3}{x^2} = \dfrac{1}{x}$

32. $3 - \dfrac{1}{x} + \dfrac{6}{x^2} = 0$

33. $\dfrac{2x - 3}{3x + 5} = \dfrac{x + 7}{3x - 4}$

34. $\dfrac{5 - 2x}{6x + 1} = \dfrac{3 + 2x}{5 - x}$

35. $x^2 + \sqrt{2}x - \sqrt{3} = 0$

36. $\sqrt{5}x^2 + 4x = \sqrt{2}$

37. $(y + 5)(y - 4) = (2y + 1)(3y - 4)$

38. $(2a - 1)(3a + 5) = (a - 3)(4a - 7)$

39. $(-2t)^2 + (4t - 3)^2 + (-1 - t)^2 = 18$

40. $(-4a)^2 + (3a - 1)^2 = (5a - 7)^2$

41. $x - \dfrac{1}{x} = 17$

42. $m + \dfrac{2}{m} = 23$

43. $x^2 - \dfrac{5}{3}x - \dfrac{4}{3} = 0$

44. $0.8t^2 - 24t + 90 = 0$

45. $2t^2 + 0.3t = 0.4$

46. $7(p - 0.7) + \dfrac{0.7(3p - 0.5)}{p} = 0$

47. $5800v^2 + 6000v - 10,000 = 0$

48. $2400m^2 = 7700m - 12,000$

49. $x = 7\sqrt{x} + 30$

50. $x - 13\sqrt{x} + 12 = 0$

51. $\sqrt{x + 5} = 3$

52. $\sqrt{x + 10} = 4$

53. $\sqrt{x + 7} + 5 = x$

54. $\sqrt{y + 5} = y - 1$

55. $\sqrt{t - 2} = 3$

56. $11 = \sqrt{4 + m}$

◆ Applications

57. The surface area of a right circular cylinder is 170 m². The radius is one-half the height. Find the dimensions of the cylinder.

58. The bottom of an extension ladder is 8 ft. from the wall on which it is leaning. If the ladder is not extended, it is 17 ft. long. As the ladder is extended to its maximum safe length, the top of the ladder moves up the wall 9 ft. To what length does the ladder extend?

59. The hypotenuse of a right triangle is 17 ft. One leg is 7 ft. longer than the other. Find the length of the legs.

60. The manufacturer of a computer printer notes that demand for the product fluctuates with the price. If the price is p dollars, then the number of sales in a year s are given by the formula

$$s = 18,000 - 20p$$

Suppose the manufacturer wishes to achieve revenues of $4,000,000 in a given year. What should be the price of the printers?

61. The formula

$$S = -16t^2 + v_0t + h$$

gives the height S, in feet, after t seconds of an object thrown upward from a height h, at an initial velocity v_0.

a. A ball is thrown upward at an initial velocity of 96 ft./sec. from the top of the Sears Tower, which is 1456 ft. tall. How long does it take the ball to reach the ground?

b. After what amount of time will the ball be at a height of 1500 ft?

62. The formula

$$S = -16t^2 + v_0t + h$$

gives the height S, in feet, after t seconds of an object thrown downward from a height h, at an initial velocity v_0.

a. An object is thrown downward from the roof of a 960-ft.-high building with an initial velocity of 64 ft./sec. How long does it take to reach the ground?

b. After what amount of time will the object be 500 ft. off the ground?

63. The more chairs a factory produces, the less it costs to make each chair. It costs $25 a chair to make 100 chairs. For each additional chair, the average cost per chair for all chairs produced goes down $0.20. How many chairs will be produced when the total cost is $2,470?

64. The more fruit trees planted per acre, the less the yield of fruit for each tree. When 30 trees are planted per acre, the yield is 50 bushels per tree. For each additional tree planted, the yield per tree decreases by 2 bushels. How many trees should be planted to yield 900 bushels per acre?

65. One can use the boiling point T of water to estimate the elevation H at which the water is boiling. The quadratic equation that can be used is

$$H = 1000(100 - T) + 580(100 - T)^2$$

where H = elevation, in meters, and T = boiling point, in degrees Celsius.

a. The boiling point is 99.4°C. Find the elevation.

b. What is the boiling point at sea level ($H = 0$)?

c. Denver, Colorado, is 5280 ft. (1609 m) above sea level. At what temperature does water boil?

d. Mt. Elbert outside Denver is 14,431 ft. (4400 m) above sea level and has a road that can be driven to the top. A family drives up Mt. Elbert and boils some water for a tailgate party. How much lower is the boiling point than in Denver?

66. There are many rules for determining the dosage of a medication for a child based on an adult dosage. The following are two such rules:

$$\text{Young's Rule:} \quad C = \frac{A}{A + 12}D$$

$$\text{Cowling's Rule:} \quad C = \frac{A + 1}{24}D$$

where A = the age of the child in years, D = adult dosage, and C = child's dosage. (*Warning!* Do not apply these formulas without consulting a physician. These are estimates and may not apply to a particular medication.)

a. An adult dosage of liquid medication is 5 cc (cubic centimeters). Find the child's dosage using each formula for a child whose age is 3 years.

b. At what age A of a child are the dosages the same?

67. What is the interest rate if interest is compounded annually and $1200 grows to $1323 in 2 years?

68. What is the interest rate if interest is compounded annually and $1532 grows to $1737.63 in 2 years?

69. What is the interest rate if interest is compounded semi-annually and $2500 grows to $2924.65 in 2 years?

70. $1400 is invested at 10.5% for 2 years, interest compounded quarterly. What interest rate, compounded annually, would this same amount of money have to be invested at in order to grow to the same amount in the same time?

71. The length of each side of an equilateral triangle is a. (See figure.)

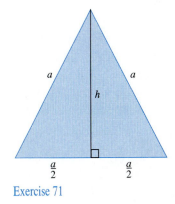

Exercise 71

a. Find the formula for the height h.

b. Find a formula for the area A.

c. Find a formula for the perimeter P.

d. The area of an equilateral triangle is $25\sqrt{3}$ cm². What is the perimeter?

72. An open box is to be folded out of a 15 in. by 25 in. piece of cardboard by cutting out square corners from the box. (See figures.) The resulting area at the bottom of the box is 231 in.^{2}. What is the length of each side of the square corners?

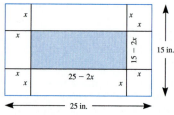

Exercise 72(a)

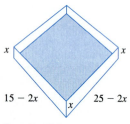

Exercise 72(b)

73. A polygon (see figure) with n sides has D diagonals, where D is given by

$$D = \frac{n(n-3)}{2}$$

a. How many diagonals does an octagon have?

b. A polygon has 54 diagonals. How many sides does it have?

c. Can there exist a polygon with 10 diagonals?

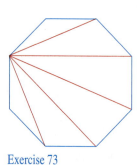

Exercise 73

74. There are n teams in a sports league. If each team plays every other team twice, then the total number of games G the league plays is given by $G = n(n-1)$.

a. There are 10 teams in a certain basketball conference. In their season, each team plays every other team twice. What is the total number of games played in a season?

b. Is it possible for a sports league to have 56 total games in its season when each team plays every other team twice?

Solve each formula for the indicated letter.

75. $G = \dfrac{n(n-1)}{2}$, for n.

76. $D = \dfrac{n(n-3)}{2}$, for n.

77. $H = -16t^2 + v_0t + h$, for t.

78. $S = 4.9t^2 + v_0t$, for t.

79. $S = 2\pi rh + 2\pi r^2$, for r.

80. $S = 2\pi r^2 + \pi rs$, for r.

Find all real solutions.

81. $2x(x+3)(5x-4) = 0$

82. $(5x+7)(x-4)(x^2-5) = 0$

83. $x^3 - 1 = 0$ 84. $8x^3 - 2 = 0$

85. $x^3 + 3x^2 - x - 3 = 0$

86. $3y^3 - 4y^2 - 12y + 16 = 0$

87. $12x^3 - 12x^2 = 0$ 88. $4x^3 - 4 = 0$

89. $5x^4 - 5 = 0$ 90. $2x - \dfrac{432}{x^2} = 0$

91. $50x - \dfrac{3200}{x^2} = 0$

92. $-(x - x^2)^{-2}(1 - 2x) = 0$

93. $\dfrac{m - \dfrac{1}{4}}{1 + \dfrac{1}{4m}} = \dfrac{\dfrac{3}{2} - m}{1 + \dfrac{3}{2}m}$

✐ In Your Own Words

94. Find another formula for solving quadratic equations by rationalizing the numerator of the quadratic formula. What restriction is placed on this formula that does not apply to the quadratic formula?

➡ Technology

95. If you have a programmable calculator, write a program to solve a quadratic equation in terms of the coefficients.

2.4 EQUATIONS SOLVABLE USING QUADRATIC EQUATIONS

In the preceding section, we learned to solve quadratic equations using any one of several methods. Moreover, we solved a number of applied problems that, when translated into algebraic language, resulted in quadratic equations. There are many equations that, although not quadratic, can be reduced to solving quadratic equations. Some examples of such equations are:

$$x^4 - 7x^2 + 6 = 0$$

$$\left(\frac{x-3}{x+4}\right)^2 + 2\frac{x-3}{x+4} - 3 = 0$$

$$\sqrt{x-1} + \sqrt{x+2} = 5$$

As we shall see shortly, the first two equations can be solved by the **method of substitution**. That is, we substitute for x a new variable so that the resulting equation is a quadratic equation with respect to the new variable. The third equation can be solved by removing the radicals using algebraic manipulations.

The Method of Substitution

Often, an equation may be put into a solvable form by changing the variable. For example, the equation

$$x^4 - 4x^2 + 4 = 0$$

is a fourth-degree equation. However, if we introduce the new variable

$$y = x^2$$

then we may write the equation in terms of the new variable as

$$y^2 - 4y + 4 = 0$$

a quadratic equation, which we can solve using the methods of the preceding section. For instance, the equation may be factored into

$$(y - 2)^2 = 0$$

which gives the solution $y = 2$. Of course, to solve the original equation, we must relate the solution in y back to values of the original variable x. To do this, substitute $y = 2$ into the equation giving the relationship between the variables:

$$y = x^2$$
$$2 = x^2$$
$$x = \pm\sqrt{2}$$

These are the possible solutions of the given equation. Because of the possibility of extraneous solutions, we must check these values before declaring them solutions:

$$(\sqrt{2})^4 - 4(\sqrt{2})^2 + 4 \quad \bigg| $$
$$= 4 - 8 + 4 \quad \bigg|$$
$$= 0 \quad \bigg| \quad 0$$
$$(-\sqrt{2})^4 - 4(-\sqrt{2})^2 + 4 \bigg|$$
$$= 4 - 8 + 4 \quad \bigg|$$
$$= 0 \quad \bigg| \quad 0$$

Thus, both values are solutions of the given equation. The method used to solve the equation is called the **method of substitution**.

We may summarize this method as follows:

METHOD OF SUBSTITUTION FOR SOLVING AN EQUATION

1. Make an appropriate substitution

$$u = [\text{expression in } x]$$

2. Solve the resulting equation for u.

3. For each solution $u = a$ of Step 2, solve the equation

$$a = [\text{expression in } x]$$

4. Check each value of x to determine which are solutions of the original equation.

The following two examples provide further practice in applying the method of substitution.

➤ **EXAMPLE 1**
Method of Substitution

Use the method of substitution to solve the following equations:

1. $x^4 - 7x^2 + 6 = 0$
2. $x - 7\sqrt{x} + 6 = 0$
3. $(x^2 + 3x)^2 - 7(x^2 + 3x) + 6 = 0$

Solution

1. Based on examination of the equation, the following substitution suggests itself:

$$y = x^2$$

In terms of y, we can rewrite the equation as

$$y^2 - 7y + 6 = 0$$

This equation is easily factorable:

$$(y - 6)(y - 1) = 0$$
$$y = 6, \ 1$$

In terms of x, this means that

$$x^2 = y = 6 \qquad \text{or} \qquad x^2 = y = 1$$

$$x = \pm\sqrt{6} \text{ or } \pm 1$$

We must check these four values of x to determine which are solutions of the original equation:

$$(\sqrt{6})^4 - 7(\sqrt{6})^2 + 6$$
$$= 36 - 7 \cdot 6 + 6$$
$$= 0 \qquad \Big| \qquad 0$$

$$(-\sqrt{6})^4 - 7(-\sqrt{6})^2 + 6$$
$$= 36 - 7 \cdot 6 + 6$$
$$= 0 \qquad \Big| \qquad 0$$

$$(1)^4 - 7(1)^2 + 6$$
$$= 1 - 7 \cdot 1 + 6$$
$$= 0 \qquad \Big| \qquad 0$$

$$(-1)^4 - 7(-1)^2 + 6$$
$$= 1 - 7 \cdot 1 + 6$$
$$= 0 \qquad \Big| \qquad 0$$

Thus, all four values are solutions of the given equation.

2. In the equation $x - 7\sqrt{x} + 6 = 0$, we can make the substitution

$$y = \sqrt{x}$$

and rewrite the equation as

$$y^2 - 7y + 6 = 0$$

This is the same equation encountered in part 1. We saw that its solutions were

$$y = 6, 1$$
$$\sqrt{x} = 6, 1$$

Squaring both sides of the last equation, we have

$$x = 36, 1$$

Let's now check these values in the original equation:

$$36 - 7\sqrt{36} + 6$$
$$= 36 - 7 \cdot 6 + 6$$
$$= 0 \qquad \Big| \qquad 0$$

$$1 - 7\sqrt{1} + 6$$
$$= 1 - 7 + 6$$
$$= 0 \qquad \Big| \qquad 0$$

Both values of x are solutions of the given equation.

3. In the equation $(x^2 + 3x)^2 - 7(x^2 + 3x) + 6 = 0$, we make the substitution

$$y = x^2 + 3x$$

Once again we arrive at the equation

$$y^2 - 7y + 6 = 0$$

From part 1, we have

$$y = 6, 1$$

That is, we have

$$x^2 + 3x = 6 \qquad \text{or} \qquad x^2 + 3x = 1$$

Using the quadratic formula, we have the solutions

$$x = \frac{-3 \pm \sqrt{33}}{2} \qquad \text{and} \qquad \frac{-3 \pm \sqrt{13}}{2}$$

Let's now check these values in the original equation. In the case of the values $x = (-3 \pm \sqrt{33})/2$, we find that

$$x^2 + 3x = 6$$
$$(x^2 + 3x)^2 - 7(x^2 + 3x) + 6$$
$$= (6)^2 - 7(6) + 6$$
$$= 0 \qquad \qquad 0$$

In a similar fashion, for the values $x = (-3 \pm \sqrt{13})/2$, we have

$$x^2 + 3x = 1$$
$$(x^2 + 3x)^2 - 7(x^2 + 3x) + 6$$
$$= 1^2 - 7(1) + 6$$
$$= 0 \qquad \qquad 0$$

Thus, all four values of x are solutions of the given equation.

➤ **EXAMPLE 2**
Substitution Leading
to Extraneous Solutions

Solve the following equation:

$$\left(\frac{x - 3}{x + 4}\right)^2 + 2\frac{x - 3}{x + 4} - 3 = 0$$

Solution

In this example, we make the substitution

$$y = \frac{x - 3}{x + 4}$$

In terms of y the equation becomes

$$y^2 + 2y - 3 = 0$$

This equation may be factored:

$$(y + 3)(y - 1) = 0$$
$$y = -3, 1$$

This means that

$$y = \frac{x - 3}{x + 4} = -3$$

or

$$y = \frac{x - 3}{x + 4} = 1$$

The first equation is equivalent to

$$x - 3 = -3(x + 4)$$
$$4x = -9$$
$$x = -\frac{9}{4}$$

Provided that $x \neq -4$, the second equation is equivalent to

$$x - 3 = x + 4$$

which leads to the contradiction

$$0 = 7$$

This last equation is **inconsistent.** So the only solution of the given equation is $x = -9/4$. Here is a check that this value is a solution:

$$\left(\frac{x - 3}{x + 4}\right)^2 + 2\frac{x - 3}{x + 4} - 3 = \left(\frac{-(9/4) - 3}{-(9/4) + 4}\right)^2 + 2\left(\frac{-(9/4) - 3}{-(9/4) + 4}\right) - 3 \qquad \Big|$$

$$= \frac{(-21/4)^2}{(7/4)^2} + 2\left(\frac{-21/4}{7/4}\right) - 3$$

$$= 9 + 2(-3) - 3$$

$$= 0 \qquad\qquad\qquad\qquad\qquad\qquad\qquad\qquad\qquad\quad\Big|\quad 0$$

> **EXAMPLE 3**
Raising an Equation
to a Power

Solve the equation:

$$\sqrt[3]{7x^2 - 15} = -2$$

Solution

When you encounter a radical equation, you should attempt to remove the radical by raising both sides of the equation to an appropriate power. In this case, we raise both sides to the third power:

$$\left[\sqrt[3]{7x^2 - 15}\right]^3 = (-2)^3$$
$$7x^2 - 15 = -8$$
$$7x^2 = 7$$
$$x^2 = 1$$
$$x = \pm 1$$

Let's now check these solutions.

$$\sqrt[3]{7(\pm 1)^2 - 15} \quad = \sqrt[3]{-8} \quad \Big|$$
$$= -2 \quad \Big| \quad -2$$

So the solution set of the given equation is $\{+1, -1\}$.

➤ **EXAMPLE 4**
Several Radicals
in One Equation

Solve the following equation:

$$\sqrt{x-1} + \sqrt{x+2} = 5$$

Solution

As in the preceding example, we must remove the radicals. To do so, move one radical to the right and square both sides:

$$\sqrt{x-1} = 5 - \sqrt{x+2}$$
$$(\sqrt{x-1})^2 = (5 - \sqrt{x+2})^2$$
$$x - 1 = 25 - 10\sqrt{x+2} + (x+2)$$
$$\sqrt{x+2} = -\frac{14}{15}$$
$$x + 2 = \left(-\frac{14}{15}\right)^2 = \frac{196}{25}$$
$$x = \frac{146}{25}$$

To check this solution, we substitute it into the given equation:

$$\sqrt{x-1} + \sqrt{x+2} = \sqrt{\frac{146}{25} - 1} + \sqrt{\frac{146}{25} + 2}$$
$$= \sqrt{\frac{121}{25}} + \sqrt{\frac{196}{25}}$$
$$= \frac{11}{5} + \frac{14}{5}$$
$$= 5 \qquad\qquad\qquad 5$$

More Applied Problems

In our discussion above, we learned to solve equations that could be reduced to quadratic equations. Let's now illustrate how such equations arise in applied problems involving relative motion and job planning.

➤ **EXAMPLE 5**
Relative Motion

A bicyclist rides 15 miles on a straight road, then turns around and returns to the starting point. The time for the total ride is 3 hours. Suppose that throughout the trip there is a 5 mile-per-hour wind blowing in the initial direction of the cyclist. The wind increases the speed of the cyclist by 5 miles per hour in the initial direction and slows her down by 5 miles per hour in the reverse direction. How fast can the cyclist ride in still air?

Solution

Let r denote the speed of the cyclist in still air. On the first half of the trip, she travels 15 miles at a speed of $r + 5$. From the formula for distance, Distance = Rate · Time, we can calculate the time for the first segment of the trip:

$$\text{Time 1} = \frac{\text{Distance}}{\text{Rate}}$$

$$\text{Time 1} = \frac{15}{r+5}$$

For the second half of the trip, the cyclist travels 15 miles at a speed of $r - 5$. The time for this segment of the trip is

$$\text{Time 2} = \frac{15}{r - 5}$$

Since the time for the total trip is 3 hours, we have

$$\text{Time 1} + \text{Time 2} = 3 \text{ hours}$$

$$\frac{15}{r + 5} + \frac{15}{r - 5} = 3$$

To solve this equation, we clear the fractions by multiplying both sides by the least common denominator, $(r + 5)(r - 5)$. For $r \neq 5, -5$, this gives us

$$15(r - 5) + 15(r + 5) = 3(r - 5)(r + 5)$$

$$30r = 3r^2 - 75$$

$$3r^2 - 30r - 75 = 0$$

$$r^2 - 10r - 25 = 0$$

By the quadratic formula, with $a = 1$, $b = -10$, and $c = -25$, we have

$$r = \frac{10 \pm \sqrt{100 + 100}}{2}$$

$$= \frac{10 \pm \sqrt{200}}{2}$$

$$= \frac{10 \pm 10\sqrt{2}}{2}$$

$$= 5 \pm 5\sqrt{2}$$

$$= 12.0711, -2.0711$$

The negative solution makes no sense in the context of the problem. So the cyclist can go 12.0711 miles per hour in still air. We can check this solution by substituting $5 + 5\sqrt{2}$ for r in the original equation:

$$\frac{15}{r + 5} + \frac{15}{r - 5} = \frac{15}{(5 + 5\sqrt{2}) + 5} + \frac{15}{(5 + 5\sqrt{2}) - 5}$$

$$= \frac{15}{10 + 5\sqrt{2}} + \frac{15}{5\sqrt{2}}$$

$$= \frac{3}{2 + \sqrt{2}} + \frac{3}{\sqrt{2}}$$

$$= \frac{3}{2 + \sqrt{2}} \cdot \frac{2 - \sqrt{2}}{2 - \sqrt{2}} + \frac{3}{\sqrt{2}} \cdot \frac{\sqrt{2}}{\sqrt{2}}$$

$$= \frac{6 - 3\sqrt{2}}{2} + \frac{3\sqrt{2}}{2}$$

$$= 3 \qquad\qquad\qquad 3$$

➤ **EXAMPLE 6**

Job Planning

Joe and Sam have a painting business. Joe can paint a particular room in one hour less than Sam. Together they can paint the room in 5 hours. How long would it take Sam to paint the room alone?

Solution

For each of the men, we have the relationship

Rate of working $\times$ Time worked = Fractional part of job done

Let t denote the total amount of time (in hours) that Sam requires to paint the room by himself. Then his rate of working is $1/t$ rooms per hour. So by the above equation, in 5 hours the part of the job he accomplishes equals $5/t$. Joe paints the room in one hour less than Sam, or $t-1$ hours. So Joe's rate of working is $1/(t-1)$ rooms per hour. Therefore, in 5 hours, the part of the job he accomplishes equals

$$5 \cdot \frac{1}{t-1} = \frac{5}{t-1}$$

Together Joe and Sam can paint the entire room in 5 hours, which we can express as

Sam fractional part + Joe fractional part = Whole job

$$\frac{5}{t} + \frac{5}{t-1} = 1$$

We can clear the fraction in this equation by multiplying by the least common denominator $t(t-1)$. This gives us

$$5(t-1) + 5t = t(t-1)$$
$$10t - 5 = t^2 - t$$
$$t^2 - 11t + 5 = 0$$

By the quadratic formula, we have

$$t = \frac{11 \pm \sqrt{(-11)^2 - 4(1)(5)}}{2}$$
$$= \frac{11 \pm \sqrt{101}}{2}$$
$$= 10.5249, 0.4751$$

We reject the solution $(11 - \sqrt{101})/2$ since, although it is a solution of the quadratic equation, it leads to a proportion $1/(t-1)$ for Joe's work that is negative. So t has the value $(11 + \sqrt{101})/2$. That is, it takes Sam about 10.5 hours to paint the room by himself.

Exercises 2.4

Solve for real solutions.

1. $\dfrac{1}{5} + \dfrac{1}{7} = \dfrac{2}{t}$

2. $\dfrac{2}{x} - \dfrac{3}{x} = 1$

3. $\dfrac{x-1}{3} + \dfrac{2x+1}{4} = \dfrac{1}{2}$

4. $\dfrac{3x+2}{3} - \dfrac{x+5}{2} = \dfrac{x+1}{3}$

5. $\dfrac{10}{x} - \dfrac{1}{x} = 3 - \dfrac{7}{x}$

6. $\dfrac{16}{t} - \dfrac{16}{t+5} = \dfrac{8}{t}$

7. $\dfrac{2}{a} + \dfrac{1}{a+2} = \dfrac{4}{a^2 + 2a}$

8. $\dfrac{2}{y-2} + \dfrac{2y}{y^2 - 4} = \dfrac{3}{y+2}$

9. $\dfrac{4}{y+3} + \dfrac{7}{y^2 - 3y + 9} = \dfrac{108}{y^3 + 27}$

10. $\dfrac{11}{m^2 - m + 1} + \dfrac{8}{m+1} = \dfrac{24}{m^3 + 1}$

11. $\dfrac{2x}{x-3} - \dfrac{6}{x} = \dfrac{18}{x^2 - 3x}$

12. $\dfrac{4x}{x+5} + \dfrac{20}{x} = \dfrac{100}{x^2+5x}$

13. $\sqrt{x^2+3x-2} = x+3$

14. $\sqrt{2x^2+7x+10} = x+4$

15. $\sqrt{2x+3} - \sqrt{x+4} = 1$

16. $\sqrt{20+x} + \sqrt{20-x} = 5$

17. $t^{-1/4} = 2$ 18. $s^{-1/3} = -\dfrac{1}{3}$

19. $\sqrt[3]{3x+7} = 4$ 20. $\sqrt[3]{x-3} = -2$

21. $\sqrt{t-1} + \sqrt{t+4} = 5$

22. $\sqrt{m+10} = 10 + \sqrt{m-10}$

23. $\sqrt{y} + \sqrt{y+8} = 2$

24. $\sqrt{x+1} + 1 = \sqrt{2x+3}$

25. $\sqrt[4]{y^2-6y} = 2$ 26. $\sqrt[5]{(y-4)(y-8)} = 2$

27. $x^{3/2} = 27$ 28. $t^{2/3} = 9$

29. $x - 7\sqrt{x} + 12 = 0$ 30. $x + 4\sqrt{x} - 5 = 0$

31. $x^4 + 5x^2 - 36 = 0$ 32. $3x^4 + 2 = 7x^2$

33. $(x^2-3x)^2 - 2(x^2-3x) - 8 = 0$

34. $(t^2-7t)^2 = 2(t^2-7t) + 48$

35. $(\sqrt{y}-10)^2 + 30 = 11(\sqrt{y}-10)$

36. $(3-\sqrt{t})^2 - 7(3-\sqrt{t}) + 18 = 0$

37. $\left(\dfrac{x}{x-1}\right)^2 + \left(\dfrac{x}{x-1}\right) - 20 = 0$

38. $\left(\dfrac{2x+3}{x+1}\right) + 12 = \left(\dfrac{2x+3}{x+1}\right)^2$

39. $7z^{1/2} - 3 = 2z$ 40. $3x^{2/3} = 20 - 7x^{1/3}$

41. $\sqrt[3]{(x^2+2x)^2} - \sqrt[3]{x^2+2x} - 2 = 0$

42. $(\sqrt{2x+1}-2)^2 - 4(\sqrt{2x+1}-2) = 5$

43. $\sqrt{3+2x} - (3-2x)^{1/2} = \sqrt{2x}$

44. $\sqrt[6]{2x+2} = \sqrt[3]{3x-1}$ 45. $(3x+1)^{3/2} = 8$

46. $\dfrac{196}{t^2-7t+49} - \dfrac{2t}{t+7} = \dfrac{2058}{t^3+343}$

♦ **Applications**

47. Marcia can deliver the papers on her route in 2 hours. Louis can deliver the same route in 3 hours. How long would it take them to deliver the papers if they worked together?

48. A draftsman can draw a schematic diagram in 3 hours. It takes a trainee 5 hours to draw the same diagram. How long would it take if they worked together?

49. A can mow the lawn in 1/3 hr., B can do it in 1/4 hr., and C in 2/5 hr. How long would it take them if they worked together?

50. A can do a job in 0.4 hr. Working with B, the job takes 0.25 hr. After C is hired, they can do the job in 0.18 hr. How long would it take B working alone to do the job?

51. It takes Madonna 9 hours longer to build a wall than Sean. If they work together, they can build the wall in 20 hours. How long would it take Sean to build the wall by himself?

52. A boat travels 15 miles downstream and then turns around and travels the same 15 miles upstream in a total time of 4 hours. The speed of the stream is 5 mph. Find the speed of the boat in still water.

53. The student drove 2/3 of the duration of a trip at 60 mph. At what speed would that student have to drive the rest of the trip so that his average speed would be 54 mph?

54. The distance M, in miles, that a person can see to the horizon from a height h, in feet, is given by $M = 1.2\sqrt{h}$.

   a. A sailor is standing on a mast of a ship 150 ft. above sea level. How far can the sailor see to the horizon?

   b. From an airplane window a traveler can see 207.85 miles to the horizon. How high is the airplane?

   c. Solve the formula for h.

55. An airplane can travel 500 mph in still air. It travels 2080 miles with a tailwind and 1440 with a headwind in a total time of 7 hours. Find the speed of the wind.

56. Two cyclists move away from each other at a right angle at speeds of 12 ft./sec. and 16 ft./sec. After what amount of time will they be 90 ft. apart?

57. The diameter of a circular ventilating system is d. (See figure.) The system is joined to a square duct system with side s. To ensure smooth air flow, the areas of the circle and the square section must be the same.

   a. Find the formula for s in terms of d.

   b. Find s when $d = 202$ mm.

   c. Find d when $s = 400$ mm.

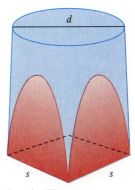

Exercise 57

This is an example of a classical problem in mathematics known as *squaring the circle*.

58. The time T it takes a yo-yo (see figure) of radius R to return to the hand after going down length L with no initial velocity is given by

$$T = \frac{2}{r}\sqrt{\frac{L(2r^2 + R^2)}{9.8}}$$

where r is the distance between the two halves of the yo-yo. If the length is fixed at 0.5 m, and $r = 1$ cm, what diameter yo-yo should be chosen to get a return time of 5 sec.?

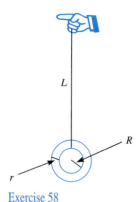

Exercise 58

59. A tank contains water H m deep. (See figure.) A hole is pierced in a wall of the tank h m below the water surface. The stream of water coming from that hole will hit the floor x m from the edge of the tank, where x is given by

$$x = 2\sqrt{h(H - h)}$$

A tank of water 10 m deep springs a leak near the top. The stream of water hits the ground 8 m from the bottom of the tank. Can a person standing on the rim of the tank reach inside the tank and plug the hole with his finger without getting his head wet? Explain.

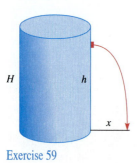

Exercise 59

60. $2000 is deposited at the beginning of the year. At the beginning of the next year, $1200 is deposited in another bank at the same interest rate. At the beginning of the third year, there is a total of $3573.80 in both accounts. If interest is compounded annually, what is the interest rate?

61. A right triangle contains as many square centimeters in area as it has centimeters in perimeter. Its longer leg is twice as long as its shorter leg. What is the length of the shorter leg?

62. An electrical circuit contains two resistors in parallel. One resistor is 5 ohms more than the other. The total resistance is 13 7/11. What is the resistance of each of the two resistors? (The formula for resistors in parallel is $1/R = 1/R_1 + 1/R_2$, where R is the total resistance.)

63. A brick is dropped into a well. The sound of the splash is heard 3 seconds later. How deep is the well? (Assume that sound travels at 1142 ft./sec.)

64. A student falls off a deck during a party. The sound of the student striking the ground reaches the top of the deck in 1 second. How high is the deck above the ground?

Determine all real solutions for x.

65. $x^3 + 3x^2 + 3x + 1 = 0$

66. $x^3 - 6x^2 + 12x = 8$

67. $\dfrac{x^3 + x^2 + 2x + 2}{x^2 + 4x + 3} = \dfrac{2x^3 + 4x^2 + x + 2}{x^2 + 5x + 6}$

68. $\dfrac{x^4 - 1}{x^4 - 16} = \dfrac{x^2 - 1}{x^2 - 4}$

69. $\dfrac{(x + a)^t}{(2x + b)^n} = \dfrac{(x + a)^{t+1}}{(2x + b)^{n+1}}$

70. $\dfrac{x + 4}{x + 6} = \dfrac{x + 2}{x + 4}$

71. $\dfrac{x + 3}{x + 5} = \dfrac{x + 1}{x + 3}$

72. $2x - \dfrac{2}{x^3} = 0$

73. $3x^2 - \dfrac{3(36)^3}{x^4} = 0$

74. $\dfrac{(x - 1)(2x) - x^2}{(x - 1)^2} = 0$

75. $\dfrac{(y^4 + 1)(-12y^2) + 32y^6}{(y^4 + 1)^3} = 0$

✍ In Your Own Words

76. Why does the method of substitution to solve a polynomial using quadratic equation equations not work for a polynomial with an x^3 term?

2.5 INEQUALITIES

In Chapter 1, we introduced the inequality relations $<$, $>$, $\leq$, and $\geq$ among real numbers. In many applications of algebra, particularly in calculus, inequalities are used extensively. In this section, we learn to solve inequalities that involve one variable.

Linear Inequalities

An inequality may involve one or more variables. For example, the inequality

$$3x + 1 > -x - 4$$

involves the variable x. The inequality is valid for some values of x (such as $x = 4$) and is invalid for other values of x (such as $x = -5$). Determining the set of values of the variable for which an inequality holds is called **solving the inequality**.

In many respects, solving an inequality is analogous to solving an equation. We use the laws of inequalities to obtain equivalent inequalities, that is, inequalities that have the same solutions. The next few examples provide some practice in solving simple inequalities that involve linear expressions.

➢ **EXAMPLE 1**
Solving an Inequality

Solve the inequality $3x + 1 > -x - 4$.

Solution

We perform allowable operations on the inequality as we would on an equation. First add quantities to both sides to bring all terms involving x to one side of the inequality and all constant terms to the other:

$$3x + x > -4 - 1$$
$$4x > -5$$

Now multiply both sides of the inequality by 1/4:

$$\frac{1}{4}(4x) > \frac{1}{4}(-5)$$

$$x > -\frac{5}{4}$$

At each step, the inequality is replaced by an equivalent one. So the original inequality is satisfied precisely by those x satisfying $x > -5/4$. A description of all the solutions is given, using set builder notation, as

$$\left\{ x : x > -\frac{5}{4} \right\}$$

This set is called the **solution set** of the inequality.

➢ **EXAMPLE 2**
Merchandising

A local sporting goods store earns a profit of $30 per tennis racket. As part of a promotion, it spends $5000 on advertising. How many tennis rackets must be sold if the profit is to be at least $10,000?

Solution

Let q denote the quantity of tennis rackets sold. Then the result to $30q - 5000$. The condition that the profit must be at least expressed by the inequality:

$$\text{Profit} \geq 10{,}000$$

$$30q - 5000 \geq 10{,}000$$

We may solve this last inequality for q:

$$30q \geq 15{,}000$$

$$q \geq 500$$

In other words, if the profit is to be at least \$10,000, the quantity sold must be at least 500.

Many applications involve pairs of inequalities of the form

$$a < b \qquad \text{and} \qquad b < c$$

which can also be written as

$$a < b < c$$

For instance, the interest rate I quoted on 30-year fixed-rate mortgages by different banks may be greater than or equal to 9.5%, and less than 10.75%. This corresponds to the double inequality

$$0.095 \leq I < 0.1075$$

The solution set of a double inequality consists of the numbers that satisfy both inequalities. That is, the solution set consists of the numbers in common to the two solution sets of the two inequalities. For example, the solution set of the double inequality $-1 < x < 3$ consists of the numbers in common to the two solution sets: $\{x : x > -1\}$ and $\{x : x < 3\}$.

If A and B are sets, then the set that consists of the elements in common is called the **intersection** of A and B and is denoted $A \cap B$. In terms of intersections, the solution set of the double inequality may be written as

$$\{x : x > -1\} \cap \{x : x < 3\}$$

To solve a double inequality, we must solve two independent inequalities and form the intersection of the respective solution sets. The following example illustrates the procedure.

➤ **EXAMPLE 3**
A Three-Termed Inequality

Solve the inequality

$$-5x < 3x + 1 < 2x + 4$$

Solution

We must determine the solution sets of the inequalities:

$$-5x < 3x + 1 \qquad \text{and} \qquad 3x + 1 < 2x + 4$$

The first inequality may be solved as follows:

$$-5x < 3x + 1$$

$$-5x - 1 < 3x$$

$$-1 < 3x + 5x$$

$$-1 < 8x$$

$$-\frac{1}{8} < x$$

The solution set of the first inequality is thus

$$\left\{ x : x > -\frac{1}{8} \right\}$$

The second inequality may be solved as follows:

$$3x + 1 < 2x + 4$$
$$3x - 2x + 1 < 4$$
$$x < 4 - 1$$
$$x < 3$$

The solution set of this inequality is thus

$$\left\{ x : x < 3 \right\}$$

The solution set of the given double inequality consists of the set of numbers in common to the two solution sets just found. That is, the intersection

$$\left\{ x : x > -\frac{1}{8} \right\} \cap \left\{ x : x < 3 \right\}$$

This intersection consists of all numbers x that are greater than $-1/8$ and less than 3, so the solution set is

$$\left\{ x : -\frac{1}{8} < x < 3 \right\}$$

In solving inequalities, it is often useful to break the solution into two or more cases. A solution must arise as a solution in one of the cases. If A and B are sets, then the set consisting of the elements in either A or B or both is called the **union** of A and B and is denoted

$$A \cup B$$

The next example provides an illustration of an inequality whose solution may be expressed as a union.

➤ **EXAMPLE 4**
Polynomial Inequality

Solve the inequality:

$$(x + 1)(x + 5) > 0$$

Solution

The inequality requires that the product of two factors be positive. Either both factors must be positive or both factors must be negative. It is simplest to determine this set using a picture like Figure 1. We draw a number line for each of the factors, and we draw $+$ where the factor is positive and $-$ where the factor

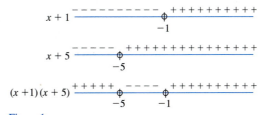

Figure 1

is negative. On a third coordinate axis, we indicate where both factors have the same sign. The solution set is

$$\{x : x > -1\} \cup \{x : x < -5\}$$

This set consists of all x that are either greater than -1 or less than -5.

> ### EXAMPLE 5
> Rational Inequality

Solve the following inequality:

$$\frac{3x + 7}{4x - 5} > 0$$

Solution

In order for the inequality to hold, the numerator and denominator must have the same sign. Reason geometrically as in the previous example. (See Figure 2.) The solution is the set

$$\left\{x : x > \frac{5}{4}\right\} \cup \left\{x : x < -\frac{7}{3}\right\}$$

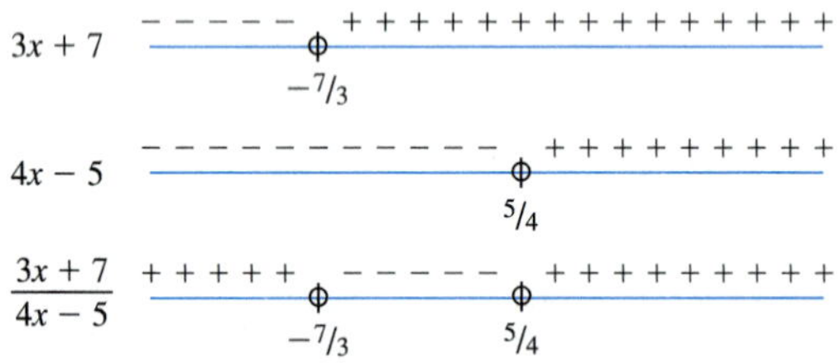

Figure 2

Warning: You might be tempted to solve the inequality by multiplying both sides by $4x - 5$ to clear the denominator. However, this is not a useful way to proceed, since the inequality sign may face in either direction after multiplying, depending on the sign of $4x - 5$.

Intervals

The subset of the number line determined by a single or double inequality is called an **interval**. In the preceding examples, we determined solution sets of inequalities that turned out to be intervals, intersections of intervals, or unions of intervals. We classify intervals on the basis of whether their endpoints are included, according to the following definition.

> ### Definition 1
> ### Intervals
>
> Let a and b be real numbers. Then the **open interval** (a, b) is the set of real numbers
>
> $$(a, b) = \{x : a < x < b\}$$
>
> The **closed interval** $[a, b]$ is the set of real numbers
>
> $$[a, b] = \{x : a \le x \le b\}$$
>
> The **half-open intervals** $[a, b)$ and $(a, b]$ are defined, respectively, as the sets of real numbers
>
> $$[a, b) = \{x : a \le x < b\}$$
> $$(a, b] = \{x : a < x \le b\}$$

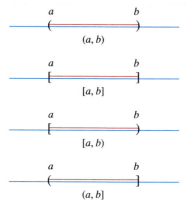

Figure 3

An interval can be represented as a line segment on the number line. An open interval includes neither endpoint of the segment, a closed interval includes both endpoints, and a half-open interval includes one endpoint. The various types of intervals are illustrated in Figure 3. Note that a bracket at the end of a line segment indicates that the endpoint is included. A parenthesis at the end of a line segment indicates that the endpoint is not included.

Intervals that extend indefinitely far to the right or left on the number line are called **infinite intervals.** The solution sets of the inequalities $x > a$ and $x < a$ are the open infinite intervals denoted (a, ∞) and $(-\infty, a)$, respectively. The solution set of the inequalities $x \geq a$ is $[a, \infty)$, and $x \leq a$ is $(-\infty, a]$. The symbol ∞ is read "infinity." The symbol $-\infty$ is read "negative infinity." The infinity symbols indicate intervals extending "infinitely far" to the right and left, respectively. The various types of infinite intervals are shown in Figure 4.

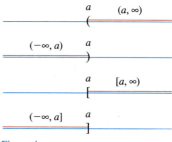

Figure 4

➤ **EXAMPLE 6**
Sketching Intervals

Sketch the graphs of the following intervals:

    1. $(-1, 3)$ 2. $(-\infty, 5)$ 3. $[0, 1]$ 4. $[-2, \infty)$

Solution

    1. This is an open interval, so neither endpoint is included. See Figure 5.

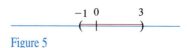

Figure 5

    2. This is an open infinite interval. The right endpoint is not included, and the interval extends indefinitely far to the left. See Figure 6.

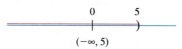

Figure 6

    3. This is a closed interval. Both endpoints are included. See Figure 7.

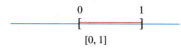

Figure 7

4. This is a closed infinite interval. The left endpoint is included, and the interval extends indefinitely far to the right. See Figure 8.

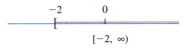

Figure 8

> **EXAMPLE 7**
Graphing Intersections
and Unions

Graph the sets on the number line described as follows:

1. $(-1, 3) \cap (2, 4)$ 2. $[0, 5] \cup [5, 7]$ 3. $(0, 1) \cup (2, \infty)$

Solution

Figure 9

Figure 10

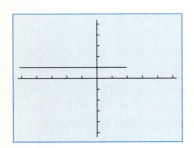

Figure 11

1. The graph consists of the points in common to the intervals $(-1, 3)$ and $(2, 4)$. In Figure 9, we have sketched both intervals. The points in common are those that are shaded twice. These points constitute the interval $(2, 3)$. Note that the endpoints are not included, since 2 does not belong to the interval $(2, 4)$ and 3 does not belong to $(-1, 3)$.

2. The graph consists of points that belong to either of the intervals $[0, 5]$ and $[5, 7]$. In Figure 10, we have shaded each of these intervals. The graph of the union consists of the points shaded at least once. It includes the number 5, which is shaded twice. So the indicated union is the interval $[0, 7]$.

3. The graph consists of points belonging to either of the intervals $(0, 1)$ and $(2, \infty)$. Figure 11 shows each of these intervals as shaded. The union consists of the points shaded at least once.

Solving Inequalities via Graphics Calculator

Many graphics calculators allow you to graph inequalities. Just enter the inequalities as if they were equations. For instance, suppose that you wish to graph the inequality

$$3x + 1 > 5x - 3$$

Call up the equation entry screen of your graphics calculator. (This is the screen that allows you to enter equations of the form **Y= . . . .**) For the first equation, enter the inequality by keying in the symbols from the calculator keyboard, and then press the key that shows the graph. The calculator will display a graph like the one shown in Figure 12. Note that the line indicating the infinite interval solution is shown 1 unit above the x-axis. It is hard for a calculator to show the line along the x-axis as in our previous figures.

Be careful in interpreting calculator graphs. You may set the x and y range to be shown. Depending on the x range you select, you may see only part of the graph or nothing at all. For instance, if you select the x range to be from 3 to 10, then you will see the graph shown in Figure 13. Don't misinterpret this graph by saying that the inequality has no solutions!

You may determine the simultaneous solutions of several inequalities by writing the inequalities in sequence, each surrounded in parentheses. For example to determine the graph of the inequality

$$1 < 2x + 1 < 3$$

Figure 12

we would write it as the simultaneous solutions of the two inequalities

$$1 < 2x + 1 \quad \text{and} \quad 2x + 1 < 3$$

which would then be entered in the calculator as

$$(1 < 2x + 1)(2x + 1 < 3)$$

The resulting graph is shown in Figure 14.

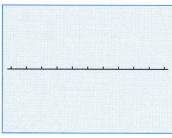

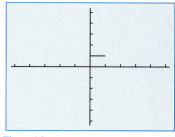

Figure 13   Figure 14   ◄

Polynomial Inequalities

An interesting application of intervals arises in solving polynomial inequalities, that is, inequalities of the form

$$p(x) \geq 0 \quad \text{or} \quad p(x) \leq 0$$

where $p(x)$ is a polynomial in the variable x. Let's consider the particular case in which the polynomial may be factored into factors of the form

$$p(x) = A(x - a_1)(x - a_2) \cdots (x - a_n)$$

where the real numbers $a_1, a_2, \ldots, a_n$ are arranged in increasing size:

$$a_1 \leq a_2 \leq \cdots \leq a_n$$

Associated to such a polynomial, we have the collection of intervals

$$(-\infty, a_1), (a_1, a_2), \ldots, (a_{n-1}, a_n), (a_n, \infty)$$

These intervals are called **test intervals** for $p(x)$.

Each of the factors is of constant sign in each of the test intervals. Therefore, we have the following result.

CONSTANT SIGN THEOREM

The polynomial $p(x) = A(x - a_1)(x - a_2)\ldots$ is of constant sign in each of the intervals $(-\infty, a_1), (a_1, a_2), \ldots, (a_n, \infty)$. That is, if c and d are two numbers in the same interval, then the numbers $p(c)$ and $p(d)$ are either both positive or both negative.

According to the theorem, if one number in a test interval is a solution to the inequality, then so is every other number in the interval. Therefore, the solution set of a polynomial inequality consists of a union of test intervals. To determine which test intervals to include in the union, it suffices to test a single number in each interval. The procedure is illustrated by the following example.

> **EXAMPLE 8**
Cubic Inequality

Solve the inequality $2x^3 + x^2 - x < 0$. Graph the solution.

Solution

We begin by factoring the polynomial:

$$2x^3 + x^2 - x = x(2x^2 + x - 1) = x(2x - 1)(x + 1)$$

$$= 2[x - (-1)](x - 0)\left(x - \frac{1}{2}\right)$$

In the last factorization above, the factors have been written so that the coefficient of x in each factor is 1 and the constants in the linear factors are in increasing order $(-1 < 0 < 1/2)$. The test intervals for this polynomial are

$$(-\infty, -1), \ (-1, 0), \left(0, \frac{1}{2}\right), \left(\frac{1}{2}, \infty\right)$$

We choose a number in each to act as a test case for that interval: $-2, -1/2, 1/4,$ 1. We tabulate the value of the polynomial for each of these numbers and determine the sign of the polynomial using the factored form. For example, at $x = -2$, the first factor (x) is negative, the second $(2x - 1)$ is positive, as is the third $(x + 2)$. So the product is $(-)(+)(+) = -$. The data may be summarized in the following table.

Interval	Test Point x	Sign of $2x^3 + x^2 - x$
$(-\infty, -1)$	-2	$-$
$(-1, 0)$	$-1/2$	$+$
$(0, 1/2)$	$1/4$	$-$
$(1/2, \infty)$	1	$+$

The first and the third intervals have test points that yield negative values for the polynomial. So the solution consists of the union of the first and third intervals:

$$(-\infty, -1) \cup \left(0, \frac{1}{2}\right)$$

Figure 15

The solution is graphed in Figure 15. You could also solve the inequality using a graphing calculator entering the inequality as

$$2x\wedge3 + x\wedge2 - x < 0$$

The graph shown would resemble the one shown in Figure 16. Note that the tick marks on the x-axis are 1/2 unit apart. Be careful in interpreting graphs on a graphing calculator. Since the tick marks are not labeled, misinterpretations are easy to make.

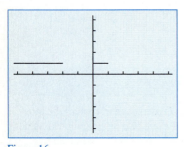

Figure 16

The method used in Example 8 may be used to solve any inequality of the form $p(x) > 0$, where $p(x)$ is a product of linear factors. Note that if we are given an inequality of the form $p(x) > c$, where c is nonzero, then, in order to apply the method, we must transform the inequality into one where the right side is 0 and the left is a product of linear factors.

Exercises 2.5

Translate each of the following to inequalities.

1. c is positive.
2. t is negative.
3. a is nonnegative.
4. q is nonpositive.
5. -3 is less than or equal to m.
6. 9 is greater than b.
7. a is at most b.
8. p is at least q.

Solve the following inequalities and write the solution set in interval notation. Sketch the solution set.

9. $-2 \leq x < 3$
10. $5 < t \leq 7$
11. $-4 \leq x \leq 5$
12. $-5 \leq t \leq -3$
13. $-2 \leq x < 3$
14. $5 < t \leq 17$
15. $x \geq -2$
16. $t < 3$

Describe each of the following using set notation.

17. $[-1, 3)$
18. $(-3, -1)$
19. $(-2, \infty)$
20. $(-\infty, 5]$
21. $(-\infty, 2]$
22. $[5, \infty)$

Solve the following inequalities and write the solution sets in interval notation.

23. $-2 \leq 5x + 3$
24. $3t + 2 < 17$
25. $-3 - 4x > 0$
26. $6 - 2t < 0$
27. $x + 3 \geq 2x - 1$
28. $5x - 8 < -2(x - 3)$
29. $3 - x \leq 4x + 7$
30. $10x - 3 \geq 13x - 8$
31. $3 - (4 - x) \leq 3x - (x + 5)$
32. $3 - x - x^2 \geq 2x - x(x + 1)$
33. $-0.3x > 0.2x + 1.3$
34. $-2.6x < 1.2 + 0.04x$
35. $-8 \leq 2x \leq 10$
36. $-9 < -3t \leq -6$
37. $-5 \leq 8 - 3x \leq 5$
38. $-7 \leq 2x - 3 \leq 7$
39. $-11 < 2x - 1 \leq -5$
40. $3 \leq 4y - 3 < 19$
41. $\dfrac{2}{5}x - 3 \leq \dfrac{4}{5}x - 13$
42. $\dfrac{3}{5}x \geq \dfrac{1}{3}x - \dfrac{1}{10}$
43. $-6 < \dfrac{2 - 3x}{2} \leq 13$
44. $1 \leq \dfrac{4 + 3t}{4} < 9$
45. $(3x - 2)^2 < 3(x - 2)(3x + 4)$
46. $4(x - 2)(x + 3) \geq (2x + 3)(2x - 3)$
47. $\dfrac{1}{2x + 3} < 0$
48. $\dfrac{-2}{4 - 5x} > 0$
49. $(x + 5)(x - 3) > 0$
50. $(x + 4)(x - 1) < 0$
51. $x^2 - 7x + 12 \geq 0$
52. $x^2 - x \leq 20$
53. $t^2 < 25$
54. $12x^2 > 4$
55. $36x^2 - 24x > 0$
56. $3x^2 + 6x + 3 > 0$
57. $(x + 3)(x - 2)(x + 1) > 0$

58. $(x - 4)(x - 5)(x + 2) \leq 0$
59. $x^3 + 3x^2 - 10x \leq 0$
60. $x^4 - 7x^3 + 12x^2 > 0$
61. $t^3 + 2t^2 - t - 2 < 0$
62. $2x^3 - x^2 - 8x + 4 \geq 0$
63. $\dfrac{x - 3}{x + 1} > 0$
64. $\dfrac{x + 5}{x - 2} < 0$
65. $\dfrac{2x - 3}{x - 4} \leq 0$
66. $\dfrac{3 - 2x}{3x + 5} \geq 0$
67. $\dfrac{(x + 1)(x - 2)}{x + 4} < 0$
68. $\dfrac{x + 7}{(x - 3)(x - 2)} > 0$
69. $12x^2 > 0$
70. $x^2 + 1 < 0$

♦ Applications

71. The base of a triangle is 12 in. For what values of the height h is the area at most 200 in.^{2}?

72. A car is traveling at a speed of 65 mph. For what times t can the car travel so that it goes at least 700 miles?

73. A student receives grades of 67, 83, and 78 on three math tests. What scores S can the student score on the fourth test and obtain an average greater than 70 and less than 80?

74. A pitcher's earned run average, ERA, is given by ERA $= 9R/I$, where R is the number of earned runs given up pitching I innings. A pitcher gives up 50 earned runs. What number of innings can be pitched so that the pitcher's ERA is at most 3.00?

75. The following are expressions for total revenue and total cost from the sale of x units of a product: $R(x) = 80x$, and $C(x) = 20x + 10,000$. Total profit $P(x)$ is given by $P(x) = R(x) - C(x)$.

    a. Find the number of units x for which $P(x) > 0$.

    b. Find the number of units x for which $P(x) \geq 0$.

76. The formula $S = -16t^2 + 64t + 940$ gives the height S of an object thrown upward from the roof of a 960-ft. building at an initial velocity of 64 ft./sec.

    a. For what times t will the height be greater than 992 ft.?

    b. For what times t will the height be less than 960 ft.?

77. A polygon with n sides has D diagonals, where D is given by

$$D = \frac{n(n-3)}{2}$$

For what values of n is $20 \leq D \leq 119$?

78. There are n teams in a sports league. If each team plays every other team twice, then the total number of games G played is given by $G = n(n-1)$. For what values of n is $G \geq 380$?

In Exercises 79–80, determine for what values of p the quadratic equation will have at least one real number solution.

79. $px^2 + 4x + 8 = 0$ 80. $12x^2 + px + 3 = 0$

Solve the inequalities. Write interval notation for the solution sets.

81. $-12x + 12x - x^2 < 0$

82. $\dfrac{(x^2 + 1) - x(2x)}{(x^2 + 1)^2} > 0$

83. $(x - 2)^3 2(x + 1) + (x + 1)^2 3(x - 2)^2 > 0$

84. $\dfrac{(x - 1)^2(2x - 2) - (x^2 - 2x)(2x - 2)}{(x - 1)^4} > 0$

85. $\dfrac{2x/3}{(x^2 - 8)^{2/3}} > 0$

86. Prove or disprove: For any number a and b, if $a < b$, then $a^2 < b^2$.

➡ **Technology**

Use a graphing calculator to solve the following inequalities.

87. $\dfrac{(x + 1)(x - 2)}{x + 4} < 0$

88. $\dfrac{x + 7}{(x - 3)(x - 2)} > 0$

89. $(x + 3)(x - 2)(x + 1) > 0$

90. $(x - 4)(x - 5)(x + 2) \leq 0$

91. $-11 < 2x - 1 \leq -5$ 92. $3 \leq 4y - 3 < 19$

93. $-6 < \dfrac{2 - 3x}{2} \leq 13$ 94. $1 \leq \dfrac{4 + 3t}{4} < 9$

✍ **In Your Own Words**

95. Describe in words the procedure for solving a polynomial inequality using test invervals.

96. Describe in words the set of simultaneous solutions to three given inequalities.

2.6 ABSOLUTE VALUE EQUATIONS AND INEQUALITIES

As we have seen in Chapter 1, the absolute value may be used to calculate the distance between points on the number line. In many problems, it is necessary to consider points whose distances from other points are restricted by inequalities, say, the points whose distance from 5 is less than 2. The algebraic statement of such problems gives rise to inequalities involving absolute values. Such inequalities also arise in dealing with numerical approximations, say, in approximating the area of a field within a measurement error of at most 1 square foot. In this section, we discuss techniques for solving equations and inequalities that involve absolute values.

Equations Involving Absolute Values

The main point to remember in dealing with absolute values is that they have a two-part definition. Most absolute value problems must be broken into two cases: the case where the expression within the absolute value bars is positive or zero, and the case where it is negative. The next two examples demonstrate how this is typically handled.

➤ **EXAMPLE 1**
Absolute Value Equation

Solve the equation:

$$|2x - 1| = 5$$

Solution

Since $|2x - 1|$ equals either $2x - 1$ or $-(2x - 1)$, the equation is equivalent to
$$2x - 1 = 5 \quad \text{or} \quad 2x - 1 = -5$$
The first equation has the solution 3, and the second has the solution -2. So the given equation has two solutions, 3 and -2.

➤ **EXAMPLE 2**
Quadratic Absolute Value Equation

Determine all solutions to the equation:
$$x|x| = 3x - 1$$

Solution

We break the solution into two cases.

    Case 1: $x \geq 0$

    In this case, we have $|x| = x$ and the equation reads
$$x \cdot x = 3x - 1$$
$$x^2 - 3x + 1 = 0$$
This quadratic equation may be solved using the quadratic formula:
$$x = \frac{3 \pm \sqrt{5}}{2} = 2.618, 0.382$$

Both solutions are positive and so satisfy the condition of case 1. So the solutions in case 1 are
$$x = \frac{3 + \sqrt{5}}{2}, \qquad x = \frac{3 - \sqrt{5}}{2}$$

    Case 2: $x < 0$

    In this case, we have $|x| = -x$, so that the equation reads
$$x(-x) = 3x - 1$$
$$-x^2 - 3x + 1 = 0$$
$$x^2 + 3x - 1 = 0$$

Solving this equation using the quadratic formula, we obtain the solutions
$$x = \frac{-3 + \sqrt{13}}{2} \quad \text{or} \quad x = \frac{-3 - \sqrt{13}}{2}$$
Since $x = (-3 + \sqrt{13})/2 = 0.303$ is positive, which violates our assumption that x is negative, the only solution in case 2 is
$$x = \frac{-3 - \sqrt{13}}{2} = -3.303$$

Inequalities Involving Absolute Values

Let's now turn to inequalities that involve absolute values. Here are some examples of such inequalities:
$$|2x + 5| < 4$$
$$|x| > 6$$

The key to solving inequalities involving absolute values is the following fundamental result.

> ## FUNDAMENTAL ABSOLUTE VALUE INEQUALITY
>
> Let a be a nonnegative real number. Then x satisfies the inequality $|x| < a$ if and only if $-a < x < a$.

Proof   Recall that $|x|$ is the distance of x from the origin. The inequality $|x| < a$ says that the distance of x to the origin is less than a. However, the numbers satisfying this property are exactly those satisfying the inequality

$$-a < x < a$$ ♦

The next several examples illustrate how this rule may be used to solve inequalities involving absolute values.

➤ **EXAMPLE 3**
Absolute Value Inequality

Solve the inequality:

$$3|2 - 5x| + 1 < 10$$

Solution

We may simplify the inequality by subtracting 1 from each side and then dividing each side by 3. This gives the inequality

$$|2 - 5x| < 3$$

Convert this last inequality to a double inequality:

$$-3 < 2 - 5x < 3$$
$$-5 < -5x < 1 \qquad \textit{Subtracting 2 from both sides}$$
$$1 > x > -\frac{1}{5} \qquad \textit{Multiplying by } -1/5$$
$$-\frac{1}{5} < x < 1 \qquad \textit{Reversing the inequality}$$

The solution set of the given inequality is the interval $(-1/5, 1)$.

➤ **EXAMPLE 4**
Calculus

In calculus, one meets inequalities of the form $|x - a| < \delta$, where a is a real number and δ is positive. Solve this inequality and graph its solution.

Solution

The given inequality is equivalent to the double inequality

$$-\delta < x - a < \delta$$
$$-\delta + a < x < \delta + a$$

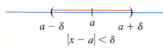

$|x - a| < \delta$

Figure 1

In other words, the solutions of the inequality consist of all x in the interval $(a - \delta, a + \delta)$. This interval is illustrated in Figure 1.

In the preceding examples, we considered inequalities in which an absolute value was less than a constant. In some applications, it is necessary to consider inequalities of the form

$$|x| > a$$

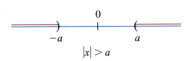

Figure 2

where a is a positive number. This inequality states that the distance from x to the origin is greater than a. The values of x for which this holds fall into two categories: either $x > a$, or $x < -a$. That is, the solution set of the inequality is

$$\{x : x < -a\} \cup \{x : x > a\}$$

The graph of this solution set is shown in Figure 2.

➤ **EXAMPLE 5**
Solution Set of a Union

Solve the inequality

$$|4x - 1| > 1$$

Graph the solution set.

Solution

According to our earlier discussion, this inequality may be replaced by the two inequalities

$$4x - 1 > 1 \qquad \text{or} \qquad 4x - 1 < -1$$

Solving these inequalities, we have:

$$
\begin{array}{ll}
4x - 1 > 1 & 4x - 1 < -1 \\
4x > 1 + 1 & 4x < -1 + 1 \\
4x > 2 & 4x < 0 \\
x > \dfrac{1}{4} \cdot 2 & x < 0 \\
x > \dfrac{1}{2} &
\end{array}
$$

The solution set consists of those x that satisfy either of the above inequalities. That is, the solution set is the following:

$$\left\{x : x > \frac{1}{2}\right\} \cup \{x : x < 0\} = (-\infty, 0) \cup \left(\frac{1}{2}, \infty\right)$$

Figure 3

The graph of the solution set is sketched in Figure 3.

We may summarize the two types of absolute value inequalities as follows:

INEQUALITIES INVOLVING ABSOLUTE VALUE			
Inequality	Solution	Distance Interpretation	Solution in Interval Notation
$|x| < a$	$-a < x < a$	The set of x whose distance from the origin is less than a.	$(-a, a)$
$|x| > a$	$x > a$ or $x < -a$	The set of x whose distance from the origin is greater than a.	$(-\infty, a) \cup (a, \infty)$

Exercises 2.6

Solve each equation or inequality. Describe your answers, where appropriate, using interval notation.

1. $|x| = 3$
2. $|x| = 6$
3. $|x + 3| = 2$
4. $|5 - x| = 7$
5. $|3x| = 12$
6. $4|x - 1| = 16$
7. $|2x - 3| = 15$
8. $|5 + 6x| = 41$
9. $|x| \leq 23$
10. $|x| > 23$
11. $|x| > 14$
12. $|x| < 14$
13. $|x - 2| > 4$
14. $|x + 9| \geq 7$
15. $|6 - 3x| < 9$
16. $\left|1 - \dfrac{2}{3}x\right| \leq 7$

17. $\left|\dfrac{3x - 2}{4}\right| \leq 5$
18. $\left|\dfrac{6 + 7x}{4}\right| < 26$

19. $|x| = -10$
20. $|x| < -10$
21. $|x| > -10$
22. $|3x + 2| \leq 0$
23. $|2 - 3x|^{-1} > 0$
24. $|2 - 3x|^{-2} \geq 0$
25. $4|5 + 6x| - 7 < 19$
26. $6|2 - 5x| + 12 > 36$
27. $|2x + 1| = x - 4$
28. $|3 - 2x| \geq 1 + 2x$
29. $4 \leq |2 - x| \leq 5$
30. $|x - 1| > |x - 5|$
31. $|2x - 1| < -2$
32. $|3x + 1| > 0$
33. $|x + 1| + |x - 1| = 2$
34. $|t + 3| + |t - 3| = 6$
35. $|m + 3| + |m - 3| = 4$
36. $|m + 3| + |m - 3| < 8$
37. $|y - 3| + |y| \leq 4$
38. $\left|\dfrac{x + 1}{2x + 3}\right| < 0$
39. $|x|^2 - 2|x| + 1 \leq 9$
40. $|t|^2 + 6|t| + 9 > 25$
41. $\left|\dfrac{x + 3}{x - 2}\right| < 1$
42. $\left|\dfrac{2x - 3}{3x + 4}\right| \geq 2$
43. $|t^2 - 5t - 1| > 5$
44. $|m^2 + m - 4| \leq 2$
45. $|x^2 - 1| \leq 3$

Solve for x.

46. $|x - a| \leq \delta$
47. $|x - x_0| < \delta$

48. Use the triangle inequality to show the following. Such derivations can occur in calculus.

   a. If $|x - 4| < 0.17$ and $|y + 4| < 0.23$, then $|x + y| < 0.4$.

   b. If $|x - a| < e/2$ and $|y + a| < e/2$, then $|x + y| < e$.

49. Prove: $|a - b| \leq |a| + |b|$
50. Prove: $a^2 \leq b^2$ if and only if $|a| \leq |b|$

Each of the following is a solution set of an absolute value inequality. Find the inequality.

51. $[5 - r, 5 + r]$
52. $\left(-2, \dfrac{2}{3}\right)$
53. $(-\infty, -4] \cup [8, \infty)$
54. $\left(-\infty, -\dfrac{3}{4}\right) \cup \left(\dfrac{13}{4}, \infty\right)$

♦ **Applications**

55. A company's profits P for a certain year are estimated to satisfy the inequality

$$|P - \$3,000,000| \leq \$225,000$$

Determine the interval over which the company's profits will vary.

56. During a strange illness a patient's temperature T varied according to the inequality

$$|T - 98.6°| \leq 2$$

Determine the interval over which the patient's temperature varied.

57. A weight is bobbing on a spring (see figure) in such a way that its distance d, in inches, from the top satisfies the inequality

$$|d - 10| < 4$$

Determine the interval over which the distance varies.

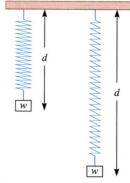

Exercise 57

2.7 COMPLEX NUMBERS

Definition of the Complex Numbers

The need to perform arithmetic operations has motivated the introduction of larger and more comprehensive number systems. The first number system used by humans consisted of the counting numbers 1,2,3,... However, to perform certain subtraction operations among such numbers, say, $5 - 5$ or $12 - 28$, mathematicians created zero and the negative numbers, which, along with the counting numbers, form the set of integers **I**.

The integers are closed under addition, subtraction, and multiplication. However, certain division operations, such as 5/3 or 128/9, result in numbers that are not integers. The desire to perform division among all integers motivated mathematicians to extend the set of integers to the set of rational numbers **Q**. The rational numbers are closed under division in the sense that if a and b are rational numbers with b nonzero, then a/b is also a rational number.

As we pointed out in Chapter 1, the decimal expressions of all rational numbers are terminating or repeating. To deal with numbers with nonterminating and nonrepeating decimal expressions, mathematicians defined the set of real numbers **R**, which contains rational and irrational numbers. The real numbers are closed under some arithmetic operations with respect to which the rational numbers are not. For example, if a is a rational number, $\sqrt{a}$ is not always a rational number, as in the case of $\sqrt{2}$. But if a is a positive real number, then $\sqrt{a}$ is a real number.

As comprehensive as the set of real numbers appears to be, certain elementary arithmetic operations among real numbers are not possible using real numbers. For example, there is no real number whose square is -1. To put this another way, it is impossible to compute $\sqrt{-1}$ within the real numbers because the square of any real number is nonnegative and so cannot be -1.

Just as the real numbers do not contain a square root of -1, they do not contain the square root of any negative number. This is a serious shortcoming of the real numbers and has motivated mathematicians to create a larger number system, called the **complex numbers,** in which taking square roots is always possible. We define the complex numbers as follows.

Definition 1
Complex Numbers

The **imaginary unit,** denoted i, is a number whose square is -1. That is,
$$i^2 = -1, i = \sqrt{-1}$$
A **complex number** is a number of the form
$$a + bi,$$
where a and b are real numbers, and $i^2 = -1$. The set of complex numbers is denoted **C.**

Here are some examples of complex numbers:
$$2 + 3i, \qquad \frac{1}{2} - 4i, \qquad 5 + 0i, \qquad \sqrt{2}i$$

If $a + bi$ is a complex number, then a is called its **real part** and bi is called its **imaginary part**.

Definition 2
Equality of Complex Numbers

Two complex numbers $a + bi$ and $c + di$ are **equal** if and only if $a = c$ and $b = d$.

The definition of equality of complex numbers is useful in solving equations involving complex numbers, as we will see later in this section.

It is customary to write a complex number $a + 0i$ as a and to consider it the same as the real number a. In this way, the complex numbers contain all the real numbers as a subset.

Figure 1 shows in pictorial fashion the relationships between the complex numbers and the various number systems we introduced earlier.

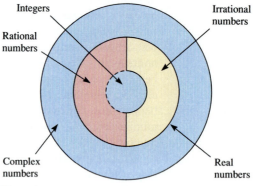

Figure 1

Arithmetic Involving Complex Numbers

Definition 3 Addition and Multiplication of Complex Numbers	Let $\alpha = a + bi$ and $\beta = c + di$ be complex numbers. Their sum and product are the numbers defined by the formulas: $$\alpha + \beta = (a + c) + (b + d)i$$ $$\alpha \cdot \beta = (ac - bd) + (ad + bc)i$$

For example, we have

$$(5 - 3i) + (4 + 2i) = (5 + 4) + (-3 + 2)i = 9 - i$$
$$(2 + 4i)(3 - 7i) = 2 \cdot 3 + 2 \cdot (-7i) + (4i)(3) + (4i)(-7i)$$
$$= 6 - 14i + 12i - 28i^2$$
$$= 6 - 2i - 28(-1)$$
$$= 6 - 2i + 28$$
$$= 34 - 2i$$

The definition of addition is about what you might expect. But the definition of multiplication of complex numbers may seem strange. However, this is just the definition we get assuming that the distributive law holds, and using the fact that $i^2 = -1$. Indeed, we have the following motivation for the definition of the product of two complex numbers:

$$\alpha \cdot \beta = (a + bi)(c + di)$$
$$= ac + (ad)i + (bc)i + (bd)i^2$$
$$= ac + (-bd) + (ad + bc)i \qquad \textit{Since } i^2 = -1$$
$$= (ac - bd) + (ad + bc)i$$

Addition and multiplication of complex numbers have the same fundamental properties as addition and multiplication of real numbers. Namely, they obey the commutative, associative, and distributive laws: If α, β, γ are complex numbers, then

$$\alpha + \beta = \beta + \alpha \qquad \textit{Commutative law of addition}$$
$$(\alpha + \beta) + \gamma = \alpha + (\beta + \gamma) \qquad \textit{Associative law of addition}$$
$$\alpha\beta = \beta\alpha \qquad \textit{Commutative law of multiplication}$$
$$\alpha(\beta\gamma) = (\alpha\beta)\gamma \qquad \textit{Associative law of multiplication}$$
$$\alpha(\beta + \gamma) = (\alpha \cdot \beta) + (\alpha \cdot \gamma) \qquad \textit{Distributive law}$$

Complex numbers have an **additive identity,** namely, 0 (or $0 + 0i$), with the property

$$\alpha + 0 = \alpha$$
$$a + bi + 0 = a + bi$$

A complex number $\alpha = a + bi$ has an **additive inverse**, namely, $-\alpha = -a - bi$, with the property

$$\alpha + (-\alpha) = 0$$
$$(a + bi) + (-a - bi) = 0$$

Subtraction of complex numbers is defined, as for real numbers, in terms of the additive inverse. Namely:

$$\alpha - \beta = \alpha + (-\beta)$$

Thus, for instance,

$$(5 - 4i) - (3 - 3i) = (5 - 4i) + (-3 + 3i) = 2 - i$$

Addition and subtraction of complex numbers is consistent with the corresponding operations for real numbers. Indeed, if we consider two real numbers a and c, then

$$(a + 0i) + (c + 0i) = (a + c) + 0i + 0i = a + c$$
$$(a + 0i) - (c + 0i) = (a + 0i) + (-c + 0i) = (a - c) + 0i + 0i = a - c$$

The number 1 (or $1 + 0i$) is the **multiplicative identity**:

$$\alpha \cdot 1 = \alpha$$
$$(a + bi) \cdot 1 = a + bi$$

➤ **EXAMPLE 1**
Arithmetic of Complex Numbers

Perform the following calculations involving complex numbers:

1. $2(3 + i) - i(2 - 4i)$ 2. $(1 + i)^2$

Solution

1. $2(3 + i) - i(2 - 4i) = (6 + 2i) + (-2i + 4i^2)$
$$= (6 + 2i) + (-2i - 4) \qquad \textit{Since } i^2 = -1$$
$$= 2 + 0i$$

2. $(1 + i)^2 = (1 + i)(1 + i)$
$$= (1 - 1) + (1 + 1)i \qquad \textit{Definition of multiplication}$$
$$= 2i$$

Just as with the real number system, the only way for a product of complex numbers to be zero is for at least one of the factors to be zero. That is,

$$\alpha\beta = 0 \text{ implies } \alpha = 0 + 0i \text{ or } \beta = 0 + 0i$$

We will find this fact of great use in determining solutions of polynomial equations in the complex number system.

Powers of i

The complex number i has the property that its square is -1. Using this fact and the laws of exponents, we can easily calculate higher powers of i. Here are the first few:

$$i^3 = i^2 \cdot i = (-1) \cdot i = -i$$
$$i^4 = i^2 \cdot i^2 = (-1)(-1) = 1$$

Using the values of the first four powers of i, we may compute higher powers, as follows:

$$i^5 = i^4 \cdot i = 1 \cdot i = i$$
$$i^6 = i^4 \cdot i^2 = 1 \cdot (-1) = -1$$
$$i^7 = i^4 \cdot i^3 = 1 \cdot (-i) = -i$$
$$i^8 = i^4 \cdot i^4 = 1 \cdot 1 = 1$$

You should note that the powers of i repeat each time the exponent is increased by 4. Moreover, if the exponent is divisible by 4, then the power of i is 1. We can use these facts to rapidly calculate any power of i. For example, let's calculate

$$i^{57}$$

First, divide the exponent by 4 and write it in the form

$$57 = 14 \cdot 4 + 1$$

Substitute this expression for the exponent of i:

$$i^{57} = i^{14\cdot4+1}$$
$$= i^{14\cdot4} \cdot i^1$$
$$= (i^4)^{14} \cdot i$$
$$= (1)^{14}i$$
$$= i$$

Square Roots of Negative Real Numbers

Mathematicians introduced the complex numbers to create a number system that contains square roots of negative numbers. Suppose that p is a positive real number. Then, as we have observed, there is no real number whose square is $-p$. However, there is a complex number with this property, namely,

$$\sqrt{p}i$$

For we have

$$(\sqrt{p}i)^2 = (\sqrt{p})^2 i^2$$
$$= p(-1)$$
$$= -p$$

The quantity $\sqrt{p}i$ is called the **principal square root** of $-p$.

Historical Note:
Origin of Complex Numbers

Square roots of negative numbers were used for centuries in calculating solutions to equations of degrees 2, 3, and 4. However, mathematicians regarded them with great suspicion. They realized, of course, that they were not real numbers, but they were not quite sure of the legitimacy of working with them. Accordingly, square roots of negative numbers were called **imaginary numbers** or just **imaginaries**. The German mathematician Karl Friedrich Gauss in 1801 gave a formal definition of the complex numbers that provided them with a logical foundation and the basis for legitimate proofs of theorems regarding them.   ▮▮

➤ **EXAMPLE 2**
Square Roots of a Negative Number

Determine two square roots of -4.

Solution

One square root of -4 is given by the principal square root, $\sqrt{4}\,i = 2i$. A second square root is given by the negative of this quantity, $-2i$, since

$$(-2i)^2 = (-2)^2 i^2$$
$$= 4 \cdot (-1) = -4$$

➤ **EXAMPLE 3**
Complex Products

Compute the product

$$(x - \sqrt{-1})(x + \sqrt{-1})$$

Solution

Apply the distributive law:

$$(x - \sqrt{-1})(x + \sqrt{-1}) = x^2 + \sqrt{-1}\,x - \sqrt{-1}\,x - (\sqrt{-1})^2$$
$$= x^2 - (-1) \qquad \textit{Using } (\sqrt{-1})^2 = -1$$
$$= x^2 + 1$$

The preceding example can also be read in reverse—not as a multiplication example, but as a factorization. Starting with our solution, a sum of squares,

$$x^2 + 1$$

we can arrive at the factorization

$$(x - i)(x + i)$$

In a similar fashion, we can factor the sum of squares

$$x^2 + y^2$$

into the linear factors

$$(x - yi)(x + yi)$$

To verify this fact, we just multiply the linear factors as in Example 3. Let's record this fact for future reference.

FACTORING A SUM OF SQUARES
$x^2 + y^2 = (x + yi)(x - yi)$

Note that the sum of two squares is not factorable using real numbers, yet we can factor it using complex numbers.

Quadratic Equations Revisited

The quadratic equation

$$ax^2 + bx + c = 0$$

has solutions given by the quadratic formula

$$x = \frac{-b \pm \sqrt{b^2 - 4ac}}{2a}$$

As we have seen, in case the discriminant $b^2 - 4ac$ is negative, the equation has no real solutions. However, in this case, the formula gives two distinct complex solutions.

> ## EXAMPLE 4
Quadratic with Complex Zeros

Determine all solutions of the equation $x^2 + x + 1 = 0$.

Solution

By the quadratic formula, the solutions are given by

$$\begin{aligned}
x &= \frac{-b \pm \sqrt{b^2 - 4ac}}{2a} \\
&= \frac{-1 \pm \sqrt{1^2 - 4 \cdot 1 \cdot 1}}{2 \cdot 1} \\
&= \frac{-1 \pm \sqrt{-3}}{2} \\
&= \frac{-1 \pm \sqrt{3}i}{2}
\end{aligned}$$

Conjugation

Definition 4	Let $a + bi$ be a complex number. Its conjugate is the number $a - bi$ and is
Conjugate of a Complex Number	denoted $\overline{a + bi}$.

Here are some examples of conjugates:

$$\overline{2 + 3i} = 2 - 3i, \qquad \overline{5} = 5, \qquad \overline{\frac{1}{2}i} = -\frac{1}{2}i, \qquad \overline{-8 - 7i} = -8 + 7i$$

The following formulas are very useful.

PROPERTIES OF CONJUGATES

Let $\alpha = a + bi$ be a complex number.

Conjugate of a conjugate: $\overline{\overline{\alpha}} = \alpha$

Sum of conjugates: $\alpha + \overline{\alpha} = 2a$

Difference of conjugates: $\alpha - \overline{\alpha} = 2bi$

Product of conjugates: $\alpha\overline{\alpha} = a^2 + b^2$

Each nonzero complex number $\alpha = a + bi$ has a multiplicative inverse α^{-1} such that

$$\alpha\alpha^{-1} = 1$$

In fact, we can use conjugates to explicitly calculate the multiplicative inverse, as given in the following result.

MULTIPLICATIVE INVERSE OF A COMPLEX NUMBER

Let $\alpha = a + bi$ be a nonzero complex number. Then

$$\alpha^{-1} = \frac{\overline{\alpha}}{a^2 + b^2}$$

Proof   We have

$$\alpha^{-1} = \frac{1}{\alpha} = \frac{1 \cdot \overline{\alpha}}{\alpha \cdot \overline{\alpha}} = \frac{\overline{\alpha}}{a^2 + b^2} \qquad\qquad \blacklozenge$$

Division of complex numbers is defined in terms of the multiplicative inverse, just as it is for real numbers. Assume that α and β are complex numbers with $\beta \neq 0$. The quotient α/β is then defined as

$$\frac{\alpha}{\beta} = \alpha\beta^{-1}$$

➤ **EXAMPLE 5**
Inverse and Division

Express the following complex numbers in the form $a + bi$.

1. $(5 - 4i)^{-1}$ 2. $\dfrac{3 + i}{1 + 2i}$

Solution

1. We may write the complex number as $1/(5 - 4i)$. Multiply the numerator and denominator by the conjugate $5 + 4i$:

$$\frac{1}{5 - 4i} = \frac{1}{5 - 4i} \cdot \frac{5 + 4i}{5 + 4i} = \frac{5 + 4i}{5^2 + 4^2} = \frac{5}{41} + \frac{4}{41}i$$

2. We may calculate the quotient by multiplying numerator and denominator by the conjugate of the denominator:

$$\frac{3 + i}{1 + 2i} = \frac{3 + i}{1 + 2i} \cdot \frac{1 - 2i}{1 - 2i}$$

$$= \frac{3 \cdot 1 + (-2)i^2 + (1 - 3 \cdot 2)i}{1^2 + 2^2}$$

$$= \frac{5 - 5i}{5}$$

$$= 1 - i$$

More Examples Involving Complex Numbers

In many problems, it is necessary to deal with polynomials with complex coefficients. The next two examples provide some practice in calculating with such polynomials.

➤ **EXAMPLE 6**
Polynomial with Complex
Coefficients

Calculate the product of the following binomials with complex coefficients.

$$\left(x - \frac{1 + \sqrt{3}i}{2}\right)\left(x - \frac{1 - \sqrt{3}i}{2}\right)$$

Solution

The product equals

$$\left[\left(x - \frac{1}{2}\right) - \frac{\sqrt{3}}{2}i\right]\left[\left(x - \frac{1}{2}\right) + \frac{\sqrt{3}}{2}i\right]$$

$$= \left(x - \frac{1}{2}\right)^2 + \left(\frac{\sqrt{3}}{2}\right)^2$$

$$= x^2 - x + \frac{1}{4} + \frac{3}{4}$$

$$= x^2 - x + 1$$

➤ **EXAMPLE 7**
Solving an Equation with
Complex Coefficients

Find all solutions (x, y) of the following equation:

$$(2x + 1) + (-y + 3)i = 6 + \sqrt{2}i$$

Solution

By the definition of equality for complex numbers, in order for two complex numbers to be equal, their corresponding real and imaginary parts must be equal. In this case,

$$2x + 1 = 6 \qquad \text{and} \qquad -y + 3 = \sqrt{2}$$

$$x = \frac{5}{2} \qquad \text{and} \qquad y = 3 - \sqrt{2}$$

Exercises 2.7

Express in terms of i.

1. $\sqrt{-3}$ 2. $\sqrt{-4}$ 3. $\sqrt{-81}$ 4. $\sqrt{-27}$

5. $\sqrt{-98}$ 6. $-\sqrt{-18}$ 7. $-\sqrt{-49}$

8. $-\sqrt{-125}$ 9. $4 - \sqrt{-60}$ 10. $6 - \sqrt{-84}$

11. $\sqrt{-4} + \sqrt{-12}$ 12. $-\sqrt{-76} + \sqrt{-125}$

Simplify.

13. i^7 14. i^{11} 15. i^{24} 16. i^{35}

17. i^{42} 18. i^{64} 19. i^9 20. $(-i)^{71}$

Simplify to the form $a + bi$.

21. $7 + i^4$ 22. $-18 + i^3$ 23. $i^4 - 26i$

24. $i^5 + 37i$ 25. $i^2 + i^4$ 26. $5i^5 + 4i^3$

27. $i^5 + i^7$ 28. $i^{84} - i^{100}$

29. $1 + i + i^2 + i^3 + i^4$ 30. $i - i^2 + i^3 - i^4 + i^5$

31. $5 - \sqrt{-64}$ 32. $\sqrt{-12} + 36i$

33. $\dfrac{8 - \sqrt{-24}}{4}$ 34. $\dfrac{9 + \sqrt{-9}}{3}$

35. $\dfrac{\sqrt{-16}}{\sqrt{-25}}$ 36. $\dfrac{\sqrt{-9}}{\sqrt{-36}}$

37. $(3 + 2i) + (2 + 4i)$ 38. $(2 - 5i) + (3 + 6i)$

39. $(3 + 4i) + (3 - 4i)$ 40. $(2 + 5i) + (-2 - 5i)$

41. $(9 + 12i) - (7 + 8i)$ 42. $(10 - 4i) - (6 + 2i)$

43. $6i - (7 - 4i)$ 44. $45 - (23 + 5i)$

45. $(1 - 3i)(2 + 4i)$ 46. $(-2 + 3i)(6 - 7i)$

47. $(2 - 5i)(2 + 5i)$ 48. $(-5 + 7i)(-5 - 7i)$

49. $2i(4 - 3i)$ 50. $5i(-8 + 6i)$ 51. $(5 - 2i)^2$

52. $(3 + i)^2$ 53. $\dfrac{1 + 2i}{3 - i}$ 54. $\dfrac{2 + i}{2 - i}$

55. $\dfrac{\sqrt{3} - i}{\sqrt{3} + i}$ 56. $\dfrac{\sqrt{2} + i}{\sqrt{2} - i}$ 57. $\dfrac{5 - 3i}{i}$

58. $\dfrac{\sqrt{2} - i}{i}$ 59. $\dfrac{i}{1 - i}$ 60. $\dfrac{4}{3 + 10i}$

61. $\dfrac{2 + i}{(2 - i)^2}$ 62. $\dfrac{5 - i}{(5 + i)^2}$

63. $(1 - 2i)^{-1}$ 64. $(1 + i)^{-1}$

65. $\dfrac{(1 + i)(2 - i)}{(4 - 2i)(5 - 3i)}$ 66. $\dfrac{(2 - 3i)(5 - 6i)}{(9 + 2i)(4 + 3i)}$

67. $\dfrac{1 + i}{1 - i} + \dfrac{2 - i}{2 + i}$ 68. $\dfrac{5 - 2i}{3 + 2i} - \dfrac{7 - i}{4 + i}$

69. $(1 + i)^{-2}$ 70. $(\sqrt{3} - 2i)^{-2}$

Determine two square roots of each of the following.

71. -5 72. -9 73. -64 74. -17

Calculate each product.

75. $(x - 2i)(x + 2i)$ 76. $(x + 5i)(x - 5i)$

77. $\left(x - \dfrac{1 - \sqrt{2}i}{3}\right)\left(x - \dfrac{1 + \sqrt{2}i}{3}\right)$

78. $\left(x - \dfrac{\sqrt{3} + 2i}{4}\right)\left(x - \dfrac{\sqrt{3} - 2i}{4}\right)$

Factor.

79. $x^2 + 4$ 80. $x^2 + 25$ 81. $x^2 + 3$

82. $x^2 + 5$ 83. $a^2 + b^2$ 84. $x^2 + 16y^2$

Find all solutions (x, y) of the following equations, where x and y are real numbers.

85. $5x + 8i = -9 + yi$

86. $(-2x + y)i - 8 = 3x - 4y + 9i$

87. $x^2 + 4yi = 6i + 9$

88. $y^3 + 5xi = 40i - 8$

Let $\alpha = a + bi$ be any complex number. Prove each of the following.

89. $\alpha + \overline{\alpha} = 2a$ 90. $\alpha - \overline{\alpha} = 2bi$

91. $\alpha\overline{\alpha} = a^2 + b^2$

Suppose α and β are any complex numbers. Prove the following.

92. $\overline{\alpha + \beta} = \overline{\alpha} + \overline{\beta}$ (The conjugate of a sum is the sum of the conjugates.)

93. $\overline{\alpha \cdot \beta} = \overline{\alpha} \cdot \overline{\beta}$ (The conjugate of a product is the product of the conjugates.)

94. $\overline{\alpha^n} = (\overline{\alpha})^n$, where n is a positive integer. (The conjugate of a power is the power of the conjugate.)

95. $\overline{\alpha - \beta} = \overline{\alpha} - \overline{\beta}$ (The conjugate of a difference is the difference of the conjugates.)

96. $\overline{\alpha} = \alpha$ if α is a real number. (The conjugate of a real number is the same number.)

97. Prove that the sum of a complex number and its conjugate is a real number.

Simplify each of the following conjugates of polynomials in the complex number α.

98. $\overline{\alpha^3 + \alpha^2 + \alpha + 1}$

99. $\overline{4\alpha^2 - 25}$

100. $\overline{5\alpha^4 - 6\alpha^3 + 2\alpha^2 + 3\alpha - 23}$

✍ In Your Own Words

101. Describe in words a procedure for determining if a quadratic equation has nonreal solutions.

102. What is the conjugate of the conjugate of a complex number?

2.8 CHAPTER REVIEW

Important Concepts, Properties, and Formulas—Chapter 2

Linear equation	$ax + b = 0$	p. 57
Compound interest formula	$A = P\left(1 + \dfrac{I}{n}\right)^{nt}$	p. 73
Complex number	$a + bi$, where a and b are real numbers and $i = \sqrt{-1}$.	p. 113
Quadratic equation	An equation of the form $ax^2 + bx + c = 0$ where $a \neq 0$.	p. 75
Quadratic formula	$x = \dfrac{-b \pm \sqrt{b^2 - 4ac}}{2a}$	p. 78

Absolute value inequalities	For a positive number a, $|x| < a$ if and only if $-a < x < a$.	p. 110
	For a positive number a, $|x| > a$ if and only if $x < -a$ or $x > a$.	p. 110

Cumulative Review Exercises—Chapter 2

Solve these equations and inequalities. Where appropriate, describe your answers using interval notation.

1. $2(2x + 4 - 3x) = 8 - 2x$

2. $4(3x + 5) - 25 - 6(4x - 3) + 8x - 90$
   $\quad = 34 - 3x + 6 - 8(x + 4)$

3. $(2x - 5)(3x + 4) = (x + 2)(6x - 2)$

4. $\dfrac{x}{3} - \dfrac{3x - 5}{2} = 8$ 5. $4x^2 + 5x = 1$

6. $4x^2 = 3 + x$ 7. $10t^2 - t = 100$

8. $1 - \dfrac{1}{6x} + \dfrac{2}{3x^2} = 0$ 9. $\dfrac{5}{2x + 6} = \dfrac{1 - 2x}{4x} + 2$

10. $2x^2 + 10x = 14$

Solve by completing the square. Show your work.

11. $(2a - 3)^2 = (a + 1)^2$ 12. $y^2 + \sqrt{3}y - 2 = 0$

13. $\sqrt{x^2 + 7} = x - 1$ 14. $15 - 3\sqrt[3]{2x + 1} = 0$

15. $x^2 - 9x^{3/2} + 20x = 0$ 16. $\sqrt{\dfrac{x + 1}{4x - 1}} = \dfrac{1}{2}$

17. $\dfrac{2}{x - 2} + \dfrac{2x}{x^2 - 4} = \dfrac{3}{x + 2}$

18. $\dfrac{5}{y^2 + 5y} - \dfrac{1}{y^2 - 5y} = \dfrac{1}{y^2 - 25}$

19. $\dfrac{4x - 20}{x - 9} - \dfrac{16}{x} = \dfrac{144}{x^2 - 9x}$

20. $\dfrac{6t}{t - 5} - \dfrac{300}{t^2 + 5t + 25} = \dfrac{2250}{t^3 - 125}$

21. $m^4 - 7m^2 + 10 = 0$ 22. $6t^{-2} = 6t^{-1} + 36$

23. $-3\left(\dfrac{x + 2}{x - 3}\right)^2 - 4\left(\dfrac{x + 2}{x - 3}\right) + 4 = 0$

24. $t^{2/3} = 16$

25. $x^3 + 4x^2 - 9x - 36 = 0$

26. $x - 10\sqrt{x} + 16 = 0$

27. $3x - 4 < 5x + 6$ 28. $5 - \dfrac{x - 3}{9} < \dfrac{2 + x}{6}$

29. $3 < 6 - \dfrac{1}{3}x \le 9$ 30. $|2x - 5| = 7$

31. $|2x - 5| < 7$ 32. $|3x + 12| \ge 9$

33. $x^2 + 2x > 35$ 34. $4x^2 - 5 < x$

35. $\dfrac{2x + 6}{3x - 1} > 1$ 36. $\left|1 - \dfrac{2}{x}\right| < 3$

37. $|m - 1| + |m + 1| = 5$ 38. $|x| + |x - 2| = -3$

Solve for the indicated letter.

39. $Q = mn^2$, for n, $n > 0$.   40. $Q = mn^3$, for n.

41. $v = \dfrac{1}{2}\sqrt{1 + \dfrac{T}{L}}$, for L.   42. $r = \sqrt[3]{\dfrac{3w}{4\pi d}}$, for w.

43. $S = \dfrac{n(n + 1)}{2}$, for n.   44. $S = -16t^2 + v_0 t$, for t.

45. $y = x + \dfrac{1}{x}$, for x.

♦ Applications

46. What percent of $225 is $175?

47. A wind is blowing from the west at 50 km/hr. A plane flies east 375 km and then back in a total time of 4 hours. How fast does the plane fly in still air?

48. Three pipes can fill a swimming pool in one hour. Alone, pipe A can fill the pool in one third the time pipe B can, and pipe C can fill the pool in one-half the time pipe B can. How long would it take each pipe to fill the pool by itself?

49. A family has budgeted $30 per month for city water. With one month left in the year, the average per month they have spent is $32.50. If water costs $0.85 per hundred cubic feet, how much water can they use in the last month to average $30 for the year?

50. Suppose $50,000 is invested at 8.4%. After 5 years, how much is in the account if interest is compounded (a) annually, (b) semiannually, (c) monthly, (d) daily (use 360 days per year)?

51. $50,000 is invested for 2 years at an interest rate that is compounded annually. It grows to $59,405. What is the interest rate?

52. To promote business, bank *A* offers to give a grandfather clock worth $595 to anyone who deposits over $5000 at 9% for 10 years, where interest is compounded annually. Bank *B* offers no free gift, but compounds interest quarterly, also at 9%. Which bank has the better deal for a $5000 deposit? For a deposit of $10,000?

53. The *period* of a pendulum *T*, in seconds, is the time it takes to move in one direction and then back, and is given by the formula

$$T = 2\pi\sqrt{\frac{L}{9.8}}$$

where *L* is the length of the pendulum in meters. (See figure.)

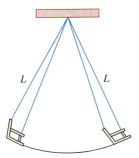

Exercise 53

a. How long is a pendulum that has a period of $3\pi/2$ second?

b. How long is a pendulum that has a period of 1 second? Use 3.14 for π.

c. Solve the formula for *L*.

54. One leg of a right triangle is 2 cm longer than the other. The hypotenuse is 2 cm longer than the longer leg. What is the perimeter of the triangle?

55. $3500 is deposited at the beginning of the year. At the beginning of the next year, $4400 is deposited in another bank at the same interest rate. At the beginning of the third year, there is a total of $8172.60 in both accounts. If interest is compounded annually, what is the interest rate?

56. A chemist has one solution of sulfuric acid and water that is 20% acid and a second that is 75% acid. How many liters of each should be mixed together to get 240 liters of a solution that is 40% acid?

57. The chemical mercury is a liquid at Celsius temperatures *C* satisfying the inequality

$$-39° < C < 357°$$

To convert from Celsius to Fahrenheit we use the formula

$$C = \frac{5}{9}(F - 32)$$

Find an inequality involving Fahrenheit temperatures *F* for which mercury is a liquid.

58. A brick is dropped into a well. The sound of the splash is heard 5 seconds later. How deep is the well?

59. Two investments are made totaling $120,000. For a certain year these investments yield $8850 in simple interest. Part of the $120,000 is invested at 9% and the rest at 6%. Find the amount invested at each rate.

60. The *Highway Code in Great Britain* uses the following quadratic equation

$$D = \text{thinking distance} + \text{braking distance} = r + \frac{r^2}{20}$$

to estimate, under normal driving conditions, the stopping distance *D* (in feet) of a car traveling *r* mph.

a. How many feet would it take to stop if you were driving 25 mph? 55 mph? 65 mph?

b. It takes a driver 120 ft. to stop a car. How fast was the car moving?

c. For what speeds *r* is the stopping distance *D* such that $75 \leq D \leq 175$?

d. Solve the equation for *r*.

61. The formula

$$S = -16t^2 + v_0 t + h$$

gives the height *S*, in feet, after *t* seconds of an object thrown upward from a height *h*, at an initial velocity v_0.

a. A ball is thrown upward at an initial velocity of 48 ft./sec. from a roof that is 1120 ft. tall. How long does it take the ball to reach the ground?

b. At what time will the ball be at a height of 1216 ft.?

Simplify to the form $a + bi$.

62. i^{90}

63. i^{95}

64. $-\sqrt{-5}\sqrt{-36}$

65. $\dfrac{16 + \sqrt{-40}}{8}$

66. $(8 - i)(2 + 5i)$

67. $(1 + 3i)^2 - (4 - 8i)$

68. $(2 - 5i)^{-1}$

69. $\dfrac{2 + 3i}{5 - 4i}$

70. $(7 - 4i)(7 + 4i)$

71. $i + i^2 + i^3 - i^4$

72. $\dfrac{1 + i}{1 - i} - \dfrac{2 + i}{2 - i}$

Determine two square roots of each of the following.

73. -25

74. -40

Factor.

75. $x^2 + 9$

76. $x^2 + 7$

Find all solutions (x, y) of each of the following.

77. $(3x - y)i - 19 = 4x + 2y - 8i$

78. $x^2 - 6yi = 64 + 12i$

Calculate each product.

79. $(x - 3i)(x + 3i)$

80. $\left(x - \dfrac{2 + \sqrt{3}i}{2}\right)\left(x - \dfrac{2 - \sqrt{3}i}{2}\right)$

In Exercises 81 and 82, determine for what values of p the quadratic equation will have (a) at least one real number solution, (b) two complex nonreal solutions.

81. $x^2 + x + p = 0$

82. $2x^2 + px + 4 = 0$

Solve.

83. $2x - \dfrac{4(108)}{x^2} = 0$

84. $\dfrac{8x^3 - 51x^2}{(2x - 8.6)^2} = 0$

85. $\dfrac{x - \dfrac{1}{x}}{1 + x\left(\dfrac{1}{x}\right)} = 1$

86. $\dfrac{3}{2} = \dfrac{(y - 7)(2y) - y^2}{(y - 7)^2}$

87. $\dfrac{\dfrac{4x}{3}}{(x^2 - 4)^{1/3}} > 0$

FUNCTIONS AND GRAPHS

*Graphs of stock market prices provide examples of functions that
arise in many everyday financial transactions.*

The notion of a function is central in modern mathematics. It arose historically as a generalization of the notion of a "formula," which expresses one variable in terms of another variable. Such relationships arise naturally in applications.

For example, the graph of Figure 1 expresses the relationship between elapsed time and the number of people who have heard a rumor. The graph shows how the rumor is slow to get started and then picks up steam as it spreads. As time passes, most of the population has heard the rumor, and the graph flattens out, indicating the slower spread of the rumor among the few remaining people. The graph is a particular instance of a **logistic curve**, which arises in many mathematical models in such disparate fields as ecology and epidemiology.

Figure 2 shows the relationship between elapsed time and amount remaining of a sample of radioactive carbon-14. The data contained in this graph are used by archaeologists in applying the process of **carbon dating** to determine the ages of ancient artifacts such as mummies and papyrus scrolls.

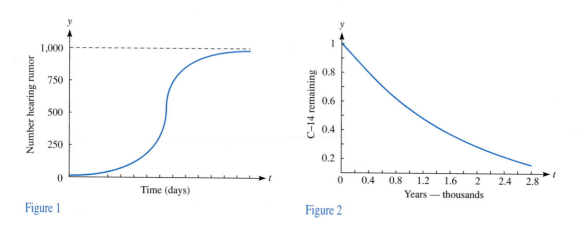

Figure 1

Figure 2

Each of the two graphs depicts a relationship between the variable time, t, and another variable (the number N of people having heard the rumor or the amount A of carbon-14 remaining). For each value of t, the relationship determines the corresponding value of N or A. Such relationships are examples of **functions** and are the subject of this chapter.

3.1 TWO-DIMENSIONAL COORDINATE SYSTEMS

In Chapter 1 we introduced the notion of a one-dimensional coordinate system. Using such a system allowed us to identify a point on a line with a real number, its coordinate. We now introduce two-dimensional coordinate systems, which allow us to assign coordinates to points in the plane. Each point of the plane will be identified with an ordered pair (a, b) of real numbers. The precise method for carrying out this identification can be very simply described.

Coordinate Systems in a Plane

Start with two perpendicular lines in the plane. On each line introduce a one-dimensional coordinate system, as described in Chapter 1. Recall that this involves choosing a positive direction and a unit distance on each line. This pair of one-dimensional coordinate systems is called a **two-dimensional Cartesian coordinate system** in the plane. (See Figure 3.)

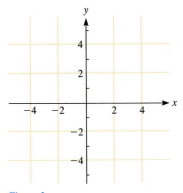

Figure 3

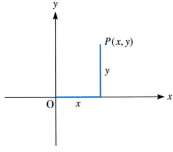

Figure 4

The perpendicular lines are called the **axes**. Usually, the horizontal axis is labeled the *x*-axis and the vertical axis the *y*-axis. However, in applications the axes may have applied meanings. For example, suppose that we wish to explore the relationship between time, *t*, and the velocity *v* of a race car. In this case, we would label the horizontal axis *t* and the vertical axis *v*. The point where the axes intersect is called the **origin** and is customarily denoted by the letter *O*. (See Figure 4.)

Given a Cartesian coordinate system, each point *P* may be identified with a pair of real numbers (x, y); *P* is reached by starting at the origin and moving horizontally a signed distance *x* and vertically a signed distance *y*, where *signed distance* means that movements to the right correspond to positive values of *x*, and movements to the left correspond to negative values of *x*; similarly, vertical movements upward correspond to positive values of *y*, and vertical movements downward correspond to negative values of *y*. The pair of numbers (x, y) is called the **coordinate representation** of *P*. The number *x* is called the *x*-**coordinate**, or **abscissa**, of *P*. The number *y* is called the *y*-**coordinate**, or **ordinate**, of *P*.

Note: The coordinate representation (x, y) should not be confused with the open interval from *x* to *y*. Even though the same notation is used, it will be clear from the context which is meant.

Figure 5 shows the coordinate representations of various points in the plane. Here are the coordinates of some special points:

1. Origin has both coordinates 0 : $(0, 0)$.
2. Point on *x*-axis has *y*-coordinate 0 : $(x, 0)$.
3. Point on *y*-axis has *x*-coordinate 0 : $(0, y)$.

The coordinate axes divide the plane into four regions called **quadrants**. The quadrants are numbered counterclockwise starting from the positive *x*-axis, as shown in Figure 6. In a particular quadrant, the abscissa and ordinate are of constant sign:

Quadrant	*x*-coordinate	*y*-coordinate
I	Positive	Positive
II	Negative	Positive
III	Negative	Negative
IV	Positive	Negative

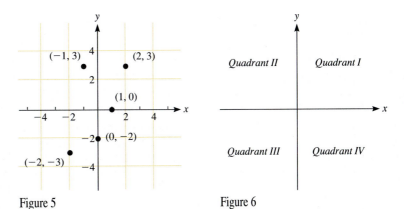

Figure 5   Figure 6

The Distance Formula

We may compute the distance between two points in terms of their coordinates using the following formula:

DISTANCE FORMULA

Let $P(x_1, y_1)$ and $Q(x_2, y_2)$ be two points. Then the distance between them is given by

$$d(P,Q) = \sqrt{(x_2 - x_1)^2 + (y_2 - y_1)^2}$$

See Figure 7.

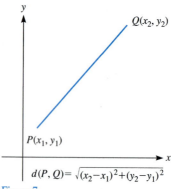

Figure 7

Proof Draw the line segment connecting P and Q and put it in a triangle with sides parallel to the horizontal and vertical axes, as shown in Figure 8. From examining the figure, we see that the horizontal side has length $|x_2 - x_1|$ and that the vertical side has length $|y_2 - y_1|$. Since the triangle PQR is a right triangle, we may apply the Pythagorean theorem to compute the length of the hypotenuse:

$$d(P,Q) = \sqrt{(RP)^2 + (RQ)^2}$$
$$= \sqrt{|x_2 - x_1|^2 + |y_2 - y_1|^2}$$
$$= \sqrt{(x_2 - x_1)^2 + (y_2 - y_1)^2}$$

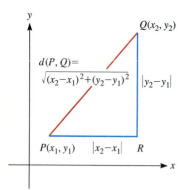

Figure 8

➤ EXAMPLE 1
Calculating Distances

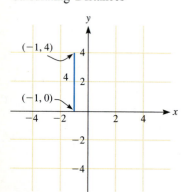

Figure 9

1. Determine the distance between the points $(-1, 3)$ and $(-2, -1)$.
2. Determine the closest distance between $(-1, 4)$ and the x-axis.

Solution

1. By the distance formula, the distance is equal to

$$\sqrt{[-2 - (-1)]^2 + (-1 - 3)^2} = \sqrt{1 + 16} = \sqrt{17}$$

2. The point on the x-axis closest to $(-1, 4)$ is the point on the axis with the same x-coordinate, namely, $(-1, 0)$. The desired distance is shown in Figure 9 and, by the distance formula, equals

$$\sqrt{[-1 - (-1)]^2 + (4 - 0)^2} = 4$$

> **EXAMPLE 2**
A Point Specified
by an Equation

A point R on the positive y-axis is twice as far from the point $P(2,1)$ as from the point $Q(0,0)$. What are the coordinates of the point?

Solution

The condition on R is that

$$d(R,P) = 2d(R,Q)$$

R has x-coordinate 0 because it is on the y-axis, so we can assign it coordinates $(0, y)$. We can write the preceding equation as

$$\sqrt{(2-0)^2 + (1-y)^2} = 2\sqrt{(0-0)^2 + (y-0)^2} \quad \text{(distance formula)}$$
$$2^2 + (1-y)^2 = 4(0^2 + y^2) \quad \text{(squaring both sides)}$$
$$4 + (1 - 2y + y^2) = 4y^2$$
$$3y^2 + 2y - 5 = 0$$

Solve the last equation by factoring:

$$(3y + 5)(y - 1) = 0$$
$$3y = -5 \qquad \text{or} \qquad y = 1$$
$$y = -\frac{5}{3} \qquad \text{or} \qquad y = 1$$

Since we are given that R is on the positive y-axis, the negative value of y does not apply, and the desired point is $R(0, 1)$.

The Midpoint Formula

In many problems, especially those arising in geometry and physics, it is necessary to determine the coordinates of the midpoint of a line from the coordinates of the endpoints. The following result asserts that the coordinates of the midpoint may be obtained by averaging the respective coordinates of the endpoints.

MIDPOINT FORMULA

The midpoint of the line segment with endpoints $P(x_1, y_1)$ and $Q(x_2, y_2)$ is the point

$$M\left(\frac{x_1 + x_2}{2}, \frac{y_1 + y_2}{2}\right)$$

See Figure 10.

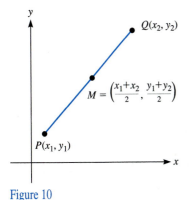

Figure 10

Proof Let's prove the result by establishing the equalities:

$$d(M,P) = \frac{1}{2}d(P,Q) \tag{1}$$

$$d(M,Q) = \frac{1}{2}d(P,Q) \tag{2}$$

These equalities state that M is equidistant from P and Q. Moreover, we can deduce from them that

$$d(P,M) + d(M,Q) = \frac{1}{2}d(P,Q) + \frac{1}{2}d(P,Q)$$
$$= d(P,Q)$$

That is, the distance from P to M plus the distance from M to Q equals the distance from P to Q. From an elementary result in geometry, this implies that M lies on the line segment $\overline{PQ}$. To prove formula (1), we apply the distance formula:

$$d(M,P) = \sqrt{\left(\frac{x_1 + x_2}{2} - x_1\right)^2 + \left(\frac{y_1 + y_2}{2} - y_1\right)^2}$$

$$= \sqrt{\left(\frac{x_2 - x_1}{2}\right)^2 + \left(\frac{y_2 - y_1}{2}\right)^2}$$

$$= \sqrt{\frac{(x_2 - x_1)^2 + (y_2 - y_1)^2}{4}}$$

$$= \frac{1}{2}\sqrt{(x_2 - x_1)^2 + (y_2 - y_1)^2}$$

$$= \frac{1}{2}d(Q,P) \qquad \text{(distance formula)}$$

$$= \frac{1}{2}d(P,Q) \qquad \text{(the distance from P to Q is equal to the distance from Q to P)}$$

The proof of the equality (2) is similar and will be omitted. This completes the proof of the theorem. ◆

➤ **EXAMPLE 3**
Calculating a Midpoint

Determine the midpoint of the line segment with endpoints $P(4, 1)$ and $Q(8, -5)$.

Solution

By the midpoint formula, the coordinates of the midpoint of the line segment are

$$\left(\frac{x_1 + x_2}{2}, \frac{y_1 + y_2}{2}\right) = \left(\frac{4 + 8}{2}, \frac{1 + (-5)}{2}\right) = (6, -2)$$

Exercises 3.1

Plot the following points on a Cartesian coordinate system.

1. *a.* $(3, 2)$
   b. $(3, -1)$
   c. $(0, 2)$
   d. $(-3, 0)$

2. *a.* $(0, -4)$
   b. $(-2, 3)$
   c. $(5, 0)$
   d. $(3, 4.5)$

3. For each point in Exercise 1, determine the quadrant in which it lies, or state the axis it belongs to.

4. For each point in Exercise 2, determine the quadrant in which it lies, or state the axis it belongs to.

Determine the distance between the points in Exercises 5–16.

5. $(2, 2)$ and $(-4, -3)$

6. $(6, 10)$ and $(-1, 5)$

7. $(0, -5)$ and $(2, -3)$

8. $(-3, -3)$ and $(3, 3)$

9. $(-2, 5)$ and $(4, 5)$

10. $(-3, 4)$ and $(-3, 8)$

11. $(2a, 3)$ and $(-a, 5)$

12. $(x, 0)$ and $(-2, -3)$

13. (a, b) and $(0, 0)$

14. $(\sqrt{3}, -2)$ and $(0, \sqrt{5})$

15. $(\sqrt{a}, \sqrt{b})$ and $(0, 0)$

16. $(a + b, a - b)$ and $(a - b, a - b)$

17. Determine the distance between $(-1, 4)$ and the y-axis.

18. Determine the distance between $(-2, 3)$ and the x-axis.

19. Determine the distance between the point (a, b) and the origin.

20. Determine the distance between the point (a, b) and the x-axis.

Determine the midpoints of the segments having the following endpoints.

21. $(2, 2)$ and $(-4, -3)$

22. $(6, 10)$ and $(-1, 5)$

23. $(0, -5)$ and $(2, -3)$

24. $(-3, -3)$ and $(3, 3)$

25. $(-2, 5)$ and $(4, 5)$

26. $(-3, 4)$ and $(-3, 8)$

27. $(2a, 3)$ and $(-a, 5)$

28. $(x, 0)$ and $(-2, -3)$

29. (a, b) and $(0, 0)$

30. $(\sqrt{3}, -2)$ and $(0, \sqrt{5})$

31. $(\sqrt{a}, \sqrt{b})$ and $(0, 0)$

32. $(a + b, a - b)$ and $(a - b, a - b)$

33. A point R on the negative y-axis is twice as far from the point $P(2, 1)$ as from the point $Q(0, 0)$. What are the coordinates of point R?

34. A point W on the positive x-axis is three times as far from the point $P(-2, 5)$ as from the point $Q(-1, 2)$. What are the coordinates of the point W?

Consider the points $A(0, 0)$ and $B(2, 5)$ in Exercises 35 and 36.

35. Find the coordinates of a point C on the x-axis such that $\angle ABC$ is a right angle.

36. Find the coordinates of a point C on the y-axis such that $\angle ABC$ is a right angle.

37. The point $(x, 3)$ is at a distance of 7 units from the point $(-2, 7)$. Find x.

38. The point $(-4, y)$ is at a distance of 5 units from the point $(0, 9)$. Find y.

39. Find the area of the triangle with vertices $A(2, 3)$, $B(-2, -3)$, and $C(5, -3)$.

40. Find the area of the parallelogram with vertices $P(-1, -7), Q(-1, 2), R(3, 5)$, and $S(3, -4)$.

41. Find the point on the y-axis equidistant from the points $(-2, 3)$ and $(-5, -7)$.

42. Find the point on the x-axis equidistant from the points $(-2, 3)$ and $(-5, -7)$.

Use the distance formula to determine whether each set of three points is on a straight line. (*Hint:* For three points to be on a straight line, the sum of the distances from the outer points to the middle one equals the distance between the outer points.)

43. $(-3, -4), (2, 3)$, and $(4, 7)$

44. $(-3, -4), (6, 1)$, and $(10, 6)$

Use the distance formula to determine whether the three points are vertices of a right triangle.

45. $(-4, 7), (6, 3)$, and $(-8, -3)$

46. $(-3, 6), (2, 4)$ and $(6, 14)$

47. Find an equation that must be satisfied by any point (x, y) that is equidistant from $(-2, 3)$ and $(5, -4)$.

48. Find an equation that must be satisfied by any point (x, y) that is 5 units from the point (h, k).

49. Find the center and radius of the circle that is inscribed in the square with vertices $(7, 5), (2, 10), (7, 15)$, and $(12, 10)$.

Use the distance formula and/or the midpoint formula to prove each of the following. Locate the polygons on a coordinate system. Put one of the points at the origin to ease the proof.

50. The midpoint of the hypotenuse of any right triangle is equidistant from each of the vertices of the triangle.

51. The diagonals of a rectangle have the same length.

52. The line segments joining the midpoints of the sides of any quadrilateral form a parallelogram.

53. The diagonals of any parallelogram bisect each other.

54. If the diagonals of a parallelogram have the same length, then the parallelogram is a rectangle.

55. Find all points $(a, 0)$ whose distance from $(0, 3)$ is greater than $\sqrt{19}$.

✍ In Your Own Words

56. Think of an example of a real life object that can be thought of as a two-dimensional coordinate system. How would you define distance in this system?

▌ 3.2 GRAPHS OF EQUATIONS

In the preceding section, we learned to plot individual points in a Cartesian coordinate system, given their coordinates. In many applications, it is necessary to deal not with individual points, but with sets of points. This is the case, for example, in studying equations in two variables from a graphical viewpoint. In this section, we turn our attention to graphing sets of points and, in particular, to graphing equations.

Graphing Equations

Suppose we are given an equation in two variables x and y, such as

$$3x + 5y = 7$$

$$3x^2 + 4y^3 = 1$$

$$y = 3x^4 + 1$$

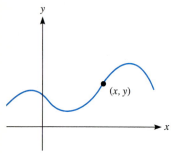

Figure 1

A **solution** of such an equation is an ordered pair (c, d) for which the equation is satisfied when $x = c$ and $y = d$. For example, $(0, 7/5)$ is a solution of the first equation, since if $x = 0$, $y = 7/5$, the equation is satisfied:

$$3(0) + 5\left(\frac{7}{5}\right) = 7$$

Given an equation in x and y, we may plot all of its solutions as points in the plane. The set of points plotted will generally form some sort of curve, which is called the **graph of the equation**. (See Figure 1.) The graph of an equation is a way of representing the algebraic properties of the equation geometrically. In the remainder of this section, we will discuss how to sketch the graphs of various sorts of equations and how to read information from graphs.

> ➤ **EXAMPLE 1**
> Points with Specific
> x- or y-Coordinates

Figure 2 shows the graph of an equation.

1. Determine all points on the graph for which $x = 1$.
2. Determine all points on the graph for which $y = 2$.

Solution

1. Draw a vertical line through the point $x = 1$ on the x-axis. (See Figure 3.) The points on this line all have x-coordinate 1. This line intersects the graph at the point $(1, 0)$, which is the only point on the graph for which $x = 1$.
2. Draw a horizontal line through the point $y = 2$. (See Figure 4.) The points on this line all have y-coordinate 2. This line intersects the graph in the points $(0, 2)$ and $(3, 2)$, which are the only points on the graph for which $y = 2$.

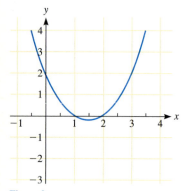

Figure 2

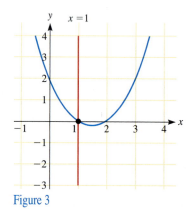

Figure 3

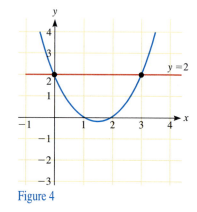

Figure 4

Graphing Linear Equations

The simplest equations in two variables have the form

$$cx + dy + e = 0$$

where c, d, and e are real numbers, with c and d not both zero. Such equations are called **linear equations in two variables**. The graph of such an equation is a straight line, and every straight line is the graph of some linear equation in two

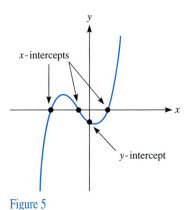

Figure 5

variables. We won't prove these last statements but rather will use them as the definition of a straight line.

To graph a linear equation in two variables, it suffices to determine two points on the line. The places where the line crosses the coordinate axes are particularly easy to determine and thus can be used in drawing the graph.

A point at which a graph crosses the x-axis is of the form $(a, 0)$. The number a is called an x-**intercept of the graph**. A point at which a graph crosses the y-axis is of the form $(0, b)$. The number b is called a y-**intercept of the graph**. (See Figure 5.) Note that the graph of an equation may, in general, have many x- and y-intercepts.

DETERMINING INTERCEPTS

To determine the intercepts of the graph of an equation:

1. x-intercepts: Set $y = 0$ in the equation and solve for x. For each solution $x = a$, the point $(a, 0)$ is an x-intercept.
2. y-intercepts: Set $x = 0$ in the equation and solve for y. For each solution $y = b$, the point $(0, b)$ is a y-intercept of the graph.

Figure 6

A graph's intercepts often provide valuable geometric information. Intercepts often provide information that has meaning in the context of a given application. For example, if we graph the price for a commodity (on the vertical axis) against its demand (on the horizontal axis), the y-intercept corresponds to the price at which 0 units of the commodity are demanded. (See Figure 6.)

In the following example, we show how to graph a linear equation by determining the intercepts.

➤ **EXAMPLE 2**
Graphing a Line
Using Intercepts

1. Determine the x- and y-intercepts of the graph of the linear equation
$$2x + 3y = 1$$
2. Graph the equation.

Solution

1. To obtain the y-intercept, we substitute $x = 0$ in the given equation:
$$2(0) + 3y = 1$$
$$y = \frac{1}{3}$$

So the y-intercept is 1/3. To obtain the x-intercept, we set y equal to 0 in the equation:
$$2x + 3(0) = 1$$
$$x = \frac{1}{2}$$

So the x-intercept is 1/2.

2. To obtain the graph, we use the intercepts to plot the points where the graph intercepts the axes and draw a line through them. See Figure 7.

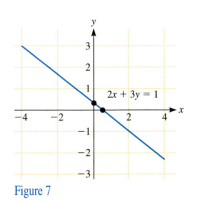

Figure 7

> **EXAMPLE 3**
Vertical and Horizontal Lines

Graph the following linear equations.

    1. $x = 1$ 2. $y = 3$

Solution

1. Since y does not appear in the equation, it may assume any value, but the value of x is always 1. Therefore, the graph consists of all points $(1, t)$, where t is any real number. The graph is a vertical line with x-intercept 1. See Figure 8.

2. Since x does not appear in the equation, it may assume any value, but the value of y is always 3. Therefore, the graph consists of all points $(t, 3)$, where t is any real number. The graph is a horizontal line with y-intercept 3. See Figure 9.

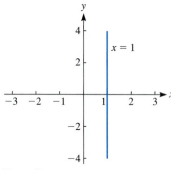

Figure 8

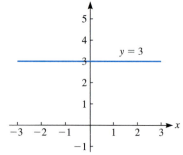

Figure 9

> **EXAMPLE 4**
Demand for a Commodity

A supermarket observes the following relationship between the weekly sales of frozen orange juice q (measured in cans) and the price per can p (measured in dollars):

$$q = -3000p + 12,000$$

1. Graph the linear equation representing the relationship.
2. Give an interpretation for the p-intercept of the graph.

Solution

1. In this example, the horizontal axis is labeled p and the vertical axis q. (See Figure 10.) The q-intercept is the value of q at which p equals 0. We obtain this value of q by setting p equal to 0 and solving the equation:

$$q = (-3000)0 + 12,000 = 12,000$$

So the q-intercept is 12,000. The p-intercept is the value of p for which the value of q is 0:

$$0 = -3000p + 12,000$$
$$p = 4$$

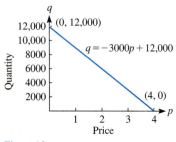

Figure 10

So the p-intercept is 4. To graph the equation, we plot the two points at which the graph intersects the axes and draw a line through the plotted points. See Figure 10.

2. The p-intercept is the value of p where the line crosses the horizontal axis, that is, the point where the value of q is 0; this value of p is 4. A value of zero for q means that no sales are being made. Since the value of p, price,

is $4 when q is zero, this graph can be interpreted to mean that the public will cease to buy orange juice when the price is $4 per can.

➤ **EXAMPLE 5**

Graphing an Absolute Value Equation

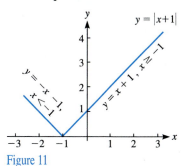

Figure 11

Determine the graph of the equation $y = |x + 1|$.

Solution

According to the definition of absolute value, we have

$$|x + 1| = \begin{cases} x + 1 & \text{if } x + 1 \geq 0 \\ -(x + 1) & \text{if } x + 1 < 0 \end{cases}$$

Therefore, the graph consists of two portions corresponding, respectively, to values of x satisfying $x \geq -1$ or $x < -1$, namely the graphs of the equations

$$y = x + 1 \quad \text{and} \quad x \geq -1$$
$$y = -x - 1 \quad \text{and} \quad x < -1$$

Each of these equations is linear. However, the values of x are restricted. The graphs are shown in Figure 11.

Graphing by Plotting Points

The simplest method for graphing an equation is to plot a number of points that lie on the graph and connect the plotted points with a smooth curve. The next two examples illustrate the procedure for doing this.

➤ **EXAMPLE 6**

Graphing a Quadratic Equation

Graph the equation

$$y = x^2$$

Solution

We determine some points on the graph by using small values of x and determining the corresponding values of y, as shown in the following table. Since the equation expresses y in terms of x, we choose a representative set of values for x and calculate the corresponding values of y. In Figure 12, we plotted the points and drew a smooth curve connecting the points. Note that as $|x|$ increases, so does y. Geometrically, this is reflected in the fact that the graph rises as we proceed either right or left from the origin.

Figure 12

x	y	(x, y)
-3	9	$(-3, 9)$
-2	4	$(-2, 4)$
-1	1	$(-1, 1)$
0	0	$(0, 0)$
1	1	$(1, 1)$
2	4	$(2, 4)$
3	9	$(3, 9)$

➤ **EXAMPLE 7**

Graphing Where x Is Expressed in Terms of y

Plot the graph of the following equation

$$y^2 = x$$

Solution

In this example, since x is given in terms of y, it is simpler to choose representative values of y and compute the corresponding values of x. The results of the calculations are contained in the following table. These points are plotted and the resulting graph sketched in Figure 13. Note that as $|y|$ increases, so does the value of x. Geometrically, this means that the graph heads to the right as we move either up or down from the origin.

y	x	(x, y)
-3	9	$(9, -3)$
-2	4	$(4, -2)$
-1	1	$(1, -1)$
0	0	$(0, 0)$
1	1	$(1, 1)$
2	4	$(4, 2)$
3	9	$(9, 3)$

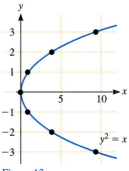

Figure 13

> **EXAMPLE 8**
An Equation Yielding
a Pair of Equations to Graph

Graph the equation

$$x^2 + y^2 = 1$$

Solution

This equation does not allow immediate computation of either variable from the other. To obtain a set of points on the graph, it is easiest to first solve the equation for one variable in terms of the other. Solving for y in terms of x, we obtain

$$y^2 = 1 - x^2$$
$$y = \pm \sqrt{1 - x^2}$$

Using this equation, it is a simple matter to compute a table of points on the graph. The following table gives points corresponding to selected values of x :

$$x = 0, \pm 0.2, \pm 0.4, \pm 0.6, \pm 0.8, \pm 1$$

The points are plotted in Figure 14, where we see that the graph appears to be a circle.

x	y
-1	± 0
-0.8	± 0.6
-0.6	± 0.8
-0.4	± 0.91652
-0.2	± 0.97980
0	± 1
0.2	± 0.97980
0.4	± 0.91652
0.6	± 0.8
0.8	± 0.6
1	0

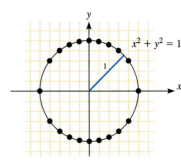

Figure 14

> **EXAMPLE 9**
An Equation with an
Undefined x-Value

Sketch the graph of the equation

$$y = \frac{1}{x}, \quad x \neq 0$$

Solution

Whenever you sketch an equation by plotting points, it is necessary to plot a set of points that are representative of the various geometric features the graph possesses. Any kind of algebraic irregularity will likely lead to some interesting geometric feature that should be investigated in some detail. In this example, the expression $1/x$ is not defined for x equal to 0. This suggests that the graph of the equation will have some interesting behavior for points (x, y) with x near 0. Therefore, in tabulating a set of points to plot, it is wise to use a number of points with x near 0. The following table includes points for which x is equal to

$$\pm 3, \pm 2, \pm 1, \pm 0.5, \pm 0.2, \pm 0.1$$

The points are plotted and the graph sketched in Figure 15.

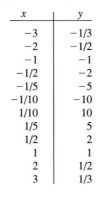

x	y
-3	$-1/3$
-2	$-1/2$
-1	-1
$-1/2$	-2
$-1/5$	-5
$-1/10$	-10
$1/10$	10
$1/5$	5
$1/2$	2
1	1
2	$1/2$
3	$1/3$

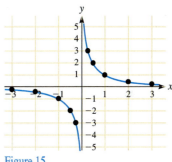

Figure 15

Based on the four examples just worked, we can state the following strategy for graphing an equation by plotting points.

GRAPHING AN EQUATION BY PLOTTING POINTS

1. Solve the equation for x in terms of y or for y in terms of x, whichever is easier.
2. If y is expressed in terms of x, choose a representative set of values for x; if x is expressed in terms of y, choose a representative set of values for y.
3. Calculate the other value corresponding to each of the representative values chosen.
4. Plot the points corresponding to the pairs (x, y) calculated.
5. Draw a smooth curve through the points, making sure that the curve has breaks corresponding to undefined values of the variables.

Graphing an equation is a tedious process. In order to obtain an accurate graph, it is often necessary to plot a large number of points. The calculations involved in determining these points can be quite gruesome.

We should note that plotting points will not always yield, unambiguously, the correct shape of a graph. Given a finite number of points, it is always possible to fit a number of different curves to the data. For instance, in Figure 16, we have sketched several different curves that fit the same set of points. The only way to tell which of the graphs is correct is to further examine the algebraic properties of the equation that gave rise to the points. Let's consider several pieces of information that allow us to infer geometric properties of the graph and thereby supplement the graphing of points in determining the correct form of a graph.

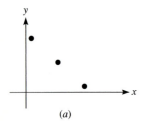
(a)

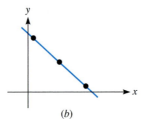
(b)

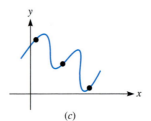
(c)

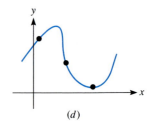
(d)

Figure 16

Spreadsheets and Plotting

As you can see from the preceding discussion, graphing equations by plotting points can be quite tedious. Much of the tedium can be eliminated by allowing a computer program to tabulate the data and plot the points. Many programs are capable of performing these tasks. However, almost all personal computer users have one type of program that can be easily used, namely, a spreadsheet.

A **spreadsheet** is a program that turns your computer into a sheet of accounting paper. That is, the screen displays a rectangular grid of cells, as shown in Figure 17. A cell is identified by a number (designating the row) and a letter (designating the column). Thus, D2 indicates the cell in the second row, fourth column. Into each cell you can type a number, text, or a formula.

Suppose that you wish to enter data into a spreadsheet to plot the graph of $y = x^2$ for $-2 \le x \le 2$, using data points with $x = -2.0, -1.8, -1.6, \ldots,$ 1.6, 1.8, 2.0. Here are the commands for entering the data points using the most popular spreadsheet program, 1–2–3 from Lotus Development. Use the **Data Fill** command to enter the x-values into the first column beginning in A1. This command enters data with a fixed increment between consecutive data items. To enter the desired x-values, give the command:

```
/ Data Fill
```

(Type the first letter of each command word.) The program will ask you to enter the fill range. Type

```
A1..A21
```

and press **ENTER**. This indicates that we wish to fill 21 cells in the first column with data. The program then asks for the start. Type

```
-2.0
```

and press **ENTER**. Next the program asks for the step. Type

```
0.2
```

and press **ENTER**. Finally, the program asks for the stop value. Type

```
2.0
```

The cells A1–A21 will now be filled with the desired data, as shown in Figure 18.

Next we enter into the cells B1–B21 the y-values corresponding to the values of x. Move the cursor to cell B1. Type the formula

```
+A1^2
```

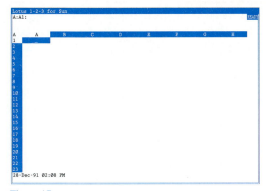

Figure 17

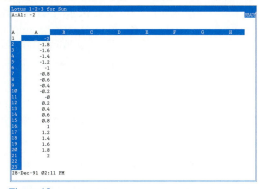

Figure 18

and press **ENTER**. This formula enters into cell B1 the square of the value in A1, or 4. The spreadsheet now looks like Figure 19. Now copy the formula in B1 into the cells B2 through B21 by giving the command

/ Copy

The program asks for the range to copy from. Type

B1

and press **ENTER**. Next, the program asks for the range to copy to. Type

B2..B21

and press **ENTER**. The program now copies the formulas into the cells B2–B21 and evaluates them. The spreadsheet now looks like Figure 20. You could just as easily ask the program to generate 100, 200, or even 500 points.

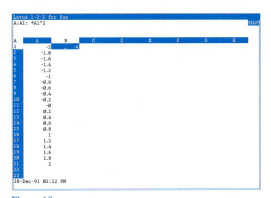

Figure 19

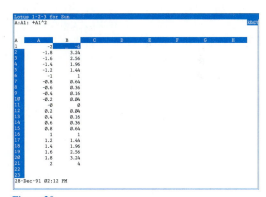

Figure 20

1–2–3 has a built-in graphing utility for graphing data contained in the spreadsheet. For instance, here is how to plot the points we have just generated. Give the command

/ Graph

The program will display the graphing menu shown in Figure 21. Select the command **Type** and from the list of graph types displayed, select **XY**. The graph menu is now redisplayed. Next specify the x-coordinates by choosing **X** from the menu. The program asks for the **X** range. Type

A1..A21

and press **ENTER**. The graph menu is again redisplayed. Next, specify the *y*-coordinates by selecting **A**. (The letter A indicates the first data set to be graphed. The program allows graphing of up to six sets of data, A–F, simultaneously.) The program asks for the **A** data set. Type

`B1..B21`

and press **ENTER**. The graph menu is redisplayed. Now view the graph by selecting **View**. You will see the graph shown in Figure 22.

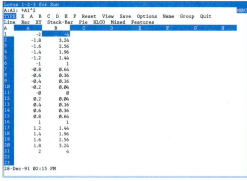

Figure 21

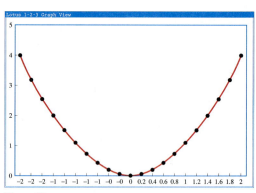

Figure 22

Circles and Their Equations

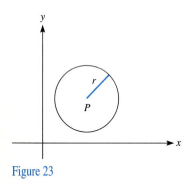

Figure 23

Certain curves occur so often in applications that it is useful to know their equations in advance. One such family of curves are the circles.

A circle with center P and radius r is the set of all points whose distance from P equals r. (See Figure 23.) Let's determine the equation of such a circle. Let $Q(x, y)$ be a typical point on the circle and let $P(h, k)$ be the center. Then, by the distance formula, the distance from P to Q equals

$$d(P,Q) = \sqrt{(x - h)^2 + (y - k)^2}$$

so that the condition

$$d(P,Q) = r$$

may be rewritten in the form:

$$\sqrt{(x - h)^2 + (y - k)^2} = r$$
$$(x - h)^2 + (y - k)^2 = r^2$$

This proves the following result:

EQUATION OF A CIRCLE

Suppose that a circle has center (h, k) and radius r. Then the circle is the graph of the equation

$$(x - h)^2 + (y - k)^2 = r^2$$

See Figure 24.

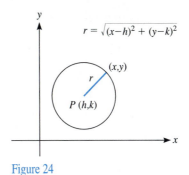

Figure 24

> **EXAMPLE 10**
Graphing a Circle

Describe the graph of the equation

$$(x - 1)^2 + (y + 5)^2 = 25$$

Draw the graph.

Solution

The equation may be written in the form

$$(x - 1)^2 + [y - (-5)]^2 = 5^2$$

This is the equation of a circle with center C at $(1, -5)$ and radius 5. We can draw the graph by plotting the point $(1, -5)$ and then using a compass to draw the circle of the indicated radius with the plotted point as center. (Refer to Figure 25.)

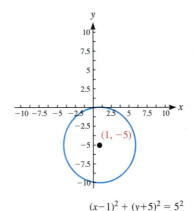

$$(x-1)^2 + (y+5)^2 = 5^2$$

Figure 25

> **EXAMPLE 11**
Completing the Square
to Graph a Circle

Describe the graph of the equation

$$4x^2 + 24x + 4y^2 - 8y = 41$$

Draw the graph.

Solution

The equation has terms x^2 and y^2 with the same coefficient. This suggests that the equation can be transformed into an equivalent equation that is the equation of a circle. We begin by dividing the equation by 4:

$$(x^2 + 6x) + (y^2 - 2y) = \frac{41}{4}$$

Now complete the square for the expressions within the parentheses:

$$(x^2 + 6x + 9) + (y^2 - 2y + 1) = \frac{41}{4} + 9 + 1$$

$$(x + 3)^2 + (y - 1)^2 = \frac{81}{4}$$

To recognize this as the equation of a circle, we write it in the form:

$$[x - (-3)]^2 + (y - 1)^2 = \left(\frac{9}{2}\right)^2$$

This is the equation of a circle with center at $(-3, 1)$ and radius 9/2. See Figure 26.

y
x
9/2
(−3, 1)
10
7.5
5
2.5
−2.5
−5
−7.5
−10
−10 −7.5 −5 −2.5
2.5 5 7.5 10

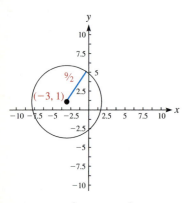

$4x^2 + 24x + 4y^2 - 8y = 41$

Figure 26

Optional

Graphing via Calculator

Many hand-held calculators have the capability to graph equations. By using such calculators, you can apply technology to perform many of the graphical tasks of algebra with minimal effort. The following short discussion is an introduction to graphing calculators using the Texas Instruments TI-81 calculator to provide specific keystrokes. If you have another brand of graphing calculator, its operation will be similar, but you will need to refer to its reference manual for the specific keystrokes.

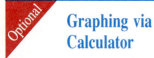

Graphing Equations

To graph an equation using the TI-81, you must first enter the equation. This is done by pressing the key **Y=**, which results in a display like the one shown in Figure 27. The screen shown allows you to enter up to four equations to graph simultaneously. For instance, suppose you wish to graph the equation

:Y1 =
:Y2 =
:Y3 =
:Y4 =

Figure 27

$$y = x^2$$

You would move the cursor to the **Y1 =** line and type

X^2

You will find the X on the key **X|T**.

You may edit equations using the arrow keys to move the cursor, **Ins** to enable insertion at the cursor, and **Del** to delete the character at the cursor.

Once all equations are entered to your satisfaction, you must set the x-range and y-range you wish to view. For the moment, let's use the default ranges of the calculator, which are

$$-10 \le x \le 10, \qquad -10 \le y \le 10$$

Using the defaults means that we don't need to set the ranges and can proceed to view the graph.

To view the graph, press the key **Graph**. The calculator will now display the graph of the specified equations in the specified ranges. You will see a graph as shown in Figure 28.

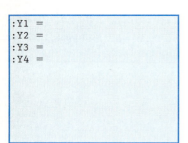

Figure 28

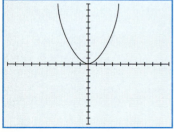

Setting Ranges

To set ranges, press the key **Range**. You will see a display as shown in Figure 29. Use the cursor movement and editing keys to enter the desired minimum and maximum values of x (denoted XMIN and XMAX,

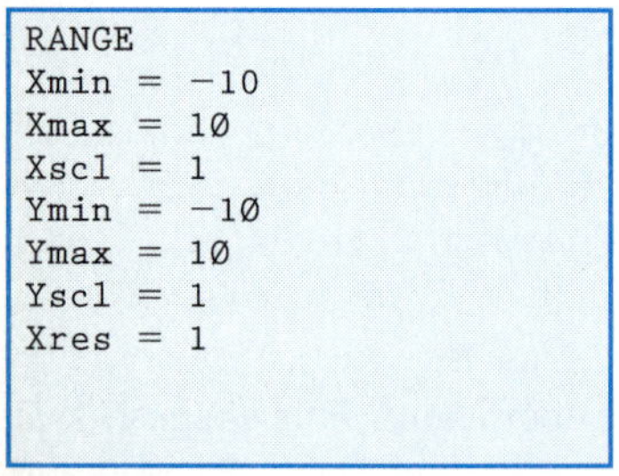

RANGE
Xmin = −10
Xmax = 10
Xscl = 1
Ymin = −10
Ymax = 10
Yscl = 1
Xres = 1

Figure 29

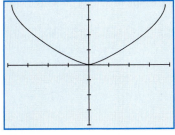

Figure 30

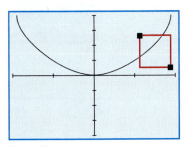

$X = .77894737$ $Y = .606759$

Figure 31

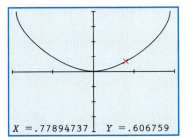

Figure 32

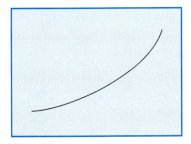

Figure 33

respectively) and the desired minimum and maximum values of y (denoted YMIN and YMAX, respectively). For instance, to graph

$$y = x^2$$

in the range

$$-2 \leq x \leq 2, \qquad -4 \leq y \leq 4$$

we would set:

$$XMIN = -2$$
$$XMAX = 2$$
$$YMIN = -4$$
$$YMAX = 4$$

Pressing **Graph** again, we would see the graph shown in Figure 30.

Tracing   You may move a point along a displayed graph and have the calculator display the numerical coordinates of the point. This process is called **tracing** and is useful in calculating the coordinates of particular points that you can locate visually (say, the intersection point of two graphs).

To trace a graph, first display the graph and press the key **Trace**. The calculator will display a point on the graph. On the bottom of the display will be the x- and y-coordinates of the point. You may move the point along the curve using the left and right arrow keys. Each keystroke moves the point one dot to the left or right. Each time you move the point, the coordinates of the new point are displayed. (See Figure 31.)

If several curves are displayed, you may move the point from curve to curve using the up and down arrow keys.

Zooming   You may control the level of detail in a graph by zooming. By zooming out, you "step away" from the graph and thereby expand the range. By zooming in, you "step toward" the graph and thereby contract the range. There are a number of methods of zooming, provided from a menu displayed in response to the key **Zoom**. Perhaps the most useful method of zooming is specifying a box on the graph. The portion of the graph in the box specifies the new range.

Here is how to zoom using a box. Press **Zoom**. Select the option **Box** from the menu displayed. The calculator will display the graph and a cursor. Using the arrow keys, move the cursor to one corner of the desired box and press the key **Enter**. Then move the cursor to the opposite corner of the box. As you move the cursor, you will see the box expand or contract. (See Figure 32.) When the cursor indicates the desired opposite corner, press the key **Enter**. The calculator now redisplays the graph with the designated box as the range. (See Figure 33.)   ◄

Symmetry of Graphs

One of the easiest-to-recognize geometric properties is the symmetry of the graph of an equation. Intuitively, a graph is symmetric with respect to a line if the graph looks the same on either side of the line. More precisely, if the line is regarded as a mirror, then the graph is symmetric with respect to the line provided that the part of the graph on one side of the line is a mirror image of the part on the other side. (See Figure 34, where the graph is symmetric with respect to the y-axis.)

Let's assume that the line of symmetry is the y-axis. Then a graph is symmetric with respect to this line provided that whenever a point (x, y) is on the graph, then so is its mirror image, $(-x, y)$. (See Figure 35.) This provides us with a way to express the symmetry algebraically.

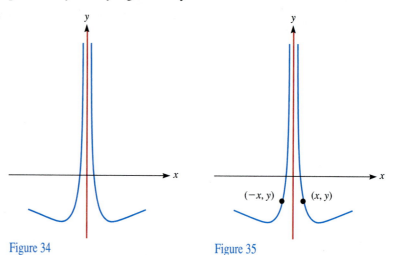

Figure 34   Figure 35

Definition 1
Symmetry with Respect to y-Axis

The graph of an equation is said to be **symmetric with respect to the y-axis** provided that we obtain an equivalent equation when x is replaced throughout by $-x$.

For example, the graph of
$$y = x^4 + 3x^2 + 1$$
is symmetric with respect to the y-axis, since if we substitute $-x$ for x in the equation, we obtain
$$y = (-x)^4 + 3(-x)^2 + 1$$
$$= x^4 + 3x^2 + 1$$
which is the original equation. Figure 36 shows the graph of this equation.

On the other hand, the graph of $y = x^3$ is not symmetric with respect to the y-axis, since if we substitute $-x$ for x throughout the equation, we get
$$y = (-x)^3$$
$$y = -x^3$$
which is not equivalent to the original equation. (See Figure 37.)

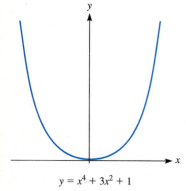

$y = x^4 + 3x^2 + 1$

Figure 36

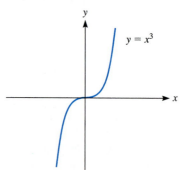

Figure 37

In a similar fashion, we define symmetry with respect to the x-axis.

Definition 2 **Symmetry with Respect** **to x-Axis**	The graph of an equation is said to be **symmetric with respect to the x-axis** provided that we obtain an equivalent equation when y is replaced throughout by $-y$.

For example, the graph of

$$y^2 + \frac{1}{y^2} = x$$

is symmetric with respect to the x-axis, since if we replace y by $-y$ throughout, we obtain the equation

$$(-y)^2 + \frac{1}{(-y)^2} = x$$

$$y^2 + \frac{1}{y^2} = x$$

which is the same as the original equation. (See Figure 38.)

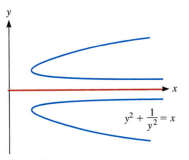

$$y^2 + \frac{1}{y^2} = x$$

Figure 38

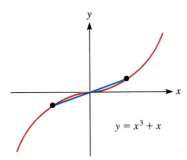

$$y = x^3 + x$$

Figure 39

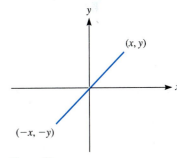

(x, y)

$(-x, -y)$

Figure 40

A third type of symmetry occurs when the graph is its own reflection in the origin. (See Figure 39.) Since the reflection of (x, y) is $(-x, -y)$ (see Figure 40), this suggests the following:

Definition 3 **Symmetry with Respect** **to Origin**	The graph of an equation is said to be **symmetric with respect to the origin** provided that we obtain an equivalent equation when x is replaced throughout by $-x$ and y by $-y$.

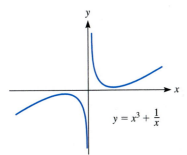

$$y = x^3 + \frac{1}{x}$$

Figure 41

For instance, the graph of $y = x^3 + (1/x)$ is symmetric with respect to the origin since if we replace x by $-x$ and y by $-y$, we have

$$-y = (-x)^3 + \frac{1}{-x}$$

$$-y = -x^3 - \frac{1}{x}$$

$$y = x^3 + \frac{1}{x}$$

which is the original equation. (See Figure 41.)

➤ **EXAMPLE 12**
Determining Symmetry
of a Graph

Determine the symmetry of the graphs shown in the following figures.

1.

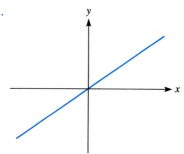

2.

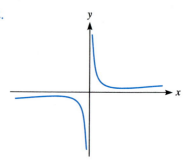

3.

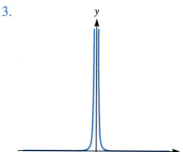

4.

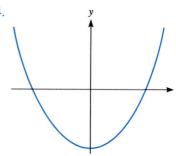

5.

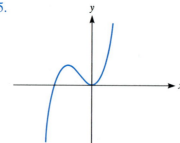

6.

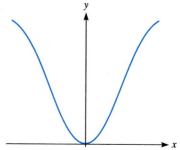

Solution

1. Symmetric with respect to the origin.
2. Symmetric with respect to the origin.
3. Symmetric with respect to y-axis.
4. Symmetric with respect to y-axis.
5. No symmetry.
6. Symmetric with respect to y-axis.

Translating Graphs

Given one graph, it is possible to derive others using the geometric operation of **translation**, which rigidly moves a graph in a horizontal and/or a vertical direction. Consider the original graph shown in Figure 42 and let h and k be positive numbers. We may translate the graph by moving it horizontally h units and vertically k units. The result of this movement is the translated graph shown in Figure

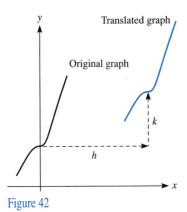

Figure 42

42. Note that the shape of the graph is not distorted as a result of the movement. This is what we mean when we say that the movement is **rigid**.

In Figure 42, we have shown the movement corresponding to positive values of h and k. In general, h and k may be any real numbers. The corresponding translation of the graph is h units in the horizontal direction and k units in the vertical direction. The horizontal movement is to the right if h is positive and to the left if h is negative. The vertical movement is upward if k is positive and downward if k is negative. We say that the graph is **translated by** (h, k). The equation of a translated graph is obtained as follows:

> ### EQUATION OF A TRANSLATED GRAPH
>
> The equation of a graph translated by (h, k) is obtained by replacing x by $x - h$ and y by $y - k$.

➤ **EXAMPLE 13**
Equation of a Translated Graph

Suppose the graph of the equation

$$y = x^2$$

is translated 5 units horizontally and 3 units vertically. Determine the equation of the translated graph. (See Figure 43.)

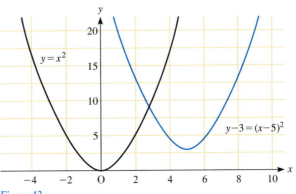

Figure 43

Solution

Replace x by $x - 5$ and y by $y - 3$ throughout the equation. This gives us $y - 3 = (x - 5)^2$, the equation of the translated graph.

➤ **EXAMPLE 14**
Graphing a Translated Equation

Sketch the graph of the equation

$$y + 1 = (x + 4)^2$$

Solution

The form of the equation suggests that its graph may be obtained by translating. To see just which translation, we must write it in terms of variables $x - h$ and $y - k$, for appropriate h and k. To do this, we write the equation in the form

$$y - (-1) = (x - (-4))^2$$

This equation may be obtained by replacing $x - (-4)$ by x and $y - (-1)$ by y in the equation

$$y = x^2$$

So we obtain the desired graph by translating the graph of $y = x^2$ by $(-4, -1)$. That is, we translate 4 units to the left and 1 unit downward. See Figure 44.

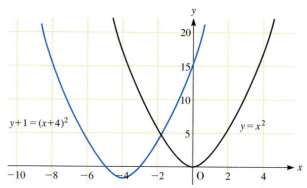

Figure 44

Exercises 3.2

1. Determine whether $(2, -1)$ is a solution of $-2x - 16y = 12$.

2. Determine whether $(5, -5)$ is a solution of $7x + 5y = 5$.

3. Determine whether $(1, -4)$ is a solution of $2x - 3y^2 = 46$.

4. Determine whether $(1, -1)$ is a solution of $2x^2 - y^2 = 3$.

5. Determine whether $(4, -2)$ is a solution of $-y = \sqrt{x}$.

6. Determine whether $(0, 2)$ is a solution of $y = (1/4)(x - 2)^2$.

7. Determine whether $(-2, 15)$ is a solution of $(x - 3)^2 + 2(x - 3) = y$.

8. Determine whether $(1/8, 16)$ is a solution of $x^{-2/3} = y^{1/2}$.

Plot the graph of each equation.

9. $y = 2x + 3$ 10. $y = 4 - x$ 11. $y = -2$

12. $x = 4$ 13. $2x - y = 4$ 14. $x + y = 1$

15. $-4x - 3y = 12$ 16. $2x + 5y = 10$

17. $y = x^2 + 1$ 18. $y = 1 - x^2$ 19. $x = y^2 + 3$

20. $x = -y^2$ 21. $y = -\dfrac{1}{x}$ 22. $y = \dfrac{2}{x}$

23. $y = |x|$ 24. $y = |2x - 1|$ 25. $y = x^3$

26. $x = y^3$ 27. $x = |y + 3|$ 28. $x = |y|$

29. $xy = 1$ 30. $xy = -0.25$

31. $y = \dfrac{1}{x^2}$ 32. $y = \dfrac{1}{|x|}$

33. $x^2 + y^2 = 4$ 34. $x^2 + y^2 = 25$

♦ Applications

35. **Spread of an organism**. A certain kind of dangerous organism is released by accident over an area of 3 square miles. (See figure.) After time t, in years, it spreads over area A, in square miles, where A and t are related by the equation

$$A = 1.8t + 3$$

a. Graph the linear equation representing the relationship between elapsed time and area.

b. Give an interpretation of the t-intercept of the graph. Interpret the A-intercept.

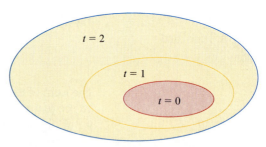

Exercise 35

36. **Driving costs**. The cost C, in cents per mile, of driving a car is related to the number of years t since 1980 by the equation

$$100C - 7t = 2320$$

where $t = 0$ corresponds to 1980, $t = 1$ to 1981 , and so on.

   a. Graph the linear equation representing the relationship between cost and elapsed time. Consider time t on the x-axis and cost C on the y-axis.

   b. Give an interpretation of the C-intercept of the graph.

Describe and graph each equation.

37. $(x - 4)^2 + y^2 = 7$

38. $x^2 + (y + 1)^2 = 5$

39. $(x + 3)^2 + (y - 7)^2 = 13$

40. $(x - 6)^2 + (y + 2)^2 = 11$

41. $x^2 + y^2 + 12x + 19 = -1$

42. $x^2 + 6x + y^2 - 16y + 48 = 0$

43. $x^2 + x + y^2 - y - 5 = 0$

44. $x^2 + y^2 = 4y$

Find an equation in the standard form of a circle satisfying the following conditions.

45. Center $(-6, 1)$, radius 2

46. Center $(3, -4)$, radius $\sqrt{3}$

47. Center at the origin, radius $\sqrt{39}$

48. Center $(0, a)$, radius 0.1

Find an equation in the standard form of a circle satisfying the following conditions.

49. Tangent to the y-axis, center $(-2, 3)$

50. Tangent to the x-axis, center $(9, 10)$

51. Tangent to both axes, center in the fourth quadrant, radius 11

52. Tangent to both axes, center in the third quadrant, radius $2/3$

53. A diameter whose endpoints have coordinates $(-5, 4)$ and $(3, 8)$

54. A diameter whose endpoints have coordinates $(-3, -5)$ and $(2, -7)$

55. Center $(-2, -3)$, area 9π square units

56. Center $(-2, -3)$, circumference 22π units

57. Inscribed in a square whose vertices are $(2, 10)$, $(7, 15)$, $(12, 10)$, and $(7, 5)$

For each of the following equations, find the center and radius of the circle.

58. $x^2 + y^2 + 3.5x - 4.2y - 10 = 0$

59. $x^2 + y^2 + 10x - 12y - 20 = 0$

60. $144x^2 + 144y^2 + 192x - 216y + 81 = 45$

61. $2x^2 + 2y^2 - 6x + 8y - 100 = 0$

Determine whether each of the following ordered pairs lies on the circle $x^2 + y^2 = 1$.

62. $\left(\dfrac{1}{2}, \dfrac{\sqrt{3}}{2}\right)$

63. $\left(-\dfrac{\sqrt{2}}{2}, \dfrac{\sqrt{2}}{2}\right)$

64. $\left(\sqrt{2}, \sqrt{3}\right)$

65. $(0, 2)$

66. $(1, -1)$

67. $(1, 0)$

68. $(0.9781, 0.2079)$

69. $(-0.2419, -0.9703)$

70. Describe the graph of this equation:

$$(x - 3)^2 + (y + 5)^2 = (x + 7)^2 + (y + 4)^2$$

71. Describe the solution set of this inequality:

$$(x + 3)^2 + (y + 5)^2 \le 36$$

➡ Technology

In Exercises 72–79, use a graphing calculator to graph the equations. Use the default values for the viewing rectangle.

72. $y = 3x - 5$

73. $y = -2x + 3.1$

74. $y = x^2 - x$

75. $y = 0.3x^2 + 0.151x$

76. $y = \dfrac{3}{x + 1}$

77. $y = \sqrt{3x - 1}$

78. $y = \dfrac{1}{x^2 + 1}$

79. $y = \dfrac{x - 1}{(x + 1)(x + 2)}$

80. Use the trace feature to determine the y-coordinate on the graph of $y = x^2 - 3x$ when

   a. $x = 1.2$

   b. $x = 2.1$

   c. $x = -1$

   d. $x = 0.001$

81. Use the trace feature to determine a point of the graph of $y = x/(x^2 + 1)$ at which $y = 0.1$.

82. Use the zoom feature to enlarge the graph of Exercise 72 in the range $1 \le x \le 2$.

83. Use the zoom feature to enlarge the graph of Exercise 73 in the range $0 \le x \le 1, 0 \le y \le 10$.

84. By using the zoom and trace features, determine the points at which the graph of Exercise 74 crosses the horizontal axis.

85. By using the zoom and trace features, determine, to two significant digits, the points at which the graph of Exercise 75 crosses the horizontal axis.

86. Graph the equations $y = x^2$ and $y = -2x + 3$ on the same coordinate system. Choose the viewing rectangle so that the two intersection points are visible.

87. Determine the coordinates of the intersection points of the preceding exercise.

88. Graph the equations $y = 1/x$, $y = 3x^3$ on the same coordinate system. Choose the viewing rectangle so that the intersection points are visible.

89. Determine the coordinates of the intersection points of the preceding exercise.

90. Graph the equations $y = x^2$, $y = 3x^2$, and $y = 0.6x^2$ on the same coordinate system. What do you observe about the points where they cross the horizontal axis? Can you make a conclusion about what happens to horizontal-axis intersections when a graph is scaled?

91. Graph on the same coordinate system the equation $y = 5x + 3$ and the graph translated 3 units horizontally and 2 units vertically.

92. Graph on the same coordinate system the equation $y = -x^2 + 1$ and the graph translated -1 unit horizontally and 3 units vertically.

93. Graph on the same coordinate system the equation $y = -x^2 + x - 1$ and the graph translated -2 units horizontally and -1 unit vertically.

94. Program your calculator to solve quadratic equations. The program should pause, ask for each of the coefficients of the equation, and then calculate and display the solutions.

✍ In Your Own Words

95–105. Use a spreadsheet program to plot points and graph each of the equations given in Exercises 9–18. Use 30 points in the range $-3 \le x \le 3$.

106. Explain how a spreadsheet program's cells can be thought of as the points of a two-dimensional coordinate system.

107. Explain how you might define the distance between two cells and the midpoint of the line between two cells when considering a spreadsheet as a two-dimensional coordinate system.

3.3 STRAIGHT LINES

Straight lines are the simplest curves. In many applied situations, experiments or physical laws suggest that certain variables are related to one another via a linear equation. In such circumstances, the relationship between the variables can be depicted using a straight-line graph. In this section, we take a closer look at straight lines and show how to determine equations for them from various sets of data.

Slope of a Straight Line

Let's consider a nonvertical straight line on which lie the distinct points (x_1, y_1) and (x_2, y_2). The difference $y_2 - y_1$ measures the vertical change between the two points and is called the **rise**. The difference $x_2 - x_1$ measures the horizontal difference between the two points and is called the **run**. Geometric interpretations of the rise and run are shown in Figure 1. Since the two points are assumed to be distinct and the line nonvertical, the points must have different x-coordinates. That is, $x_2 - x_1 \ne 0$ (the run is nonzero).

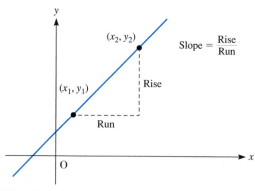

Figure 1

The ratio of the rise to the run measures the steepness of the line and is called the **slope.** That is, we have the following definition.

Definition 1 **Slope of a Line**	Suppose that L is a nonvertical line passing through the points (x_1, y_1) and (x_2, y_2). The **slope** of the line, denoted m, is defined as the ratio $$m = \frac{\text{Rise}}{\text{Run}} = \frac{y_2 - y_1}{x_2 - x_1}$$ For a vertical line, the slope is undefined.

> **EXAMPLE 1**
Calculating Slope

For the straight lines in the following graphs, determine the slope.

1. 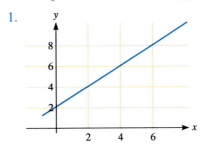   2.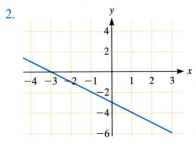

Solution

1. To calculate the slope, we need two points on the line. Examining the graph, we see that these points may be taken to be $(2, 4)$ and $(4, 6)$. So the slope of the line is

$$m = \frac{y_2 - y_1}{x_2 - x_1} = \frac{6 - 4}{4 - 2} = 1$$

2. Two points on the graph are $(0, -3)$ and $(2, -5)$. Therefore, the slope is equal to

$$m = \frac{y_2 - y_1}{x_2 - x_1} = \frac{-5 - (-3)}{2 - 0} = -1$$

There's more to the definition of slope than meets the eye! As we shall soon see, the notion of slope is one of the most important measures describing a straight line. However, before we describe any applications of slope, let's make several observations about the definition.

First, note that the value of the slope does not depend on which point is labeled with subscript 1 and which with subscript 2. Indeed, we have

$$\frac{y_2 - y_1}{x_2 - x_1} = \frac{-(y_1 - y_2)}{-(x_1 - x_2)} = \frac{y_1 - y_2}{x_1 - x_2}$$

Next, note that the value of the slope does not depend on which points we choose as (x_1, y_1) and (x_2, y_2), provided, of course, that we choose points on the line. Indeed, suppose that the straight line has the equation

$$Ax + By + C = 0$$

In this equation, A and B can't both be 0. (Otherwise neither x nor y would be involved in the equation.) Since (x_1, y_1) and (x_2, y_2) are on the line, both points satisfy the equation of the line. That is,

$$Ax_1 + By_1 + C = 0$$
$$Ax_2 + By_2 + C = 0$$

Subtracting the first equation from the second, we have

$$A(x_2 - x_1) + B(y_2 - y_1) = 0$$
$$A(x_2 - x_1) = -B(y_2 - y_1)$$
$$\frac{y_2 - y_1}{x_2 - x_1} = -\frac{A}{B}$$

The expression on the left side of the last equation equals the slope. The expression on the right side depends only on the equation of the line and not on the particular points (x_1, y_1) and (x_2, y_2). That is, the value of the slope is independent of the points used to calculate it.

Next, let's emphasize that the notion of slope is undefined for vertical lines. Indeed, in the case of a vertical line, all points have the same x-coordinate. That is, x_1 equals x_2, and the formula used to define the slope would involve division by 0. Since that is an undefined operation, we leave as undefined the slope of a vertical line.

The Greek letter Δ (delta) is often used to denote the change in a variable. In computing the slope, the change in the y variable from one point to the other is $y_2 - y_1$, or the rise. For this reason, the rise is sometimes denoted Δy (read "delta y"). Similarly, the run is denoted Δx (read "delta x"). In this notation, the slope is expressed as

$$m = \frac{\Delta y}{\Delta x}$$

➤ **EXAMPLE 2**
Calculating Slope
from the Equation

Determine the slopes of the lines having the following equations.

1. $y = 4x - 3$
2. $3x + 7y = -10$
3. $2y = 2x + 2y - 6$

Solution

1. To calculate the slope, we need two points on the line. To obtain such points, choose any two values for x, say, $x = 0$ and $x = 1$, and determine the corresponding values for y:

$$y = 4 \cdot 0 - 3 = -3$$
$$(x_1, y_1) = (0, -3) \qquad \textit{on line}$$
$$y = 4 \cdot 1 - 3 = 1$$
$$(x_2, y_2) = (1, 1) \qquad \textit{on line}$$

We now calculate the slope:

$$m = \frac{y_2 - y_1}{x_2 - x_1} = \frac{1 - (-3)}{1 - 0} = 4$$

That is, the line has slope 4, which is the coefficient of x.

2. Again we determine the slope by determining two points on the line, corresponding to $x = 0$ and $x = 1$. The corresponding values of y are

$$7y = -3x - 10$$
$$7y = -3 \cdot 0 - 10$$
$$y = -\frac{10}{7}$$
$$(x_1, y_1) = \left(0, -\frac{10}{7}\right) \quad \textit{on line}$$
$$7y = -3 \cdot 1 - 10 = -13$$
$$y = -\frac{13}{7}$$
$$(x_2, y_2) = \left(1, -\frac{13}{7}\right) \quad \textit{on line}$$

Therefore, the slope is given by:

$$m = \frac{y_2 - y_1}{x_2 - x_1} = \frac{(-13/7) - (-10/7)}{1 - 0} = -\frac{3}{7}$$

That is, the slope of the line is $-3/7$.

3. We simplify the equation by combining like terms:

$$2y = 2x + 2y - 6$$
$$0 = 2x - 6$$
$$x = 3$$

This is the equation of a vertical line. The slope of a vertical line is undefined.

➤ **EXAMPLE 3**
Calculating Slope from Points

A line passes through the points $(-1, 5)$ and $(0, 4)$. Determine the slope of the line.

Solution

Using the definition of slope, we have

$$\text{Slope} = \frac{y_2 - y_1}{x_2 - x_1} = \frac{4 - 5}{0 - (-1)} = \frac{-1}{1} = -1$$

Let's now develop an alternative method of computing the slope directly from the equation of a line. Suppose that the equation of a line is

$$Ax + By + C = 0$$

where A and B are not both 0 (that is, the equation involves at least one of x and y). If B is not 0, then we may solve the equation for y to obtain

$$y = -\frac{A}{B}x + \left(-\frac{C}{B}\right)$$

As we showed earlier, the expression $-A/B$ equals the slope m of the line. Let's denote the expression $-C/B$ by b. Then the equation of the line may be written as

$$y = mx + b$$

The value b has a geometric significance. Indeed, if we set x equal to 0, we see that $y = m(0) + b = b$. In other words, the point $(0, b)$ is on the line, and b is its y-intercept. Since the equation $y = mx + b$ expresses both the slope and the y-intercept of the line, it is called the **slope-intercept form of the equation of a line**.

The derivation of the slope-intercept form started from the assumption $B \neq 0$. On the other hand, if $B = 0$, then the equation of the line reads

$$Ax + C = 0, \quad A \neq 0$$

$$x = -\frac{C}{A}$$

This is the equation of a vertical line.

The preceding discussion may be summarized as follows.

EQUATIONS OF STRAIGHT LINES

Let $Ax + By + C = 0$ be the equation of a straight line. If $B \neq 0$, then solving the equation for y in terms of x gives an equivalent equation of the form $y = mx + b$, the slope-intercept form of the equation. Here m is the slope of the line, and b is the y-intercept. If $B = 0$, then the line is vertical. In this case, solving the equation for x yields an equivalent equation of the form $x = a$.

➤ **EXAMPLE 4**
Slope-Intercept Form

Determine the slope-intercept form, the slope, and the y-intercept of the line with equation $5x + 3y - 12 = 0$.

Solution

To determine the slope-intercept form, solve for y in terms of x.

$$5x + 3y - 12 = 0$$
$$3y = -5x + 12$$
$$y = -\frac{5}{3}x + 4$$

The last equation is the slope-intercept form. The slope of the line equals the coefficient of x, that is, $-5/3$. The y-intercept equals the constant term of the slope-intercept form, or 4.

The next law gives a geometric interpretation of the slope of a line. As we will see, this interpretation is useful in many applied problems.

GEOMETRIC INTERPRETATION OF SLOPE

Suppose that a line has slope m. Starting from any point on the line, if you move h units in the x direction, then it is necessary to move mh units in the y direction to return to the line. See Figure 2.

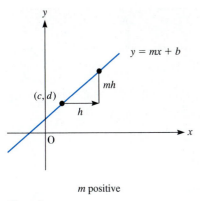

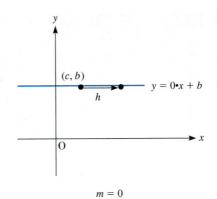

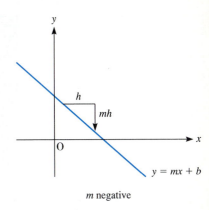

m positive   m = 0   m negative

Figure 2

Proof   Assume that the initial point on the line is (x_1, y_1). Since the point is on the line, it satisfies the equation

$$y_1 = mx_1 + b$$

If the x-coordinate is changed by h, then the new x-coordinate is $x_1 + h$. The corresponding value of y on the line can be obtained from the preceding equation:

$$y = m(x_1 + h) + b$$
$$= (mx_1 + b) + mh$$
$$= y_1 + mh$$

That is, the value of y changes from y_1 to $y_1 + mh$. The amount of the change is mh, as claimed.   ♦

➤ **EXAMPLE 5**
Application of Slope

Consider the line with equation

$$y = -2x + 9$$

1. Suppose that x is increased by 5. What is the change in the value of y?
2. Suppose that x is decreased by 100. What is the change in the value of y?

Solution

1. The slope of the line is -2. By the Geometric Interpretation of Slope, a change of h in x results in a change of mh in y. So an increase of 5 in x results in a change of

$$m \cdot 5 = -2 \cdot 5 = -10$$

in y. That is, y decreases by 10.

2. Again as a consequence of the Geometric Interpretation of Slope, a change of -100 in x results in a change of

$$mh = -2 \cdot (-100) = 200.$$

That is, y increases by 200.

➤ **EXAMPLE 6**
Manufacturing

The cost C (in dollars) of manufacturing q television sets in a production run is given by

$$C = 250q + 50,000$$

1. Give an interpretation for the slope of this linear equation.
2. Suppose that production is increased by 100 units. What will be the effect on production cost?

Solution

1. The slope of the line represented by the equation is 250. By the Geometric Interpretation of Slope, if q is increased by 1, that is, if the company produces one more television set, then the value of C, the total cost of manufacturing, changes by

$$250 \cdot 1 = 250$$

That is, the slope equals the cost of producing one additional television set. This additional cost is called **the marginal cost of production** and is used by economists in analyzing the microeconomics of production.

2. If production increases by $h = 100$ units, then the cost C changes by

$$mh = 250 \cdot 100 = 25,000$$

That is, the cost increases by $25,000.

> **EXAMPLE 7**
Distance versus Time

After t hours of driving, a trucker's distance D, in miles, from his delivery point is given by the equation

$$D = -50t + 500$$

Give a physical interpretation of the slope of the corresponding line.

Solution

The relationship connecting D and t is a linear equation expressed in slope-intercept form. The slope is -50, and the vertical-axis intercept (the D-intercept) is 500. According to the preceding discussion, for each increment of time Δt, the truck goes an additional distance ΔD. Moreover, the slope equals the quotient

$$\frac{\Delta D}{\Delta t} = \frac{\text{Change in distance}}{\text{Change in time}}$$

But according to the well-known formula, distance = rate $\times$ time. Therefore, the ratio [Change in distance]/[Change in time] is equal to the **rate** at which the distance is changing per unit time. This quantity is called the **velocity** of the truck and is expressed in units of miles per hour. The statement Velocity $= -50$ means that the distance to the delivery point is decreasing at a rate of 50 miles per hour.

Finding Equations of Straight Lines

In solving problems, it is often necessary to determine equations of straight lines from given data. A number of different forms of equations for straight lines allow us to do just that. One is the slope-intercept form, which we have been using in our discussion of slope.

➤ EXAMPLE 8
Cost Equation

The developer of an office building determines that initial development costs are $3,000,000 for land, permits, and site work. These costs don't vary with the size of the building and so are called **fixed costs.** The construction of the building itself costs $50 for each square foot of office space. The amount of this construction cost is called the **variable cost** of the building. Determine an equation that relates the total cost of the building to the number of square feet of office space it contains.

Solution

Let C denote the cost of the building and S the number of square feet it contains. From the Geometric Interpretation of Slope, the slope in the equation expressing the relationship is 50 since an increase of S by 1 square foot causes C to increase by 50. The number 3,000,000 gives the constant term in the equation, since this number gives the cost for S equal to 0. Therefore, using the slope-intercept form, we may write the equation as

$$C = 50S + 3,000,000$$

The next formula allows us to determine the equation of a line given its slope and a point on the line.

POINT-SLOPE FORMULA

Suppose a line has slope m and passes through the point (x_1, y_1). Then an equation for the line is

$$y - y_1 = m(x - x_1)$$

To verify the point-slope formula, first note the point (x_1, y_1) is on the line determined by the equation. Indeed, if we substitute x_1 for x and y_1 for y, we see that the equation is satisfied:

$$y_1 - y_1 = m(x_1 - x_1)$$
$$0 = 0$$

Writing the equation in the form

$$y - y_1 = mx - mx_1$$
$$y = mx + (y_1 - mx_1)$$
$$y = mx + b \qquad \textit{(where } b = y_1 - mx_1\text{)}$$

we see that the slope of the line is m.

➤ **EXAMPLE 9**
Point-Slope Form

Determine the equation of a line passing through the point $(3, -2)$ and having slope 4.

Solution

Using the point-slope formula, we have the equation

$$y - (-2) = 4(x - 3)$$
$$y + 2 = 4x - 12$$
$$y = 4x - 14$$

➤ **EXAMPLE 10**
Line Passing through
Two Points

Determine the equation of a line passing through the points $(5, 3)$ and $(2, -2)$.

Solution

The slope of the line is

$$m = \frac{3 - (-2)}{5 - 2} = \frac{5}{3}$$

Therefore, by the point-slope formula, the desired equation is:

$$y + 2 = \frac{5}{3}(x - 2)$$

$$y = \frac{5}{3}x - \frac{10}{3} - 2$$

$$y = \frac{5}{3}x - \frac{16}{3}$$

The following table summarizes the various equations of straight lines we have considered.

EQUATIONS OF LINES

1. General linear equations: $Ax + By + C = 0$ or $Ax + By = D$
2. Slope-intercept form: $y = mx + b$
3. Point-slope form: $y - y_1 = m(x - x_1)$
4. Vertical line: $x = a$
5. Horizontal line: $y = b$

Parallel and Perpendicular Lines

The next two statements describe the relationships between slopes of parallel lines and between slopes of perpendicular lines.

SLOPES OF PARALLEL LINES

Two distinct nonvertical lines are parallel if and only if they have the same slope and have different y-intercepts. (See Figure 3.)

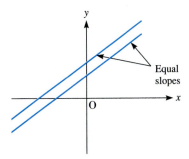

Figure 3

SLOPES OF PERPENDICULAR LINES

Two lines with slopes m_1 and m_2 are perpendicular if and only if the product of the slopes is -1, that is, if and only if

$$m_1 m_2 = -1 \quad \text{or} \quad m_1 = -\frac{1}{m_2}, \quad \text{where} \quad m_2 \neq 0$$

(See Figure 4.)

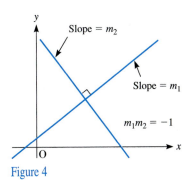

Figure 4

The proof of these results is outlined in the exercises.

➤ **EXAMPLE 11**
Perpendicular Lines

Determine whether the lines described by the following equations are perpendicular to one another.

$$-4x + 3y = 1, \quad 3x + 4y = 17$$

Solution

Let's apply the preceding result. To do so, we must first determine the slopes of the lines. The slope-intercept forms of the equations are, respectively,

$$-4x + 3y = 1$$
$$3y = 4x + 1$$
$$y = \frac{4}{3}x + \frac{1}{3}$$

and

$$3x + 4y = 17$$
$$4y = -3x + 17$$
$$y = -\frac{3}{4}x + \frac{17}{4}$$

From these equations, we see that the slopes of the lines are $m_1 = 4/3$ and $m_2 = -3/4$, respectively. We then have

$$m_1 m_2 = \frac{4}{3} \cdot -\frac{3}{4} = -1$$

Therefore, by the preceding result, the two lines are perpendicular.

➤ **EXAMPLE 12**
Parallel Lines

Determine the equation of a line passing through the point $(-1, -2)$ and parallel to the line with equation

$$y = 3x - 18$$

Solution

The line with the given equation has slope 3. By the Slopes of Parallel Lines rule, a line parallel to this line has the same slope, 3. Since the desired line must pass through the point $(-1, -2)$, its equation may be derived from the point-slope formula:

$$y - (-2) = 3[x - (-1)]$$
$$y + 2 = 3x + 3$$
$$y = 3x + 1$$

➤ **EXAMPLE 13**
Perpendicular Lines

Determine the equation of a line passing through the point $(7, 4)$ and perpendicular to the line having equation

$$3x - 4y - 12 = 0$$

Solution

The given equation has slope-intercept form

$$y = \frac{3}{4}x - 3$$

so the corresponding line has slope 3/4. By the Slopes of Perpendicular Lines rule, the slope m of a line perpendicular to it is

$$m = -\frac{1}{3/4} = -\frac{4}{3}$$

Since the perpendicular line must pass through $(7, 4)$, we may derive its equation using the point-slope formula:

$$y - 4 = -\frac{4}{3}(x - 7)$$

$$y = -\frac{4}{3}x + \frac{40}{3}$$

Exercises 3.3

For each of the following equations, determine the slope of the line.

1. $y = 0.3x - 27$

2. $y = \frac{2}{3}x + \frac{4}{5}$

3. $-3y = 18x$

4. $x = 4y$

5. $x - 2y = 6$

6. $6x + 3y = 9$

7. $y + 6 = \frac{2}{3}(x - 3)$

8. $3(x - 3) - 4(y + 5) = 10(x + 7) - 5(y - 8)$

9. Consider the line with equation

$$y = -12x + 34$$

   a. Suppose that x is increased by 3. What is the change in the value of y?

   b. Suppose that x is decreased by 50. What is the effect on y?

10. Consider the line with equation

$$y = 3.4x - 15$$

   a. Suppose that x is increased by 200. What is the effect on y?

   b. Suppose that x is decreased by 39. What is the effect on y?

♦ Applications

11. **Spread of an organism**. A certain kind of dangerous organism is released by accident over an area of 3 square miles. After time t, in years, it spreads over area A, in square miles, where A and t are related by the equation

$$A = 1.8t + 3$$

   a. Give an interpretation for the slope of this linear equation.

   b. Suppose that time is increased by 10 years. What will be the effect on the area A affected by the organism?

12. **Driving costs**. The cost C, in cents per mile, of driving a car is related to the number of years t since 1980 by the equation

$$100C - 7t = 2320$$

where $t = 0$ corresponds to 1980, $t = 1$ to 1981, and so on.

   a. Give an interpretation of the slope of this linear equation. Consider time t on the x-axis and cost C on the y-axis.

   b. Suppose that time is increased by 8 years. What will be the effect on the cost C?

13. **Revenue**. The revenue R, in dollars, from the sale of q television sets from a production run is given by

$$R = 409q$$

   a. Give an interpretation of the slope of this linear equation.

   b. Suppose that sales are increased by 2000 units. What will be the effect on revenue?

14. **Temperature conversion**. Fahrenheit temperatures F and Celsius temperatures C are related by the following linear equation.

$$F = \frac{9}{5}C + 32$$

   a. Give an interpretation of the slope of this linear equation.

   b. Suppose that a person's temperature rises 1°C during an illness. What is the change in the person's Fahrenheit temperature?

15. **Tail length of a snake**. It has been found in a study that the total length L and the tail length t, both in millimeters, of females of the snake species *Lampropeltis polyzona* are related by the linear equation

$$t = 0.143L - 1.18$$

   a. Give an interpretation of the slope of this linear equation.

   b. Suppose a snake is found whose tail is 80 mm longer than that of another. What is the difference in the total length of one snake over the other?

Determine the slope of the line passing through each pair of points.

16. $(0, 0)$ and $(3, 6)$

17. $(4, 0)$ and $(0, 2)$

18. $(1, -1)$ and $(3, -5)$

19. $(-2, -3)$ and $(-1, -6)$

20. $(2, 3)$ and $(-1, 3)$

21. $(-7, 3)$ and $(-7, 2)$

22. $(a, 2a + 3)$ and $(a + h, 2(a + h) + 3)$

23. $(k, -3k)$ and $(k, -3(k + h))$

♦ Applications

24. **Demand for orange juice.** A supermarket owner observes the following relationship between weekly sales of frozen orange juice q (measured in cans) and the price p (measured in dollars):

$$q = -3000p + 12{,}000$$

   a. Give an interpretation of the slope of this line.

   b. What is the effect on sales of lowering the price \$1?

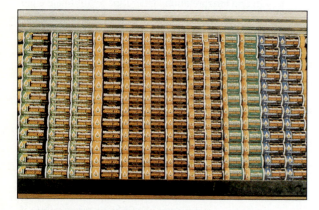

25. **Straight-line depreciation** is a way a business computes the value of an item for tax purposes. A company buys an office machine for \$9700 on January 1 of a given year. The machine is expected to last for 6 years, at the end of which time its trade-in or salvage value will be \$1300. If the company figures the decline in value to be the same each year, then the book value or salvage value V after t years, $0 \le t \le 6$, is given by

$$V = \$9700 - \$1400t$$

Give an interpretation of the slope of this line.

26. **Cost.** A clothing manufacturer is planning to market a new type of sweatshirt to college students. The company estimates that fixed costs for setting up the production line are \$43,000. Variable costs for producing each sweatshirt are \$3.50. Determine an equation that relates cost to the number of sweatshirts produced.

27. A salesperson gets a salary of \$24,000 per year plus a commission of 38% of sales. Determine an equation that relates salary to sales.

Determine the equation of the line satisfying the given conditions.

28. Through $(-1, 2)$ with slope -3

29. Through $(4, -2)$ with slope $-\dfrac{1}{2}$

30. Through $(2, -3)$ with slope undefined

31. Through $(2, -3)$ with slope 0

32. With y-intercept $(0, -1)$ and slope $-\dfrac{1}{5}$

33. With y-intercept $(0, 0)$ and slope -4

34. Through $(2, 3)$ and $(-1, 5)$

35. Through $(-2, -6)$ and $(4, -10)$

36. Through $(9, -3)$ and $(9, 7)$

37. Through $(4, -6)$ and $(3, -6)$

Find the slope and y-intercept.

38. $y = -\dfrac{2}{3}x + 1$

39. $2y + 3x = 6$

40. $5x - 3y - 9 = 0$

41. $\dfrac{2}{3}x - \dfrac{1}{5}y - \dfrac{1}{10} = 0$

42. $2y + 3 = 6$

43. $-3x = 6$

Determine whether each pair of equations represents parallel lines, perpendicular lines, or neither.

44. $2x - 3y = 4,\ -2x + 3y = -8$

45. $y = \dfrac{4}{5}x - 15,\ 4y - 5x + 5 = 0$

46. $2x - 3y = 56,\ x + 4y = 7$

47. $4x - 2y = 86,\ y = -3x + 7$

48. $x - 5y = 32,\ x + 11y = 4$

49. $x = -3,\ y = 0$

50. $x = -5y,\ x = 12$

51. $y - 4x = 34,\ y - 4x = -3$

Determine the equation of a line passing through the given point and parallel to the given line.

52. $(-3, -2),\ 2x + 2y = 10$

53. $(-3, 5),\ 9y = 3x + 1$

54. $(3, -1),\ y = x$

55. $(5, -2),\ x = 3$

56. $(0, 0),\ 4x - 6y = 2$

57. $(-3, 5),\ 9y = 18$

Determine the equation of a line passing through the given point and perpendicular to the given line.

58. $(0, -2),\ y = \dfrac{2}{3}x + 1$

59. $(-1, 3),\ 2x - 4y = 6$

60. $(0, 0),\ y = x$

61. $(4, 2),\ -4x = -24$

62. $(-2, 5),\ 4y = 2$

63. $(-3, 5),\ 2x = 9y + 18$

In each case determine (1) the equation of the line through P that is parallel to the line containing A and B and (2) the equation of the line through P that is perpendicular to the line containing A and B.

64. $A(-2, 3), B(4, 5), P(4, -6)$

65. $A(-1, -8), B(1, 10), P(6, -2)$

66. Prove that the graph of an equation in the form

$$\frac{x}{a} + \frac{y}{b} = 1$$

has x-intercept $(a, 0)$ and y-intercept $(0, b)$.

67. Use the result of Exercise 66 to find the x-intercept and y-intercept of each of the following.

   a. $\dfrac{x}{4} + \dfrac{y}{-2} = 1$

   b. $4x - 3y = 12$

68. Find the equation of the perpendicular bisector of the line segment with endpoints $(2, 3)$ and $(6, -7)$.

69. Find the equation of the perpendicular bisector of the line segment with endpoints $(0, 1)$ and $(-4, -5)$.

70. Determine whether the points $A(-1, 5), B(-5, 1)$, $C(5, 1)$, and $D(1, 5)$ are vertices of a rectangle. Use slopes.

71. Find k such that the line containing $(-2, k)$ and $(3, 8)$ is parallel to the line containing $(5, 3)$ and $(1, -3)$.

72. Find k such that the line containing $(-2, k)$ and $(3, 8)$ is perpendicular to the line containing $(5, 3)$ and $(1, -3)$.

♦ **Applications**

73. In 1983 there were 2.2 million telephone-answering machines sold. In 1986 there were 4.9 million sold.

   a. Use the two-point equation to find a linear equation relating the number sold N to the year Y.

   b. Use the equation to predict sales in 1996.

74. Highway records show that about 34% of drivers aged 20 and 22% of those aged 25 will be involved in at least one driving accident within a one-year period.

   a. Use the two-point equation to find a linear equation describing the percentage P of drivers who will have an accident at age A.

   b. Predict the percentage of those of age 30 who will have an accident.

75. **Computer diskettes.** A computer supply catalog recently listed a box of ten 5¼-inch diskettes to sell at $1.49 each if 10 diskettes are purchased. If 60 diskettes are purchased the price is reduced to $1.29 each.

   a. Use the two-point equation to find a linear equation relating the price P, per diskette, to the number purchased, N.

   b. Predict the price per diskette if 100 diskettes are purchased. The catalog listed the price to be $1.19 if 100 are purchased. How does this compare to the prediction?

76. **Charles's law.** In 1787 Jacques Charles, a French scientist, noticed that gases expand when heated and contract

Jacques Charles

when cooled according to a linear equation, assuming the pressure remains constant.

   a. Suppose a particular gas has a volume $V = 500$ cc when $T = 27°C$ and a volume $V = 605$ cc when $T = 90°C$. Use the two-point equation to find a linear equation relating V and T.

   b. At what temperature does $V = 0$ cc? This is an estimate of what is known as *absolute zero*, the coldest temperature possible.

77. **Slopes of Perpendicular Lines rule: Outline of proof**. To prove a sentence of the type "P if and only if Q," one has two statements to prove: (1) If P, then Q. (2) If Q, then P.

   "P if and only if Q" means that P and Q are equivalent statements. Another way to prove "P if and only if Q" is to produce a string of equivalent statements, starting with P and ending with Q. We do so for this proof. Consider the accompanying figure showing two nonvertical lines L and M with respective equations $y = m_1x + b_1$ and $y = m_2x + b_2$. Lines L and M are perpendicular if and only if ΔPQR is a right triangle, and ΔPQR is a right triangle if and only if $PR^2 + RQ^2 = PQ^2$, from the Pythagorean theorem. Now let (x_1, y_1) and (x_2, y_2) be coordinates of two points on lines L and

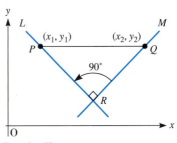

Exercise 77

M, respectively. Use these points and the result of the Pythagorean theorem to complete the proof.

➡ **Technology**

Use a graphing calculator to graphically approximate the slope and *y*-intercept of the following lines.

78. $y = -5.1x + 2.3$

79. $y = 1.17(x + 3.11) + 2x$

80. $5y + 3 = 4x - 7$

81. $5(y + 2.9) = 4(x - 1.1) + 31.7$

✍ **In Your Own Words**

82. There is more than one way to define the distance function. For instance, consider the distance function $d(x, y) = |x_1 - x_2| + |y_1 - y_2|$.

   a. How is this definition of distance different from the definition normally used?

   b. Why do you think this definition is called the taxicab distance function?

   c. Describe another model that has another definition for distance.

3.4 THE CONCEPT OF A FUNCTION

Definition of a Function

The notion of a function is crucial to all of modern mathematics. All fields of modern mathematics use the terminology of functions to express ideas and results. The notion of a function was not formulated overnight. It took a number of reformulations and generalizations before the modern concept of a function arose. To motivate the idea of a function, let's start with one of the older formulations. Suppose that a formula expresses the variable *y* in terms of the variable *x*. Then we say that *y* **is a function of** *x*. For example, consider the formula

$$y = 3x^2$$

For each value of *x*, this formula determines a corresponding value of *y*. For instance, if $x = 1$, the corresponding value of *y* is

$$y = 3 \cdot 1^2 = 3$$

We say that 3 is the **value of the function** at $x = 1$. It is customary to assign letters to denote functions. For instance, the above function can be denoted *f*, and we may write the function in the form:

$$y = f(x)$$

(read "*y* equals *f* of *x*"). This notation indicates that the value of *y* is determined by the value of *x* using the function *f*. We say that *x* is the **independent variable** and *y* is the **dependent variable**.

A function may have more than one independent variable. For example, the volume *V* of a right circular cylinder of radius *r* and height *h* is given by the formula

$$V = \pi r^2 h$$

This formula defines a function $C(r, h)$ of the two independent variables *r* and *h*. For given values of *r* and *h*, the value of the function is given by the formula

$$C(r, h) = \pi r^2 h \quad (r > 0, h > 0)$$

Note that the variables *r* and *h* are restricted to be positive because of the physical interpretation of the variables, as the radius and height of a cylinder, respectively.

The previous function *f* can be viewed as a correspondence that associates the real number *x* with the real number $3x^2$. In this view, the function is a device for associating elements of one set with elements of a second set. Namely, it associates with each *x* in **R** the number $3x^2$ in **R**. The function *C* associates with each point (r, h) in the first quadrant an element of **R**.

In many applications of mathematics, it is necessary to consider **correspondences** or **mappings** between sets that have nothing to do with numbers. For example, consider the set of all dots of light on a computer screen. Each dot of light is called a **pixel.** For each image portrayed on the screen, there is a correspondence associating with each pixel its color. We may visualize the correspondence as shown in Figure 1. On the left is the set P of pixels on the screen, and on the right is the set C of possible colors. The correspondence or mapping associates with each pixel p a color c. We give the correspondence the letter S as a name, and we say that S **associates** c **with** p and write

$$c = S(p)$$

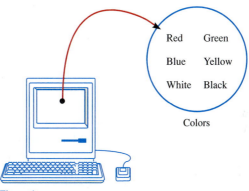

Figure 1

Pictorially, we indicate that c is associated with p by drawing an arrow from p to $S(p)$, as in Figure 2. We may view the correspondence S as a general "formula" that indicates, for each pixel p, the color c. To be sure, this "formula" is not an algebraic one like $y = 3x^2$. However, associating a color with a pixel is no different conceptually from associating the value $3x^2$ with x.

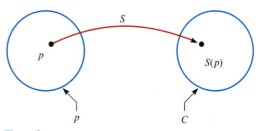

Figure 2

In a similar fashion, we may consider a correspondence F between the set of all people in your algebra class and the set of months of the year. The correspondence F associates with each person the month in which he or she was born. We might have, for example,

$$F(\text{Jane Smith}) = \text{December}$$

Again, the mapping F can be viewed as a rule that tells you, for each person in your algebra class, the month of birth.

The preceding examples of correspondences lead to the following definition.

| Definition 1
Function as a
Correspondence | Let A and B be sets. A **function** (or **mapping**) f from A to B is a correspondence that associates with each element of A one and only one element of B. If x is an element of A, then the element of B associated with x under the correspondence f is denoted $f(x)$, read "f of x," and is called the **value** of f at x. |

If f is a function from A to B, then we write

$$f : A \rightarrow B$$

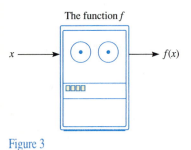

The function f

The set A is called the **domain** of f. That is, the domain of f consists of all x for which the function has a value defined. The subset of B consisting of all values assumed by f is called the **range** of f. If x is in the domain of f, we say that f is **defined at** x. Similarly, if x is not in the domain of f, we say that f **is not defined at** x.

You should view a function as a rule for associating to each x in A one and only one value $f(x)$ in B. You can also think of a function as a machine (say, a computer) into which you feed a value x from A. The machine outputs a unique value $f(x)$ from B. See Figure 3.

Figure 3

Applications of Functions

As we saw in Chapter 2, many applied problems give rise to equations involving variables x and y. Many of these equations can be written in an equivalent form in which y is expressed as a function $f(x)$, such as $y = x^2 + 1$.

Any such equation can be regarded as defining a function specifying the value of y in terms of the expression in x. The examples that follow will provide some practice in determining functions that arise in typical applied situations.

➤ **EXAMPLE 1**
Falling Object

A ball is thrown vertically upward from the roof of a skyscraper. Let $h(t)$ denote the height of the ball from the ground t seconds after it is thrown. Then, from the laws of physics, $h(t)$ is given by an expression of the form

$$h(t) = -16t^2 + v_0 t + h_0$$

where v_0 is the initial velocity of the ball and h_0 is the initial height of the ball. Suppose that the building is 1200 feet tall. Furthermore, suppose that an observer sees the ball pass by a window 800 feet from the ground 10 seconds after the ball was thrown. Determine a formula for the function $h(t)$.

Solution

We are given the data $h(0) = 1200$, the height of the ball when it was thrown (at time $t = 0$); and $h(10) = 800$, the height of the ball 10 seconds after it was thrown ($t = 10$). Inserting $t = 0$ into the expression for $h(t)$ gives us

$$h(0) = 1200 = -16(0)^2 + v_0(0) + h_0 = h_0$$

This gives us the value $h_0 = 1200$. Substituting $t = 10$ into the expression for $h(t)$ gives us

$$h(10) = 800 = -16(10)^2 + v_0(10) + 1200$$
$$1200 = 10v_0$$
$$v_0 = 120$$

So we can express h as a function of t as follows:

$$h(t) = -16t^2 + 120t + 1200$$

➤ **EXAMPLE 2**
Storage Tank

Large tanks to store industrial chemicals are built in a cylindrical shape 10 feet tall. The paint used for the tanks costs \$50 per gallon, and one gallon is sufficient to cover 200 square feet of the tank's surface. Calculate the cost $C(r)$ of painting a tank as a function of its radius r. (Both the top and the sides of the tank must be painted.)

Solution

Let h denote the height of the tank. (See Figure 4.) From the formulas for the area of a circle and the surface area of a cylinder, the area to be painted equals

$$\text{Area of top} + \text{Area of sides} = \pi r^2 + 2\pi r h$$

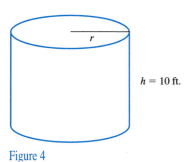

Substituting $h = 10$, we have

$$\text{Total area} = \pi r^2 + 2\pi r(10)$$
$$= \pi r^2 + 20\pi r$$

Figure 4

Since one gallon of paint covers 200 square feet, the number of gallons required to paint the entire cylinder is equal to the total area divided by 200:

$$\frac{\pi r^2 + 20\pi r}{200} = \frac{1}{200}\pi r^2 + \frac{1}{10}\pi r$$

We find the cost by multiplying this expression by 50 (the cost of one gallon) and arrive at an expression for the cost C in terms of the radius r:

$$C(r) = \frac{1}{4}\pi r^2 + 5\pi r$$

➤ **EXAMPLE 3**
Distance between Moving Objects

Two cars leave an intersection at the same time, proceeding in perpendicular directions. One car is moving at 45 miles per hour, and the other is moving at 30 miles per hour. Determine an expression for the function $D(t)$ that gives the distance between the two cars as a function of time.

Solution

In t hours, the first car goes $45t$ miles and the second car goes $30t$ miles. The positions of the cars at time t is shown in Figure 5. The distance between the two cars is given by the hypotenuse of a right triangle. So, by the Pythagorean theorem, we have

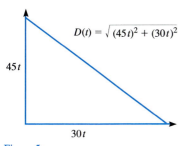

$$D(t) = \sqrt{(45t)^2 + (30t)^2}$$
$$= \sqrt{2025t^2 + 900t^2}$$
$$= \sqrt{(2025 + 900)t^2}$$
$$= \sqrt{2025 + 900} \cdot \sqrt{t^2}$$
$$= \sqrt{2925}\,t$$

Figure 5

since time t is considered to be nonnegative.

Examples of Functions, Domains, and Ranges

Let's now explore some of the simplest functions, namely, those that have real values and are specified by formulas involving a single variable.

Suppose that a and b are real constants. We may define the function

$$f(x) = ax + b$$

where x is a real number. Such a function is called a **linear function**. Similarly, a function of the form

$$f(x) = ax^2 + bx + c, \quad a \neq 0$$

is called a **quadratic function**. A function of the form

$$f(x) = ax^3 + bx^2 + cx + d, \quad a \neq 0$$

is called a **cubic function**. In general, a function in which $f(x)$ is specified as a polynomial expression in x is called a **polynomial function**.

➤ **EXAMPLE 4**
Calculating Domain and Range

Let f denote the linear function that associates with each real number x the real number $f(x) = 3x + 1$.

1. What is the domain of f?
2. Determine $f(2)$ and $f(-1)$.
3. Determine $f(a)$, $f(a + 1)$, and $f(2t - 1)$, where a and t are real numbers.
4. What is the range of f?

Solution

1. Since x may be any real number, the domain consists of the set of all real numbers **R**. Any value of x is a permissible substitution for x in the formula $3x + 1$.
2. We have

$$f(2) = 3(2) + 1 = 7$$
$$f(-1) = 3(-1) + 1 = -2$$

3. The function may be evaluated at any expression, as long as the value of the expression is contained within the domain. In this case, a and t are presumed to be real numbers. The given expressions are also real numbers and, hence, are in the domain. In each case, we substitute the given expression for x in the definition of the function. In the first case, we replace each occurrence of x by a:

$$f(a) = 3a + 1$$

In the second case, we replace each occurrence of x by $a + 1$:

$$f(a + 1) = 3(a + 1) + 1 = 3a + 4$$

In the third case, we replace each occurrence of x by $2t - 1$:

$$f(2t - 1) = 3(2t - 1) + 1 = 6t - 2$$

4. To determine the range, we must determine those real numbers c that are assumed as values of the function for some x, namely,

$$f(x) = c$$

This equation is equivalent to

$$3x + 1 = c$$

$$x = \frac{1}{3}(c - 1)$$

That is, for this value of x, the function f assumes the value c. Since x may be any real number, f assumes all real numbers as values, and its range is **R**, the set of all real numbers.

If we don't specify the domain of a function, we understand the domain to consist of all values of the variable for which the defining expression makes sense. For instance, since the expression $2x^3 - 1$ is defined for all real numbers x, the domain of the function

$$f(x) = 2x^3 - 1$$

consists of all real numbers x. Similarly, since the expression $1/x$ is defined for $x \neq 0$, the domain of the function

$$f(x) = \frac{1}{x}$$

consists of all nonzero real numbers x. Since the expression $\sqrt{x}$ is defined only for nonnegative real numbers, the domain of the function $f(x) = \sqrt{x}$ consists of the nonnegative real numbers x.

The domain is often specified in parentheses after the expression defining the function. For example, the preceding three functions may be defined as follows:

$$f(x) = 2x^3 - 1 \quad \text{(all real } x)$$

$$f(x) = \frac{1}{x} \quad (x \neq 0)$$

$$f(x) = \sqrt{x} \quad (x \geq 0)$$

In some cases, we wish to specify a domain that is smaller than the domain of the defining expression. This may be done by stating the conditions for defining the domain in parentheses after the defining expression. For instance, the function

$$f(x) = 3x \quad (3 < x \leq 5)$$

has domain the interval $(3, 5]$.

➤ **EXAMPLE 5**
Evaluating a Function

Suppose that we are given the real-valued function
$$f(x) = \sqrt{x - 1}$$

1. Determine the domain of f.
2. Determine the value of $f(10)$ and $f(-1)$.
3. Determine the value of $f(w^2)$.
4. Determine the range of f.

Solution

1. The expression $\sqrt{x - 1}$ produces a real number provided that
$$x - 1 \geq 0$$
$$x \geq 1$$

So the domain is

$$\{x : x \geq 1\} \quad \text{or} \quad [1, \infty)$$

2. For $f(10)$, we have
$$f(10) = \sqrt{10 - 1} = \sqrt{9} = 3$$
For $f(-1)$, we have
$$f(-1) = \sqrt{-1 - 1} = \sqrt{-2}$$
But this is a complex, nonreal number. Since we are considering only functions with real numbers as the domain and range, -1 is not in the domain, so $f(-1)$ is undefined.

3. We substitute w^2 for x in the definition of the function:
$$f(w^2) = \sqrt{w^2 - 1}$$
The expression on the right gives the value of f at w^2.

4. As x takes values in the domain of f, the expression $x - 1$ assumes as values all nonnegative real numbers, so $\sqrt{x - 1}$ assumes the same set of values as the square root function, namely, all nonnegative real numbers.

➤ **EXAMPLE 6**
Determining Domain
of a Quotient

Let h be the function defined by
$$h(x) = \frac{\sqrt{x}}{(x - 1)(x - 2)}$$

1. What is the domain of h?
2. Determine $h(4)$.

Solution

1. The expression on the right requires a number of restrictions on x for it to be defined. First, the radical in the numerator is defined only for nonnegative numbers, that is, for
$$\{x : x \geq 0\}$$
Also, the expression is not defined at x for which the denominator is 0. This requires the exclusions $x \neq 1, 2$. Combining these restrictions, we arrive at the domain of h:
$$\{x : x \geq 0 \quad \text{and} \quad x \neq 1, 2\}$$

2. We substitute the given value for x into the expression defining h:
$$h(4) = \frac{\sqrt{4}}{(4 - 1)(4 - 2)} = \frac{2}{3 \cdot 2} = \frac{1}{3}$$

In the next example, we evaluate an expression typical of those you will encounter when calculating derivatives in calculus.

➤ **EXAMPLE 7**
Calculus

Let $f(x) = x^2$. Evaluate and simplify the expression
$$\frac{f(a + h) - f(a)}{h}$$
where a and h are real numbers and $h \neq 0$.

Solution

Substituting in the given expression for $f(x)$, we have

$$f(a + h) = (a + h)^2 = a^2 + 2ah + h^2$$

Therefore, we have

$$\frac{f(a + h) - f(a)}{h} = \frac{(a^2 + 2ah + h^2) - a^2}{h}$$

$$= \frac{2ah + h^2}{h}$$

$$= \frac{h(2a + h)}{h}$$

$$= 2a + h$$

since $h \neq 0$

➤ **EXAMPLE 8**
Square Root of a Square

Let f be the function defined by

$$f(x) = \sqrt{x^2}$$

1. Determine the domain of f.
2. Find an equivalent expression for f.

Solution

1. For any real number x, the expression x^2 is always nonnegative, so the expression $\sqrt{x^2}$ produces a real number for any real number x.
2. If x is nonnegative, then $\sqrt{x^2} = x$. However, since $(-x)^2 = x^2$, we see that for negative x

$$\sqrt{x^2} = \sqrt{(-x)^2} = -x$$

Thus, we see that

$$\sqrt{x^2} = x \quad (x \geq 0)$$
$$= -x \quad (x < 0)$$

On the right side, we recognize the definition of absolute value. So we have

$$\sqrt{x^2} = |x|$$

as an equivalent expression for f. (See Properties of Radicals on page 20.)

$$\boxed{\sqrt{x^2} = |x|}$$

Exercises 3.4

For $f(x) = 4x - 5$, find each of the following.

1. $f(0)$
2. $f(-2)$
3. $f(-2a)$
4. $f(a + h)$

For $g(x) = 3x^2 + x - 1$, find each of the following.

5. $g(10)$
6. $g(-1)$

7. $g(a - h)$
8. $g(-4t)$

For $f(x) = |5 - x^2|$, find each of the following.

9. $f(0)$
10. $f(-4)$
11. $f(1.2)$
12. $f(a + h)$

For $g(x) = \dfrac{1}{x^2} + \dfrac{1}{x} - 1$, find each of the following, if possible.

13. $g(0)$ 14. $g(-2)$ 15. $g\left(\dfrac{1}{3}\right)$ 16. $g\left(\dfrac{1}{a}\right)$

For $g(x) = 2x^3 - 3x^2$, find each of the following.

17. $g(1)$ 18. $g(-2)$

19. $g(2 - h)$ 20. $g(a + h)$

For $f(x) = (1 + x)^3$, find each of the following.

21. $f(2)$ 22. $f(-1)$

23. $f(a + h)$ 24. $f\left(\dfrac{1}{t}\right)$

For $g(x) = (2x + 1)^2$, find each of the following.

25. $g(0)$ 26. $g\left(-\dfrac{1}{2}\right)$

27. $g(a + h)$ 28. $g(t + h)$

For $f(x) = 12$, find each of the following.

29. $f\left(-\dfrac{3}{2}\right)$ 30. $f(0)$

31. $f(a + h)$ 32. $f(t + h)$

Determine the domain and range of each function.

33. $f(x) = 3x + 7$ 34. $g(x) = 5 - 4x$

35. $f(x) = 12$ 36. $g(x) = -4$

37. $f(x) = \dfrac{1}{x - 3}$ 38. $f(x) = -\dfrac{2}{3x + 7}$

39. $g(x) = \sqrt{3 - 2x}$ 40. $g(x) = \sqrt{1 + 4x}$

41. $g(x) = x^3$ 42. $g(x) = 4x^3 + 5$

43. $f(x) = |x|$ 44. $g(x) = |5 - 3x|$

45. $g(x) = \sqrt{4 - x^2}$ 46. $g(x) = \sqrt{3 - x^2}$

47. $g(x) = \sqrt{x^2 - 5x + 6}$ 48. $f(x) = \sqrt{3 + 2x}$

For each function, evaluate and simplify the expressions

$$\dfrac{f(a + h) - f(a)}{h}, \quad \dfrac{f(a + h) - f(a - h)}{h}, \quad h \neq 0$$

49. $f(x) = 2x - 3$ 50. $f(x) = 4 + 5x$

51. $f(x) = x^2 - 5x + 7$ 52. $f(x) = x^3$

53. $f(x) = \sqrt{2x - 3}$ 54. $f(x) = x^{-2}$

55. $f(x) = -10$ 56. $f(x) = -x$

57. $f(x) = x^4$ 58. $f(x) = x^5$

Determine the domain of each function.

59. $f(x) = \dfrac{|x|}{x}$ 60. $f(x) = 2x - 4x^{1/2} + 2$

61. $f(x) = \dfrac{1}{x^2 + 3x + 2}$ 62. $g(x) = \dfrac{\sqrt{x}}{x^2 - 16}$

63. $g(x) = \sqrt{4x^2}$ 64. $f(x) = \sqrt{25x^2}$

65. $f(x) = \sqrt[3]{x}$ 66. $f(x) = \dfrac{x^2 - 1}{2x^2 - 3x + 1}$

67. $f(x) = \dfrac{|1 - x|}{x - 1}$ 68. $f(x) = x^{3/2}$

69. $f(x) = \dfrac{1}{|4x - 5|}$ 70. $f(x) = \dfrac{3}{5x + 9}$

71. $f(x) = \dfrac{1}{\sqrt{x^2 + 5}}$ 72. $f(x) = \sqrt{\dfrac{5x - 5}{3x - 3}}$

◆ Applications

73. Two cars leave an intersection at the same time, proceeding in perpendicular directions. One car is moving at 65 mph, and the other is moving at 55 mph. Let D denote the distance between the cars. Determine a formula in one variable for $D(t)$, where t is the time the cars travel.

74. From a thin piece of cardboard 20 in. by 20 in., square corners of length x are cut out so the sides can be folded up to make a box. Let V denote the volume of the box. Determine a formula in one variable for $V(x)$.

75. A container company is constructing an open-top, rectangular metal tank with a square base of length x that will have a volume of 300 cubic ft. Let S denote the surface area of the tank. Determine a formula in one variable for $S(x)$.

76. The owner of a 30-unit motel, by checking records of occupancy, knows that when the room rate is $50 a day, all units are occupied. For every increase of x dollars in the daily rate, x units will be left vacant. Each unit occupied costs $10 per day to service and maintain. Let R denote the total daily profit. Determine a formula in one variable for $R(x)$.

77. A Palladian window is a rectangle with a semicircle on top, as shown in the figure. Suppose the perimeter of a particular Palladian window is to be 28 ft. Let A denote the total area of the window. Determine a formula in one variable for $A(x)$.

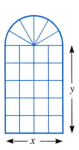

Exercise 77

78. Let A denote the area of a circle with diameter d. Determine a formula in one variable for $A(d)$.

79. A power line is to be constructed from a power station at point A to an island at point Q, which is 1 mile out in the water from a point B on the shore. (See the accompanying figure.) Point B is 3 miles downshore from the power station at A. It costs \$6000 per mile to lay the line under the water and \$4000 per mile to lay the line underground. Let C denote the total cost of the power line. Determine a formula for $C(x)$.

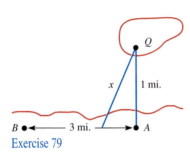

Exercise 79

80. A 28-in. piece of string is to be cut into two pieces. One piece is to be used to form a circle and the other to form a square. Let A denote the total area. Determine a formula for $A(x)$.

81. A ball is thrown vertically downward from the roof of a giant new skyscraper. Let S denote the height of the ball from the ground t seconds after being thrown. Then, $S(t)$ is given by an expression of the form

$$S(t) = -16t^2 + v_0 t + h$$

where v_0 represents the initial velocity and h represents the height of the building. Suppose that the building is 1800 feet tall. Furthermore, suppose that an observer sees the ball pass by a window 700 feet from the ground 8 seconds after the ball was thrown. Determine a formula for $S(t)$.

82. An airplane can travel 500 mph in still air. It travels 2080 miles with a tailwind of w mph and 1440 miles against the same tailwind. Let t denote the total time of travel. Find a function in one variable for $t(w)$.

✍ In Your Own Words

83. Think about the following relation that assigns values to the cells of the first column and first row in a spreadsheet:

$$f(x) = \begin{cases} rownumber & \text{if } x \text{ is in column A} \\ -1 & \text{if } x \text{ is in row 1} \end{cases}$$

Is this a function in the domain for which it is defined? Why or why not?

3.5 THE GRAPH OF A FUNCTION

Let $y = f(x)$ be a function with independent variable x and dependent variable y, where x and y both assume real values. The graph of the equation $y = f(x)$ is called the **graph of the function** f. This graph is generally a curve of some sort depicting the relationship between x and y.

A point (x, y) is on the graph precisely when x and y satisfy the equation—that is, provided $y = f(x)$. So the graph consists of the points

$$(x, f(x))$$

where x is in the domain of f. See Figure 1.

From the graph of a function, the function value $f(x)$ may be determined for any value of the independent variable x. Just start at x on the horizontal axis and proceed vertically upward or downward until you get to the graph. The y-coordinate of the corresponding point is the value of $f(x)$. Given the graph of a function, you know the function, in the sense that you can (at least approximately) determine the function value for any x. To put it succinctly: The graph defines the function. For instance, consider the graph of Figure 1 at the beginning of this chapter, which plots the number of people N who have heard a rumor (y-axis) against elapsed time t (t-axis). For each value of t, the graph specifies the number of people who had heard the rumor by time t. That is, the graph specifies N as a function of t.

The domain of a function may be visualized in terms of its graph as a subset of the x-axis. A point a is in the domain provided that there is a point on the graph directly above or below a. For example, in Figure 2, note that a_1 is in the domain but a_2 is not.

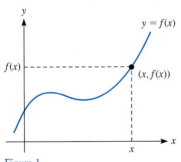

Figure 1

The range of a function may be visualized as a subset of the y-axis. A point b on the y-axis is in the range provided that there is a point on the graph at the same height as b. For example, in Figure 3, note that b_1 is in the range but b_2 is not.

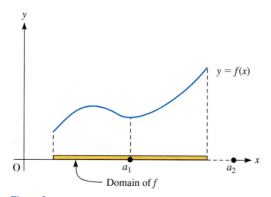

Figure 2

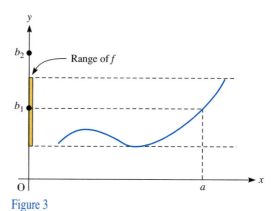

Figure 3

> **EXAMPLE 1**
> Determining Domain
> and Range from the Graph

Each of the following figures shows the graph of a function. From the graph, determine the domain and range of the corresponding function.

1.
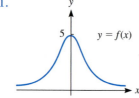

2.
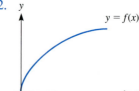

3.
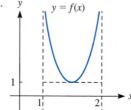

4.
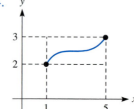

Solution

1. The domain consists of the values of x over which lie points on the graph. In this case, there is a point over every value of x on the x-axis, so the domain is the infinite interval $(-\infty, \infty)$. The range consists of all values y on the y-axis for which there are points on the graph at the same height. In this case, the range is the interval $(0, 5]$.

2. Domain $= [0, \infty)$, range $= [0, \infty)$

3. Domain $= (1, 2)$, range $= [1, \infty)$

4. Domain $= [1, 5]$, range $= [2, 3]$

➤ **EXAMPLE 2**
Graphing a Function

Graph the function

$$f(x) = x^2 \quad (-2 \le x \le 2)$$

From the graph determine the domain and range of f.

Solution

We graphed the equation

$$y = x^2$$

earlier in the chapter. In Figure 4, we have drawn this graph. The graph of the function consists of the points with x-coordinates satisfying the restriction $-2 \le x \le 2$. The domain and range are indicated on the x- and y-axes, respectively. The domain is the interval $[-2, 2]$, and the range is the interval $[0, 4]$.

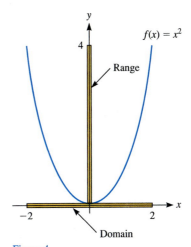

Figure 4

➤ **EXAMPLE 3**
Graphing

Sketch the graph of the function

$$f(x) = \sqrt{1 - x^2}$$

Solution

In order for $f(x)$ to be defined, the quantity within the radical must be nonnegative. That is,

$$1 - x^2 \ge 0$$
$$x^2 \le 1$$
$$\sqrt{x^2} \le \sqrt{1}$$
$$|x| \le 1$$
$$-1 \le x \le 1$$

So the domain is the set

$$\{x : -1 \le x \le 1\}$$

Next, we must graph the equation

$$y = \sqrt{1 - x^2}$$

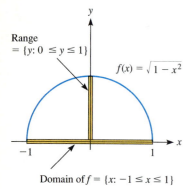

Range
= {y: 0 ≤ y ≤ 1}

$f(x) = \sqrt{1-x^2}$

Domain of f = {x: −1 ≤ x ≤ 1}

Figure 5

We square both sides of this equation to obtain

$$y^2 = 1 - x^2 \quad \text{and} \quad y \geq 0$$
$$x^2 + y^2 = 1 \quad \text{and} \quad y \geq 0$$

The last equation has as its graph a circle with center $(0, 0)$ and radius 1. The condition $y \geq 0$ restricts the graph to the upper semicircle. See Figure 5.

Let f be a function whose domain and range are subsets of the real numbers. With each x in the domain of f, the function associates a single real number $f(x)$. This means that on a graph of $f(x)$, for a given value of x there is only one point (x, y). To put this another way, a vertical line passing through x on the horizontal axis intersects the graph in exactly one point, namely, the point $(x, f(x))$. See Figure 6. Not *every* vertical line intersects the graph, of course.

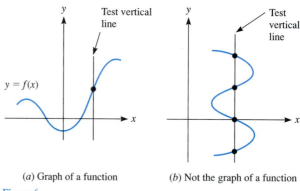

(a) Graph of a function   (b) Not the graph of a function

Figure 6

Vertical lines through various points on the x-axis tell us much about a graph. If x is not in the domain of a function f, then there is no point of the form (x, y) on the graph, and thus a vertical line through x will not intersect the graph. See Figure 6(a). Furthermore, if a vertical line passing through x intersects the graph in more than one point, the graph does not have a single y associated with x and is not the graph of a function. See Figure 6(b).

We summarize these observations in the following test, which allows us to determine whether a graph in the xy-plane is the graph of a function.

VERTICAL LINE TEST

A set of points in the xy-plane is the graph of a function if and only if each vertical line intersects the graph in at most one point.

➤ **EXAMPLE 4**
Using the Vertical Line Test

Determine which of the graphs in Figure 7 are the graphs of functions.

Solution

1. In Figure 8(a), any vertical line either does not intersect the graph or intersects it in a single point. This graph is the graph of a function by the vertical line test.

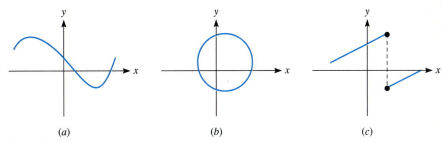

Figure 7

2. In Figure 8(*b*), we have drawn a vertical line that intersects the graph in more than one point. According to the vertical line test, the graph is not the graph of a function.

3. In Figure 8(*c*), we have drawn a vertical line that intersects the graph more than once. According to the vertical line test, the graph is not the graph of a function.

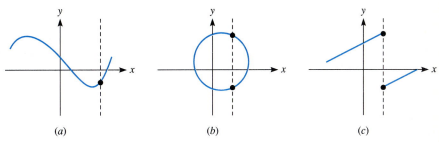

Figure 8

Horizontal and Vertical Translations and Scaling

Up to this point, we have described only one method for sketching the graphs of functions, namely, by plotting points. This method is tedious and produces an accurate graph only if you use enough points and take care that these points represent all of the various geometric features of the graph. There are a number of additional methods for sketching graphs of functions. Some of these belong to the province of algebra; others overlap into calculus. In the remainder of this section, we present several methods for sketching graphs using known graphs as a starting point.

The first method we present involves functions whose graphs are obtained by translating known graphs vertically or horizontally.

Suppose that $f(x)$ is a given function and that c is a real number. The graph of the function is the graph of the equation

$$y = f(x)$$

Suppose that we translate this graph vertically $|c|$ units (upward if $c > 0$ and downward otherwise). We have seen in Section 3.2 that the equation of the resulting graph can be obtained by replacing y by $y - c$. That is, the equation is

$$y - c = f(x)$$

Or, solving for y, we have the equation

$$y = f(x) + c$$

This equation represents a function, which we will call $g(x)$; that is,

$$g(x) = f(x) + c$$

which can be evaluated for each x by adding c to the value of $f(x)$. As we have just seen, the graph of $g(x)$ can be obtained by vertically shifting the graph of $f(x)$. See Figure 9. Rephrasing this geometric fact gives us the following result.

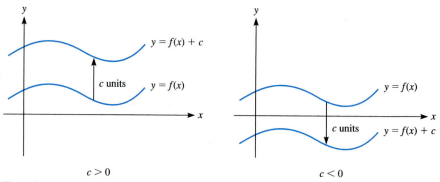

Figure 9

PRINCIPLE OF VERTICAL TRANSLATION

Let $f(x)$ be a function. Then the graph of the function g, defined by

$$g(x) = f(x) + c$$

is obtained by vertically translating the graph of $f(x)$ by c units. If $c > 0$, the shift is upward. If $c < 0$, the shift is downward.

➤ **EXAMPLE 5**
Applying Translation

Sketch the graph of the function

$$f(x) = \sqrt{1 - x^2} + 4$$

Solution

We have already seen (Example 3) that the graph of the function

$$g(x) = \sqrt{1 - x^2}$$

consists of the upper half of the circle of radius 1 centered at the origin. According to the principle of vertical shifting, the graph of the function $f(x)$ can be obtained by shifting the graph of $g(x)$ upward 4 units. See Figure 10.

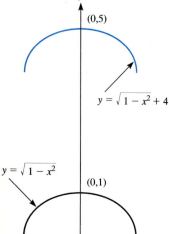

Figure 10

We can also sketch graphs by translating known graphs horizontally. Start with a function $f(x)$. Its graph is the graph of the equation

$$y = f(x)$$

As we have seen (Section 3.2), if we replace x by $x - c$, the resulting equation has a graph that can be obtained by shifting the graph of f. More precisely, we can state the following result.

PRINCIPLE OF HORIZONTAL TRANSLATION

Let $g(x)$ be a given function. Then the graph of the function $g(x - c)$ can be obtained by shifting the graph of $g(x)$ by c units horizontally. If $c > 0$, the shift is to the right. If $c < 0$, the shift is to the left. See Figure 11.

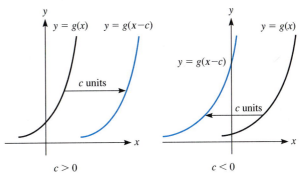

Figure 11

In addition to horizontal and vertical translation, we may apply **scaling** to known graphs. Suppose that $f(x)$ is a given function and that c is a positive number. Consider the graph of the function $cf(x)$. A typical point on this graph is

$$(x, cf(x))$$

where x is in the domain of f. This point can be obtained by replacing the point $(x, f(x))$ with a point with the same x-coordinate but with the y-coordinate scaled to be c times as large. See Figure 12.

Scaling results from multiplying the y-values of a function by a positive number. If the positive number is less than 1, then each y-value shrinks; in this case, the scaling is called **shrinking**. If the positive number is greater than 1, each y-value grows; in this case, the scaling is called **stretching**.

If we multiply a function f by -1, we obtain the function $-f$, whose graph is obtained by reflecting the graph of f in the x-axis. See Figure 13. More generally, if c is *any* number, then the graph of cf can be obtained by first scaling by a factor of $|c|$ and then, if $c < 0$, reflecting the resulting graph in the x-axis.

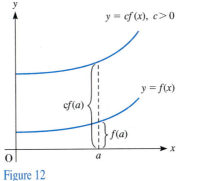

Figure 12   Figure 13

➤ **EXAMPLE 6**

Applying Translation and Scaling

Sketch the graphs of the following functions.

1. $f(x) = (x - 2)^2$ 2. $f(x) = (x + 3)^2$ 3. $f(x) = 2x^2$

4. $f(x) = -2x^2$ 5. $f(x) = (x - 1)^2 + 3$

Solution

1. By the principle of translation, the graph of $f(x)$ can be obtained by translating the graph of $g(x) = x^2$ to the right 2 units. (See Figure 14(a).)

2. By the principle of translating, the graph of $f(x)$ can be obtained by translating the graph of $g(x) = x^2$ to the left 3 units (since $c = -3$ is less than 0). See Figure 14(b).

3. The graph of $f(x) = 2x^2$ can be obtained by scaling the graph of $g(x) = x^2$ by 2. See Figure 14(c).

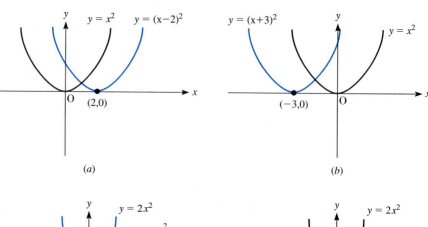

(a)   (b)

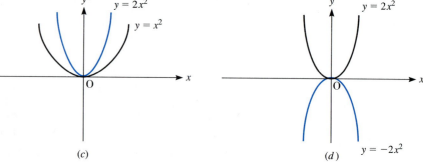

(c)   (d)

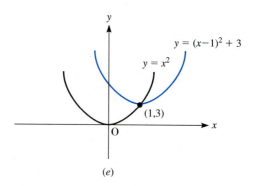

(e)

Figure 14

4. The graph of $f(x) = -2x^2$ can be obtained by first scaling the graph of $g(x) = x^2$ by $|-2| = 2$. Because the scaling factor (-2) is less than 0, we must then reflect the graph in the x-axis, yielding the graph in Figure 14(d).

5. The graph of $f(x) = (x - 1)^2 + 3$ can be obtained by translating the graph of $g(x) = x^2$ to the right 1 unit and 3 units upward. See Figure 14(e).

Increasing and Decreasing Functions

Up to this point in the section, we have concentrated on developing techniques for sketching graphs of functions. In many applied problems, however, we are given graphs of functions and must describe their features and interpret these features in terms of the applications. One of the most important features is whether a graph is increasing or decreasing for values of x in a particular interval. Specifically, we say that a function $f(x)$ is **increasing on the interval** (a, b) provided that the values of $f(x)$ increase as x moves from left to right through the interval, that is, provided $f(x_1) < f(x_2)$ whenever $a < x_1 < x_2 < b$. Figure 15 shows a graph that is increasing on the interval (a, b). Similarly, we say that $f(x)$ is **decreasing on the interval** (a, b) provided that the values of $f(x)$ are decreasing as x moves from left to right through the interval, that is, provided $f(x_1) > f(x_2)$ whenever $a < x_1 < x_2 < b$. Figure 16 shows a graph of a function that is decreasing on the interval (a, b).

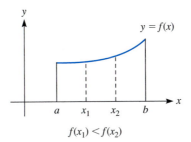

Figure 15

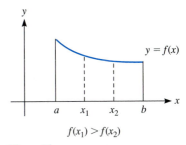

Figure 16

> **EXAMPLE 7**
Income Function

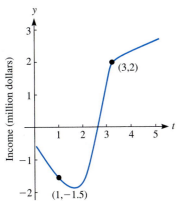

Figure 17

Figure 17 shows the graph of a function that represents the income generated by a young biotechnology firm t years after its founding.

1. Determine the increase in revenue from year 1 to year 3.

2. In what intervals is the graph increasing, and in what intervals is it decreasing?

Solution

1. Reading the graph, we see that the ends of years 1 and 3 correspond, respectively, to the points $(1, -1.5)$ and $(3, 2)$. The change from year 1 to year 3 is given by the difference in y-coordinates, namely, $2 - (-1.5) = 3.5$. The positive sign of the difference indicates that revenue increased from year 1 to year 3, and the amount of increase was \$3.5 million.

2. Examining the graph, we see that it is decreasing in the interval $(0, 2)$ and increasing in the interval $(2, 5)$.

Symmetry of Graphs

Early in the chapter, we introduced the notion of symmetry of the graph of an equation with respect to either the x- or the y-axis. Let's now connect that discussion with symmetry in the case of graphs of functions.

▌Definition 1
Even and Odd Functions

Let f be a function. We say that f is an **even function** provided that

$$f(x) = f(-x)$$

for all x in the domain of f. We say that f is an **odd function** provided that

$$f(-x) = -f(x)$$

for all x in the domain of f.

➤ **EXAMPLE 8**
Evenness and Oddness

Determine which of the following functions are even, which are odd, and which are neither.

1. $f(x) = x^3 - 2x$ 2. $f(x) = \dfrac{1}{x^2 + 1}$

3. $f(x) = x^2 + 3$ 4. $f(x) = (x + 1)^2$

Solution

1. We substitute $-x$ for x in the expression for f:

$$f(-x) = (-x)^3 - 2(-x)$$
$$= -x^3 + 2x$$
$$= -\left(x^3 - 2x\right)$$

We see that this last expression equals $-f(x)$. So f is an odd function.

2. Again we substitute $-x$ for x:

$$f(-x) = \frac{1}{(-x)^2 + 1} = \frac{1}{x^2 + 1}$$

Since the last expression is the same as the expression for $f(x)$, $f(x)$ is an even function.

3. Again we substitute $-x$ for x in the expression for f:

$$f(-x) = (-x)^2 + 3 = x^2 + 3$$

But the last expression is just the expression for $f(x)$. In this case, $f(x)$ is an even function.

4. Proceeding as in the previous parts, we have

$$f(-x) = (-x + 1)^2 = x^2 - 2x + 1$$

Since

$$f(x) = (x + 1)^2 = x^2 + 2x + 1$$

we see that $f(-x)$ is equal to neither $f(x)$ nor $-f(x)$ for all x in the domain of f. Thus, f is neither even nor odd.

Suppose $f(x)$ is an even function. If $(x, f(x))$ is a point on the graph of f, then so is $(-x, f(-x)) = (-x, f(x))$. These two points are located symmetri-

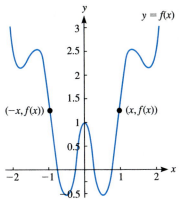

Figure 18

cally with respect to the y-axis. This is illustrated in Figure 18. In other words, an even function has a graph that is symmetric with respect to the y-axis.

As examples of graphs of even functions, consider the functions in parts 2 and 3 of Example 8. Their graphs are symmetric with respect to the y-axis as shown in Figures 19 and 20.

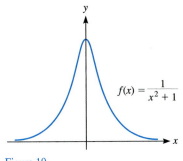

Figure 19

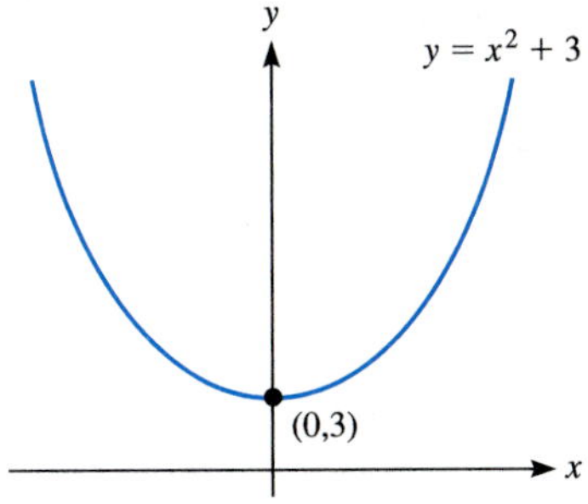

Figure 20

Suppose f is an odd function. Then for each x in the domain of f, the points $(x, f(x))$ and $(-x, -f(x))$ are on the graph. These two points are located symmetrically with respect to the origin. See Figure 21.

As an example of the graph of an odd function, consider the function f defined in part 1 of Example 8. Its graph is illustrated in Figure 22.

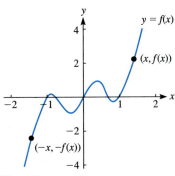

Figure 21

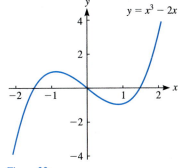

Figure 22

Piecewise-Defined Graphs

All of the functions that we have considered so far have been defined using a single formula. However, it is possible to define a function using several formulas, one for each of several parts of the domain. For example, consider the function

$$f(x) = \begin{cases} x & (x < 0) \\ -3x & (x \geq 0) \end{cases}$$

For $x < 0$, we use the first formula to determine $f(x)$; for $x \geq 0$, we use the second formula. For instance, for $x = -3$ and $x = 2$, we have

$$f(-3) = -3, \qquad f(2) = -3(2) = -6$$

A function that is defined using different formulas is said to be **piecewise defined**. The graph of a piecewise-defined function is drawn by separately sketching the graph corresponding to each formula. For example, the graph of the preceding function is shown in Figure 23.

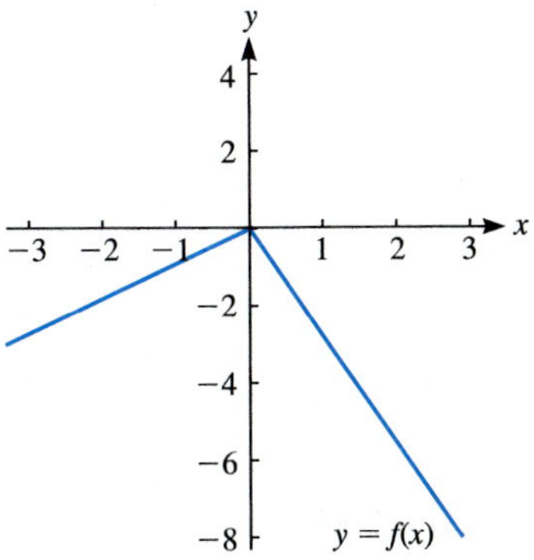

Figure 23

Piecewise-defined graphs are often used in applications, as the next example illustrates.

> **EXAMPLE 9**
Engineering

The voltage on a certain pin of a semiconductor device is equal to either $+5$ volts or -5 volts. Let $f(t)$ denote the voltage on the pin at time t, where $t = 0$ corresponds to the beginning of a certain circuit test. Graph the function f assuming that the voltage is initially -5 volts for $t < 0$, equal to $+5$ volts for t between 0 and 2 milliseconds inclusive, and equal to -5 volts thereafter.

1. Write formulas defining $f(t)$.
2. Sketch the graph of $f(t)$.

Solution

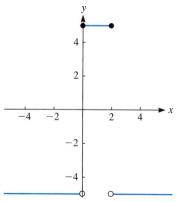

Figure 24

1. Using the data given, $f(t)$ is piecewise defined, requiring three defining formulas, as follows:

$$f(t) = \begin{cases} -5 & (t < 0) \\ 5 & (0 \le t \le 2) \\ -5 & (t > 2) \end{cases}$$

2. We sketch the graph of $f(t)$ by separately considering the function over the intervals

$$t < 0, \qquad 0 \le t \le 2, \qquad t > 2$$

In each interval, we use the appropriate definition of $f(t)$ to produce the graph. The result is the graph shown in Figure 24.

A piecewise-defined function that appears often in applied problems is the **greatest integer function**, denoted by $[x]$. This function is defined for all real numbers x, and its value is given by

$$[x] = \text{greatest integer} \le x$$

Here are the values of the greatest integer function for selected values of x:

$$[5.1] = 5, \qquad [0.17] = 0, \qquad [111] = 111, \qquad [-1.1] = -2$$

Note that if $0 \le x < 1$, the largest integer less than or equal to x is 0, so that

$$[x] = 0 \qquad (0 \le x < 1)$$

For $1 \le x < 2$, the largest integer less than or equal to x is 1, so that

$$[x] = 1 \qquad (1 \le x < 2)$$

Similarly, for any nonnegative integer n, if $n \le x < n+1$ the largest integer less than or equal to x is n, so that

$$[x] = n \qquad (n \le x < n + 1)$$

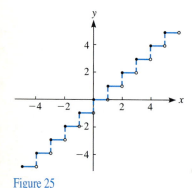

Figure 25

The greatest integer function has a separate formula for each interval $n \le x < n + 1$. The graph of the function for x in this interval is a constant. The total graph resembles a series of steps, with a jump occurring at each integer. See Figure 25.

> **EXAMPLE 10**
Cost Function

A cargo service charges by the weight of a package. It charges a flat fee of $2 plus $1 for each pound or fraction thereof. Determine a formula for the cost $C(x)$ of sending a package weighing x pounds.

Solution

The function $C(x)$ is defined for $x > 0$. The cost of \$1 for each pound or fraction thereof can be expressed in terms of the greatest integer function as

$$[x + 1]$$

dollars. Indeed, for amounts less than 1 pound, $x + 1$ is at least 1 and less than 2, so that $[x + 1] = 1$; for weights at least 1 pound but less than 2 pounds, $x + 1$ is at least 2 and less than 3, so $[x + 1] = 2$; and so forth. The function $[x + 1]$ counts 1 for each whole pound. Adding in the flat fee, the desired cost function is given by

$$C(x) = 2 + [x + 1]$$

Exercises 3.5

Determine whether each of the following is the graph of a function.

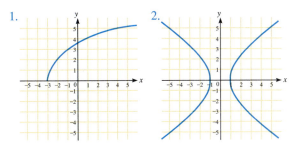

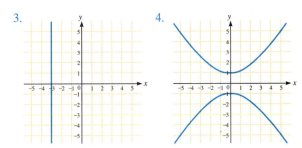

For each graph of a function, find $f(-2)$, $f(0)$, $f(3)$, and $f(5)$.

5.
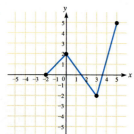

6.
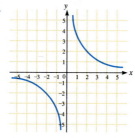

7.
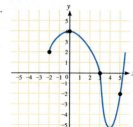

8.
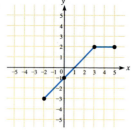

Determine the domain and range of the following functions.

9. See Exercise 5.

10. See Exercise 6.

11. See Exercise 7.

12. See Exercise 8.

Graph each function and from the graph determine the domain and range.

13. $f(x) = 4 - 8x$ $(-1 \le x \le 1)$

14. $f(x) = 2x - 5$ $(-3 \le x \le 3)$

15. $g(x) = x^2$ $(-1 \le x \le 1)$

16. $g(x) = x^2 - 3$ $(-2 \le x \le 2)$

17. $f(x) = x^2 - 3$

18. $g(x) = x^3 - 1$

19. $g(x) = x^3 - 1$ $(-2 \le x \le 2)$

20. $g(x) = 5 - x^2$ $(-3 \le x \le 3)$

21. $f(x) = \sqrt{4 - x^2}$ $(-2 \le x \le 2)$

22. $f(x) = \sqrt{-9 + x^2}$ $(|x| \ge 3)$

23. $f(x) = \dfrac{1}{x}$ $(x > 0)$

24. $f(x) = (x - 1)^2$ $(-1 \le x \le 3)$

Consider the following graph of a function $y = f(x)$ for Exercises 25–34.

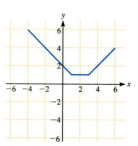

Exercises 25–34

Graph each of the following.

25. $y = f(x) + 1$

26. $y = f(x) - 1$

27. $y = f(x - 1)$

28. $y = f(x + 3)$

29. $y = f(x + 1) - 4$

30. $y = f(x - 2) + 3$

31. $y = -2f(x - 2) + 2$

32. $y = 2f(x) - 5$

33. $y = -\dfrac{1}{2}f(x) + 1$

34. $2y - 6 = \dfrac{1}{2}f(x - 4)$

35. a. Graph $f(x) = \sqrt{x}$, and then use that graph to graph parts (b)–(f).

b. $g(x) = \sqrt{x - 3}$

c. $g(x) = 2 + \sqrt{x + 1}$

d. $g(x) = 4\sqrt{x}$

e. $g(x) = -\sqrt{2x}$

f. $g(x) = -1 - \dfrac{1}{2}\sqrt{x - 1}$

36. a. Graph $y = |x|$, and then use that graph to graph parts (b)–(f).

b. $y = |x + 3|$

c. $y = |x - 1| - 2$

d. $y = |-2x|$

e. $y = -\dfrac{1}{2}|x| + 2$

f. $2y - 6 = |x + 1|$

Determine whether each graph is symmetric with respect to the x-axis or the y-axis.

37. $x^2 + y^2 = 4$

38. $(x - 4)^2 - (y - 2)^2 = 25$

39. $x = y$

40. $y = |x + 2| + |x - 2|$

41. $3x + 3y = 5$

42. $3x^4 - y = 10$

43. $x + 3 = |y|$

44. $|x| + 8 = 2y$

Determine whether each graph is symmetric with respect to the origin.

45. $x^2 - 3y^2 = 4$

46. $(x + 3)^2 + (y - 5)^2 = 16$

47. $2y = \dfrac{5}{3x}$ 48. $3x^2 - 7y^2 = 8x$

49. $\dfrac{1}{x} + \dfrac{2}{y} = 3$ 50. $|y| = |-2x|$

51. $y = x^7 + x^5 + x^3 + x$ 52. $y = x^6 + x^4 + x^2 + 5$

Determine whether each of the following functions is even, odd, or neither.

53. $f(x) = 3x - 2$ 54. $f(x) = -5x$

55. $f(x) = 2x^3 - 2x^2 + x$ 56. $f(x) = -2x^2$

57. $f(x) = \dfrac{3}{x^2}$ 58. $f(x) = \dfrac{5}{x^3 - x}$

59. $f(x) = (x - 3)^2$ 60. $f(x) = x^2 + 2x - 4$

61. $f(x) = \left(\dfrac{1}{x}\right)^{-3}$ 62. $f(x) = -\left|\dfrac{2}{3}x\right|$

63. $f(x) = 4x^{2/3}$ 64. $f(x) = (x^2 + 8)^2$

Graph each piecewise-defined function.

65. $f(x) = \begin{cases} x & (x \le 1) \\ -x & (x > 1) \end{cases}$

66. $f(x) = \begin{cases} -1 & (x < 0) \\ 0 & (x = 0) \\ 1 & (x > 0) \end{cases}$

67. $f(x) = \begin{cases} x - 1 & \text{if } x < -3 \\ x^2 & \text{if } -3 \le x \le 3 \\ 1 - x & \text{if } x > 3 \end{cases}$

68. $f(x) = \begin{cases} |x| & (x < 2) \\ -x^2 & (2 \le x < 3) \\ 2x - 1 & (x \ge 3) \end{cases}$

69. $f(x) = \begin{cases} \dfrac{x^2 - 9}{x - 3} & (x \ne 3) \\ 2 & (x = 3) \end{cases}$

70. $f(x) = \begin{cases} \dfrac{x^2 - 9}{x + 3} & (x \ne -3) \\ 6 & (x = -3) \end{cases}$

71. $f(x) = \begin{cases} -x - 3 & (x \le -2) \\ 3 - x^2 & (-2 < x < 2) \\ x - 3 & (x \ge 2) \end{cases}$

72. $f(x) = \begin{cases} -3 & (x \text{ not an integer}) \\ 3 & (x \text{ an integer}) \end{cases}$

♦ Applications

73. A parking garage charges $1.00 for the first hour or part of an hour, and $0.50 for each hour or part of an hour thereafter, up to a maximum of $5.00 per day. Graph this function for 0 to 10 hours.

74. It costs $20 to connect to a mainframe computer and $20 per minute or part of a minute of CPU time for the first 30 minutes. After that the cost drops to $10 per minute or part of a minute. Graph this function for the use of 0 to 1 hour of CPU time.

The greatest integer function is given by $f(x) = [x]$, or, in computer language, $f(x) = \text{INT}(x)$. Use the graph of this function to construct the following graphs.

75. $g(x) = [x - 1]$

76. $g(x) = 2[x]$

77. $g(x) = \text{INT}(x)$

78. $f(x) = 3\text{INT}(x + 2) - 1$

79. We define a *decimal place function* as follows: Given a number x, find its decimal notation. Then $f(x)$ is the digit in the fourth decimal place. For example,

$$f\left(\frac{2}{5}\right) = f(0.400000) = 0$$

$$f\left(\frac{18}{19}\right) = f(0.947368\dots) = 3$$

$$f\left(-\frac{39}{14}\right) = f(-2.785714\dots) = 7$$

and so on.

  a. Find $f(3/4)$, $f(14/23)$, $f(23/13)$, and $f(\sqrt{2})$.

  b. Determine the domain and range of the function.

  c. Describe the graph of this function.

✍ In Your Own Words

80. For what values of c is the equation $ax + by = c$ symmetric with respect to the origin?

81. For what values of a, b, and c is the equation $ax + by = c$ symmetric with respect to the x-axis? To the y-axis?

➡ Technology

Use a graphing calculator to determine the domain and range of each of the following functions.

82. $f(x) = \sqrt{1 - 4x^2}$ 83. $f(x) = \dfrac{x}{x^2 + 1}$

84. $f(x) = |x(x + 1)|$ 85. $f(x) = \sqrt{x + 9}$

3.6 OPERATIONS ON FUNCTIONS

It is possible to perform algebraic operations on functions similar to the operations we discussed for algebraic expressions, namely, addition, subtraction, multiplication, and division. We can also perform a totally different operation called *composition of functions*. In this section, we introduce these operations and indicate some of their applications.

Sums and Differences of Functions

Let f and g be functions with domains A and B, respectively. We define the sum of f and g, denoted $f + g$, to be the function such that

$$(f + g)(x) = f(x) + g(x)$$

That is, the value of $f + g$ at x is obtained by adding the values of f and g at x. The domain of $f + g$ consists of those x that belong to both A and B. That is, the domain of $f + g$ is the intersection

$$A \cap B$$

➤ **EXAMPLE 1**
Domain of a Sum

Suppose that

$$f(x) = \frac{x}{x - 1}, \qquad g(x) = \frac{2}{x - 2}$$

Determine the domain of the sum $f + g$.

Solution

The domains of f and g are given by

$$\text{Domain}(f) = \{ x : x \neq 1 \}, \qquad \text{Domain}(g) = \{ x : x \neq 2 \}$$

Therefore,

$$\text{Domain}(f + g) = \{ x : x \neq 1, 2 \}$$

The graph of $f + g$ can be described simply in terms of the separate graphs of f and g. For each x in the domain of $f + g$, the height of the point

$$(x, (f + g)(x))$$

above the x-axis is obtained by adding the corresponding y-coordinates of the graphs of f and g, namely, $f(x)$ and $g(x)$. See Figure 1.

Parallel to our definition for sum, we define the difference of f and g by the formula

$$(f - g)(x) = f(x) - g(x)$$

That is, the value of the difference $f - g$ at x equals the value of f at x minus the value of g at x. The domain of $f - g$ consists of those values x that belong to both the domain of f and the domain of g.

Sums and differences of functions appear often in applications. For example, suppose that $R(x)$ and $C(x)$ denote the revenue and cost, respectively, of producing x bushels of corn for a certain agricultural corporation. Then the difference

$$R(x) - C(x)$$

represents the profit from producing x bushels of corn.

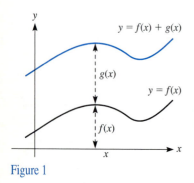

Figure 1

Products and Quotients of Functions

The product of f and g, denoted $f g$, is the function defined by the formula

$$(f g)(x)$$

As with the sum and the difference, the product is defined for values of x that are in the domains of both f and g. That is, the domain of $f g$ is the intersection

$$A \cap B$$

where A and B are the domains of f and g, respectively.

➤ **EXAMPLE 2**
Domain of Product

Suppose we have the functions

$$f(x) = x + 1, \qquad g(x) = x - 1$$

Determine the domain of and an expression for the product $f g$.

Solution

Each of the functions f and g has as its domain the set **R** of all real numbers, so $f g$ also has as its domain the set **R** of real numbers. Moreover,

$$(f g)(x) = (x + 1)(x - 1) = x^2 - 1$$

➤ **EXAMPLE 3**
Product of Two Functions

Suppose we have the functions

$$f(x) = \sqrt{x}, \qquad g(x) = \frac{x - 1}{\sqrt{x}}$$

Determine the domain of and an expression for the product $f g$.

Solution

We have

$$\text{Domain}(f) = \{ x : x \geq 0 \}, \qquad \text{Domain}(g) = \{ x : x > 0 \}$$

The domain of $f g$ is therefore $\{ x : x > 0 \}$, and

$$(f g)(x) = (\sqrt{x}) \left(\frac{x - 1}{\sqrt{x}} \right) = x - 1$$

Note that even though after simplification the expression for the product is defined for all real numbers x, the product $f g$ is defined only for positive values of x.

A very common mistake is to compute a product, simplify it, and then substitute a value of x that is not in the domain of the product. This sort of error can lead to mathematical nonsense, such as statements of the form $0 = 1$. To avoid such an error, be careful to note the domains of any functions you compute.

The quotient of f divided by g, denoted f / g, is the function defined by the formula

$$(f/g)(x) = \frac{f(x)}{g(x)}$$

For the expression on the right to be defined, x must be in both the domain of f and the domain of g, and $g(x)$ must be nonzero. (Otherwise division by 0 would result.)

➤ **EXAMPLE 4**
Quotient of Two Functions

Suppose

$$f(x) = \frac{x^2 - 1}{x}, \qquad g(x) = \frac{x + 1}{\sqrt{x}}$$

Determine the domain of and an expression for the quotient f/g.

Solution

We have

$$\text{Domain}(f) = \{x : x \neq 0\}, \qquad \text{Domain}(g) = \{x : x > 0\}$$

and we have

$$(f/g)(x) = \frac{\dfrac{x^2 - 1}{x}}{\dfrac{x + 1}{\sqrt{x}}}$$

$$= \frac{x^2 - 1}{x} \cdot \frac{\sqrt{x}}{x + 1}$$

$$= \frac{x - 1}{\sqrt{x}}$$

The quotient f/g has domain

$$\{x : x > 0\}$$

➤ **EXAMPLE 5**
Calculating a Quotient
Function

Suppose

$$f(x) = \frac{x}{x^2 + 1}, \qquad g(x) = (x - 1)(x - 2)$$

Determine an expression for the quotient f/g. What is its domain?

Solution

Here

$$\text{Domain}(f) = \mathbf{R}, \text{Domain}(g) = \mathbf{R}$$

The quotient is defined by the formula

$$\left(\frac{f}{g}\right)(x) = \frac{x}{(x^2 + 1)(x - 1)(x - 2)}$$

Moreover, $g(x)$ is equal to 0 for $x = 1, 2$. These values must be excluded from the domain of the quotient, which is therefore given by

$$\text{Domain}\left(\frac{f}{g}\right) = \{x : x \neq 1, 2\}$$

In working with quotients, you must exercise the same care as in working with products with regard to the domain. Even though you may be able to simplify the quotient to an expression with a larger domain, the quotient is defined only for the domain consisting of values common to the domains of the numerator and denominator and for which the denominator is nonzero.

Composition of Functions

In many applications, it is necessary to substitute an expression for the variable of a function $g(x)$. Such substitutions are conveniently described in terms of **composition of functions**, defined as follows:

Definition 1 **Composition of Functions**	Suppose that f and g are functions. The composite $$g \circ f$$ is the function defined by the formula $$(g \circ f)(x) = g(f(x))$$ That is, to compute the value of the composite at x, we first evaluate f at x and then use that value to evaluate g at $f(x)$. See Figure 2.

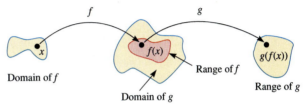

Figure 2

The machine concept of functions gives us a convenient way of viewing a **composite**. Suppose that f is viewed as one machine and g as a second. Then the machine corresponding to the composite is a serial combination of the first machine followed by the second.

Let's now determine the domain of a composite. Suppose that x is a value in the domain of $f(x)$. For the value $g(f(x))$ to make sense, the value $f(x)$ must be contained in the domain of g. This is necessary if we are to be able to evaluate $g(z)$ at the value $z = f(x)$. In other words,

$$\text{Domain}(g \circ f) = \{ x : f(x) \text{ is in Domain}(g) \}$$

Note that the domain of $g \circ f$ is a subset of the domain of f.

In forming composites, it helps to think in terms of an "inner function" and an "outer function." In the expression

$$g(f(x))$$

$f(x)$ is the inner function, and $g(x)$ is the outer function.

➤ **EXAMPLE 6**
Calculating a Composite

Suppose

$$f(x) = x + 1, \qquad g(x) = x^2$$

Determine the composite $g \circ f$.

Solution

To form the composite of g by f, we substitute the expression for f wherever x appears in the expression for g. We have $g(x) = x^2$. Therefore, to obtain an expression for the composite, we replace x by $x + 1$ in the expression for g. We have

$$(g \circ f)(x) = g\,(f(x))$$
$$= g\,(x + 1)$$
$$= (x + 1)^2$$

➤ **EXAMPLE 7**
Calculating a Composite

Suppose that

$$f(x) = \frac{1}{\sqrt{x}}, \qquad g(x) = x^2 + 1$$

Determine the composite of g by f.

Solution

The expression for the composite of g by f is obtained by substituting $f(x)$ for x in the expression for g:

$$(g \circ f)(x) = g(f(x)) = g\left(\frac{1}{\sqrt{x}}\right)$$

$$= \left(\frac{1}{\sqrt{x}}\right)^2 + 1$$

$$= \frac{1}{x} + 1$$

Note that the order of forming composites is important. For instance, if we let f and g be as in the last example, we may form the composite of f by g by substituting $g(x)$ for x in $f(x)$, with the following result:

$$(f \circ g)(x) = f(g(x)) = f(x^2 + 1) = \frac{1}{\sqrt{x^2 + 1}}$$

Thus, in this case, $f \circ g$ is not equal to $g \circ f$.

➤ **EXAMPLE 8**
Domain of a Composite

Suppose that

$$f(x) = x^3 - 1, \qquad g(x) = \sqrt{x} - 1$$

1. Determine the composite

$$g \circ f$$

2. What is the domain of the composite?

Solution

1. To compute the composite, we replace x by $f(x)$ in the expression for $g(x)$:

$$f(x) = x^3 - 1$$
$$(g \circ f)(x) = g\,(f(x)) = g(x^3 - 1)$$
$$= \sqrt{x^3 - 1} - 1$$

2. To determine the domain of the composite, the value of $f(x)$ must be in the domain of g, which in this case consists of the nonnegative real numbers. The domain of the composite thus consists of all x for which

$$x^3 - 1 \geq 0$$
$$x^3 \geq 1$$
$$x \geq 1$$

That is, the domain of the composite is

$$\{x : x \geq 1\}$$

➤ **EXAMPLE 9**
Writing a Function
as a Composite

Let h be the function defined by

$$h(x) = \sqrt[3]{x^2 + 5}$$

Express h as the composite of two functions.

Solution

Suppose that we define

$$g(x) = \sqrt[3]{x}, \qquad f(x) = x^2 + 5$$

Then we have

$$g(f(x)) = h(x)$$

In other words, h is the composite of g by f. Note that there is no unique way to express a function as a composite. For example, there are other functions we can use instead of f and g as we defined them here, such as $g(x) = \sqrt[3]{(x + 5)}$ and $f(x) = x^2$. However, the functions we chose were the "obvious" ones that do the job.

Operations
on Functions
Using Graphing
Technology

If you are using a graphing calculator or graphing software, you will find that most models can graph sums, differences, products, and quotients of functions as well as the composition of functions.

For example, suppose you are using the TI-81 and wish to graph the functions $f(x) = 2x$ and $g(x) = x^2$ and their sum $f(x) + g(x)$. One way to do this would be to enter each of these as separate functions, graph them, and then add them together algebraically and graph the result. To carry out the details, begin by setting an appropriate range for the graph. Then using the **Y=** key, enter $2x$ in Y_1 and x^2 in Y_2. Graph each of the functions separately as in Figure 3. You can see

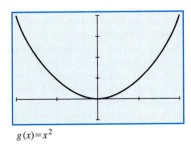

$g(x) = x^2$

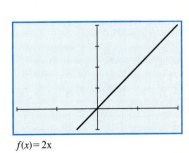

$f(x) = 2x$

Figure 3

the graphs one at a time as follows: To turn off the graph of one equation, display the list of equations, position the cursor over the equal sign of the equation, and press **ENTER**. Using this procedure, you may view the graphs one at a time. By repeating the procedure, you may redisplay a graph.

Next, let's graph the sum. Return to the **Y=** menu. Use the **2nd** [Y − VAL] key to enter $Y_1 + Y_2$ for Y_3. Then graph Y_3 without graphing the other two equations. The graph should look like Figure 4.

The method for graphing a composite function on the TI-81 is similar. Consider the functions

$$f(x) = \frac{1}{\sqrt{x}} \text{ and } g(x) = x^2 + 1$$

as given in Example 7. The composite function we are looking for is $g(f(x))$. To graph this, first enter $f(x)$ in Y_1 in the **Y=** menu. Now move to Y_2 and enter $Y_1{}^2 + 1$. Then graph Y_2. The results should resemble Figure 5.

The power of a graphing utility is the ease with which we can graph many functions. For instance, by entering $g(x)$ in Y_1 and $1/\sqrt{Y_1}$ in Y_2 and graphing Y_2, we have the graph of $f(g(x))$ as shown in Figure 6. The statement in Example 7 that $f(g(x)) \neq g(f(x))$ is clearly true by looking at the two graphs.

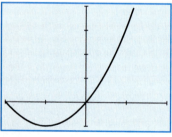

Figure 4

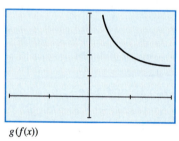

$g(f(x))$

Figure 5

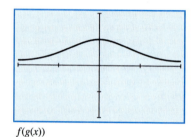

$f(g(x))$

Figure 6 ◀

Exercises 3.6

For each of the following functions find the domain of $f, g, f + g, f - g, fg, ff, f/g, g/f, f \circ g$, and $g \circ f$. Then find $(f + g)(x), (f - g)(x), (fg)(x), (ff)(x), (f/g)(x), (g/f)(x), (f \circ g)(x)$, and $(g \circ f)(x)$.

1. $f(x) = x - 4, g(x) = x + 5$

2. $f(x) = x^2 - 4, g(x) = 2x + 3$

3. $f(x) = 2x^2 + x - 3, g(x) = x^3$

4. $f(x) = \sqrt{x}, g(x) = x^2$

Let $f(x) = x^2 - 1$ and $g(x) = 2x + 3$. Find each of the following.

5. $(f - g)(3)$

6. $(f + g)(-1)$

7. $(f - g)(x)$

8. $(f + g)(x)$

9. $(fg)(3)$

10. $\left(\dfrac{f}{g}\right)(-1)$

11. $\left(\dfrac{g}{f}\right)(-1)$

12. $\left(\dfrac{f}{g}\right)(-1.5)$

13. $(fg)(x)$

14. $\left(\dfrac{f}{g}\right)(x)$

15. $\left(\dfrac{g}{f}\right)(x)$

16. $(f \circ f)(x)$

17. $(f \circ g)(x)$

18. $(f \circ f \circ f)(x)$

19. $(g \circ g)(x)$

20. $(g \circ f)(x)$

Find $(f \circ g)(x)$ and $(g \circ f)(x)$.

21. $f(x) = \dfrac{2}{3}x, g(x) = \dfrac{3}{2}x$

22. $f(x) = x + 5, g(x) = x - 5$

23. $f(x) = 2x + 1, g(x) = \dfrac{x - 1}{2}$

24. $f(x) = \dfrac{4}{3}x + \dfrac{16}{3}, g(x) = \dfrac{3}{4}x - 4$

25. $f(x) = x^3 - 5, g(x) = \sqrt[3]{x + 5}$

26. $f(x) = \sqrt[7]{x + 1}, g(x) = x^7 - 1$

27. $f(x) = \sqrt{x - 2}, g(x) = x^2 + 2$

28. $f(x) = \sqrt[4]{x}, g(x) = x^4$

29. $f(x) = \dfrac{1}{x+1}, g(x) = \dfrac{1-x}{x}$

30. $f(x) = \dfrac{x^2+1}{x^2-1}, g(x) = \dfrac{2x-5}{3x+4}$

31. $f(x) = 3, g(x) = -5$

32. $f(x) = -4, g(x) = 3x + 1$

In each of the following, find $f(x)$ and $g(x)$ such that $h(x) = (f \circ g)(x)$. Answers may vary, but try to choose the most obvious answer, but not $g(x) = x$.

33. $h(x) = (4x^3 - 1)^5$

34. $h(x) = \sqrt[3]{x^2 + 1}$

35. $h(x) = \dfrac{1}{(x+5)^4}$

36. $h(x) = \dfrac{1}{\sqrt{7x+2}}$

37. $h(x) = \dfrac{x^3+1}{x^3-1}$

38. $h(x) = |4x^2 - 3|$

39. $h(x) = \left(\dfrac{1+x^3}{1-x^3}\right)^4$

40. $h(x) = (\sqrt{x} + 3)^4$

41. $h(x) = \sqrt{\dfrac{x+2}{x-2}}$

42. $h(x) = \sqrt{2 + \sqrt{2+x}}$

43. $h(x) = (x^5 + x^4 + x^3 - x^2 + x - 2)^{63}$

44. $h(x) = (x - 7)^{2/3}$

In Exercises 45–48, graph each equation. Then graph $y = (f + g)(x)$ by adding y-coordinates.

45. $f(x) = x^2, g(x) = 2x - 3$

46. $f(x) = x, g(x) = \dfrac{1}{x}$

47. $f(x) = \sqrt{x}, g(x) = 3 - x^2$

48. $f(x) = 2 - x^2, g(x) = x^2 - 2$

◆ Applications

49. An airplane is 200 ft. from the control tower at the end of the runway. It takes off at a speed of 125 mph. (See figure.)

   a. Let a be the distance the plane travels down the runway. Find a formula for a in terms of the time t the plane travels. That is, find an expression for $a(t)$.

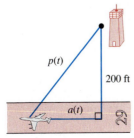

Exercise 49

   b. Let P be the distance of the plane from the control tower. Find a formula for P in terms of the distance a. That is, find an expression for $P(a)$.

   c. Find $(P \circ a)(t)$. Explain the meaning of this function.

50. A tanker had an oil spill. Oil is moving away from the tanker in a circular pattern such that the radius is increasing at the rate of 10 ft. per hour. (See figure.)

   a. Find a function $r(t)$ for the radius in terms of time t.

   b. Find a function $A(r)$ for the area of the oil spill in terms of the radius r.

   c. Find $(A \circ r)(t)$. Explain the meaning of this function.

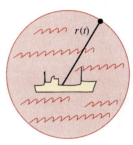

Exercise 50

51. A baseball diamond is a square 90 ft. on a side. A batter runs to first base at the rate of 3 ft./sec. (See figure.)

   a. Find a function $b(t)$ for the distance b of the runner from first base in terms of time t.

   b. Find a function $h(b)$ for the distance of the runner from second base. This distance is to be expressed in terms of the distance b. Note that as shown in the drawing, this is the straight-line distance to second base.

   c. Find $(h \circ b)(t)$. Explain the meaning of this function.

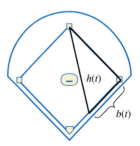

Exercise 51

In Exercises 52 and 53 compute the total profit function $P(x) = R(x) - C(x)$.

52. $R(x) = 50x, C(x) = 2x^3 - 12x^2 + 40x + 10$

53. $R(x) = 50x - 0.5x^2, C(x) = 10x + 3$

54. Consider $f(x) = 3x - 2$ and $g(x) = 5x + b$. Find b such that $(f \circ g)(x) = (g \circ f)(x)$ for all real numbers x.

55. Consider $f(x) = 3x - 2$ and $g(x) = mx + b$. Find m and b such that $(f \circ g)(x) = (g \circ f)(x) = x$ for all real numbers x.

56. Prove that the composition of two even functions is even.

57. Prove that the sum of two even functions is even.

58. Prove that the sum of two odd functions is odd.

59. Prove that the composition of two odd functions is odd.

60. Prove that the product of an even function and an odd function is odd.

61. Form a conjecture and prove a result about the product of two odd functions.

62. Form a conjecture and prove a result about the product of two even functions.

Exercises 63–74 refer to the graph of the function $f(x)$ shown in the following figure. In each exercise draw the graph of the indicated function.

63. $f(x) + 1$ 64. $f(x) + 3$ 65. $f(x + 2)$

66. $f(x + 1)$ 67. $f(x - 1)$ 68. $f(x - 5)$

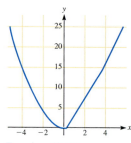

Exercises 63–74

69. $-f(x)$ 70. $-f(x - 1)$ 71. $f(-x)$

72. $-f(-x + 1)$ 73. $f(-x + 2)$ 74. $1 - f(x)$

➡ **Technology**

Use a graphing calculator to graph on a single coordinate system $f(x)$, $g(x)$, and $f(x)/g(x)$. What do you observe about the behavior of the graph of the quotient at zeros of the denominator?

75. $f(x) = x + 1$, $g(x) = x$

76. $f(x) = x^2 - x$, $g(x) = (x + 1)^2$

77. $f(x) = x^2 - 4x + 4$, $g(x) = x - 2$

78. $f(x) = \sqrt{x + 1}$, $g(x) = x^2$

3.7 INVERSE FUNCTIONS

One of the important tasks of algebra is in defining functions that are suitable for solving the widest variety of real-world problems. Throughout the book, we will be introducing new functions required to describe various events and processes. One general method for constructing new functions from known ones is by using inverse functions. In this section, we define inverse functions and develop an initial facility in dealing with them. Many of the functions we will encounter later in the book (such as the inverse trigonometric functions and logarithmic functions) are constructed as inverses of known functions.

One-to-One Functions

Suppose that f is a function with domain A and range B. Then for every element c in B, there is some x in the domain A such that $f(x) = c$. However, there may be more than one x for which f has the value c. For instance, consider the function $f(x) = x^2$. In this case, there are two values of x for which $f(x) = 4$, namely, $x = 2$ and $x = -2$. On the other hand, for some functions, each value c in the range corresponds to exactly one value x for which $f(x) = c$. Let's assign such functions a name.

Definition 1
One-to-One Function

We say that f is a **one-to-one function** provided that for each c in the range, there is precisely one x in the domain such that $f(x) = c$.

When the domain and range are both subsets of the set of real numbers, we can use a simple geometric test to determine whether a function is one-to-one. On the graph of $f(x)$, the values of x for which $f(x) = c$ are the y-coordinates of the points at which the graph intersects the horizontal line $y = c$. Generally, there may be several such values of x, that is, many such points of intersection. (See Figure 1(a).) However, if $f(x)$ is one-to-one, then **every** line $y = c$ intersects the graph of $f(x)$ in at most one point. (See Figure 1(b).) Thus, we have the following geometric criterion for one-to-one functions:

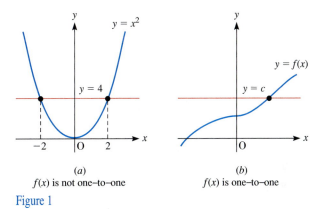

(a)
$f(x)$ is not one–to–one

(b)
$f(x)$ is one–to–one

Figure 1

HORIZONTAL LINE TEST FOR ONE-TO-ONE FUNCTIONS

A function f is one-to-one if and only if each horizontal line intersects the graph of f in at most one point.

As an example of a one-to-one function, consider the function $f(x) = \sqrt{x}$ $(x \geq 0)$, whose graph is shown in Figure 2. Note that if a horizontal line intersects the graph, it intersects it in precisely one point. (Of course, not every horizontal line intersects the graph.) So the function f is one-to-one.

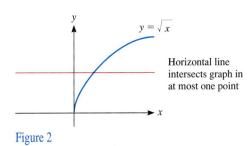

Horizontal line intersects graph in at most one point

Figure 2

➤ **EXAMPLE 1**
Determining If a Function
Is One-to-One

Determine which of the following functions are one-to-one.

1. $f(x) = -2x + 5$
2. $f(x) = \sqrt{1 - x^2}$ $(0 \leq x \leq 1)$
3. $f(x) = x^2$

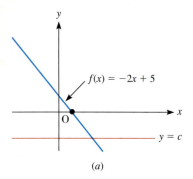

Figure 3(a)

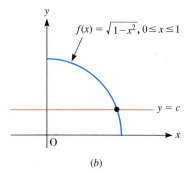

Figure 3(b)

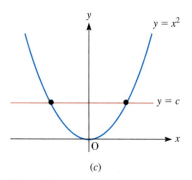

Figure 3(c)

Inverse of a Function

Solution

1. Suppose that we are given a value y in the range of f (which is the set of all real numbers). A particular value of x satisfies $f(x) = y$ precisely if

$$-2x + 5 = y$$

Solving this equation for x in terms of y, we obtain

$$x = -\frac{1}{2}(y - 5)$$

This equation shows that for a particular value of y, there is a single value of x such that $f(x) = y$. Thus, f is one-to-one. The graph of f is shown in Figure 3(a). From the horizontal line test, we obtain a second proof that the function is one-to-one.

2. Again suppose that y belongs to the range of f. The condition $f(x) = y$ is equivalent to

$$\sqrt{1 - x^2} = y$$
$$1 - x^2 = y^2 \quad \text{and } x \text{ in } [0, 1]$$
$$x^2 = 1 - y^2$$
$$x = \pm\sqrt{1 - y^2}$$

At first glance, it appears that for a given value of y, there are two possible values of x, one for each choice of sign. However, since x is in $[0, 1]$, the negative sign cannot hold, and there is a single choice for x, namely,

$$x = \sqrt{1 - y^2}$$

Since a single value of x corresponds to a given value of y, the function f is one-to-one. In Figure 3(b), we have sketched the graph of f. It is easy to verify that the horizontal line test holds in this case.

3. We have sketched the graph of f in Figure 3(c). It is clear that the horizontal line test fails. For instance, the figure shows a horizontal line that intersects the graph in two places. This is reflected in the algebra as follows: If we solve the equation $f(x) = y$ for x in terms of y, we have

$$y = x^2$$
$$x = \pm\sqrt{y}$$

For all $x \neq 0$, the numbers $\sqrt{y}$ and $-\sqrt{y}$ are different (one is positive and the other negative). Therefore, for $x \neq 0$ there are two distinct values of x corresponding to y. This is confirmed by examining the graph.

Let f be a one-to-one function with domain A and range B. Then for each y in B, there is exactly one x in A such that $f(x) = y$. We can define a function from B to A whose value at y is x. This function is called the **inverse** of f and is denoted f^{-1}. Note that -1 is not an exponent in this notation but rather stands for the inverse function.

Recall that we can represent a function $f : A \rightarrow B$ as an arrow from the element x of A to the element $f(x)$ of B. The inverse function f^{-1} can be viewed as represented by the reverse arrow, which goes from $f(x)$ to x. See Figure 4.

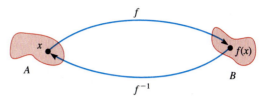

Figure 4

The following facts about inverse functions follow directly from their definition.

INVERSE FUNCTIONS

1. The inverse function f^{-1} is defined only if the function f is one-to-one.
2. The domain of the inverse function is the range of f.
3. The range of the inverse function is the domain of f.
4. The inverse function satisfies the property

$$f^{-1}(y) = x \text{ if and only if } f(x) = y$$

Suppose that we start with a number x in the domain of the function f. The function value $f(x)$ lies in the range of f (the domain of the inverse function). The definition of the inverse function says

$$f^{-1}(f(x)) = x \tag{1}$$

Similarly, suppose that y is a number in the domain of f^{-1} (i.e., in the range of f). Then the definition of the inverse function says that

$$f(f^{-1}(x)) = x \tag{2}$$

The last two formulas can be phrased in terms of composition of functions:

$$(f^{-1} \circ f)(x) = x \qquad x \text{ in domain } (f) \tag{3}$$
$$(f \circ f^{-1})(x) = x \qquad x \text{ in domain } (f^{-1}) \tag{4}$$

These formulas state important relationships between a function and its inverse. We will be using these formulas in many special cases to perform algebraic manipulations in expressions that involve both a function and its inverse.

Examples of Inverse Functions

The algebraic reasoning we used in Example 1 to determine whether a given function was one-to-one also allows us to determine the inverse function. Consider, for example, the function

$$f(x) = -2x + 5$$

We showed that if y is a real number, then $f(x) = y$ is equivalent to the equation

$$x = -\frac{1}{2}(y - 5)$$

Or, in terms of the inverse function, we have

$$f^{-1}(y) = -\frac{1}{2}(y - 5)$$

The name of the variable we use doesn't make any difference. And since we are accustomed to writing the variables of our functions as x, the inverse function may be described as

$$f^{-1}(x) = -\frac{1}{2}(x - 5)$$

where x is in the range of f, that is, x is any real number.

As another example, consider the function

$$f(x) = \sqrt{1 - x^2} \quad (0 \le x \le 1)$$

In Example 1 we determined that the function f is one-to-one and that the equation $f(x) = y$ is equivalent to the equation

$$x = \sqrt{1 - y^2} \quad (0 \le y \le 1)$$

This means that the inverse function of f is given by

$$f^{-1}(y) = \sqrt{1 - y^2} \quad (0 \le y \le 1)$$

Or, replacing the variable y by the customary variable x, we have

$$f^{-1}(x) = \sqrt{1 - x^2} \quad (0 \le x \le 1)$$

Note that in this particular example, the inverse of the function is the function itself.

We can condense the preceding computations into the following method for determining the inverse of a one-to-one function f.

DETERMINING THE INVERSE FUNCTION OF f

1. Determine that $f(x)$ has an inverse by determining that it is one-to-one.
2. Write the equation $f(x) = y$.
3. Solve the equation for x in terms of y:

$$x = \text{expression in } y$$

4. Write

$$f^{-1}(y) = \text{expression in } y$$

5. Replace y throughout with x to obtain the inverse function:

$$f^{-1}(x) = \text{expression in } x$$

➤ **EXAMPLE 2**
Calculating Inverses

Determine the inverse functions of the following functions. Determine the domain and range of the inverse.

1. $f(x) = 3x + 5$ 2. $f(x) = \sqrt{x}, \quad x \ge 0$

Solution

1. The graph of f is a nonvertical straight line. By the horizontal line test, f is one-to-one and so has an inverse function. The domain and range of f

both consist of the set **R** of all real numbers. Therefore, the domain and range of f^{-1} are also the set **R**. To obtain a formula for f^{-1}, we write

$$y = 3x + 5$$

We now solve this equation for x in terms of y:

$$y - 5 = 3x$$

$$x = \frac{1}{3}(y - 5)$$

We now interchange x and y to obtain

$$y = \frac{1}{3}(x - 5)$$

$$f^{-1}(x) = \frac{1}{3}(x - 5)$$

2. Figure 5 shows the graph of $y = \sqrt{x}$. By inspecting the graph, we see that it passes the horizontal line test, so the function is one-to-one. Therefore, f has an inverse function. Both the domain and range of f are the set of all nonnegative real numbers. Therefore, both the domain and range of the inverse function are the set of all nonnegative real numbers. We now obtain a formula for the inverse function by first solving the equation for x in terms of y:

$$y = \sqrt{x}$$
$$x = y^2 \quad \text{and} \quad y \geq 0$$

Interchanging x and y, we have

$$y = x^2 \quad \text{and} \quad x \geq 0$$
$$f^{-1}(x) = x^2$$

That is, the inverse function is $f^{-1}(x) = x^2 \quad (x \geq 0)$.

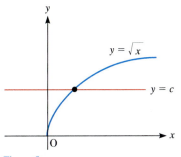

Figure 5

The Graph of the Inverse Function

From the definition of the inverse function, we can see that the point (a, b) is on the graph of f if and only if (b, a) is on the graph of the inverse of f. Indeed, we have

$$b = f(a) \text{ if and only if } a = f^{-1}(b)$$

The points (a, b) and (b, a) are located symmetrically with respect to the line $y = x$, which bisects the first and third quadrants. (See Figure 6.)

To better understand the relationship of the points (a, b) and (b, a), think of the line $y = x$ as a mirror that reflects images on one side of it to the other side, preserving all distances. Then the point (a, b) is the reflection of the point (b, a) and vice versa. This fact provides us with a simple method for sketching the graph of the inverse function.

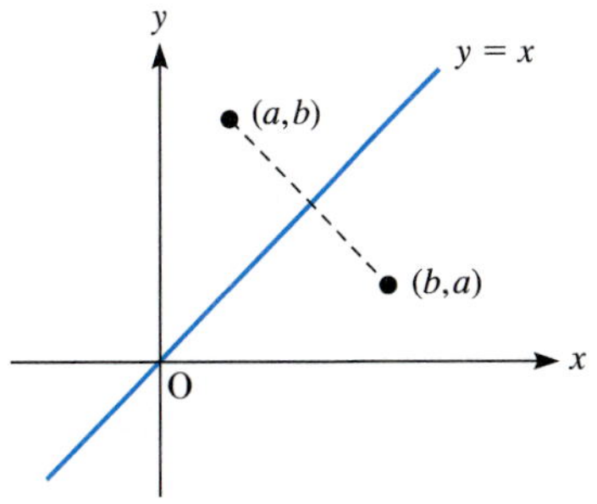

Figure 6

GRAPH OF AN INVERSE FUNCTION

The graph of the inverse function $f^{-1}(x)$ is the reflection of the graph of f in the line $y = x$. See Figure 7.

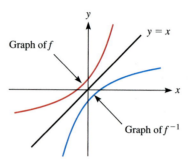

Figure 7

> **EXAMPLE 3**
Graphing Inverses

Sketch the graph of the inverse function of $f(x)$.

1. $f(x) = -2x + 5$
2. $f(x) = \sqrt{1 - x^2}$ $(0 \le x \le 1)$

Solution

1. We first sketch the graph of $f(x)$, as in Figure 8. To obtain the graph of the inverse, we reflect this graph in the line $y = x$.
2. We proceed as in part 1. We first graph f and then reflect that graph in the line $y = x$. (See Figure 9.) Note that in this case, the reflected graph is the same as the original. This is a geometric confirmation of the fact that we discovered earlier in this section by computation, namely, that the inverse of the function

$$f(x) = \sqrt{1 - x^2} \quad (0 \le x \le 1)$$

is the function $f(x)$ itself.

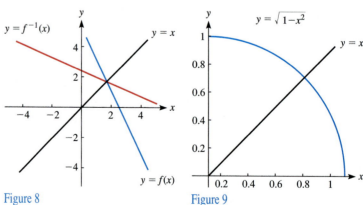

Figure 8   Figure 9

Exercises 3.7

The following are graphs of functions. Determine which functions have inverse functions. For those that do, draw the graph of the inverse function.

1.
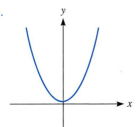

2.
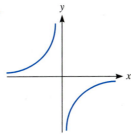

3.
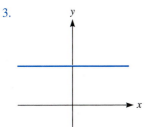

4.
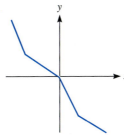

5.
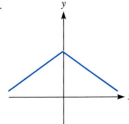

6.
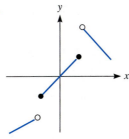

Graph each of the following functions. Determine which functions have an inverse. For those that do, find a formula for $f^{-1}(x)$ and draw its graph.

7. $f(x) = 5 - 8x$

8. $f(x) = 4$

9. $f(x) = -8$

10. $f(x) = \dfrac{x - 2}{4}$

11. $f(x) = x^2 - 3$

12. $f(x) = \sqrt{x} \quad (x \geq 0)$

13. $f(x) = \sqrt{4 - x^2} \quad (0 \leq x \leq 2)$

14. $f(x) = \dfrac{1}{x} \quad (x \neq 0)$

15. $f(x) = \sqrt{4 - x^2} \quad (-2 \leq x \leq 2)$

16. $f(x) = \sqrt{4 - x^2} \quad (-2 \leq x \leq 0)$

In each of the following exercises, determine whether the functions are inverses of each other by calculating $f \circ g(x)$ and $g \circ f(x)$.

17. $f(x) = \dfrac{2x}{3}, \qquad g(x) = \dfrac{3x}{2}$

18. $f(x) = x + 5, \qquad g(x) = x - 5$

19. $f(x) = 2x + 1, \qquad g(x) = \dfrac{x - 1}{2}$

20. $f(x) = \dfrac{4x}{3} + \dfrac{16}{3}, \qquad g(x) = \dfrac{3x}{4} - 4$

21. $f(x) = x^3 - 5, \qquad g(x) = \sqrt[3]{x + 5}$

22. $f(x) = \sqrt[7]{x + 1}, \qquad g(x) = x^7 - 1$

23. $f(x) = \sqrt{x - 2}, \qquad g(x) = x^2 + 2$

24. $f(x) = \sqrt[4]{x} \quad (x \geq 0), \qquad g(x) = x^4$

25. $f(x) = \dfrac{1}{x + 1}, \qquad g(x) = \dfrac{1 - x}{x}$

26. $f(x) = \dfrac{x^2 + 1}{x^2 - 1}, \qquad g(x) = \dfrac{2x - 5}{3x + 4}$

27. $f(x) = 3, \qquad g(x) = \dfrac{1}{3}$

28. $f(x) = -4, \qquad g(x) = 3x - 1$

For each of the following functions, find a formula for $f^{-1}(x)$.

29. $f(x) = 2x - 3$

30. $f(x) = 7x + 2$

31. $f(x) = \dfrac{2x}{3} + \dfrac{3}{5}$

32. $f(x) = 1.6x - 6.4$

33. $f(x) = x^2 \quad (x \geq 0)$

34. $f(x) = 4x^2 \quad (x \leq 0)$

35. $f(x) = \sqrt{x + 2} \quad (x \geq -2)$

36. $f(x) = \sqrt{7x - 2} \quad \left(x \geq \dfrac{2}{7}\right)$

37. $f(x) = x^5$

38. $f(x) = x^3 - 2$

39. $f(x) = \dfrac{2}{x}$

40. $f(x) = -\dfrac{4}{7}x$

41. $f(x) = \dfrac{2x - 3}{3x + 1}$

42. $f(x) = \dfrac{7x + 2}{8x - 3}$

43. $f(x) = \sqrt{25 - x^2} \quad (-5 \leq x \leq 0)$

44. $f(x) = \sqrt{49 - x^2} \quad (0 \leq x \leq 7)$

45. $f(x) = \sqrt[3]{x + 8}$

46. $f(x) = \sqrt[5]{6 - 3x}$

47. For $f(x) = 23{,}457x - 3456$, find $(f \circ f^{-1})(739)$ and $(f^{-1} \circ f)(5.00023)$.

48. For $f(x) = 677x^3$, find $(f \circ f^{-1})(958.34)$ and $(f^{-1} \circ f)(765/577)$.

49. The following formulas for conversion between Fahrenheit and Celsius temperatures have been considered several times up to now in the text:

$$F = \frac{9}{5}C + 32, \qquad C = \frac{5}{9}(F - 32)$$

Show that these functions are inverses of one another.

50. Find a formula for the area A of a circle in terms of its radius r. Then find a formula for the radius r in terms of the area A. Show that the functions defined by these formulas are inverses of one another.

51. Let $f(x) = mx + b$, $m \neq 0$. Find a formula for $f^{-1}(x)$.

52. For what integers n does $f(x) = x^n$ have an inverse?

53. Determine whether or not an even function has an inverse.

54. Find examples of odd functions that are inverses of one another.

55. Suppose that f and g both have inverses. Show that $(f \circ g)^{-1}(x) = (g^{-1} \circ f^{-1})(x)$.

56. The function f is increasing over an interval provided that $a < b$ for a and b in the interval always implies that $f(a) < f(b)$. Prove that if f is increasing over an interval, then it has an inverse.

➡ **Technology**

For each of the following functions $f(x)$, use a graphing calculator to determine whether or not it has an inverse function. If the inverse exists, calculate $f^{-1}(2)$.

57. $f(x) = x^2 - 3.5x + 1$ 58. $f(x) = -3.9x^3 + 9.6$

59. $f(x) = \dfrac{x}{4x + 3}, \quad x > 0$

60. $f(x) = \dfrac{1}{x^2 - 17}$ 61. $f(x) = |3x - 11|$

62. $f(x) = \sqrt{x^2 - x}, \quad x \geq 1$

✍ **In Your Own Words**

63. Write a paragraph defining the concept of a one-to-one function and providing examples of functions that are and functions that are not one-to-one. Use complete sentences.

64. Explain the procedure for determining the inverse of a one-to-one function.

65. What happens when you attempt to apply the procedure of Exercise 64 to a function that is not one-to-one?

66. Does an inverse function always exist for a decreasing function? Explain your reasoning.

3.8 VARIATION AND ITS APPLICATIONS

Many applications involve equations in which one variable is expressed in terms of other variables. Such an equation expresses a **functional relationship** between the variables. In this section, we will explore some of the most common types of functional relationships.

Direct Variation

Let's begin with the simplest type of functional relationship between two variables:

Definition 1 **Direct Variation**	We say that y **varies directly as** x (or that y is **proportional to** x) provided that $$y = kx$$ for some constant k. The constant k is called the **proportionality factor**. When y varies directly as x, any increase in x results in a proportional increase in y, and any decrease in x results in a proportional decrease in y.

➤ **EXAMPLE 1**
Direct Variation

Suppose that y varies directly as x. Moreover, suppose that the value of y is 100 when $x = 4$.

1. Determine the functional relationship between y and x.
2. What is the value of y when $x = 5$?

Solution

1. Since y varies directly as x, we have

$$y = kx$$

for some positive constant k. We may determine the constant k using the given data. Substitute the values $y = 100$ and $x = 4$ into this equation to obtain

$$100 = k(4)$$
$$k = 25$$

So the functional relationship between x and y is given by

$$y = 25x$$

2. Using the functional relationship just derived, we see that when $x = 5$, we have

$$y = 25x = 25(5) = 125$$

➤ **EXAMPLE 2**

Hooke's Law

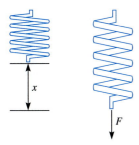

Figure 1

Hooke's law from physics states that the distance a spring stretches from its rest position is directly proportional to the amount of force applied. Suppose that a spring is stretched 8 in. by a force of 50 lbs. Suppose that a force of 60 lbs. is applied. How far will the spring be stretched? (See Figure 1.)

Solution

Denote the force acting on the spring by F and the distance the spring stretches by x. Then Hooke's law states that F varies directly as x. Therefore, we have

$$F = kx$$

for some constant k. Inserting the given data into this equation, we find that

$$F = kx$$
$$50 = k(8)$$
$$k = \frac{25}{4}$$
$$F = \frac{25}{4}x$$

Therefore, when F is equal to 60 pounds, we have

$$60 = \frac{25}{4}x$$
$$\frac{60 \cdot 4}{25} = x$$
$$x = \frac{48}{5} \text{ in.}$$

When a force of 60 lbs. is applied, the spring is stretched 48/5 or 9.6 inches.

In many applications, functional relationships of the form

$$y = kx^m$$

are required, where m and k are constants. This is a generalization of the formula for direct variation. In the case of a functional relationship of this form, we say that y **varies directly as** x^m.

Inverse Variation

> **Definition 2**
> **Inverse Variation**
>
> We say that y **varies inversely as** x^m (or that y is **inversely proportional to** x^m, provided that
>
> $$y = \frac{k}{x^m}$$
>
> for some constant k.

Here are some examples of functional relationships in which y varies inversely as some power of x:

$$y = \frac{5}{x^2}$$

$$y = \frac{0.3}{x}$$

$$y = -\frac{5}{x^4}$$

➤ **EXAMPLE 3**
Inverse Variation

Suppose that y varies inversely as x. Furthermore, suppose that when $x = 0.1$, the value of y is 20. Determine the functional relationship between x and y.

Solution

Since y varies inversely as x, we have

$$y = \frac{k}{x}$$

for some constant k. We may determine the value of k from the given data by substituting $x = 0.1$ and $y = 20$ into the equation to obtain

$$20 = \frac{k}{0.1}$$

$$k = 2$$

Replacing k by this value, 2, we have the functional relationship

$$y = \frac{2}{x}$$

➤ **EXAMPLE 4**
Electrical Resistance

The electrical resistance of an electrical wire of a certain length varies inversely with the square of the radius of the wire. Suppose that a certain wire has a radius of 0.05 in., and the resistance is measured to be 1000 ohms. What will be the resistance for a wire of radius 0.1 in.? (See Figure 2.)

Solution

Let R denote the resistance and r the radius of the wire. Then we have

$$R = \frac{k}{r^2}$$

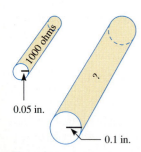

0.05 in.

0.1 in.

Figure 2

for some constant k. To determine k, we substitute $R = 1000$ $r = 0.05$ into the equation, obtaining

$$1000 = \frac{k}{(0.05)^2}$$

$$k = 2.5$$

Therefore, we have the functional relationship

$$R = \frac{2.5}{r^2}$$

In particular, if $r = 0.1$ in., then

$$R = \frac{2.5}{(0.1)^2} = 250 \text{ ohms}$$

➤ **EXAMPLE 5**

Newton's Law of Gravitation

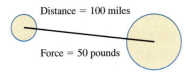

Distance = 100 miles

Force = 50 pounds

Figure 3

The force of gravitational attraction between two bodies varies inversely as the square of the distance between them. Suppose that when two space rocks are 100 miles apart, the force of gravitational attraction between them is 50 lbs. What is the gravitational force between the rocks when they are 30 miles apart? (See Figure 3.)

Solution

Let F denote the force of gravitational attraction and d the distance between the rocks. Then we are given that

$$F = \frac{k}{d^2}$$

for some constant k. Substituting $F = 50$ and $d = 100$ into the equation, we have

$$50 = \frac{k}{100^2}$$

$$k = 500,000$$

This means that the gravitational force between the rocks is given by the functional relationship

$$F = \frac{500,000}{d^2}$$

To determine the amount of the force when the rocks are 30 miles apart, we substitute $d = 30$ into this equation to obtain

$$F = \frac{500,000}{30^2} = 555.55 \text{ lbs.}$$

Other Forms of Variation

Other types of variation correspond to other types of relationships that occur in applications. A commonly encountered form of variation is joint variation, which is defined as follows.

Definition 3
Joint Variation

We say that y **varies jointly as** x **and** t (or that y **is proportional to** x **and** t) provided that

$$y = ktx$$

for some constant k.

➢ **EXAMPLE 6**
Joint Variation

Suppose that y varies jointly as x and t. What is the effect if both x and t are increased by 50%?

Solution

We are given that

$$y = kxt$$

If x and t are each increased by 50%, then x is replaced by $1.5x$ and t is replaced by $1.5t$. This means that y is replaced by

$$k(1.5x)(1.5t) = 2.25(kxt)$$

But kxt is the original value of y. So the original value of y is multiplied by 2.25.

Other types of variation are obtained by combining the types of variation already introduced. For example, to say that y varies jointly as a and b and inversely as the square root of c means that y is defined by a relationship of the form

$$y = k\frac{ab}{\sqrt{c}}$$

for some constant k. Similarly, to say that y varies jointly with the square of a and the fourth power of b means that y satisfies a relationship of the form

$$y = ka^2b^4$$

for some constant k. Other types of variation are defined similarly.

Exercises 3.8

1. y varies directly as x, and the value of y is 0.04 when $x = 2.8$.

   a. Determine the functional relationship between y and x.

   b. What is the value of y when $x = 350$?

2. y varies directly as x, and the value of y is 5.6 when $x = 3.2$.

   a. Determine the functional relationship between y and x.

   b. What is the value of y when $x = 0.64$.

3. y varies directly as the square of x, and $y = 0.8$ when $x = 0.2$.

   a. Determine the functional relationship between y and x.

   b. What is the value of y when $x = 3.7$?

4. y varies inversely as x, and $y = 2.6$ when $x = 1.2$.

   a. Determine the functional relationship between y and x.

   b. What is the value of y when $x = 0.64$?

5. y varies inversely as the square of x, and $y = 2$ when $x = 0.7$.

   a. Determine the relationship between y and x, also known as an equation of variation.

   b. What is the value of y when $x = 0.1$?

6. y varies jointly as x and z, and $y = 3.4$ when $x = 4$ and $z = 5$.

   a. Determine the equation of variation between y, x, and z.

   b. What is the value of y when $x = 8.4$ and $z = 505$?

7. y varies jointly as x and z and inversely as w. It is known that $y = 52$ when $x = 2$, $z = 4$, and $w = 1/2$.

   a. Determine the relationship between y, x, z, and w.

   b. What is the value of y when $x = 12$, $z = 24$, and $w = 15$?

8. y varies jointly as x and the square of z, and $y = 152.1$ when $x = 1.8$ and $z = 6.5$.

   a. Determine the relationship between y, x, and z.

   b. What is the value of y when $x = 0.0034$ and $z = 23.4$?

9. y varies jointly as x and z and inversely as the square of w. It is known that $y = 22$ when $x = 14$, $z = 10$, and $w = 12$.

   a. Determine the relationship between y, x, z, and w.

   b. What is the value of y when $x = 12.4$, $z = 1400$, and $w = 5$?

10. y varies jointly as x and z and inversely as the product of w and p. It is known that $y = 16$ when $x = 23$, $z = 31$, $w = 42$, and $p = 19$.

   a. Determine the relationship between y, x, z, w, and p.

   b. What is the value of y when $x = 12.4$, $z = 1400$, $w = 35.8$, and $p = 200$?

11. y varies inversely as the square of x and directly as the cube of z. It is known that $y = 3$ when $x = 4$ and $z = 6$.

   a. Determine the relationship between y, x, and z.

   b. What is the value of y when $x = 6$ and $z = 3$?

12. y varies directly as the square root of x and inversely as the product of the cube of z and the square of w. It is known that $y = 25$ when $x = 16$, $z = 2$, and $w = 40$.

   a. Determine the functional relationship between y, x, z, and w.

   b. What is the value of y when $x = 25$, $z = 4$, and $w = 20$?

Exercises 13–16 each present a situation involving variation. Determine a functional relationship, list the variation constant, and describe the variation.

13. The area A of a circle and its radius r.

14. The area A of a triangle and its base b and height h.

15. The distance d a car travels at a constant speed of 65 mph in t hours.

16. The simple interest I on a principal of P dollars at interest rate r for t years.

♦ **Applications**

17. **Water in a carrot**. The amount of water A in a raw carrot varies directly as the carrot's weight w. Suppose that a 25-g carrot contains 22 g of water. How much water does a 32-g carrot contain?

18. **Radiation energy**. The amount of energy E emitted by radiation varies directly as the wavelength L of the radiation. Suppose that one type of x-ray has a wavelength of 10^{-6} cm and emits 2×10^{-10} erg. How much energy is emitted from infrared light, which has a wavelength of 7×10^{-4} cm?

19. **Temperature of a gas**. The temperature of a gas varies jointly as its pressure P and volume V. A tank contains 100 l of oxygen under 15 atm of pressure at 20°C. If the

volume of the gas is decreased to 75 l and the pressure is increased to 22 atm, what is the temperature of the gas?

20. **Period of a pendulum**. The period of a pendulum varies directly as the square root of the length of the pendulum. A 3.15-m pendulum has a period of 3.56 sec. What is the period of a 10-m pendulum?

21. **Stopping distance**. The stopping distance of a car on a certain surface varies directly as the square of the velocity before the brakes are applied. A car traveling 55 ft./sec. needs 79 ft. to stop. What is the stopping distance for a car on the same road surface at a velocity of 95 ft./sec.?

22. **Illumination.** The illumination I from a light source varies inversely as the square of the distance d from the source. A flashlight is shining at a painting on a wall 8 ft. away. At what distance should the flashlight be placed so that the amount of light is doubled? (See figure.)

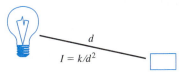

$$I = k/d^2$$

Exercise 22

23. **Speed from skid marks.** Police have used the formula $s = \sqrt{30fd}$ to estimate the speed s (in mph) at which a car was traveling if it skidded d feet. The variable f is the coefficient of friction, determined by the kind of road (concrete, asphalt, gravel, tar) and whether the road was wet or dry. Following are some values of f.

	Concrete	Tar
Wet	0.4	0.5
Dry	0.8	1.0

   a. Determine a functional relationship between s and d on a dry tar road. At 40 mph, about how many feet will a car skid? A car leaves a skid mark of about 141 ft. How fast was it going?

b. Determine a functional relationship between s and d on a wet concrete road. At 55 mph, about how many feet will a car skid? A car leaves a skid mark of about 208 ft. How fast was it going?

24. **Force of attraction.** In Newton's law of gravitation, the force F with which two masses attract each other varies directly as the product of the masses M and m and inversely as the square of the distance d between the masses. What happens to the force of attraction when the distance between the two masses is increased from 3 ft. to 12 ft.? (See figure.)

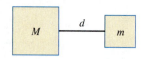

Exercise 24

25. **Gas pressure.** The pressure P of a given quantity of gas varies directly as the absolute temperature T and inversely as the volume V. At what temperature T, with $V = 200$ in.^{3}, will the pressure be three times that which exists when $T = 300°$ and $V = 500$ in.^{3}?

26. **Kepler's law** states that the time t for a planet to make a circuit of the sun is directly proportional to the three-halves power of the planet's mean distance d from the sun. The mean distances from the sun are 93 million miles for the earth and 141 million miles for Mars. In days, how long will it take Mars to make one orbit of the sun? (Use 365 days for the time it takes the earth to orbit the sun.)

27. **Accuracy of a speedometer.** If a car's speedometer is not accurate, its reading varies directly with the speed of the car. A speedometer reads 38 mph when the car is actually going 42 mph. How fast is the car going when the speedometer reads 60 mph?

28. **Power of a windmill.** Within certain limitations, the power P, in watts, generated by a windmill is proportional to the cube of the velocity V, in miles per hour, of the wind, as given by the equation

$$P = 0.015V^3$$

The constant 0.015 is a typical value that does not apply in all situations. (See figure.)

a. How much power would be generated by a continuous 6 mph wind?

b. How much power would be generated by a continuous 3 mph wind?

c. By what fraction is the power changed by cutting the wind speed in half?

Exercise 28

d. How fast would the wind speed need to be to produce 120 W of power?

(*Note:* The fact that the wind generates such little power relative to its speed is one of the frustrations of recent attempts to use windmills as an alternative energy source. Also, the functional relationship does not apply for higher and higher wind speeds. That is, the design of a windmill may not have a tripling effect on power; consider the effect of a tornado as an example.)

29. **Safe length of a beam.** The safe length S of a wooden beam varies directly as the width w and the square of the height h and inversely as the length L. An old house has beams that are 3 in. wide by 10 in. high by 18 ft. long. A remodeler wants to replace these by beams that are 2 in. wide and of the same length. What height would the new beams have to be to have twice the safe length as the old beams? (See figure.)

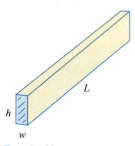

Exercise 29

30. **Wavelength of an electronic particle.** The wavelength w of a particle is inversely proportional to its momentum M, which is its mass m times its velocity v. The mass of an electron is 9.19×10^{-28} g. Traveling at one-tenth the speed of light (3×10^3 m/sec), its wavelength is 2.5×10^{-11} m. What is the wavelength of a 150-g baseball traveling at 40 m/sec.? Give your answer using scientific notation.

31. **Electrical attraction or repulsion.** The force E of electrical attraction or repulsion between two charged particles varies jointly as the magnitudes C_1 and C_2 of the two charges and inversely as the square of the distance d between them. If the force of repulsion between two

electrons, each of charge -1.6×10^{-19} coulombs and 1×10^{-8} cm apart, is 2.3×10^{-13} dyn, what is the force of attraction between an electron and a proton if a proton's charge is 1.6×10^{-19} coulombs and the particles are 8×10^{-9} cm apart? Give your answer using scientific notation.

32. If a varies jointly as c and d and inversely as the cube of b, what can be said of d in relation to a, b, and c?

33. Suppose x varies directly as the square of y.

   a. How does y vary with respect to x? (Assume that all values of x and y are positive.)

   b. If k is the variation constant for x varying with respect to y, what is the variation constant for y varying with respect to x?

34. Suppose y varies jointly as x and the square of z and inversely as w.

   a. How does x vary with respect to w, y, and z?

   b. How does w vary with respect to x, y, and z?

   c. If k is the variation constant for y varying with respect to x, z, and w, what is the variation constant for part (a)? For part (b)?

3.9 CHAPTER REVIEW

Important Concepts, Properties, and Formulas—Chapter 3

Distance formula	$d(P,Q) = \sqrt{(x_1 - x_2)^2 + (y_1 - y_2)^2}$	p. 128
Midpoint formula	$M\left(\dfrac{x_1 + x_2}{2}, \dfrac{y_1 + y_2}{2}\right)$	p. 129
Standard form of a linear equation	$ax + by = c$	p. 132
Equation of a circle	$(x - h)^2 + (y - k)^2 = r^2$	p. 140
Slope of a line	$m = \dfrac{y_2 - y_1}{x_2 - x_1}, \ x_2 - x_1 \neq 0$	p. 151
Slope-intercept formula	Suppose a nonvertical line has slope m and y-intercept $(0, b)$. Then the slope-intercept equation for the line is $$y = mx + b$$	p. 153
Point-slope formula	Suppose a line passes through (c, d) and has slope m. The equation of the line is $$y - d = m(x - c)$$	p. 157
Parallel lines	Lines that have the same slope ($m_2 = m_1$).	p. 159
Perpendicular lines	The slope of one is the negative of the multiplicative inverse of the other: $$m_1 = -\dfrac{1}{m_2}, \quad m_2 \neq 0$$	p. 159
Function	Let A and B be sets. A function f from set A to set B is a correspondence that associates each element of set A with one and only one element of set B.	p. 164
Vertical line test	A set of points in the xy plane is the graph of a function if and only if each vertical line intersects the set in at most one point.	p. 176

Vertical translation	To obtain graph of $f(x) = g(x) + c$, start with graph of $g(x)$ and $$\begin{cases} \text{shift upwards } \lvert c \rvert \text{ units} & \text{if } c > 0 \\ \text{shift downwards } \lvert c \rvert \text{ units} & \text{if } c < 0 \end{cases}$$	p. 178
Horizontal translation	To obtain graph of $f(x) = g(x - c)$, start with graph of $g(x)$ and: $$\begin{cases} \text{shift right } \lvert c \rvert \text{ units} & \text{if } c > 0 \\ \text{shift left } \lvert c \rvert \text{ units} & \text{if } c < 0 \end{cases}$$	p. 179
Vertical scaling	To obtain the graph of $f(x) = cg(x)$, start with graph of $g(x)$ and: $$\begin{cases} \text{stretch vertically } \lvert c \rvert \text{ units} & \text{if } \lvert c \rvert > 1 \\ \text{shrink vertically } \lvert c \rvert \text{ units} & \text{if } 0 < \lvert c \rvert < 1 \end{cases}$$ Then reflect the result across the y-axis if $c < 0$.	p. 179
Horizontal scaling	To obtain the graph of $f(x) = g(cx)$, start with graph of $g(x)$ and: $$\begin{cases} \text{stretch horizontally } \lvert c \rvert \text{ units} & \text{if } \lvert c \rvert > 1 \\ \text{shrink horizontally } \lvert c \rvert \text{ units} & \text{if } 0 < \lvert c \rvert < 1 \end{cases}$$ Then reflect the result across the y-axis if $c < 0$.	p. 179
Even function	f is even provided $f(x) = f(-x)$ for all x in the domain of f.	p. 182
Odd function	f is odd provided $f(-x) = -f(x)$ for all x in the domain of f.	p. 182
Operations on functions	Let f and g be functions.	p. 188ff
	Sum of f and g: $$(f + g)(x) = f(x) + g(x)$$	
	The difference of f and g: $$(f - g)(x) = f(x) - g(x)$$	
	The product of f and g: $$(fg)(x) = f(x)g(x)$$	
	The quotient of f and g: $$g : \left(\frac{f}{g}\right)(x) = \frac{f(x)}{g(x)}, \; g(x) \neq 0$$	
	The composition of f and g: $$(f \circ g)(x) = f(g(x))$$	

Inverse of a function	Exists provided that f is one-to-one. Exists if the graph passes the horizontal line test.	p. 198ff
	If f^{-1} is the inverse function, then $$f(f^{-1}(y)) = y, f^{-1}(f(x)) = x$$ for all x in the domain and y in the range of f.	

Cumulative Review Exercises—Chapter 3

1. Determine the distance between the points $(-2, -7)$ and $(3, -4)$.

2. Determine the midpoint of the segment having the endpoints $(-2, -7)$ and $(3, -4)$.

3. Determine the slope of the line through the points $(-2, -7)$ and $(3, -4)$.

4. Determine the equation of the line through $(-2, -7)$ and $(3, -4)$.

5. Find an equation in standard form of a circle having a diameter whose endpoints have coordinates $(-2, -7)$ and $(3, -4)$.

6. Find the slope and y-intercept of $3x = 6y - 24$.

7. Determine the equation of the line through $(5, -6)$ with slope $-3/2$.

Determine whether each set of lines is parallel, perpendicular, or neither.

8. $3x + 6y = 3, 6y = 10 - 4x$

9. $y = 0.32x + 16, 1000y = 3125x + 7$

10. $12x - 45y = 17, 77x - 76y = 87$

11. Find the point on the x-axis equidistant from $(7, 3)$ and $(1, -2)$.

12. Find the distance between $(2\sqrt{2}, \sqrt{5})$ and $(3\sqrt{2}, -2\sqrt{5})$.

13. Find the midpoint of the line segment with endpoints $(2\sqrt{2}, \sqrt{5})$ and $(3\sqrt{2}, -2\sqrt{5})$.

Plot the graph of each equation.

14. $y = \dfrac{1}{x - 1}$

15. $x = \dfrac{1}{y - 1}$

16. $y = -|x + 3|$

17. $y = 4 - x^2$

18. $x = |y| - 3$

19. $xy = -2$

20. $y = -2x - 5$

21. $-3x + 6y = -12$

22. $x^2 + y^2 = 9$

23. $f(x) = 3|x - 2| + 2$

24. $f(x) = -0.25\sqrt{x + 3}$ 25. $f(x) = -\dfrac{1}{2}x^3$

26. $f(x) = \begin{cases} 3 - x^2 & (0 \le x) \\ x^2 - 4 & (x < 0) \end{cases}$

27. $f(x) = \begin{cases} \dfrac{1}{2}x - 1 & (x < -2) \\ \dfrac{1}{2}x^2 & (-2 \le x < 0) \\ x + 1 & (0 \le x \le 2) \\ -3x + 5 & (x > 2) \end{cases}$

Consider the graph of $y = f(x)$ in the following figure. Graph each of the functions described in Exercises 28–31.

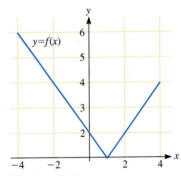

Exercises 28–31

28. $y = f(x) - 3$

29. $y = f\left(\dfrac{1}{2}x\right) + 3$

30. $y = -f(x)$

31. $y = 2f(x - 1)$

Determine whether the graphs in Exercises 32–37 are symmetric with respect to the coordinate axes and the origin.

32. $y = x^2 + 4$

33. $y = -x^{1/3}$

34. $y^2 = -x^2 + 3$

35. $y = -\dfrac{1}{2}x + 3$

36. $y = -x^2 + 4$

37. $x^3 = y^5$

Describe and graph each equation in Exercises 38–42.

38. $x^2 + (y - 3)^2 = 9$

39. $(x + 2)^2 + (y + 3)^2 = 5$

40. $x^2 + y^2 - 4x + 2y - 7 = 0$

41. $x^2 + y^2 = 12y$

42. $10x^2 + 10y^2 = 30$

Determine whether each of the following is the graph of a function.

43.

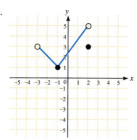

44.

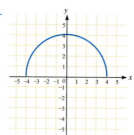

45.

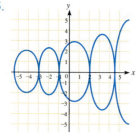

46.

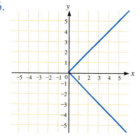

For $f(x) = x^2 + 3x - 4$, find each of the following.

47. $f(-1)$ 48. $f(5)$ 49. $f(0)$

50. $f(a + h)$ 51. $f(a - 1)$ 52. $f(0.75)$

For each function, evaluate and simplify the expression

$$\frac{f(a + h) - f(a)}{h}, \quad h \neq 0$$

53. $f(x) = x^2 + 3x - 4$ 54. $f(x) = 3\sqrt{x}$

55. $f(x) = x$ 56. $f(x) = 1 - \dfrac{1}{x}$

Find the domain and range of each function.

57. $f(x) = -5x + 8$ 58. $f(x) = x^3 - 1$

59. $f(x) = \sqrt{6 - x}$ 60. $g(x) = \dfrac{x^2 - 1}{x + 1}$

61. $g(x) = 56.7$

62. $g(x) = -\dfrac{3}{x^2 + x - 30}$

63. $f(x) = \sqrt{64 - x^2}$

Find the domain of each function.

64. $f(x) = \sqrt{24 - 5x - x^2}$

65. $f(x) = 1 - \dfrac{1}{x}$

66. $g(x) = 3x^{45} - 2x^{34} + 6$

67. $f(x) = \sqrt{|x|(x + 1)}$ 68. $g(x) = \dfrac{3}{|2x - 6|}$

69. $f(x) = \sqrt{49x^2}$ 70. $f(x) = \dfrac{1}{\sqrt{x - 1}}$

Determine whether each function is even, odd, or neither.

71. $f(x) = |4 - x|$ 72. $f(x) = (x + 3)^2$

73. $f(x) = |x + 1| + |x - 1|$

74. $f(x) = -x^2 - 2$

75. $f(x) = (x^4 + 4)(x^3 - x)$

76. $f(x) = \dfrac{x}{x + 1}$ 77. $f(x) = \sqrt{x^2 + 1}$

78. $f(x) = \sqrt[3]{x}$ 79. $f(x) = \sqrt{x^4 + 1}$

For $f(x) = x^2 - 1$ and $g(x) = 2x + 3$, find each of the following.

80. The domain of f.   81. The domain of g.

82. a. The domain of f/g.   83. a. The domain of g/f.

     b. $(f/g)(x)$ b. $(g/f)(x)$

84. a. The domain of $f + g$.   85. a. The domain of $f \circ g$.

     b. $(f + g)(x)$ b. $(f \circ g)(x)$

86. a. The domain of $g \circ f$.

     b. $(g \circ f)(x)$

Find $(f \circ g)(x)$ and $(g \circ f)(x)$.

87. $f(x) = x^2 - 2x + 1, \quad g(x) = \dfrac{3}{x^2}$

88. $f(x) = x^3 + 2, \quad g(x) = \sqrt[3]{x - 2}$

89. $f(x) = x^3, \quad g(x) = x^{11}$

90. $f(x) = \dfrac{1}{2 - x}, \quad g(x) = \dfrac{2x - 1}{x}$

91. $f(x) = \dfrac{2x + 7}{3x - 4}, \quad g(x) = \dfrac{5}{x^2}$

In each of the following, find $f(x)$ and $g(x)$ such that $h(x) = (f \circ g)(x)$. Answers may vary, but try to choose the most obvious answer, but not $g(x) = x$.

92. $h(x) = (3x^2 - 5x + 1)^{11}$

93. $h(x) = \sqrt[5]{\dfrac{x^3 + 1}{x^3 - 1}}$

94. $h(x) = (x^3 + 2)^5 + (x^3 + 2)^3 - 2(x^3 + 2)^2 - 7$

95. $h(x) = \left| \dfrac{x^2 - 5}{x^2 + 5} \right|$

96. Determine the equation of a line through $(-1, -5)$ and perpendicular to the line $3x + 4y = 6$.

97. Determine an equation of a line through $(-1, -5)$ and parallel to the line $3x + 4y = 6$.

98. Determine an equation of a line through $(-2, 3)$ and perpendicular to the line containing the points $(-3, -4)$ and $(-1, 2)$.

99. Determine an equation of a line through $(-2, 3)$ and parallel to the line containing the points $(-3, -4)$ and $(-1, 2)$.

100. y varies jointly as x and w and inversely as the cube of z, and $y = 24$ when $x = 20$, $w = 15$, and $z = 5$.

    a. Determine the relationship between y, x, w, and z.

    b. What is the value of y when $x = 10$, $w = 2.4$, and $z = 20$?

♦ **Applications**

101. **Loudness of sound.** Suppose you are sitting at a distance d from a stereo speaker. You and the speaker are outside. The loudness L of the sound is inversely proportional to the square of d. What happens to the sound if you move three times the distance d from the speaker?

102. **Safe load.** The safe load S for a rectangular beam of fixed length varies jointly as the width w and the square of the height h.

    a. Determine a relationship between S, w, and h.

    b. Which of the choices in the accompanying figure will give the strongest single beam that can be cut from a cylindrical log of diameter 25 cm?

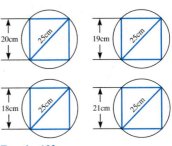

Exercise 102

103. **Growth rate of yeast cells.** The rate of growth G of yeast cells under a certain laboratory condition varies jointly as the number of cells N and $700 - N$. Determine a relationship between G, N, and $700 - N$.

104. **Temperature conversion.** Celsius temperatures C and Fahrenheit temperatures F are related by the linear equation

$$C = \frac{5}{9}(F - 32)$$

    a. Give an interpretation of the slope of this equation.

    b. Suppose a person's Fahrenheit temperature dropped 1° during an illness. What is the effect on the person's Celsius temperature?

105. **Sales commissions.** A salesperson gets a base salary of $25,000 plus a commission of 24% of sales that exceed $400,000.

    a. Determine a function that relates salary and sales.

    b. Graph this function.

106. **Boyle's law.** Boyle's law asserts that the volume V of a gas at a constant temperature is inversely proportional to the pressure P.

    a. Determine a functional relationship between V and P.

    b. A tank contains 16 cubic feet of oxygen at a pressure of 50 pounds per square inch. What volume of oxygen will be occupied if the pressure is changed to 15 pounds per square inch?

107. **Force on a batted ball.** The force F exerted by a softball bat of mass m being swung at speed v varies jointly as m and the square of v according to the equation

$$F = \frac{1}{2}mv^2$$

By what factor would a batter have to increase velocity to exert the same force with a 38-oz. softball bat as with a 35-oz. softball bat?

108. **Kelvin temperature.** Kelvin temperature K is related to Celsius temperature C by the linear equation

$$K = C + 273.15$$

a. What is the effect of a $1°$ change in Celsius temperature on the Kelvin temperature?

b. *Absolute zero* is the coldest temperature possible and occurs when $K = 0$. To what Celsius temperature does Kelvin temperature $0°$ correspond? To what Fahrenheit temperature does it correspond?

109. A rectangular box with a volume of 560 cubic feet is to be constructed with a square base and top, each with sides of length x. The cost per square foot for the bottom is $0.35, for the top is $0.20, and for the sides is $0.11. Let C denote the total cost of the box. Determine a formula in one variable for $C(x)$.

110. A boat travels 50 miles upstream and 50 miles back downstream. The speed of the stream is 2 mph. Let t denote the total time of the trip. Find a formula in one variable for $t(b)$, where b is the speed of the boat in still water.

111. A tank in the shape of an inverted cone has a cross section that is an equilateral triangle with dimensions as shown in the accompanying figure. The volume of a cone is given by

$$V = \frac{1}{3}\pi r^2 h$$

Find a function $V(h)$ in one variable for the volume in terms of the height h.

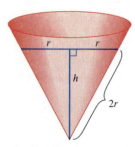

Exercise 111

112. A carpet warehouse offers discounts to buyers in quantity. If x is the number of square yards of carpet, then $C(x)$ is the cost, where

$$C(x) = \begin{cases} 12x & (0 \le x \le 50) \\ 10x & (50 < x \le 200) \\ 9x & (200 < x) \end{cases}$$

Graph this function for $0 \le x \le 300$.

113. Give a piecewise definition for the function g whose graph is shown in the accompanying figure.

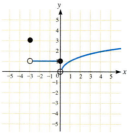

Exercise 113

114. For

$$f(x) = \frac{1}{x - 4} \qquad \text{and} \qquad g(x) = 4$$

find the domain of $f \circ g$ and $g \circ f$.

115. For the following functions, find $(f \circ g)(x)$ and $(g \circ f)(x)$.

$$f(x) = \begin{cases} x^2 & (-3 \le x \le 0) \\ 1 - x & (0 < x \le 2) \\ \dfrac{1}{x - 2} & (2 < x) \end{cases}$$

$$g(x) = \begin{cases} 2x - 7 & (-3 \le x \le 0) \\ x^3 & (0 < x \le 2) \\ \dfrac{x + 3}{x} & (2 < x) \end{cases}$$

Determine whether these functions are inverses of each other by calculating $(f \circ g)(x)$ and $(g \circ f)(x)$.

116. $f(x) = -\dfrac{4}{5}x, \quad g(x) = -\dfrac{5}{4}x$

117. $f(x) = 2x - 5, \quad g(x) = \dfrac{5}{2} + \dfrac{1}{2}x$

118. $f(x) = x^2, \quad g(x) = \sqrt{x}$

119. $f(x) = x^5, \quad g(x) = \sqrt[5]{x}$

For each of the following, find a formula for $f^{-1}(x)$.

120. $f(x) = 9x - 14$

121. $f(x) = \dfrac{3x - 4}{4 - 2x}$

122. $f(x) = \sqrt{16 - x^2} \quad (0 \le x \le 4)$

123. $f(x) = x^3 - 1$

124. For $f(x) = 45{,}677x^5 + 0.0457$, find

$$(f \circ f^{-1})(78.8999)$$

and

$$(f^{-1} \circ f)(-2{,}344{,}789)$$

POLYNOMIAL AND RATIONAL FUNCTIONS

*The symphonic shell shown has a cross-section that is a curve called a **parabola**, which is the graph of a quadratic function. The orchestra is seated in the center of the shell, and the sound it produces is reflected off the shell. Due to a geometric property of the parabola and the way sound is reflected, the sound waves are projected forward.*

I n the preceding chapter, we studied the properties of linear functions, whose graphs are straight lines. Let us now turn our attention to polynomial functions, which include the linear functions as a special case but which include many other important functions as well. We begin this chapter with a study of quadratic functions. We will provide an introduction to optimization problems, which call for maximizing or minimizing a quantity. Optimization problems are one of the main topics in a calculus course. However, by studying the properties of quadratic functions, we will be able to handle some special optimization problems. In this chapter, we will also learn to sketch the graphs of polynomial and rational functions and will develop the theory of polynomial equations.

4.1 QUADRATIC FUNCTIONS

Polynomial functions provide a rich source of mathematical models. Let's begin by defining these functions.

Definition 1
Polynomial Function

A **polynomial function** of the variable x is a function of the form
$$f(x) = a_n x^n + a_{n-1} x^{n-1} + \cdots + a_1 x + a_0$$
where n is a nonnegative integer and $a_n, a_{n-1}, \ldots, a_1, a_0$ are complex numbers.

Here are some examples of polynomial functions.
$$f(x) = 5$$
$$f(x) = -0.1x + 0.001$$
$$f(x) = 3x^2 + 100x - 310$$
$$f(x) = -x^3 + x^2 - x + 1$$
$$f(x) = 3ix^2 + 2 - 3i$$

As the first two examples illustrate, constant and linear functions are special cases of polynomial functions. Since we have already explored the properties of these functions and their graphs, let's now turn to the next most complicated polynomial functions, namely, those for which n equals 2.

Definition 2
Quadratic Function

A **quadratic function** is a function defined by an expression of the form
$$f(x) = ax^2 + bx + c$$
where a, b, and c are real numbers and a is nonzero.

Here are some examples of quadratic functions.
$$f(x) = x^2$$
$$f(x) = \frac{1}{2}x^2 + 3x - 1$$
$$f(x) = 0.001x^2 + 5.07$$

Graphs of Quadratic Functions

We have already graphed the function $y = x^2$ and shown that its graph is bowl-shaped, opening upward, with the bottom of the bowl at the origin. (See Figure 1.) The graphs of the quadratic functions
$$f(x) = ax^2, \qquad a > 0$$

have the same general appearance as that of $f(x) = x^2$. The coefficient a controls the width of the opening. The smaller the value of a, the wider the opening. Figure 2 shows the graphs of such quadratic functions for various positive numbers a.

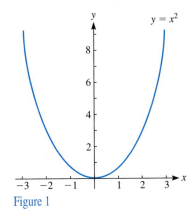

Figure 1

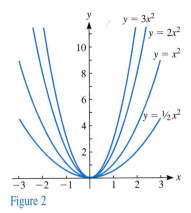

Figure 2

The graph of

$$f(x) = ax^2, \qquad a < 0$$

is obtained by reflecting the graph of $f(x) = -ax^2$ in the x-axis; thus, it is a bowl-shaped curve that opens downward. (See Figure 3.)

The graph of

$$f(x) = a(x - h)^2 + k \tag{1}$$

is obtained by translating the graph of $f(x) = ax^2$ and is therefore a bowl-shaped graph with bottom or top at (h, k). The graph opens upward if $a > 0$ and downward if $a < 0$. (See Figure 4.)

By completing the square, every quadratic function may be written in the form (1), which is called the **standard form** of the quadratic function. The bowl-shaped curve arising as the graph of a quadratic function is called a **parabola**. The bottom or top of the bowl of a parabola is called its **vertex**. In the graph of (1), the vertex is at (h, k). The vertical line through the vertex is called the **axis** of the parabola. (See Figure 5.)

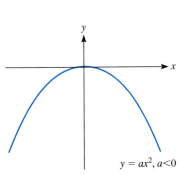

Figure 3

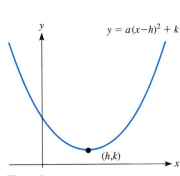

Figure 4

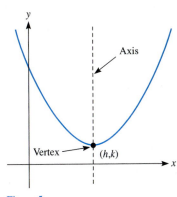

Figure 5

Let's summarize this information for easy reference:

STANDARD FORM OF A QUADRATIC FUNCTION

The quadratic function $f(x) = ax^2 + bx + c$ may be written in the form
$$f(x) = a(x - h)^2 + k$$
for suitable real numbers h and k. The graph of the function is a parabola with vertex (h, k). The parabola opens upward if $a > 0$ and downward if $a < 0$.

➤ **EXAMPLE 1**
Standard Form
for a Quadratic Function

Consider the quadratic function $f(x) = 2x^2 + 6x - 10$.

1. Write $f(x)$ in the form (1).
2. Determine the vertex and the axis of the graph of $f(x)$.

Solution

1. Complete the square:

$$f(x) = 2x^2 + 6x - 10$$
$$= 2(x^2 + 3x) - 10$$
$$= 2\left(x^2 + 3x + \left(\frac{3}{2}\right)^2\right) - 10 - 2\left(\frac{3}{2}\right)^2$$
$$= 2\left(x + \frac{3}{2}\right)^2 - \frac{29}{2}$$

This is of the form (1) with

$$a = 2, \qquad h = -\frac{3}{2}, \qquad k = -\frac{29}{2}$$

2. The vertex is at $(h, k) = (-3/2, -29/2)$. The axis is the vertical line through the vertex and thus has equation $x = -3/2$.

➤ **EXAMPLE 2**
Determining a Quadratic
Function from Geometric
Information

Determine the equation of a quadratic function whose graph passes through the point $(1, -2)$ and has vertex $(4, 5)$.

Solution

By equation (1), the quadratic function has the form
$$f(x) = a(x - 4)^2 + 5$$
for some value of a. Since the graph must pass through $(1, -2)$, we must have
$$f(1) = -2$$
$$a(1 - 4)^2 + 5 = -2$$
$$9a + 5 = -2$$
$$a = -\frac{7}{9}$$

$$f(x) = -\frac{7}{9}(x - 4)^2 + 5$$

Since the function $y = ax^2$ is even, its graph is symmetric about the y-axis. Thus, the graph of (1), which is obtained by translating the graph of $y = ax^2$, is symmetric about its axis. So to sketch the graph of a quadratic function, we may proceed as follows:

GRAPHING A QUADRATIC FUNCTION $f(x)$

1. Write the quadratic function in standard form: $f(x) = a(x - h)^2 + k$.
2. Plot the vertex (h, k).
3. Determine a point on the graph. The simplest such point to determine is usually the one corresponding to $x = h + 1$. Plot this point.
4. Draw a smooth curve from the vertex through the plotted point.
5. Complete the graph by drawing a symmetrically placed curve on the other side of the vertex.

➤ **EXAMPLE 3**
Graphing Quadratic Functions

Graph the following quadratic functions.

1. $f(x) = 3x^2 - 12x - 10$
2. $f(x) = -2x^2 + 12x$

Solution

1. As a first step, we write the function in standard form:

$$3x^2 - 12x - 10 = 3(x^2 - 4x) - 10$$
$$= 3(x^2 - 4x + 4) - 10 - 3 \cdot 4$$
$$= 3(x - 2)^2 - 22$$

The vertex is $(2, -22)$. When $x = 3$, the value of $f(x)$ is:

$$f(3) = 3 \cdot 1^2 - 22 = -19$$

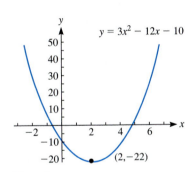

$y = 3x^2 - 12x - 10$

$(2, -22)$

Figure 6

So $(3, -19)$ is on the graph. Draw a smooth curve from $(2, -22)$ through $(3, -19)$. Now draw a symmetric curve on the left of the vertex. The completed graph is shown in Figure 6.

2. We put the function into standard form:

$$f(x) = -2x^2 + 12x$$
$$= -2(x^2 - 6x)$$
$$= -2(x^2 - 6x + 9) - (-2) \cdot 9$$
$$= -2(x - 3)^2 + 18$$

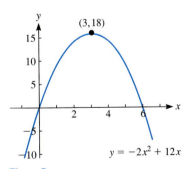

$(3, 18)$

$y = -2x^2 + 12x$

Figure 7

The vertex of the parabola in this case is $(3, 18)$. Another point on the graph is $(4, 16)$. Draw a smooth curve from the vertex through $(4, 16)$ and complete the graph via symmetry. The completed graph is shown in Figure 7.

Applications of Quadratic Functions

The graph of a quadratic function is a parabola that opens either upward or downward. In the case of a graph that opens upward, the bottom point on the graph is called the **minimum point** of the function. Its y-value is the least value that the function assumes and is called the **minimum value** of the function. (See Figure 8.)

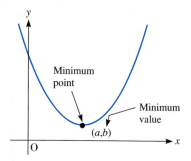

Figure 8

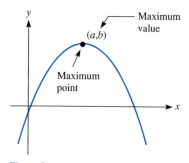

Figure 9

Analogously, in the case of a graph that opens downward, the graph has a top point, called the **maximum point** of the function. The y-value of the maximum point is the largest value that the function assumes and is called the **maximum value** of the function. (See Figure 9.)

The maximum or minimum point of the graph of a quadratic function is called the **vertex** or **extreme point** of the graph. The maximum or minimum value of the function is called the **extreme value**.

The extreme point of a quadratic function can easily be determined. Consider the quadratic function

$$f(x) = ax^2 + bx + c$$

Assume that $a > 0$, so that the graph of f opens upward. The function then has a minimum point. Write the function in the form:

$$f(x) = a\left(x^2 + \frac{b}{a}x\right) + c$$

$$= a\left(x^2 + \frac{b}{a}x + \frac{b^2}{4a^2}\right) + \left(c - \frac{ab^2}{4a^2}\right)$$

$$= a\left(x + \frac{b}{2a}\right)^2 + \frac{4ac - b^2}{4a}$$

$$= a(x - h)^2 + k$$

where

$$(h, k) = \left(-\frac{b}{2a}, \frac{4ac - b^2}{4a}\right)$$

Recall that a perfect square is nonnegative. Since we are assuming that a is positive, the expression

$$a\left(x + \frac{b}{2a}\right)^2$$

is nonnegative, being the product of a positive number and a nonnegative number. Therefore, for any value of x, we have

$$f(x) \geq k$$

In other words, the value of $f(x)$ is always at least k. Moreover, for

$$x = -\frac{b}{2a}$$

the value of f is exactly k, since

$$f\left(-\frac{b}{2a}\right) = a\left(-\frac{b}{2a} + \frac{b}{2a}\right)^2 + k$$

$$= 0 + k$$

$$= k$$

We have proved that $f(x)$ is always at least k. So we have located the minimum point of f, which is $[-b/2a, f(-b/2a)]$.

If $a < 0$, the graph opens downward, and f has a maximum point whose coordinates are given by the same formulas just derived for the minimum point.

The proof is similar to the one just given. Thus, we have proved the following important result.

EXTREME POINT OF A QUADRATIC FUNCTION

Let $f(x) = ax^2 + bx + c$ be a quadratic function. The extreme point of its graph is

$$(h, k) = \left[-\frac{b}{2a}, f\left(-\frac{b}{2a}\right) \right]$$

If $a > 0$, then the extreme point is a minimum point. If $a < 0$, then the extreme point of f is a maximum point.

We have shown that the extreme point of a parabola is at the vertex. Many applied problems can be reduced to determining the coordinates of the extreme point. Such problems are called **optimization problems** since in the applied context, the maximum or minimum point represents an optimum condition of some sort (maximum profit, minimum cost, etc.). In the next three examples, we illustrate a number of practical uses of quadratic functions.

➤ **EXAMPLE 4**
Projectile Motion

A ball is thrown vertically upward with an initial velocity of 100 feet per second.

1. Determine a function $h(t)$ that gives the height of the ball at time t seconds after it is thrown.
2. How long does it take the ball to reach its maximum height?
3. What is the maximum height the ball reaches?

Solution

1. In Chapter 3 we used the result from physics that gives the height function $h(t)$ as

    $$h(t) = -16t^2 + v_0 t + h_0$$

    where v_0 is the initial velocity of the ball (in feet per second) and h_0 is the initial height of the ball (in feet). In this case, we have the formula

    $$h(t) = -16t^2 + 100t$$

    (since the initial height h_0 equals 0).

2. The height function is a quadratic function. In Figure 10 we have drawn the path of the ball $t \geq 0$ and labeled the extreme point, which is a maximum. (This is consistent with the fact that the coefficient of the squared term is -16, which is negative.) According to the preceding proof, the maximum point occurs at

    $$t = -\frac{b}{2a} = -\frac{100}{2 \cdot (-16)} = \frac{25}{8}$$

    That is, the ball reaches its maximum height after $25/8 = 3.125$ seconds. You should note that the curve shown in Figure 10 is the physical path of the ball but does not represent the graph of the function $h(t)$.

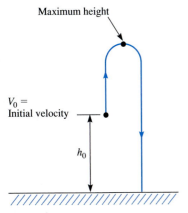

Maximum height

$V_0 = $ Initial velocity

h_0

Figure 10

3. The maximum height of the ball is given by the maximum value of the function. This maximum value equals

$$h\left(-\frac{b}{2a}\right) = h\left(\frac{25}{8}\right)$$

$$= -16\left(\frac{25}{8}\right)^2 + 100\left(\frac{25}{8}\right)$$

$$= 156.25 \text{ ft.}$$

➤ **EXAMPLE 5**
Maximizing Revenue

A company manufactures upholstered chairs. If it manufactures x chairs, then the revenue $R(x)$ per chair is given by the function

$$R(x) = 200 - 0.05x$$

How many chairs should the company manufacture to maximize total revenue?

Solution

We first find a function $T(x)$ that expresses the total revenue T received in terms of the number of chairs manufactured x. We have

$$T(x) = \text{Number of chairs} \cdot \text{Revenue per chair}$$

Therefore,

$$T(x) = x \cdot R(x)$$
$$= x(200 - 0.05x)$$
$$= -0.05x^2 + 200x$$

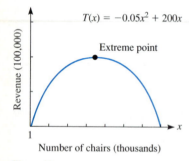

Figure 11

Thus, $T(x)$ is a quadratic function. Figure 11 shows the graph of this function. Note that the graph is a parabola opening downward. In economic terms, the graph says that initially as the number of chairs x is increased, the total revenue $T(x)$ increases. At a certain level of production, the total revenue is maximized. Increasing the number of chairs further actually decreases total revenue. The maximum occurs for

$$x = -\frac{b}{2a} = \frac{200}{2(-0.05)} = 2000$$

That is, the revenue is at a maximum when 2000 chairs are manufactured.

➤ **EXAMPLE 6**
Architecture

Suppose that an architect wishes to design a house with a fenced back yard. To save fencing cost, he wishes to use one side of the house to border the yard and to use cyclone fence for the other three sides. The specifications call for using 100 feet of fence. What is the largest area the yard can contain?

Solution

Let l denote the length of the yard and w the width. (See Figure 12.) Since the builder wishes to use 100 feet of fence, the diagram in the figure shows that

$$l + 2w = 100$$

The area of the yard, which is to be maximized, is given by the expression

$$\text{Area} = lw$$

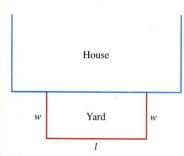

Figure 12

We may express the area as a function $A(w)$, as follows. Solve the first equation for l in terms of w:

$$l = 100 - 2w$$

Now substitute this expression for l into the expression for the area:

$$\text{Area} = lw = (100 - 2w)w = -2w^2 + 100w$$

That is, the area function $A(w)$ is given by

$$A(w) = -2w^2 + 100w$$

This is a quadratic function. The coefficient of the square term is -2, which is negative, so the extreme point is a maximum point. The value of w for which the maximum value of $A(w)$ occurs is given by

$$w = -\frac{b}{2a} = -\frac{100}{2(-2)} = 25 \text{ ft.}$$

For this value of w, the corresponding value of l is

$$l = 100 - 2w = 100 - 2(25) = 50 \text{ ft.}$$

In other words, if the area of the back yard is to be maximized, the yard should be 25 feet wide and 50 feet long. The maximum area of the yard equals

$$A(25) = -2(25)^2 + 100(25) = 1250 \text{ sq. ft.}$$

Exercises 4.1

Match each of the following functions with its graph among the graphs (a)–(f).

1. $f(x) = 2x^2$

2. $f(x) = -x^2$

3. $f(x) = 2x^2 - x$

4. $f(x) = -(x + 1)^2 + 3$

5. $f(x) = (x - 3)^2 + 1$

6. $f(x) = x(x - 1)$

(e)

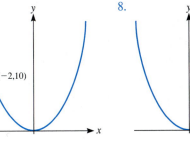
(f)

Determine the equations of each of the parabolas shown in the following graphs.

7.

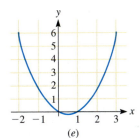

8.

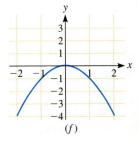
(3, 24)

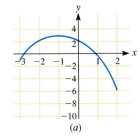
(a)

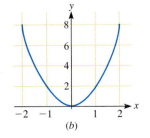
(b)

9.

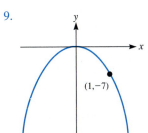
(1, −7)

10.

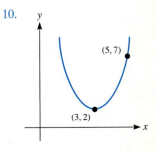
(5, 7)
(3, 2)

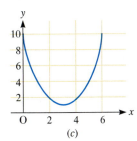
(c)

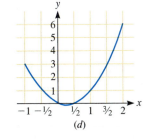
(d)

11.

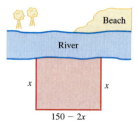

12.

For each of the following quadratic functions:

a. Find the vertex.

b. Determine whether the graph opens upward or downward.

c. Graph the function.

13. $f(x) = -2x^2$ 14. $f(x) = 2\,1/4x^2$

15. $f(x) = (x-2)^2$ 16. $f(x) = -2(x+1)^2$

17. $f(x) = 2(x-3)^2$ 18. $f(x) = -1.5(x+2)^2$

19. $f(x) = -(x+3)^2 - 2$

20. $f(x) = 1/2(x-4)^2 - 3$

For each of the following quadratic functions:

a. Determine the standard quadratic form.

b. Find the vertex.

c. Determine whether the graph opens upward or downward.

d. Find the maximum or minimum value.

e. Graph the function.

21. $f(x) = x^2 + 6x + 5$

22. $f(x) = -13 - 8x - x^2$

23. $f(x) = -1/4x^2 - 3x - 1$

24. $f(x) = 2x^2 + x - 1$

Find the maximum or minimum value.

25. $f(x) = -3(x+1)^2$ 26. $f(x) = 4(x-3)^2$

27. $f(x) = 5(x-1)^2 + 3$ 28. $f(x) = 12(x-5)^2 + 2$

29. $f(x) = -2x^2 - x + 15$ 30. $f(x) = x^2 - 6x + 9$

31. $f(x) = 3x^2 - x - 14$ 32. $f(x) = -x^2 + 1$

33. $f(x) = 21.34x^2 - 456x + 2000$

34. $f(x) = 2/3x^2 + 5/6x - 7/12$

35. $f(x) = -\sqrt{2}x^2 + 4\sqrt{3}x - \sqrt{7}$

36. $f(x) = \$240,000x^2 - \$560,000x + \$4,500,000$

Solve. (Functions for many of these exercises were developed in Exercise Set 3.4.)

37. Of all the numbers whose sum is 30, find the two that have the maximum product. That is, find x and y which

yield the maximum value of $P = xy$, where $x + y = 30$.

38. Of all the numbers whose difference is 13, find the two that have the minimum product.

39. The sum of the height and the base of triangle is 166 cm. Find the dimensions of such a triangle that has maximum area.

40. The sum of the base and height of a parallelogram is 145 yd. Find the dimensions of such a parallelogram that has maximum area.

♦ **Applications**

41. A rancher wants to fence a rectangular area next to a river, using 150 yd. of fencing, as shown in the accompanying figure.

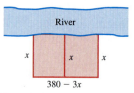

Exercise 41

a. Determine a formula in one variable for the area $A(x)$.

b. At what value of x will the area be a maximum?

c. What is the largest area that can be enclosed?

42. A rancher wants to enclose two rectangular areas with a fence, as shown in the accompanying figure, next to a river, using 380 yd. of fencing.

Exercise 42

a. Determine a formula in one variable for the enclosed area $A(x)$.

b. At what value of x will the area be a maximum?

c. What is the largest area that can be enclosed?

43. A carpenter is building a rectangular room with a fixed perimeter of 74 ft. What are the dimensions of the largest room that can be built? What is its area?

44. Of all the rectangles that have a perimeter of 98 ft., find the dimensions of the one with the largest area. What is its area?

45. A clothing firm is coming out with a new line of suits. If it manufactures x suits, then the revenue $R(x)$ per suit is given by the function

$$R(x) = 400 - 0.16x$$

How many suits should the company manufacture in order to maximize total revenue?

46. An electronics firm is marketing a new low-price stereo. It determines that its total revenue from the sale of x stereos is given by the function

$$R(x) = 280x - 0.4x^2$$

The firm also determines that its total cost of producing x stereos is given by

$$C(x) = 5000 + 0.6x^2$$

Total profit is given by

$$P(x) = R(x) - C(x)$$

a. Find the total profit.

b. How many stereos must the company produce and sell to maximize profit?

c. What is the maximum profit?

47. The owner of a 30-unit motel, by checking records of occupancy, knows that when the room rate is $50 a day, all units are occupied. For every increase of x dollars in the daily rate, x units are left vacant. Each unit occupied costs $10 per day to service and maintain. Let R denote the total daily profit.

a. Determine a formula in one variable for $R(x)$.

b. Find a value of x for which $x \geq 0$ and R is a maximum.

c. What is the maximum profit?

48. A Palladian window is a rectangle with a semicircle on top. (See figure.) Suppose the perimeter of a particular Palladian window is to be 28 ft. Let A denote the total area of the window.

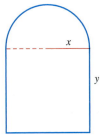

Exercise 48

a. Determine a formula in one variable for $A(x)$.

b. For what value of x will A be a maximum?

c. What is the maximum possible area?

49. A 28-in. piece of string is to be cut into two pieces. (See figure.) One piece is to be used to form a circle and the other to form a square. Let A denote the total area.

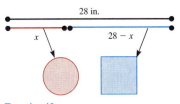

Exercise 49

a. Determine a formula for $A(x)$.

b. Find a value of x for which A is a minimum.

c. What is the minimum possible area?

50. A ball is thrown vertically upward from the roof of a skyscraper. Let S denote the height of the ball from the ground t seconds after being thrown. $S(t)$ is given by an expression of the form

$$S(t) = -16t^2 + v_0 t + h$$

where v_0 represents the initial velocity and h represents the height of the building.

Suppose that the building is 1800 ft. tall. Furthermore, suppose that an observer sees the ball pass by a window 900 ft. from the ground 8 seconds after the ball was thrown.

a. Determine a formula for $S(t)$.

b. How long does it take the ball to reach its maximum height?

c. What is the maximum height the ball reaches?

51. A soybean farmer has 15 tons of beans on hand. If he sold them now, his profit would be $300 per ton. If he waits to sell, he can add 3 tons of beans each week, but he will lose $20 per ton for each week he delays in selling his beans.

a. In how many weeks should he sell to maximize his profits?

b. What would his maximum profit be?

52. The power P, in watts, delivered in an electric circuit in which a current I is flowing is found by the formula

$$P = EI - RI^2$$

where E is the voltage and R is the resistance. Find the maximum power that can be delivered by a 200-volt circuit with resistance 5.5 ohms.

53. Find a such that

$$f(x) = ax^2 + 4x - 5$$

has a maximum value at $x = 7$.

54. Find b such that the function

$$f(x) = 2x^2 + bx - 3$$

has a minimum value of 100.

55. Find c such that the function

$$f(x) = 0.1x^2 + 7x + c$$

has a minimum value of -127.

Find the maximum or minimum value.

56. $f(x) = (1 - c)x^2 - 3cx + 12$

57. $f(x) = -(t + 2)x^2 + (t^2 - 4)x - 16$

58. Find a quadratic function that has $(-2, 3)$ as a vertex and contains the point $(-1, 8)$.

59. Find the vertex of the quadratic function

$$f(x) = px^2 - 3x + p$$

60. Graph $f(x) = |x^2 - 3|$.

61. Graph $f(x) = |6x - 5 - x^2|$.

62. Graph $x = |(y + 3)^2 - 4|$.

➡ **Technology**

Use a graphing calculator to determine the maximum or minimum value of each of the following functions. Determine the value to two significant digits.

63. $y = 5.78x^2 + 3.11x$

64. $y = -12x^2 + 13x - 11$

65. $y = \dfrac{x^2}{21} + \dfrac{11x}{2} + \dfrac{5}{23}$

66. $y = 11(x - 3.4)(x + 7.9)$

67. $y = |x^2 - 3x + 4|$ 68. $y = |1.1 - x^2|$

4.2 GRAPHS OF POLYNOMIAL FUNCTIONS OF DEGREE n

As we have already seen, the graph of a linear function is a straight line, and the graph of a quadratic function is a parabola. Many applied problems, especially those in science and engineering, require us to deal with polynomial functions of degree 3 or higher. In this section, we develop some elementary methods for sketching the graphs of such functions.

Graphs of the Functions $y = ax^n$

Among the most elementary polynomial functions are those of the form

$$y = ax^n$$

where n is a positive integer and a is a real constant. We begin our study of the graphs of polynomial functions by examining the graphs of these functions.

Consider the function

$$f(x) = x^3$$

x	$f(x)$
0	0
0.2	0.008
0.5	0.125
1	1
2	8
3	27

To sketch its graph, we first tabulate points corresponding to a set of representative positive values of x. Using these points, we plot the portion of the graph for $x > 0$.

We see that as x increases, so does $f(x)$. In fact, as x increases without bound, the value of $f(x)$ does the same. So we draw the graph heading steadily upward without bound as x increases.

Next we note that the function is odd, since

$$f(-x) = (-x)^3 = -x^3 = -f(x)$$

Therefore, its graph is symmetric with respect to the origin, as we saw in Section 3.5. So the portion of the graph for $x < 0$ is obtained by reflecting the portion for $x > 0$ in the origin. The sketch of the graph is shown in Figure 1.

Starting from the graph of Figure 1, we can sketch the graphs of many other functions using the geometric transformations of scaling, reflection, horizontal translation, and vertical translation. The next example illustrates some of the possibilities.

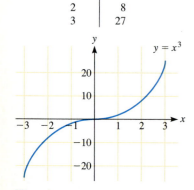

Figure 1

> **EXAMPLE 1**

Graphing Cubics

Sketch the graphs of the following functions.

1. $f(x) = 3x^3$
2. $f(x) = -x^3$
3. $f(x) = (x - 1)^3$
4. $f(x) = (x + 2)^3 + 3$
5. $f(x) = -2(x - 1)^3 - 4$

Solution

1. We scale the graph of Figure 1 using a factor of 3. See Figure 2.
2. We reflect the graph of Figure 1 in the *x*-axis because the scaling factor is negative. See Figure 3.

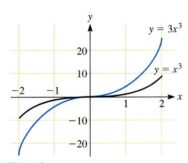

Figure 2

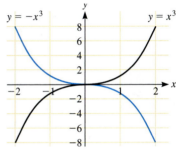

Figure 3

3. We translate the graph of Figure 1 one unit to the right. See Figure 4.
4. We translate the graph of Figure 1 two units to the left and 3 units upward. See Figure 5.
5. We perform the following geometric transformations to Figure 1: First we scale the graph using a factor of 2 . Then we reflect the graph in the *x*-axis. Finally we translate the graph 1 unit to the right and 4 units downward. See Figure 6.

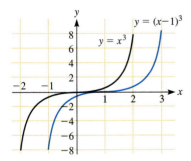

Figure 4

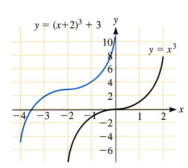

Figure 5

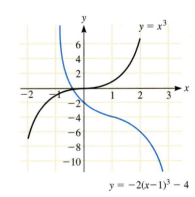

Figure 6

Let's consider now the graphs of the functions

$$y = x^n, \qquad n \geq 2$$

When n is odd, the function is odd, just as we observed above for the case $n = 3$. For n odd, the graph has the same general shape as the graph of Figure 1. In Figure 7, we show the graphs corresponding to several small odd values of n.

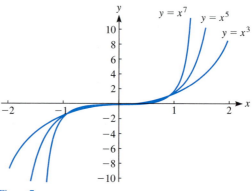

Figure 7

If n is even, the function

$$y = x^n$$

is even, like the function $y = x^2$. In this case, the graph is symmetric with respect to the y-axis and has the general shape of a parabola. In Figure 8, we have sketched the graphs corresponding to several small even values of n. Note that as n increases, the graph tends to be flatter for x between -1 and 1 and to rise more steeply for $x < -1$ and $x > 1$.

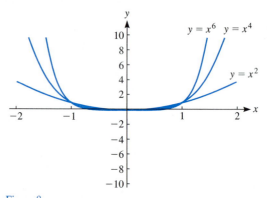

Figure 8

By using scaling, reflection, and translation with the basic graphs in Figures 7 and 8, we can sketch the graphs of all functions of the form

$$y = ax^n, \qquad n \geq 2$$

Graphs of Other Polynomial Functions

Let's now turn our attention to graphing more general polynomial functions. A full discussion of this subject really belongs to calculus, where the tools necessary to describe the various specific geometric features of these graphs are developed. However, using a few simple ideas, we can make some inroads into the problem of sketching graphs of polynomial equations and come up with rough sketches of such graphs.

Definition 1	Let $f(x)$ be a polynomial function. A complex number α_0 for which
Zero of a Polynomial	$$f(\alpha_0) = 0$$
	is called a **zero** of f.

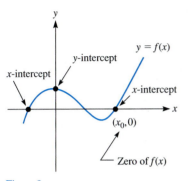

Figure 9

Suppose that x_0 is a real zero of f. Then the point $(x_0, 0)$ is an x-intercept of the the graph of f. (See Figure 9.) In general, a polynomial function $f(x)$ has a number of zeros, which can be found as solutions to the polynomial equation $f(x) = 0$. Later in the chapter, we will discuss these zeros more fully and develop techniques for determining them.

Figure 10 shows the graph of a polynomial function of degree greater than 2. Notice that the graph consists of a number of "peaks" and "valleys." A point on the graph that is either the top of a peak or the bottom of a valley is called a **turning point** or an **extreme point**. Using calculus, it is possible to prove that a polynomial of degree n has at most $n - 1$ extreme points.

Note that in the interval between two consecutive extreme points, the graph is either increasing or decreasing. (If the graph changes from increasing to decreasing or vice versa, then there must be an extreme point within the interval. The proof of this fact is part of a course in calculus.)

The following is another result that is proven in more general form in a calculus course and that is useful in sketching graphs of polynomials:

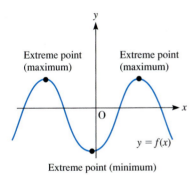

Figure 10

INTERMEDIATE VALUE THEOREM

Suppose that $f(x)$ is a polynomial with real coefficients and that a and b are real numbers with $a < b$. Suppose that one of $f(a)$ and $f(b)$ is positive and the other negative. Then $f(x)$ has a real zero in the interval (a, b).

As a consequence of the Intermediate Value theorem, we conclude that if a and b are two consecutive real zeros of f, then for x between a and b, the values of f are all of one sign, either all positive or all negative. (If not, then by the Intermediate Value theorem, there would be a zero between the supposedly consecutive zeros.) We can determine which sign prevails by evaluating f for a **test value** lying between a and b. Using the signs of f in each of the intervals determined by consecutive real zeros allows us to produce a rough sketch of the graph, as shown in the following two examples.

➤ **EXAMPLE 2**
Graphing a Polynomial
Function

Sketch the graph of the function
$$f(x) = 3(x + 2)(x - 1)(x - 4)$$

Solution

The zeros of f are determined as the solutions of the equation
$$3(x + 2)(x - 1)(x - 4) = 0$$

A product can be zero only if one of the factors is 0. In this case, that means that either $x + 2 = 0$, $x - 1 = 0$, or $x - 4 = 0$. That is, the zeros of x are -2, 1, and 4. These three real zeros divide the x-axis into four intervals, namely,

$$(-\infty, -2), (-2, 1), (1, 4), (4, +\infty)$$

In each of these intervals, we choose a test value and determine the sign of f at the test value. The results are summarized in the following table.

Interval	Test Value x	$f(x)$	Sign of f
$(-\infty, -2)$	-3	$3(-1)(-4)(-7)$	$-$
$(-2, 1)$	0	$3(2)(-1)(-4)$	$+$
$(1, 4)$	2	$3(4)(1)(-2)$	$-$
$(4, +\infty)$	5	$3(7)(4)(1)$	$+$

We now may sketch the graph. Begin by plotting the points corresponding to the x-intercepts, namely, the points $(-2, 0)$, $(1, 0)$, $(4, 0)$. Next, we consult the table of test intervals. The sign of f in a test interval determines whether the graph is above or below the x-axis in that interval. For the first interval listed, the value of f is negative, so the graph lies below the x-axis throughout the interval. As x becomes large, either positively or negatively, the values of f become arbitrarily large in absolute value. (This is a general property of polynomial functions that we will assume without proof.) Since f is negative in the first interval, the values must get arbitrarily large in the negative direction as x becomes large in the negative direction. This allows us to sketch the graph of f for x in the first interval.

In the second interval, f is positive. At the left and right endpoints of the interval, the value of $f(x)$ is 0. This means that the graph must have a turning point somewhere in the interval. Knowing this allows us to make a sketch of the graph for x in the second interval. (We have drawn the graph with one turning point in this interval. It might have more than one, but without calculus we have no way of knowing.)

In the third interval, f is negative. Again, because both endpoints give the value 0 for $f(x)$, we can use the same reasoning as in the second interval to sketch the graph corresponding to x in this interval. Finally, the last interval includes values of x that are arbitrarily large. For these values, $f(x)$ grows arbitrarily large. And since the sign of $f(x)$ is positive, the growth is in the positive direction. The final sketch of the graph is shown in Figure 11.

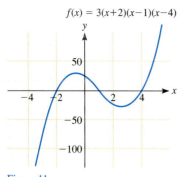

$f(x) = 3(x+2)(x-1)(x-4)$

Figure 11

> **EXAMPLE 3**
Graphing a Fourth-Degree Polynomial

Sketch the graph of the function

$$f(x) = x^4 - 4x^2$$

Solution

Factor the expression on the right to obtain

$$f(x) = x^2(x^2 - 4) = x^2(x + 2)(x - 2)$$

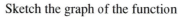

Examining the factors on the right, we see that the zeros of f are -2, 0, and 2. These divide the x-axis into four intervals, as shown in the following table. As in the preceding example, we choose a test value in each interval and determine the sign of f at the test value. The results are summarized in the following table.

Interval	Test Value x	$f(x)$	Sign of f
$(-\infty, -2)$	-3	$(-3)^2(-1)(-5)$	$+$
$(-2, 0)$	-1	$(-1)^2(1)(-3)$	$-$
$(0, 2)$	1	$(1)^2(3)(-1)$	$-$
$(2, +\infty)$	3	$(3)^2(5)(1)$	$+$

In Figure 12 we have sketched the graph of f by first plotting the zeros and then drawing a section of the graph corresponding to each of the intervals in the

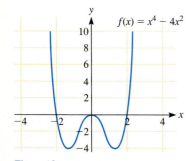

$f(x) = x^4 - 4x^2$

Figure 12

table. The accuracy of the graph is enhanced by noting that the function $f(x)$ is even, so that the graph is symmetric with respect to the *y*-axis.

Exercises 4.2

Sketch the graph of each of the following functions.

1. $f(x) = -x^2$

2. $f(x) = 0.25x^2$

3. $f(x) = \frac{1}{2}x^3$

4. $f(x) = 4x^3$

5. $f(x) = (x - 2)^3 + 1$

6. $f(x) = -(x + 1)^3 - 4$

7. $f(x) = x(x - 1)(x + 1)$

8. $f(x) = (x + 3)(x - 1)(x + 2)$

9. $f(x) = 0.25x^4$

10. $f(x) = -x^4$

11. $f(x) = -x^5$

12. $f(x) = -x^6$

13. $f(x) = -x^2 + 2$

14. $f(x) = 2.5x^2$

15. $f(x) = x^3 - 3x + 22$

16. $f(x) = x^3 - x^2 - 2x$

17. $f(x) = x^4 - 2x^2$

18. $f(x) = x^4 - 6x^2$

19. $f(x) = x^3 - 2x^2 + x - 1$

20. $f(x) = x^3 + 2x^2 - 5x - 6$

21. $f(x) = x^4 - 4x^3 - 4x^2 + 16x$

22. $f(x) = x^4 - 2x^3 + 3x - 5$

23. $f(x) = \frac{1}{3}x^3 - \frac{1}{2}x^2 - 2x + 1$

24. $f(x) = \frac{8}{3}x^3 - 2x + \frac{1}{3}$

25. $f(x) = (x - 2)^2(x + 3)$

26. $f(x) = x^2(x + 2)(x^2 + 1)$

♦ Applications

27. **Maximizing volume**. From a thin piece of cardboard 8 in. by 8 in., square corners are cut out so that the sides can be folded up to make a box. (See figure.) Let *V* denote the volume of the box.

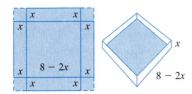

Exercise 27

a. Determine a polynomial function in one variable for $V(x)$.

b. Graph the function on the interval [0, 4].

c. Use the graph to estimate the maximum value of the function and the value of *x* at which it occurs.

28. **Deflection of a beam**. A beam rests at two points, *A* and *B*, and has a concentrated load applied to the center of the beam, as shown in the accompanying figure. Let *y* denote the deflection of the beam at a distance of *x* units measured from the beam to the left of the weight. The deflection depends on the elasticity of the board, the load, and other physical characteristics. Suppose under certain conditions that *y* is given by

$$y = \frac{1}{12}x^3 - \frac{1}{16}x$$

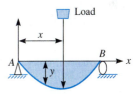

Exercise 28

a. Graph the function on the interval [0, 10].

b. Use the graph to estimate the minimum value of the function and the value of *x* at which it occurs.

➡ Technology

Graph each of the following functions using a graphing calculator. Use the resulting graph to estimate the *x*-intercepts.

29. $f(x) = x^3 - 3x^2 - 144x - 140$

30. $f(x) = x^4 - 2x^3$

31. $f(x) = 6x^5 - 24x^3$

32. $f(x) = x^2(x + 5)(x^2 + 3x + 5)^2$

33. You are given the following total revenue and total cost functions.

$$R(x) = 100x - x^2, \quad C(x) = \frac{1}{3}x^3 - 6x^2 + 89x + 100$$

a. Total profit is given by $P(x) = R(x) - C(x)$. Find $P(x)$.

b. Graph $R(x)$, $C(x)$, and $P(x)$ using the same set of axes.

c. Estimate the value or values of *x* at which $P(x)$ has a maximum. Estimate the maximum value.

♦ **Applications**

34. **Medical dosage**. The function

$$N(t) = -0.045t^3 + 2.063t + 2$$

gives the bodily concentration, in parts per million, of a certain dosage of medication after time t, in hours.

a. Graph the function.

b. Estimate the maximum value of the function and the time t at which it occurs.

c. The *minimum effective dosage* of the medication occurs for values of t for which $2 \le N(t)$. Use the graph to estimate the times for which the medication has its minimum effective dosage.

35. **Temperature during an illness**. A patient's temperature during an illness was given by

$$T(t) = 98.6 - t^2(t - 4)$$

a. Graph the function over the interval $[0, 4]$.

b. Estimate the maximum value of the function and the time t at which it occurs.

Match the following functions with the corresponding graph among graphs (a)–(f).

36. $f(x) = x^3$
37. $f(x) = x^4 + 3$
38. $f(x) = x(x + 1)(x - 2)$
39. $f(x) = x^3 - 3x + 1$
40. $f(x) = -x^4$
41. $f(x) = (x^2 - 4)(x^2 - 9)$

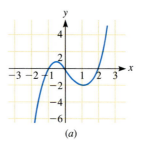

(a)

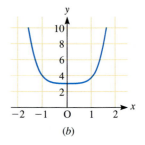

(b)

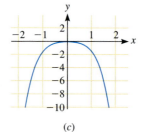

(c)

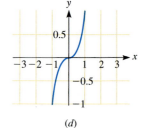

(d)

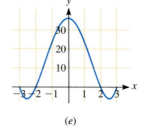

(e)

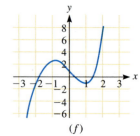

(f)

4.3 THE DIVISION ALGORITHM

In the preceding section, we saw how knowing the zeros of a polynomial function provided a first piece of information for sketching the graph. In this section and the next, we will discuss a number of theorems that provide information about the zeros of polynomial functions. In this section, we relate the zeros to the linear factors of the polynomial via a fundamental result called the division algorithm.

Long Division of Polynomials

We can divide two polynomials to produce a quotient and a remainder. As an illustration, consider the quotient

$$\frac{x^2 + 3x + 1}{x}$$

We can divide a polynomial by a monomial by dividing each term of the polynomial by the monomial:

$$\frac{x^2 + 3x + 1}{x} = \frac{x^2}{x} + \frac{3x}{x} + \frac{1}{x} = x + 3 + \frac{1}{x}$$

We may express this result purely in terms of polynomials by multiplying each term by the denominator x to obtain

$$x^2 + 3x + 1 = x(x + 3) + 1$$

The polynomial $x + 3$ is the quotient and 1 is the remainder.

You may calculate the quotient and remainder using the process of long division, which is included in first courses in algebra. The next example recalls the procedure.

➤ **EXAMPLE 1**
Long Division

Calculate the quotient and remainder when $f(x) = 2x^3 + 5x^2 - 10x - 7$ is divided by $g(x) = 2x - 1$.

Solution

Since we are seeking the quotient and remainder for the division $f(x)/g(x)$, we organize the calculation as a long division problem of $f(x)$ divided by $g(x)$.

$$
\begin{array}{r}
x^2 \quad +3x \quad -\dfrac{7}{2} \qquad \text{quotient} \\
2x - 1 \overline{\smash{\big)}\, 2x^3 \quad +5x^2 \quad -10x \quad -7} \\
\underline{2x^3 \quad -x^2} \\
6x^2 \quad -10x \\
\underline{6x^2 \quad -3x} \\
-7x \quad -7 \\
\underline{-7x \quad +\dfrac{7}{2}} \\
-\dfrac{21}{2} \qquad \text{remainder}
\end{array}
$$

Therefore, in this example, the quotient $q(x)$ and remainder $r(x)$ are:

$$
q(x) = x^2 + 3x - \frac{7}{2}, \qquad r(x) = -\frac{21}{2}
$$

Note that the degree of the denominator $g(x)$ is 1, and the degree of the remainder is 0, which is less than the degree of $g(x)$. Note also that since both $f(x)$ and $g(x)$ have all rational coefficients, so do the quotient and remainder. However, observe that even though both $f(x)$ and $g(x)$ have all integer coefficients, the quotient and remainder have some noninteger coefficients.

The division algorithm states that a computation like the one just carried out can always be carried out for computing the quotient of a polynomial $f(x)/g(x)$.

DIVISION ALGORITHM FOR POLYNOMIALS

Suppose that $f(x)$ and $g(x)$ are polynomials with complex coefficients, with $g(x)$ not the zero polynomial. Then unique polynomials $q(x)$ and $r(x)$ exist such that

1. $f(x) = g(x)q(x) + r(x)$
2. Either $r(x)$ is the zero polynomial, or the degree of $r(x)$ is less than the degree of $g(x)$.

We will omit the proof of this result.

The polynomial $q(x)$ is called the **quotient** and $r(x)$ the **remainder** when $f(x)$ is divided by $g(x)$. If $f(x)$ and $g(x)$ have real (respectively rational) coefficients, then $q(x)$ and $r(x)$ have real (respectively rational) coefficients.

Factoring and Zeros

Let's now consider the special case when a polynomial $f(x)$ is divided by a linear polynomial of the form $x - a$. By the division algorithm, we may express the division in the form

$$f(x) = (x - a)q(x) + r(x)$$

where $q(x)$ is the quotient and $r(x)$ is the remainder. From the statement of the division algorithm, either $r(x)$ is the zero polynomial or the degree of $r(x)$ is less than the degree of $x - a$, which is 1. That is, in either case, $r(x)$ is a constant polynomial.

We can determine the value of this constant by replacing x in the preceding equation with a. We then obtain

$$f(a) = (a - a)q(a) + r(a)$$
$$= 0 + r(a)$$
$$= r(a)$$

Thus, the value $r(a)$ is equal to $f(a)$. But since $r(x)$ is a constant polynomial, this means that $r(x)$ is equal to $f(a)$. This proves the following useful result.

REMAINDER THEOREM

Let $f(x)$ be a polynomial with complex coefficients and a a complex number. Then the remainder on dividing $f(x)$ by $x - a$ is the constant polynomial $f(a)$. That is

$$f(x) = (x - a)q(x) + f(a)$$

where $q(x)$ is the quotient when $f(x)$ is divided by $x - a$.

➤ **EXAMPLE 2**
Calculating the Remainder

Suppose that

$$f(x) = -x^4 + 5x^2 - 2$$

What is the remainder when $f(x)$ is divided by $x - 3$?

Solution

By the Remainder theorem, the remainder is the constant polynomial $f(3)$, which equals

$$-(3)^4 + 5(3)^2 - 2 = -81 + 45 - 2 = -38$$

That is, the remainder is the constant polynomial $r(x) = -38$.

We can draw the following simple, but important, consequence of the Remainder theorem.

FACTOR THEOREM

Let $f(x)$ be a polynomial with complex coefficients and a a complex number. Then $x - a$ is a factor of $f(x)$ if and only if a is a zero of $f(x)$.

Proof   Suppose that a is a zero of the polynomial $f(x)$. Then $f(a)$ is equal to 0 and the equation in the Remainder theorem reads

$$f(x) = (x - a)q(x)$$

That is, $x - a$ is a factor of $f(x)$. Conversely, suppose that $x - a$ is a factor of $f(x)$. Then there is a polynomial $g(x)$ such that

$$f(x) = (x - a)g(x)$$

Substituting a for x in this equation, we see that

$$f(a) = (a - a)g(a) = 0$$

That is, $f(a)$ is equal to 0 and a is a zero of $f(x)$.   ◆

Combining the factor theorem with what we already know about zeros of polynomials, we now have a number of equivalent ways of recognizing real zeros of a polynomial:

REAL ZEROS OF POLYNOMIAL FUNCTIONS

Suppose that $f(x)$ is a polynomial with real coefficients and a is a real number. Then the following are all equivalent ways of stating that a is a zero of $f(x)$.

1. a is a solution of the equation $f(x) = 0$.
2. $x - a$ is a factor of $f(x)$.
3. The graph of $f(x)$ has $(a, 0)$ as an x-intercept.

➤ **EXAMPLE 3**

Applying the Factor Theorem

Suppose that $f(x)$ is a cubic polynomial with zeros $1/2$, 1, and 3. Further, suppose that $f(0)$ is equal to 4. Determine $f(x)$.

Solution

Since $1/2$, 1, and 3 are all zeros of $f(x)$, the Factor theorem asserts that $f(x)$ has as factors each of the polynomials $x - 1/2$, $x - 1$, and $x - 3$. Therefore, $f(x)$ has as a factor the polynomial

$$\left(x - \frac{1}{2}\right)(x - 1)(x - 3)$$

And since $f(x)$ is a cubic polynomial, then

$$f(x) = c\left(x - \frac{1}{2}\right)(x - 1)(x - 3)$$

for some constant c. We are also given that $f(0) = 4$. To use this fact, we substitute 0 for x into the above equation for $f(x)$ to obtain

$$4 = f(0) = c\left(0 - \frac{1}{2}\right)(0 - 1)(0 - 3)$$

$$4 = -\frac{3}{2}c$$

Solving this equation for c gives us

$$c = -\frac{8}{3}$$

Finally, substituting this back into the formula for $f(x)$ gives us

$$f(x) = -\frac{8}{3}\left(x - \frac{1}{2}\right)(x - 1)(x - 3)$$

This answer is acceptable as it stands. Or, if you wish, you may multiply out the right-hand side to obtain the cubic polynomial expression for $f(x)$. However, this expression is preferable since it is clear just by looking at it that it satisfies the conditions specified in the problem.

➤ **EXAMPLE 4**
Application
of the Factor Theorem

Suppose that $f(x)$ is a polynomial with zeros $1 + i$ and $1 - i$. What can be said about $f(x)$?

Solution

By the Factor theorem, we conclude that $f(x)$ is divisible by the polynomials $x - (1 + i)$ and $x - (1 - i)$. Thus, $f(x)$ is divisible by

$$[x - (1 + i)][x - (1 - i)] = x^2 - 2x + 2$$

Here is an interesting consequence of the Factor theorem.

NUMBER OF ZEROS THEOREM

A polynomial of degree n has at most n distinct zeros.

Proof   If $f(x)$ has distinct zeros

$$a_1, a_2, \ldots, a_k$$

then $f(x)$ has as factors the polynomials

$$x - a_1, x - a_2, \ldots, x - a_k$$

And since all the given zeros are assumed distinct, these factors are all different, so that $f(x)$ has as a factor the polynomial

$$(x - a_1)(x - a_2) \cdots (x - a_k)$$

Since this is a polynomial of degree k (a product of k linear factors) and is a factor of $f(x)$, k is at most equal to n. This means that the number of distinct zeros of f is at most equal to n.   ◆

Synthetic Division

In the preceding discussion, we showed the significance of dividing a polynomial $f(x)$ by polynomials of the form $x - a$. The calculational procedure we have given for performing this division is often tedious and time-consuming. There is a simple algorithm, called **synthetic division**, for carrying out such calculations in a much simpler and faster manner.

Let's begin by working out an example and showing how the efficient organization of long division calculations leads naturally to synthetic division. Consider the problem of dividing

$$3x^5 - 2x^4 - 5x^3 + x^2 - x + 4$$

by $x - 2$. Here is the traditional method of performing the division:

$$
\begin{array}{r}
3x^4 + 4x^3 + 3x^2 + 7x + 13 \\
x - 2\,\overline{\big)\,3x^5 - 2x^4 - 5x^3 + x^2 - x + 4} \\
\underline{3x^5 - 6x^4} \\
4x^4 - 5x^3 \\
\underline{4x^4 - 8x^3} \\
3x^3 + x^2 \\
\underline{3x^3 - 6x^2} \\
7x^2 - x \\
\underline{7x^2 - 14x} \\
13x + 4 \\
\underline{13x - 26} \\
30
\end{array}
$$

To simplify this computation, the first thing to note is that the calculations are completely contained in the various coefficients. It is not really necessary to include the variables at all. Moreover, since we will always be dividing by a polynomial of the form $x - a$, we may omit the coefficient 1 corresponding to x. This leaves a computation of the following form:

$$
\begin{array}{r}
\;\;3\quad 4\quad 3\quad\;\; 7\quad 13 \\
-2\,\overline{\big)\;3\;\;-2\;\;-5\;\;\;\;1\;\;-1\;\;\;\;4} \\
\underline{3\;\;-6} \\
4\;\;-5 \\
\underline{4\;\;-8} \\
3\;\;\;\;1 \\
\underline{3\;\;-6} \\
7\;\;\;-1 \\
\underline{7\;\;-14} \\
13\;\;\;\;4 \\
\underline{13\;\;-26} \\
30
\end{array}
$$

Note that this array of numbers still contains a lot of duplicate information. Consider the first step of the division, to obtain the first 3 in the quotient (top row). This is just a duplicate of first 3 in the dividend (second row, inside the division sign). Moreover, the third row duplicates the 3 again. Let's simplify the computation by writing the 3 only once, in the dividend. The third row then consists of the single number -6, which is computed as 3 (from the quotient) times the -2 in the divisor (second row, outside the division sign). The next entry in the quotient, 4, is calculated by subtracting -2 minus -6. But we only write 4 once, in the quotient. The fourth row is determined by bringing down the -5 from the dividend, and the fifth row as the product 4 (quotient) times -2 (dividend), and so forth. This is how the abbreviated computation is shown:

$$
\begin{array}{r}
3 \quad\;\; 4 \quad\;\; 3 \quad\;\;\; 7 \quad\;\; 13 \\
\hline
-2\,|\,3 \quad -2 \quad -5 \quad\; 1 \quad -1 \quad\;\; 4 \\
-6 \\
\hline
-5 \\
-8 \\
\hline
1 \\
-6 \\
\hline
-1 \\
-14 \\
\hline
4 \\
-26 \\
\hline
30
\end{array}
$$

Rather than subtracting numbers at each stage, let's perform addition. This can be done by changing the sign of the -2 in the divisor. The multiplications will then result in the negatives of the previous results, so we may add the entries at each step. The computation now looks like this:

$$
\begin{array}{r}
3 \quad\;\; 4 \quad\;\; 3 \quad\;\;\; 7 \quad\;\; 13 \\
\hline
2\,|\,3 \quad -2 \quad -5 \quad\; 1 \quad -1 \quad\;\; 4 \\
6 \\
\hline
-5 \\
8 \\
\hline
1 \\
6 \\
\hline
-1 \\
14 \\
\hline
4 \\
26 \\
\hline
30
\end{array}
$$

Let's now compress the entire display into three lines as follows:

$$
\begin{array}{r}
3 \quad\;\; 4 \quad\; 3 \quad\;\; 7 \quad\; 13 \\
\hline
2\,|\,3 \quad -2 \quad -5 \quad\; 1 \quad -1 \quad\;\; 4 \\
6 \quad\;\; 8 \quad\; 6 \quad 14 \quad 26 \quad 30
\end{array}
$$

Finally, let's combine the dividend row at the top and the remainder 30 at the bottom into a single row and place it at the bottom. Moreover, tradition dictates turning the division sign inside out. (This surely doesn't affect the calculation!)

$$
\begin{array}{r|rrrrrr}
2 & 3 & -2 & -5 & 1 & -1 & 4 \\
  &   & 6 & 8 & 6 & 14 & 26 \\
\hline
  & 3 & 4 & 3 & 7 & 13 & 30
\end{array}
$$

Note how simply the bottom row may be calculated: The initial 3 in the quotient is a copy of the 3 in the dividend in the first row. Then compute 6 as 2 (divisor) times 3 (quotient). Compute 4 by adding -2 and 6. Now compute 8 as 4 (quotient) times 2 (divisor), and so forth. The quotient corresponds to the terms of the last row, except for the last, which is the remainder.

A similar procedure may be used to calculate the quotient of any polynomial $f(x)$ divided by $x - a$.

SYNTHETIC DIVISION

To divide $f(x)$ by $x - a$ using synthetic division:

1. Write the coefficients of f across the top row.
2. Start with the leftmost column. Bring down the leading coefficient into the third row.
3. Multiply the entry in the third row by a.
4. Move one column to the right and put the result of step 3 in the second row.
5. Add rows 1 and 2 of the current column and put the result in the third row.
6. Repeat steps 3–5 for each of the columns in turn.
7. The numbers in the third row are the coefficients of the quotient and remainder.

The following examples illustrate the mechanics of synthetic division and show how it may be applied to get information about a polynomial $f(x)$.

➤ **EXAMPLE 5**
Synthetic Division

Use synthetic division to determine the quotient and remainder when $f(x) = 4x^5 - 2x^4 + x^3 - 7x^2 + 3$ is divided by $x - 2$.

Solution

We use the algorithm for synthetic division. Note that we have included a zero to represent the zero coefficient of the x term. This is because there must be one column corresponding to each power of x in the divisor. If a power of x is missing, we must use 0 as a placeholder in the corresponding column. Be careful of this point. Leaving out a 0 placeholder is an easy mistake to make.

$$
\begin{array}{r|rrrrrr}
2 & 4 & -2 & 1 & -7 & 0 & 3 \\
  &   & 8 & 12 & 26 & 38 & 76 \\
\hline
  & 4 & 6 & 13 & 19 & 38 & 79
\end{array}
$$

The coefficients of the quotient are given by the first five entries in the third row of the table, making the quotient

$$q(x) = 4x^4 + 6x^3 + 13x^2 + 19x + 38$$

The last entry in the table, 79, gives the remainder.

➤ **EXAMPLE 6**
Application of Synthetic Division

Use synthetic division to determine $f(-3)$, where

$$f(x) = -x^4 - 5x^3 + 4x^2 - 9x + 10$$

Solution

By the Remainder theorem, the value of $f(-3)$ is the remainder when $f(x)$ is divided by $x - (-3)$. We use synthetic division to determine the value of this remainder.

$$
\begin{array}{r|rrrrr}
-3 & -1 & -5 & 4 & -9 & 10 \\
   &   & 3 & 6 & -30 & 117 \\
\hline
   & -1 & -2 & 10 & -39 & 127
\end{array}
$$

The last entry in the table, the remainder, provides the value of $f(-3)$. That is

$$f(-3) = 127$$

➤ **EXAMPLE 7**
**Another Application
of Synthetic Division**

Use synthetic division to prove that -2 is a zero of the polynomial

$$f(x) = x^6 - 4x^4 + 16x^2 - 64$$

Solution

According to the Remainder theorem, -2 is a zero of $f(x)$ if and only if the remainder is zero on dividing $f(x)$ by $x - (-2)$. We can determine this remainder using synthetic division.

$$
\begin{array}{r|rrrrrrr}
-2 & 1 & 0 & -4 & 0 & 16 & 0 & -64 \\
   &   & -2 & 4 & 0 & 0 & -32 & 64 \\
\hline
   & 1 & -2 & 0 & 0 & 16 & -32 & 0
\end{array}
$$

The final entry in the synthetic division table, the remainder, is 0. Therefore, $f(-2) = 0$, so -2 is a zero of $f(x)$.

Exercises 4.3

For Exercises 1–16, a polynomial $f(x)$ and a divisor $g(x)$ are given. Calculate the quotient $q(x)$ and the remainder $r(x)$.

1. $f(x) = x^5 - 2x^4 + x^3 - 5$
   $g(x) = x - 2$

2. $f(x) = 2x^5 - 3x^3 + 2x^2 - x + 3$
   $g(x) = x + 1$

3. $f(x) = 2x^3 + x^2 - 13x + 6$
   $g(x) = x + 2$

4. $f(x) = 6x^4 - 5x^3 + 9x^2 + 2x - 10$
   $g(x) = 2x - 1$

5. $f(x) = 3x^4 - x^3 + 2x - 6$
   $g(x) = 3x - 2$

6. $f(x) = 2x^5 - 3x^3 + 2x^2 - x + 3$
   $g(x) = 4x + 5$

7. $f(x) = x^5 - 2x^4 + x^3 - 5$
   $g(x) = x^2 - 2$

8. $f(x) = 2x^5 - 3x^3 + 2x^2 - x + 3$
   $g(x) = x^2 + 1$

9. $f(x) = x^3 + 125$
   $g(x) = x + 5$

10. $f(x) = x^3 - 1$
    $g(x) = x - 1$

11. $f(x) = x^4 - 3x^2 + 8$
    $g(x) = x^2 - 1$

12. $f(x) = 2x^5 - 3x^3 + 2x^2 - x + 3$
    $g(x) = x^3 + 1$

13. $f(x) = x^3 + x^2$
    $g(x) = x^2 + x - 2$

14. $f(x) = x^5 - x^4$
    $g(x) = x^4 - x^3 + 1$

15. $f(x) = x^6 - y^6$
    $g(x) = x - y$

16. $f(x) = x^5 + b^5$
    $g(x) = x + b$

17. Suppose that $f(x)$ is a quadratic polynomial with zeros -1 and 2, and $f(3) = -10$. Determine f.

18. Suppose that $f(x)$ is a quadratic polynomial with zeros 5 and -2, and $f(1) = 7$. Determine f.

19. Suppose that $f(x)$ is a quadratic polynomial with zeros i and $-i$, and $f(-2) = 6$. Determine f.

20. Suppose that $f(x)$ is a quadratic polynomial with zeros $2 - i$ and $2 + i$, and $f(1) = 7$. Determine f.

21. Suppose that $f(x)$ is a cubic polynomial with zeros -1, 0, and 1, and $f(2) = 24$. Determine f.

22. Suppose that $f(x)$ is a cubic polynomial with zeros -1, 0, and 2, and $f(1) = -24$. Determine f.

23. Suppose that $f(x)$ is a fifth-degree polynomial with zeros 0, -2, -1, i, and $-i$; and $f(1) = -24$. Determine f.

24. Suppose that $f(x)$ is a fifth-degree polynomial with zeros 1, -2, -5, i, and $-i$; and $f(1) = -24$. Determine f.

25. Suppose that $f(x)$ is a fourth-degree polynomial with zeros $1 - \sqrt{2}, 1 + \sqrt{2}, 4 \pm 3i$; and $f(0) = -2$. Determine f.

26. Suppose that $f(x)$ is a fourth-degree polynomial with zeros

$$\frac{2 + \sqrt{3}}{5}, \frac{2 - \sqrt{3}}{5}, \frac{1}{2} - \frac{3}{2}i, \frac{1}{2} + \frac{3}{2}i$$

and $f(0) = 1$. Determine f.

27. Suppose that $f(x)$ is a polynomial with zeros $2 - i$ and $2 + i$. What can be said about $f(x)$?

28. Suppose that $f(x)$ is a polynomial with zeros $1 + \sqrt{5}$ and $1 - \sqrt{5}$. What can be said about $f(x)$?

29. For $f(x) = 2x^3 - 3x^2 + x - 1$,

   a. Find $f(-2)$.

   b. Find the remainder when $f(x)$ is divided by $x + 2$.

   c. Find $f(1)$.

   d. Find the remainder when $f(x)$ is divided by $x - 1$.

30. For $f(x) = 22x^4 - x^3 + x^2 - x + 1$,

   a. Find $f(2)$.

   b. Find the remainder when $f(x)$ is divided by $x - 2$.

   c. Find $f(-1)$.

   d. Find the remainder when $f(x)$ is divided by $x + 1$.

31. For $f(x) = 2x^2 + ix + 1$,

   a. Find $f(i)$.

   b. Find the remainder when $f(x)$ is divided by $x - i$.

32. For $f(x) = 3x^2 - ix - 1$,

   a. Find $f(-i)$.

   b. Find the remainder when $f(x)$ is divided by $x + i$.

Use synthetic division to find the quotient and remainder when $f(x)$ is divided by $g(x)$.

33. $f(x) = 4x^5 + 2x^3 - x + 5$

   $g(x) = x + 2$

34. $f(x) = 2x^5 - 3x^3 + 2x^2 - x + 3$

   $g(x) = x + 1$

35. $f(x) = 2x^3 + x^2 - 13x + 6$

   $g(x) = x - 2$

36. $f(x) = 6x^4 - 5x^3 + 9x^2 + 2x - 10$

   $g(x) = x - 1$

37. $f(x) = x^3 + 125$ 38. $f(x) = x^3 - 1$

   $g(x) = x + 5$ $g(x) = x - 1$

39. $f(x) = 3x^4 - 4x^3 - 19x^2 + 8x + 12$

   $g(x) = x - \dfrac{2}{3}$

40. $f(x) = 4x^5 + 2x^3 - x + 5$

   $g(x) = x + \dfrac{3}{4}$

41. $f(x) = x^6 - y^6$ 42. $f(x) = x^5 + b^5$

   $g(x) = x - y$ $g(x) = x + b$

Use synthetic division to determine the given function values.

43. $f(x) = 2x^4 - x^3 - 3x^2 + 2$; find $f(1), f(-2), f(4)$.

44. $f(x) = x^4 - 3x^3 + 5x - 2$; find $f(-2), f(0), f(2)$.

45. $f(x) = x^3 - 4x^2 + 9$; find $f(-1), f(3), f(5)$.

46. $f(x) = 2x^3 - 5x^2 + x - 3$; find $f(-2), f(1), f(4)$.

47. $f(x) = 2x^3 - 3x^2 + 5x - 1$; find $f(3), f(-5), f(11)$, $f(2/3)$.

48. $f(x) = x^5 - 3x^4 + 2x^2 - x - 5$; find $f(-3), f(20)$, $f(-1/2)$.

Use synthetic division to determine whether each number is a zero of the given polynomial.

49. $-1, 1; f(x) = x^4 + 2x^3 + 3x^2 - 2$

50. $-1, 2; f(x) = x^4 - 2x^3 + x^2 - x - 2$

51. $i, 2i; f(x) = 2x^4 - x^3 + x^2 - x - 1$

52. $i, 0; f(x) = 3x^4 - 2x^2 + 5x$

53. $\dfrac{1}{3}, -5; f(x) = 6x^3 + 31x^2 + 4x - 5$

54. $3, -\dfrac{1}{2}; f(x) = x^3 + \dfrac{17}{2}x^2 + 19x + \dfrac{15}{2}$

Factor each polynomial $f(x)$. Then use the result of the factoring to solve the equation $f(x) = 0$.

55. $f(x) = x^4 - x^3 - 19x^2 - 11x + 30$

56. $f(x) = x^3 + 5x^2 - 12x - 36$

57. $f(x) = x^4 + 2x^3 - 13x^2 - 14x + 24$

58. $f(x) = x^3 - 4x^2 + x + 6$

59. Solve $x^4 + 2x^3 - 13x^2 - 14x + 24 < 0$.

60. Solve $x^3 - 4x^2 + x + 6 > 0$.

61. Find k such that when $2x^3 + x^2 - 5x + 2k$ is divided by $x + 1$ the remainder will be 6.

62. Find those values of k such that $x - 3$ is a factor of $kx^3 - 5x^2 - 2kx + 3$.

For Exercises 63–66, a polynomial $f(x)$ and a divisor $g(x)$ are given. Calculate the quotient $q(x)$ and the remainder $r(x)$.

63. $f(x) = (x + 1)^2 - 2(x + 1) + 2$

   $g(x) = x + 1$

64. $f(x) = (x - 2)^3 + 7(x - 2) + 4$

   $g(x) = x - 2$

65. $f(x) = (x + 1)^3 4(x - 5)^3 + (x - 5)^4 3(x + 1)^2$

    $g(x) = x + 1$

66. $f(x) = (2x^2 - 5x - 3)^3$

    $g(x) = 2x + 1$

67. *a.* Find a way to use synthetic division to divide a polynomial by a divisor of the form $ax + b$.

    b. Use the result of (a) to find the quotient and remainder when $f(x)$ is divided by $g(x)$, where
    $$f(x) = 5x^3 + 4x^2 - 3x - 7$$
    $$g(x) = 2x + 3$$

68. Prove that $x - a$ is a factor of $x^n - a^n$ for all natural numbers n.

69. Prove that $x + a$ is a factor of $x^n + a^n$ for all odd positive natural numbers n.

70. Prove that $x + a$ is a factor of $x^n - a^n$ for all even natural numbers n.

71. Show that $x^n - a^n = (x - a)(x^{n-1} + x^{n-2}a + \cdots + a^{n-1})$ for all natural numbers n.

72. Show that $x^n - a^n = (x + a)(x^{n-1} - x^{n-2}a + x^{n-3}a^2 - \cdots - a^{n-1})$ for all even natural numbers n.

73. Show that $x^n + a^n = (x + a)(x^{n-1} - x^{n-2}a + x^{n-2}a^2 - \cdots + a^{n-1})$ for all odd natural numbers n.

➡ **Technology**

74. Let $f(x) = x^3 - 5x + 1$, $g(x) = x^2 - 2x$.

    a. Calculate $q(x), r(x)$ using the division algorithm.

    b. Use a graphing calculator to graph $f(x)/g(x)$ and $q(x)$ on the same coordinate system.

    c. Zoom out the graph of (b) several times. Describe what you see.

75. Repeat the preceding exercise with $f(x) = x^4 - 2x^3 + 10$, $g(x) = 3x^2 - x$.

76. Can you generalize the graphical phenomenon exhibited in the preceding two exercises? Test your generalization using four more sets of polynomials.

4.4 THE ZEROS OF A POLYNOMIAL

In this section we introduce additional results that provide information about the zeros of a polynomial with complex coefficients.

The Fundamental Theorem of Algebra

A polynomial with real coefficients may have complex numbers as zeros. Therefore, it might be the case that some polynomials with complex coefficients have zeros belonging to some number system larger than the complex numbers. However, this is not the case, as asserted by the following theorem.

FUNDAMENTAL THEOREM OF ALGEBRA

Let $f(x)$ be a nonconstant polynomial with complex coefficients. Then $f(x)$ has at least one complex zero.

Any proof of the Fundamental Theorem of Algebra requires advanced mathematical techniques well beyond the scope of this text. However, in what follows we will assume this result without proof.

An immediate consequence of the Fundamental Theorem of Algebra is the following result.

FACTORIZATION THEOREM

Let $f(x)$ be a polynomial with complex coefficients and leading coefficient a and positive degree n. Then $f(x)$ can be written in the form
$$f(x) = a(x - \alpha_1)(x - \alpha_2) \cdots (x - \alpha_n)$$
where
$$\alpha_1, \alpha_2, \ldots, \alpha_n$$
are complex numbers.

Proof   By the Fundamental Theorem of Algebra, $f(x)$ has a complex zero α_1. By the Factor theorem, this implies that $f(x)$ has $x - \alpha_1$ as a factor.

$$f(x) = (x - \alpha_1)f_1(x)$$

If $f_1(x)$ is a constant polynomial, then $f_1(x) = a$. If not, then $f_1(x)$ has degree at least 1, and the Fundamental Theorem of Algebra implies that $f_1(x)$ has a complex zero α_2. Again applying the Factor theorem, we see that

$$f_1(x) = (x - \alpha_2)f_2(x)$$
$$f(x) = (x - \alpha_1)(x - \alpha_2)f_2(x)$$

If $f_2(x)$ is a constant polynomial, then $f_2(x) = a$. Otherwise, repeat the same reasoning again. After n repetitions of the reasoning, we will arrive at the desired factorization of f.   ◆

➤ **EXAMPLE 1**
Factoring into Linear Factors

Factor the following polynomial into linear factors with complex coefficients:

$$f(x) = 6x^4 + 6x^3 + 18x^2$$

Solution

We begin by noting that we may factor out $6x^2$ from each term to obtain

$$6x^2(x^2 + x + 3)$$

To factor the second quadratic polynomial, we use the quadratic formula to determine its zeros.

$$\frac{-1 \pm \sqrt{1^2 - 4(1)(3)}}{2} = \frac{-1 \pm \sqrt{-11}}{2} = \frac{-1 \pm \sqrt{11}i}{2}$$

Therefore, the factorization of the given polynomial is

$$6x^4 + 6x^3 + 18x^2 = 6x^2\left(x - \frac{-1 + \sqrt{11}i}{2}\right)\left(x - \frac{-1 - \sqrt{11}i}{2}\right)$$

**Historical Note:
Formulas for the Solution
of Polynomial Equations**

The quadratic formula gives us a way of calculating the zeros of a quadratic polynomial using the operations of addition, subtraction, multiplication, division, and extraction of roots. It settles, once and for all, the problem of determining the zeros of quadratic polynomials, and in a very effective manner. It is natural to expect that such formulas exist for polynomials of other degrees. In the 14th century, the Renaissance mathematician Cardano developed a formula for calculating the zeros of a cubic equation. Cardano's formula is complex, but it does allow us to calculate the zeros from the coefficients of the equation and resembles, in broad shape, the quadratic formula. Farrari, a contemporary of Cardano's, developed a similar formula for equations of the fourth degree. However, four centuries passed and, in spite of many attempts, mathematicians were unable to find an equation for the zeros of equations of the fifth degree. The reason for this failure was explained by the 19th-century mathematician Abel, who proved that there exist no formulas for equations of the fifth or higher degrees. Attempts to explain the reason for this phenomenon and to explain the general nature of the zeros of equations of higher degree led to the incredibly fertile field of mathematics called **the theory of equations**. In its turn, the theory of equations led to the field of mathematics called **group theory**, which is now a fundamental tool in the fields of crystallography, astrophysics, elementary particle theory, and quantum mechanics.

Zeros and Their Multiplicities

It can be proved that the factorization given by the Fundamental Theorem of Algebra is unique. That is, any two such factorizations of $f(x)$ differ only in the order of their linear factors. The zeros

$$\alpha_1, \alpha_2, \ldots, \alpha_n$$

may not all be different. The number of times a particular zero appears among the factors is called the **multiplicity** of the zero. For many purposes, it is convenient to count a zero of multiplicity m as m zeros. In this case, we say that *zeros are counted according to their multiplicities*. For example, consider the polynomial $f(x) = (x - 1)^3$. This polynomial has the single zero 1 of multiplicity 3, which is counted as three zeros.

Suppose that the polynomial $f(x)$ has degree n. When zeros are counted according to their multiplicities, the factors listed above consist of n zeros. That is, we have the following result.

NUMBER OF ZEROS OF A POLYNOMIAL

Suppose that $f(x)$ is a polynomial with complex coefficients and positive degree n. If zeros are counted according to their multiplicities, then $f(x)$ has exactly n zeros.

➤ **EXAMPLE 2**
Determining Multiplicities of Zeros

Determine the multiplicities of the zeros of the polynomial

$$f(x) = x^5 (x - 1)^6 (x + 7)^2 (x + i)^2 (x - i)^2$$

Solution

The zeros can be read off from the distinct factors appearing in the factorization of $f(x)$. The multiplicities can be read off from the exponents, which indicate the number of times the particular factor is present. Here are the results:

$$\alpha_1 = 0, \text{multiplicity} = 5$$
$$\alpha_2 = 1, \text{multiplicity} = 6$$
$$\alpha_3 = -7, \text{multiplicity} = 2$$
$$\alpha_4 = -i, \text{multiplicity} = 2$$
$$\alpha_5 = i, \text{multiplicity} = 2$$

Polynomials with Real Coefficients

Polynomials with real coefficients form a very important class of polynomials, since these polynomials occur most frequently in applied problems. Let's now develop some information about the zeros of these polynomials.

CONJUGATE ROOT THEOREM

Suppose that $f(x)$ is a polynomial with real coefficients. If α is a zero of $f(x)$, then so is its complex conjugate $\overline{\alpha}$.

The proof of the Conjugate Root theorem requires several important properties of complex conjugates, which we establish first. Let α and β be complex numbers. Then we have the following:

PROPERTIES OF COMPLEX CONJUGATES

1. $\overline{\alpha + \beta} = \overline{\alpha} + \overline{\beta}$
2. $\overline{\alpha\beta} = \overline{\alpha} \cdot \overline{\beta}$
3. α is real if and only if $\alpha = \overline{\alpha}$

For the proofs of these three facts, suppose that

$$\alpha = a + bi, \qquad \beta = c + di$$

for real numbers a, b, c, and d. Then we have

$$\overline{\alpha + \beta} = \overline{(a + bi) + (c - di)}$$
$$= \overline{(a + c) + (b + d)i}$$
$$= (a + c) - (b + d)i$$
$$= (a - bi) + (c - di)$$
$$= \overline{\alpha} + \overline{\beta}$$

Thus, we have proved Property 1.

The proof of Property 2 is similar and is left to the exercises.

For the proof of Property 3, note that if α is real, then $b = 0$ and

$$\overline{\alpha} = \overline{a + 0i} = a - 0i = a = \alpha$$

Conversely, if

$$\alpha = \overline{\alpha}$$

then

$$a + bi = \overline{a + bi}$$
$$a + bi = a - bi$$
$$2bi = 0$$
$$b = 0$$

That is,

$$\alpha = a + bi = a + 0i$$

is real. This proves Property 3.

Proof of the Conjugate Root Theorem   Let

$$f(x) = \alpha_n x^n + \alpha_{n-1} x^{n-1} + \cdots + a_0$$

Since all the coefficients of $f(x)$ are real, by Property 3 of complex conjugates we have

$$\overline{a_j} = a_j \quad (j = 0, 1, \ldots, n) \tag{1}$$

Since α is a zero of $f(x)$, we have

$$0 = f(\alpha) = a_n \alpha^n + a_{n-1} \alpha^{n-1} + \cdots + a_0$$

Taking the complex conjugates of the left and right sides of this equation, we have

$$\overline{0} = \overline{a_n\alpha^n + a_{n-1}\alpha^{n-1} + \cdots + a_0}$$

However, since 0 is real, it is its own conjugate. Moreover, applying Properties 1 and 2, we see that

$$0 = \overline{a_n\alpha^n + a_{n-1}\alpha^{n-1} + \cdots + a_0}$$
$$= \overline{a_n\alpha^n} + \overline{a_{n-1}\alpha^{n-1}} + \cdots + \overline{a_0}$$
$$= \overline{a_n} \cdot \overline{(\alpha)}^n + \overline{a_{n-1}} \cdot \overline{(\alpha)}^{n-1} + \cdots + \overline{a_0}$$

However, using Equation (1), the preceding equation may be written

$$a_n(\overline{\alpha})^n + a_{n-1}(\overline{\alpha})^{n-1} + \cdots + a_0 = 0$$

The left side of this last equation is just $f(x)$ with x replaced by $\overline{\alpha}$. That is, we have

$$f(\overline{\alpha}) = 0$$

This completes the proof of the theorem. ◆

An immediate consequence of Conjugate Root theorem is the following.

CONJUGATE ROOT THEOREM (ALTERNATE FORM)

The nonreal zeros of a polynomial with real coefficients occur in complex-conjugate pairs.

➤ **EXAMPLE 3**
Zeros of a Fifth-Degree Polynomial

How many real zeros can a polynomial of degree 5 with real coefficients have?

Solution

By the alternate form of the Conjugate Root theorem, the nonreal zeros occur in pairs. The number of pairs can be 0, 1, or 2. (Any more pairs would result in more than five zeros.) These three possibilities correspond to the following numbers of real zeros:

$$0 \text{ pairs:} \quad 5 - 0 = 5 \text{ real zeros}$$
$$1 \text{ pair:} \quad 5 - 2 = 3 \text{ real zeros}$$
$$2 \text{ pairs:} \quad 5 - 4 = 1 \text{ real zero}$$

➤ **EXAMPLE 4**
Determining Complex Zeros

Determine the zeros of the polynomial

$$x^6 - 1$$

Solution

We proceed by factoring the polynomial using the various factorization formulas from Chapter 1.

$$x^6 - 1 = (x^3)^2 - 1$$
$$= (x^3 - 1)(x^3 + 1)$$
$$= (x - 1)(x^2 + x + 1)(x + 1)(x^2 - x + 1)$$

From this factorization, we see that the zeros are solutions of the equations

$$x - 1 = 0, \qquad x + 1 = 0$$
$$x^2 + x + 1 = 0, \qquad x^2 - x + 1 = 0$$

The first two equations give the zeros $1, -1$. To solve the last two equations, we use the quadratic formula:

$$x^2 + x + 1 = 0$$

$$x = \frac{-1 \pm \sqrt{(1)^2 - 4(1)(1)}}{2(1)}$$

$$= \frac{-1 \pm \sqrt{3}i}{2}$$

In a similar fashion, the last equation yields the zeros

$$x = \frac{1 \pm \sqrt{3}i}{2}$$

So the six zeros of the given polynomial are

$$\pm 1, \frac{\pm 1 \pm \sqrt{3}i}{2}$$

where all possible combinations of signs are allowed in the final expression.

The Fundamental Theorem of Algebra says that a polynomial with real co-efficients can be factored in the form

$$f(x) = a(x - \alpha_1)(x - \alpha_2) \cdots (x - \alpha_n)$$

where the complex numbers

$$\alpha_1, \alpha_2, \ldots, \alpha_n$$

are the zeros of $f(x)$. According to the alternate form of the Conjugate Root theorem, the nonreal zeros occur in complex-conjugate pairs. If such a pair of zeros is

$$\beta, \overline{\beta}$$

then the product

$$(x - \beta)(x - \overline{\beta})$$

is contained in the factorization of $f(x)$ and is equal to

$$x^2 - (\beta + \overline{\beta})x + \beta\overline{\beta}$$

The coefficients of this polynomial are real because if we let $\beta = a + bi$, then

$$\beta + \overline{\beta} = (a + bi) + (a - bi) = 2a$$

and

$$\beta\overline{\beta} = (a + bi)(a - bi) = a^2 + b^2$$

and both of the expressions on the right of these two equations are real numbers. This shows that the factor corresponding to a complex-conjugate pair of zeros, namely,

$$(x - \beta)(x - \overline{\beta})$$

is a real polynomial. On the other hand, if a zero α is real, the factor $x - \alpha$ has real coefficients. This proves the following important result about the factorization of polynomials with real coefficients.

POLYNOMIAL FACTORIZATION OVER THE REALS

Let $f(x)$ be a polynomial of positive degree with real coefficients. Then $f(x)$ can be factored into a product of linear and quadratic polynomials having real coefficients, where the quadratic factors have no real zeros.

➤ **EXAMPLE 5**
Factoring into Real Factors

Factor the following polynomial into polynomials with real coefficients.
$$x^4 + x^3 + 5x^2 + 4x + 4$$

Solution

Note that the coefficient 4 appears twice. This suggests that we group terms together so that we can factor out a 4. In addition to grouping the last two terms together, we can rewrite the middle term as
$$x^2 + 4x^2$$

So now we can factor two groups of terms:
$$(x^4 + x^3 + x^2) + (4x^2 + 4x + 4) = x^2(x^2 + x + 1) + 4(x^2 + x + 1)$$
$$= (x^2 + 4)(x^2 + x + 1)$$

Notice that the zeros of each of the two factors are nonreal complex numbers, since their respective discriminants are negative (-16 for the first and -3 for the second). Therefore, the polynomials cannot be factored further into real polynomials.

Note, however, that by using the quadratic formula to determine the zeros of each of the factors, we see that the given polynomial may be factored into complex polynomials
$$(x + 2i)(x - 2i)\left(x - \frac{-1 + \sqrt{3}i}{2}\right)\left(x - \frac{-1 - \sqrt{3}i}{2}\right)$$

Descartes' Rule of Signs

It is possible to derive information about the zeros of a polynomial $f(x)$ with real coefficients by examining the number of changes of sign in the sequence of coefficients. Consider the polynomial
$$5x^6 - 3x^4 + x^3 + 0.5x^2 - 10x + 4$$

The sequence of signs in this polynomial is
$$+ - + + - +$$

The number of changes in this sequence is four: the change from the first to the second, from the second to the third, from the fourth to the fifth, and from the fifth to the sixth. Note that *no change in signs is recorded for terms with zero coefficients*. The number of changes in the sequence of signs is related to the number of real zeros of the polynomial by the following famous result of Descartes.

DESCARTES' RULE OF SIGNS

Let $f(x)$ be a polynomial with real coefficients and with nonzero constant term.

1. The number of positive real zeros is at most equal to the number of changes m in the sequence of signs of $f(x)$ and differs from m by an even integer.
2. The number of negative real zeros is at most equal to the number of changes n in the sequence of signs for $f(-x)$ and differs from n by an even integer.

➤ **EXAMPLE 6**

Number of Real Zeros

Apply Descartes' Rule of Signs to the polynomial

$$5x^6 - 3x^4 + x^3 + 0.5x^2 - 10x + 4$$

Solution

The polynomial has a nonzero constant term, so we may apply Descartes' Rule of Signs. As we have already seen, the sequence of signs for this polynomial is $+ - + + - +$, and there are four changes in this sequence, so $m = 4$. By Descartes' Rule of Signs, the polynomial has at most four positive real zeros, and the actual number differs from 4 by an even integer. Thus, the number of positive zeros is either 4, 2, or 0.

To analyze the possibilities for negative zeros, we form $f(-x)$:

$$5(-x)^6 - 3(-x)^4 + (-x)^3 + 0.5(-x)^2 - 10(-x) + 4$$
$$= 5x^6 - 3x^4 - x^3 + 0.5x^2 + 10x + 4$$

The sequence of signs for $f(-x)$ is

$$+ - - + + +$$

There are two changes in sign, so $n = 2$. Descartes' Rule of Signs then states that the polynomial has at most two negative real zeros, and the number of negative real zeros is either 2 or 0.

Thus, Descartes' Rule of Signs shows that there are six possibilities, as shown in the following table:

Positive Real Zeros	Negative Real Zeros	Nonreal Zeros	Total Zeros
4	2	0	6
4	0	2	6
2	2	2	6
0	0	6	6
0	2	4	6
2	0	4	6

➤ **EXAMPLE 7**

Study of Real Zeros

Use Descartes' Rule of Signs to analyze the zeros of the equation

$$x^3 + 2x^2 - x + 1$$

Solution

There are two changes in the sequence of signs of $f(x)$. By Descartes' Rule of Signs, this means that there are at most two positive zeros. Moreover, the number of positive zeros differs from 2 by an even integer. This means that the number of positive zeros is either 2 or 0.

To analyze the negative zeros, we form $f(-x)$:

$$f(-x) = (-x)^3 + 2(-x)^2 - (-x) + 1 = -x^3 + 2x^2 + x + 1$$

There is a single change in sign. Again appealing to Descartes' Rule of Signs, this means that there is at most one negative zero. Since the actual number of negative zeros differs from 1 by an even integer, there is only one possibility: a single negative zero.

Combining the various possibilities, we see that there are either two positive zeros and one negative zero or no positive zeros, one negative zero and two nonreal zeros.

Descartes' Rule of Signs assumes that the constant term of the polynomial is nonzero. However, this is not much of a restriction when it comes to analyzing zeros, for a polynomial with a zero constant term has a power of x as a factor. By factoring out this power of x, we can arrive at a polynomial with a nonzero constant term to which Descartes' Rule can be applied. For example, consider the polynomial

$$f(x) = x^5 - 3x^3 + 12x^2$$

It has zero constant term but can be written in the form

$$f(x) = x^2(x^3 - 3x + 12)$$

The first factor on the right corresponds to a zero with multiplicity 2. We may investigate the remaining zeros by applying Descartes' Rule of Signs to the polynomial

$$x^3 - 3x + 12$$

Exercises 4.4

Factor each of the following polynomials into linear factors with complex coefficients.

1. $f(x) = 6x^4 + 5x^3 + 5x^2$

2. $f(x) = 3x^6 + 6x^5 - 3x^4$

3. $f(x) = (x^2 + 1)(x^2 + 2x + 3)$

4. $f(x) = (x^2 + 4)(x^2 - 4x - 5)$

5. $f(x) = x^3 - 1$

6. $f(x) = x^3 + 8$

7. $f(x) = x^3 + 3x^2 + 5x + 15$ (*Hint:* Consider factoring by grouping.)

8. $f(x) = x^3 - 5x^2 + 7x - 35$ (*Hint:* Consider factoring by grouping.)

9. $f(x) = (x^2 + x + 3)^2$

10. $f(x) = (x^2 - x - 5)^2$

Determine the zeros and their multiplicities of the following polynomials.

11. $f(x) = 3x(x - 2)^2(2x - 5)^3$

12. $f(x) = (2x - 3)(x - 4)^2(x + 1)^2$

13. $f(x) = (x^2 - 4)^2(x + 2)^3$

14. $f(x) = (2x^2 - 5x - 3)^3$

15. $f(x) = (x - 2)^3(x + 1)$

16. $f(x) = (x^2 - x - 6)^2(x^2 + 2x - 3)^2$

17. $f(x) = x(x - 3)^2(x^2 + 1)$

18. $f(x) = x^2(x + 5)(x^2 + 3x + 5)^2$

19. $f(x) = x^5(x^2 + x - 1)^3$

20. $f(x) = x^4(x^2 + 5)^3(x^2 - x)$

21. How many real zeros can a polynomial of degree 3 with real coefficients have?

22. How many real zeros can a polynomial of degree 4 with real coefficients have?

23. How many real zeros can a polynomial of degree 2 with real coefficients have?

24. How many real zeros can a polynomial of degree 7 with real coefficients have?

Factor each polynomial into polynomials with real coefficients and then into polynomials with complex coefficients. Then find the zeros.

25. $x^6 - 64$ 26. $t^6 - 1$

27. $x^3 - 1$ 28. $x^3 + 1$

29. $m^3 + 8$ 30. $x^3 - 8$

31. $y^4 - 4$ 32. $y^4 - 9$

33. $r^7 + 2r^5 + r^3$ 34. $p^6 + 2p^4 + p^2$

35. $x^4 - x^3 + 3x^2 - 4x - 4$

36. $x^4 - 2x^3 + 7x^2 - 18x - 18$

37. $x^4 - 2x^2 - 3$ 38. $x^4 - 2x^2 - 35$

39. $r^3 - r^2 - 8r + 12$ 40. $t^3 - 4t^2 - 12t + 48$

Use Descartes' Rule of Signs to analyze the zeros of each polynomial.

41. $5x^4 + x^3 - 2x^2 - 3x + 5$

42. $8x^6 - 3x^5 - 2x^3 + x^2 - 3x + 8$

43. $11x^{10} + 8x^3 + x^2 - x + 2$

44. $x^8 - x^7 + 3x^6 + 2x^2 - 1$

45. $5x^4 + x^3 - 2x^2 - 3x + 5$

46. $8x^6 - 3x^5 - 2x^3 + x^2 - 3x + 8$

47. $x^5 - x^4 - 3x^3 + 2x^2 - x - 1$

48. $10x^4 - 3x^3 + 2x^2 + x + 4$

49. $11x^{10} + 8x^3 + x^2 - x + 2$

50. $x^8 - x^7 + 3x^6 + 2x^2 - 1$

51. $10x^6 - 3x^5 + 2x^4 + x^3 + 4x^2$

52. $x^8 - x^7 - 3x^6 + 2x^5 - x^4 - x^3$

For each polynomial, certain zeros are given. Find the remaining zeros.

53. $x^3 + 5x^2 + 9x + 5; -2 - i$

54. $3x^3 + 2x^2 + 3x + 2; i$

55. $x^4 - 3x^3 - 2x^2 + 6x; 3$

56. $x^4 - 4x^3 + 8x^2 - 8x - 5; 1 + 2i$

57. $x^4 - 6x^2 - 8x - 3; -1$ is a zero of multiplicity 2

58. $x^4 - 5x^3 + 6x^2 + 4x - 8; 2$ is a zero of multiplicity 3

59. $x^5 - 12x^3 + 46x^2 - 85x + 50; -5, 1 + 2i$

60. $12x^5 + 4x^4 + 7x^3 + 14x^2 - 34x + 12; -\sqrt{2}i, -\dfrac{3}{2}$

61. Prove: $\overline{\alpha\beta} = \overline{\alpha} \cdot \overline{\beta}$

4.5 CALCULATING ZEROS OF REAL POLYNOMIALS

In the preceding section, we obtained a great deal of information about the zeros of polynomials. We discussed the Fundamental Theorem of Algebra and related theorems, which state that a polynomial of degree n with complex coefficients has n complex zeros, provided that zeros are counted according to their multiplicities. We then discussed the special features of real polynomials and their zeros. In this section, we continue the discussion by considering the rational zeros of polynomials with integer coefficients. Then we discuss the approximation of real zeros of real polynomials.

Polynomials with Rational Coefficients

Suppose that $f(x)$ is a polynomial with rational coefficients and that we wish to determine the zeros of $f(x)$. The first point to notice is that by multiplying $f(x)$ by a common denominator of the coefficients, we can obtain a polynomial with the same zeros but with integer coefficients. So for the sake of our discussion, we can assume that the coefficients are all integers.

The simplest zeros of $f(x)$ are the rational zeros, that is, the zeros of the form c/d, where c and d are integers and d is nonzero. We can determine all such zeros using the following result.

RATIONAL ZERO THEOREM

Let
$$f(x) = a_n x^n + a_{n-1} x^{n-1} + \cdots + a_1 x + a_0$$
be a polynomial with integer coefficients. Suppose that c/d is a rational zero of $f(x)$, where c and d are integers, with d nonzero and the fraction c/d in lowest terms. Then c is a factor of the constant coefficient a_0, and d is a factor of the leading coefficient a_n.

The proof of this theorem belongs to either a course in abstract algebra or a course in the theory of numbers and so will be omitted here.

Using the Rational Zero theorem, the problem of determining the rational zeros of $f(x)$ is reduced to testing a finite number of possibilities obtained by factoring the leading and constant coefficients of $f(x)$ and forming all possible fractions c/d. The next two examples illustrate how to organize the calculations.

➤ **EXAMPLE 1**
Rational Zeros

Find all rational zeros of the polynomial
$$x^4 - 5x^3 + 13x^2 - 35x + 42$$

Solution

In this case, the leading and constant coefficients are given respectively by
$$a_n = 1, \qquad a_0 = 42$$
By the Rational Zero theorem, if c/d is a rational zero, then c is a factor of 42 and d is a factor of 1. If necessary, we may multiply both c and d by -1, so we may assume that d is positive. This means that $d = 1$. The choices for c are as follows:
$$\pm 1, \pm 2, \pm 3, \pm 6, \pm 7, \pm 14, \pm 21, \pm 42$$
This leads to 16 possibilities for rational zeros. We now test each of them in turn to determine which ones work. One possible method of testing them is to use synthetic division to evaluate the remainder when $f(x)$ is divided by $x - c$. After tackling the computations, taking each value of c in turn, we find zero remainders for $c = 2, 3$ and for no other values of c. This means that the only rational zeros of $f(x)$ are $c/d = 2/1 = 2$ and $3/1 = 3$.

➤ **EXAMPLE 2**
Determine Rational Zeros

Determine all rational zeros of the polynomial
$$2x^3 - 3x^2 - 10x + 15$$

Solution

In this case, the leading and constant coefficients are given respectively by
$$a_n = 2, \qquad a_0 = 15$$
By the Rational Zero theorem, if c/d is a rational zero, then c is a factor of 15 and d is a factor of 2. If necessary, we may multiply both c and d by -1, so we may assume that d is positive. This means that $d = 1$ or 2. The choices for c are $\pm 1, \pm 3, \pm 5, \pm 15$. This leads to 16 possibilities for rational zeros. We now test each of them in turn to determine which ones work. The only possibility that is a zero is $c/d = 3/2$. So the only rational zero of the given polynomial is 3/2.

➤ **EXAMPLE 3**
Factoring from Zeros

Factor the polynomial
$$f(x) = x^3 - 5x^2 - 3x - 18$$
into linear factors.

Solution

We first look for rational zeros to obtain any simple linear factors of $f(x)$. According to the Rational Zero theorem, any rational zeros can be written in the form c/d, where c is a factor of -18 and d is a factor of 1. This leads to the following choices for c/d:

$$\pm 1, \pm 2, \pm 3, \pm 6, \pm 9, \pm 18$$

As in the preceding example, we determine which of these are zeros using synthetic division. We get nonzero remainders for c/d equal to the first six choices. However, for the choice $+6$, synthetic division yields

$$
\begin{array}{r|rrrr}
6 & 1 & -5 & -3 & -18 \\
  &   & 6 & 6 & 18 \\
\hline
  & 1 & 1 & 3 & 0
\end{array}
$$

This shows that division by $x - 6$ yields a zero remainder, so $x - 6$ is a factor of $f(x)$. Moreover, from the final row of the synthetic division, we read off the factorization

$$f(x) = (x - 6)(x^2 + x + 3)$$

The quadratic polynomial $x^2 + x + 3$ has a negative discriminant and hence has two complex zeros, which may be determined by the quadratic formula. They are

$$\frac{-1 \pm \sqrt{-11}}{2} = \frac{-1 \pm \sqrt{11}i}{2}$$

This leads to the desired factorization

$$f(x) = (x - 6)\left(x - \frac{-1 + \sqrt{11}i}{2}\right)\left(x - \frac{-1 - \sqrt{11}i}{2}\right)$$

Historical Note:
Evariste Galois

The theory of polynomial equations is a vast branch of mathematics called **Galois theory**, named after its founder, Evariste Galois, who made his ingenious discoveries in the early 19th century. Galois was an unrecognized prodigy who failed the entrance examination to college because his examiners didn't understand what he was talking about. He submitted a paper for a prize given by the French Academy, only to have it lost by Augustin Cauchy, one of the most important mathematicians of the age. Totally ignored, Galois delved into radical politics and was challenged to a duel by a monarchist (the French monarchy had recently been restored after Napoleon had been exiled to Elba). Anticipating his death in the duel, Galois spent the night before hastily recording his mathematical discoveries. The next day he was killed, just short of his 21st birthday. Mathematicians have spent more than a century building on Galois' fertile ideas.   ▌▐

➤ **EXAMPLE 4**
Architecture

The cost of building an office building of n floors is equal to

$$100n^3 + 3000n^2 + 50{,}000$$

Suppose that the total cost of construction is \$450,000. How many stories is the building?

Solution

We must solve the following equation for n:

$$100n^3 + 3000n^2 + 50{,}000 = 450{,}000$$
$$100n^3 + 3000n^2 - 400{,}000 = 0$$
$$n^3 + 30n^2 - 4000 = 0$$

Let's apply the Rational Zero theorem to the polynomial

$$f(x) = x^3 + 30x^2 - 4000$$

We are looking for a positive zero. Since the leading coefficient is 1, the Rational Zero theorem asserts that the zero must be a factor of 4000. If we factor 4000, we see that

$$4000 = 2^5 5^3$$

So the factors of 4000 are obtained by multiplying a certain number of factors of 2 and a certain number of factors of 5. These are 1, 2, 4, 5, 8, 10, 16, 20, 32, ... Using synthetic division, we may test these factors one by one to determine which is a zero of $f(x)$. After a bit of calculation, we see that the only one that works is 10. So 10 is a solution of the equation, and the building has 10 stories.

Bounds on the Zeros of Real Polynomials

As we have just seen, the rational zeros of a polynomial with integer coefficients may be determined by examining a finite list of possibilities. Often, however, this list can be long, resulting in a large number of possibilities to check. The number of possibilities can be narrowed by using the following result.

UPPER AND LOWER BOUNDS FOR ZEROS

Let $f(x)$ be a polynomial with real coefficients and positive leading coefficient.

1. Let $a > 0$ be chosen so that the third row in synthetic division of $f(x)$ by $x - a$ has all positive or zero entries. Then all real zeros of $f(x)$ are less than or equal to a. The number a is called an **upper bound** for the zeros of f.
2. Let $b < 0$ be chosen so that the third row in the synthetic division of $f(x)$ by $x - b$ has alternating signs. Then all real zeros of $f(x)$ are greater than or equal to b. The number b is called a **lower bound** for the zeros of f.

Proof

1. Since the third row of synthetic division of $f(x)$ by $x - a$ has all non-negative entries, we see that

   $$f(x) = (x - a)q(x) + c$$

   where c and the coefficients of $q(x)$ are all positive or zero. Assume that $t > a$. Then

   $$f(t) = (t - a)q(t) + c > 0$$

   so that t is not a zero of f. That is, any zero of f must be less than or equal to a.
2. The proof is similar to that of part 1 and is left as an exercise.   ◆

The next example will give you an idea of how this result may be applied to obtain information about the zeros of a real polynomial.

➤ **EXAMPLE 5**
Calculating Upper and Lower Bounds for Zeros

Let

$$f(x) = x^5 - x^3 + 2x^2 - 2x + 3$$

Determine upper and lower bounds for the real zeros of $f(x)$.

Solution

In a search for an upper bound for the zeros, we may try the various positive integers in turn and perform synthetic division. For the integer $a = 1$, we have

$$\begin{array}{r|rrrrrr} 1 & 1 & 0 & -1 & 2 & -2 & 3 \\ & & 1 & 1 & 0 & 2 & 0 \\ \hline & 1 & 1 & 0 & 2 & 0 & 3 \end{array}$$

Since all entries in the last row are nonnegative, we see that 1 is an upper bound for the zeros of f. To obtain a lower bound, we test negative integers and look for third rows with alternating signs. For $a = -1$, we have

$$\begin{array}{r|rrrrrr} -1 & 1 & 0 & -1 & 2 & -2 & 3 \\ & & -1 & 1 & 0 & 2 & 0 \\ \hline & 1 & -1 & 0 & 2 \end{array}$$

The signs don't alternate. Next we try $a = -2$:

$$\begin{array}{r|rrrrrr} -2 & 1 & 0 & -1 & 2 & -2 & 3 \\ & & -2 & 4 & -6 & 8 & -12 \\ \hline & 1 & -2 & 3 & -4 & 6 & -9 \end{array}$$

The signs alternate, so -2 is a lower bound for the zeros of f.

Approximating Real Zeros

For many applications, it is sufficient to determine a real zero with an accuracy of a certain number of decimal places. We will present a method for determining such approximations based on the Intermediate Value theorem, stated in Section 4.2. Recall that this result says that if there are two points at which the graph of a polynomial lies on opposite sides of the x-axis, then the graph crosses the x-axis at some point in between. (See Figure 1.)

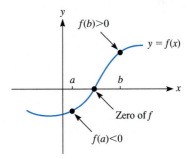

Figure 1

➤ EXAMPLE 6
Isolating a Zero

Show that the polynomial

$$f(x) = 3x^3 - x + 1$$

has a zero in the interval $(-1, 0)$.

Solution

We note that

$$f(-1) = 3(-1)^3 - (-1) + 1 = -1 < 0$$
$$f(0) = 3(0)^3 - 0 + 1 = 1 > 0$$

Therefore, $f(-1)$ and $f(0)$ have opposite signs. The Intermediate Value theorem states that the given polynomial has a zero in the interval $(-1, 0)$.

We can place a real zero in an interval as small as we wish by applying the Intermediate Value theorem to successively smaller intervals. Suppose that we determine from using the Intermediate Value theorem that $f(x)$ has a zero within a certain interval. We may divide this interval into subintervals of equal length. For one of these subintervals, the values of $f(x)$ at the endpoints are of opposite signs. This means that a real zero lies within that subinterval. We may now break this subinterval into equal parts and apply the same procedure again. In this way we can approximate the zero to any desired degree of accuracy. This method of approximating real zeros is called the **method of bisection**. The next example provides the details of the method in a particular instance.

➤ **EXAMPLE 7**
Approximating a Zero

Determine the real zero of the function

$$f(x) = 3x^3 - x + 1$$

from the preceding example with an accuracy of one tenth.

Solution

In the preceding example, we showed that for this function

$$f(-1) < 0, \qquad f(0) > 0$$

so a zero lies in the interval $(-1, 0)$. We can divide this interval into the 10 subintervals:

$$(-1, -0.9), (-0.9, -0.8), \ldots, (-0.1, 0)$$

We calculate the values of $f(x)$ at the endpoints of each of these intervals. As a result of these calculations, we determine that

$$f(-0.9) = -0.287 < 0, \qquad f(-0.8) = 0.264 > 0$$

Thus, the zero lies between -0.9 and -0.8. To pin down the zero to the nearest tenth, evaluate the function at the midpoint of the interval $(-0.9, -0.8)$, to obtain

$$f(-0.85) = 0.00763 > 0$$

Therefore, the zero lies in the interval $(-0.9, -0.85)$. That is, to the nearest tenth the zero equals -0.9.

➤ **EXAMPLE 8**
Approximation of Zero
to Specified Accuracy

Determine the real zero of $f(x) = x^4 - 2x^2 + 3x - 1$, which lies between 0 and 1. Calculate the zero to within 0.001.

Solution

We begin by tabulating the values of $f(x)$ for x between 0 and 1, at intervals of 0.1. We have used the computer program *Mathematica* to prepare the following table:

x	$f(x)$
0	-1
0.1	-0.7199
0.2	-0.4784
0.3	-0.2719
0.4	-0.0944
0.5	0.0625
0.6	0.2096
0.7	0.3601
0.8	0.5296
0.9	0.7361
1.0	1.0

You may verify the entries using your calculator. According to the shaded rows, the zero lies between 0.4 and 0.5, since the value of $f(x)$ changes sign in this interval. We now tabulate the values of the function for x between 0.4 and 0.5 at intervals of 0.01. We arrive at the following table:

x	$f(x)$
0.4	-0.0944
0.41	-0.0779424
0.42	-0.061683
0.43	-0.045612
0.44	-0.029719
0.45	-0.0139938
0.46	0.00157456
0.47	0.0169968
0.48	0.0322842
0.49	0.047448
0.5	0.0625

The shaded rows show that the zero lies between 0.45 and 0.46. Finally we tabulate the function values for x between 0.45 and 0.46 at intervals of 0.001. The following table shows that the zero lies between 0.458 and 0.459:

x	$f(x)$
0.45	-0.0139938
0.451	-0.01243
0.452	-0.0108679
0.453	-0.00930727
0.454	-0.00774819
0.455	-0.00619065
0.456	-0.00463462
0.457	-0.0030801
0.458	-0.00152706
0.459	0.0000244838
0.46	0.00157456

To three significant digits, the zero is therefore 0.459.

Exercises 4.5

Determine all rational zeros of the following polynomials.

1. $2x^3 - 11x^2 + 17x - 6$
2. $6x^4 - 5x^2 + 1$
3. $3x^3 - 2x^2 - 3x - 12$
4. $x^3 - 3x^2 - 3x - 4$
5. $5x^5 - 4x^4 - 50x^3 + 40x^2 + 125x - 100$
6. $x^5 - 2x^3 + 5x^2 - 10$
7. $x^4 - 25$
8. $z^6 - 1$

Determine upper and lower bounds for the real zeros of $f(x)$.

9. $f(x) = 2x^4 - x^3 + 3x^2 - 5$

10. $f(x) = 3x^3 + 4x^2 + 12$

11. $f(x) = x^5 + 3x^4 - x^2 + 14$

12. $f(x) = 2x^4 - 3x^2 - 9x + 1$

13. $f(x) = 16x^{12} - 11x^{10} + 2x^9 - 3x^8 + 4x - 6$

14. $f(x) = 4x^4 + x^3 - 6x^2 + 10$

15. $f(x) = x^4 - 25$ 16. $f(x) = x^6 - 1$

Find the rational zeros of the following polynomials.

17. $9x^3 - 18x^2 + 11x - 2$ 18. $3x^3 + 4x^2 + 12$

19. $4x^3 - x^2 - 100x + 25$ 20. $2x^4 - 3x^2 - 9x + 1$

21. $4x^4 - 21x^2 - 25$ 22. $81x^4 - 16$

23. $x^4 - 25$ 24. $x^6 - 1$

25. $\frac{3}{2}x^3 + \frac{11}{4}x^2 - \frac{3}{4}x - \frac{1}{2}$ 26. $\frac{3}{4}x^3 - \frac{2}{3}x^2 - \frac{5}{6}x - \frac{1}{2}$

Show that each of the following polynomials has a zero in the given interval.

27. $f(x) = x^3 + x + 1; (-1, 0)$

28. $f(x) = x^3 - x + 1; (-2, -1)$

29. $f(x) = x^3 - 3x^2 + 4x - 5; (2, 3)$

30. $f(x) = x^3 + x - 3; (1, 2)$

31. $f(x) = x^3 + 2x^2 + 8x - 2; (0, 1)$

32. $f(x) = x^3 - 3x^2 - 5; (3, 4)$

33. $f(x) = x^4 - x^2 - 3; (-1, -2)$ and $(1, 2)$

34. $f(x) = x^4 - x^3 - 1; (-1, 0)$ and $(1, 2)$

Approximate the real zeros of each of the following polynomial functions with an accuracy of 0.0001 using the method of bisection.

35. $f(x) = x^3 + x + 1$ 36. $f(x) = x^3 - x + 1$

37. $f(x) = x^3 - 3x^2 + 4x - 5$

38. $f(x) = x^3 + x - 3$

39. $f(x) = x^3 + 2x^2 + 8x - 2$

40. $f(x) = x^4 - x^2 - 3$ 41. $f(x) = x^4 - x^3 - 1$

42. $f(x) = x^4 - 3x^3 - 2x - 3$

43. $f(x) = x^5 - 3x^4 + 5x^3 - 7x^2 + 6$

44. $f(x) = -x^5 + 4x^4 - 2x^3 + x^2 - 4x + 1$

♦ Applications

45. **Cost of a building.** The cost of building an office building of n floors is equal to

$$100n^3 + 3000n^2 + 50,000$$

a. Suppose that the total cost of construction is equal to $3,487,500. How many stories is the building?

b. Suppose that the total cost of construction is about $776,000. Use the method of bisection to estimate how many stories are in the building. Give your answer correct to the nearest one.

46. **Volume of a box.** An open box of volume 108 in.^{3} can be made from a piece of cardboard that is 10 in. by 15 in. by cutting a square from each corner and folding up the sides. (See figure.) What is the length of a side of the squares? Use the method of bisection to find an approximate answer, if appropriate.

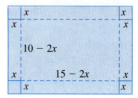

Exercise 46

47. **Deflection of a beam.** A beam rests at two points A and B and has a concentrated load applied to the center of the beam, as shown in the figure. Let y denote the deflection of the beam at a distance of x units from the left end of the beam to the left edge of the weight. The deflection depends on the elasticity of the board, the load, and other physical characteristics. Suppose under certain conditions that y is given by

$$y = \frac{1}{12}x^3 - \frac{1}{16}x$$

Use the method of bisection to estimate the number of units x at which the deflection will be 40 units.

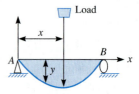

Exercise 47

48. Break-even points. Given the following total revenue and total cost functions:

$$R(x) = 100x - x^2, \qquad C(x) = \frac{1}{3}x^3 - 6x^2 + 89x + 100$$

a **break-even** point occurs at a nonnegative value of x for which $R(x) = C(x)$. Find all the break-even points for the functions. Use the method of bisection if appropriate.

49. Medical dosage. The function

$$N(t) = 0.045t^3 + 2.063t + 2$$

gives the bodily concentration in parts per million of a certain dosage of medication after time t, in hours. Find all times in the interval $[0, 10]$ for which the concentration is 4 parts per million. Use the method of bisection if appropriate.

50. Temperature during an illness. A patient's temperature during an illness was given by

$$T(t) = 98.6 - t^2(t - 4)$$

Find the times t at which the patient's temperature was 100°. Use the method of bisection where appropriate.

Find all the zeros. Use any procedures studied in this chapter.

51. $x^3 - 3x - 2$

52. $x^3 + 2x^2 + 9$

53. $x^3 - x^2 - 3x + 2$

54. $x^3 + 4x^2 + 7x + 6$

55. $x^4 - x^3 - 5x^2 + 7x - 2$

56. $x^4 + x^3 + x^2 - x - 2$

57. $6x^4 + x^3 + 4x^2 + x - 2$

58. $4x^4 - 4x^3 - 5x^2 + x + 1$

59. $4x^5 - 24x^4 + 25x^3 + 39x^2 - 38x - 24$

60. $x^5 + x^4 - 9x^3 - 5x^2 + 16x + 12$

61. $3x^4 - 2x^2 + 18$ 62. $5x^4 - 3x^2 + 2$

Solve. Use any procedures studied in this chapter.

63. $x^3 - 4x^2 + x + 6 = 0$

64. $24x^3 + 10x^2 - 13x - 6 = 0$

65. $x^4 + 3x^2 = 20$

66. $2x^3 + 15x^2 = -20x - 3$

67. $4x^5 + 4x = 17x^3$

68. $x^5 + 60x + 72 = 15x^3 + 10x^2$

69. Prove that $\sqrt{2}$ is irrational by considering the zeros of the polynomial function $f(x) = x^2 - 2$.

70. Prove that $\sqrt{5}$ is irrational by considering the zeros of the polynomial function $f(x) = x^2 - 5$.

➡ **Technology**

71. Use a graphing calculator to approximate the zeros in Exercises 51–62. Do this by zooming.

72. Use a graphing calculator to approximate all real zeros of the function $f(x) = 3x^4 + x - 1$.

73. **Cardano's formula.** Any cubic polynomial equation can be reduced to one of the form

$$y^3 + by^2 + cy + d = 0$$

by dividing on both sides by the leading coefficient. Further, we can obtain a reduced form by making the substitution

$$y = x - \frac{1}{3}b$$

This will yield an equation of the type

$$x^3 + px + q = 0$$

where

$$p = c - \frac{b^2}{3} \qquad \text{and} \qquad q = d - \frac{bc}{3} + \frac{2b^3}{27}$$

To solve the latter equation we first make the following computations:

$$a = \sqrt[3]{-\frac{q}{2} + \sqrt{\frac{q^2}{4} + \frac{p^3}{27}}}$$

$$b = \sqrt[3]{-\frac{q}{2} - \sqrt{\frac{q^2}{4} + \frac{p^3}{27}}}$$

Then the three cube roots of the reduced cubic equation are

$$x_1 = a + b$$

$$x_2 = \frac{-a + b}{2} + \frac{a - b}{2}\sqrt{3}i$$

$$x_3 = \frac{-a + b}{2} - \frac{a - b}{2}\sqrt{3}i$$

Use the preceding results to find the roots of the following cubics for which you approximated the real zeros in the preceding exercises. Then find approximations using your calculator or computer and compare.

    a. $f(x) = x^3 + x + 1$

    b. $f(x) = x^3 - x + 1$

    c. $f(x) = x^3 - 3x^2 + 4x - 5$

    d. $f(x) = x^3 + x - 3$

    e. $f(x) = x^3 + 2x^2 + 8x - 2$

74. Use the method of bisection to approximate all the real zeros of the function

$$f(x) = 1 + \sqrt{x - 2} - \sqrt[3]{x + 6}$$

on the interval $(2, 6)$.

4.6 RATIONAL FUNCTIONS

Recall that a rational function is a function given by an expression of the form

$$f(x) = \frac{p(x)}{q(x)}$$

where $p(x)$ and $q(x)$ are polynomials with real coefficients. The domain of such a function consists of all real numbers x for which $q(x)$ is nonzero. In this section, we will learn to sketch the graphs of rational functions. As we will see, the significant new feature is the behavior of the functions for x near a zero of the denominator $q(x)$.

To simplify our discussion, we will initially assume that $p(x)$ and $q(x)$ have no factors in common.

Vertical Asymptotes

We begin with a graph of a typical rational function.

➤ EXAMPLE 1
Graph of a Rational Function

Sketch the graph of the rational function

$$f(x) = \frac{1}{x^2}$$

Solution

First note that $f(x)$ is not defined for $x = 0$, the value at which the denominator is 0. Values of $f(x)$ for x near 0 therefore display special characteristics. Note that

$$f(-x) = \frac{1}{(-x)^2} = \frac{1}{x^2}$$

so $f(x)$ is even, and its graph is symmetric with respect to the y-axis. Because the graph is symmetric with respect to an axis, we can restrict ourselves at first to the positive values of x. Here is a table of the values of $f(x)$ at some representative values that decrease through positive numbers and get increasingly close to 0. We say that these values of x **approach zero from the right.**

x	$f(x)$
1	1
0.1	100
0.01	10,000
0.001	1,000,000

The data in the table suggest that as x approaches zero from the right, the values of $f(x)$ increase without bound. Geometrically, this means that, as x approaches 0 from the right, the graph of $f(x)$ rises without bound and approaches arbitrarily close to the y-axis without ever touching it. We say that the y-axis is a **vertical asymptote** of the graph. Using the data in the table, we can make a sketch of the graph for $x > 0$. By symmetry, the portion of the graph corresponding to $x < 0$ can be obtained by reflection in the y-axis. (See Figure 1.)

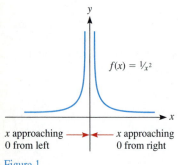

$f(x) = 1/x^2$

x approaching 0 from left ⟶ ⟵ x approaching 0 from right

Figure 1

Suppose that the rational function $f(x)$ is undefined for $x = a$ (that is, the value a would make the denominator of the function equal to zero). As we did in Example 1 for $a = 0$, we can examine the behavior of $f(x)$ as x approaches any value a from the right. If the values of $f(x)$ increase without bound as x

approaches *a* from the right, we say that $f(x)$ *approaches positive infinity from the right* and write $f(x) \to +\infty$ as $x \to a^+$. [See Figure 2(*a*).] If, as *x* approaches *a* from the right, the values of $f(x)$ decrease without bound (say, -1, -100, $-10{,}000$, etc.), then we say that $f(x)$ *approaches negative infinity from the right* and write $f(x) \to -\infty$ as $x \to a^+$. [See Figure 2(*b*).]

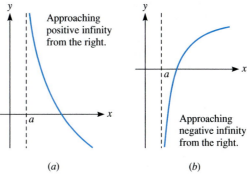

Approaching positive infinity from the right.

Approaching negative infinity from the right.

(*a*) (*b*)

Figure 2

In a similar fashion, we can consider *x* approaching *a* from the left, that is, through values smaller than *a*. If the values of $f(x)$ increase without bound when *x* approaches *a* from the left, we say that $f(x)$ *approaches positive infinity from the left* and write $f(x) \to +\infty$ as $x \to a^-$. [See Figure 3(*a*).] If the values of $f(x)$ decrease without bound when *x* approaches *a* from the left, then we say that $f(x)$ *approaches negative infinity from the left* and write $f(x) \to -\infty$ as $x \to a^-$. [See Figure 3(*b*).]

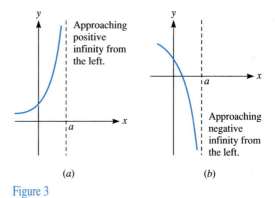

Approaching positive infinity from the left.

Approaching negative infinity from the left.

(*a*) (*b*)

Figure 3

Definition 1 Vertical Asymptote	If, as *x* approaches *a*, the graph of a rational function $f(x)$ approaches positive infinity or negative infinity from either the right or left, then we say that the line $x = a$ is a **vertical asymptote** of the graph.

Graphs of rational functions (assumed to be written in lowest terms) have a vertical asymptote corresponding to each zero of the denominator. It is necessary to allow *x* to approach such values from both the right and the left to determine the behavior of the graph in the vicinity of the asymptote. The next two examples illustrate how to do this.

> **EXAMPLE 2**

Graph with Vertical Asymptote

Vertical asymptote $x = -1$

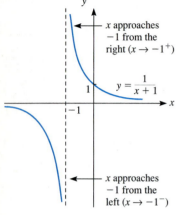

Figure 4

Sketch the graph of the rational function

$$y = \frac{1}{x + 1}$$

Solution

The function is undefined for $x = -1$, the value at which the denominator equals 0. This means that the line $x = -1$ is a vertical asymptote of the graph. To determine the behavior of the graph in the vicinity of this asymptote, we allow x to approach -1 from both the right and the left.

As x approaches -1 from the right, $x > -1$, so the value of $f(x)$ is positive. Moreover, as x approaches -1 from the right, the values of $f(x)$ increase without bound. That is, they approach positive infinity. This information allows us to sketch the graph for $x > -1$. We first sketch the asymptote $x = -1$. Then we draw in the portion of the graph to the right of the asymptote. (See Figure 4.)

Next, we consider values of x when x approaches -1 from the left. These values for x are less than -1, so $x + 1$ is negative, as are the values of $f(x)$. Moreover, as x approaches -1 from the left, the values of $f(x)$ decrease without bound. That is, $f(x)$ approaches negative infinity. (See Figure 4.)

Horizontal Asymptotes

In drawing an accurate sketch of a graph, it is helpful to indicate how the function behaves for values of x that are very large in either the positive or the negative direction. These values lie far to the right or left on the number line.

If x increases without bound through positive numbers, then we say that *x approaches positive infinity*. We may visualize this concept by imagining the value of x as represented by a point on the number line moving indefinitely far to the right. If x increases without bound through negative numbers, then we say that *x approaches negative infinity*. We may visualize this concept by imagining the value of x as represented by a point on the number line moving indefinitely far to the left.

As x approaches either positive or negative infinity, the values of a rational function $f(x)$ may exhibit several different behaviors. On the one hand, the value of $f(x)$ may approach positive infinity or negative infinity. These possibilities are shown in Figure 5(a) and (b), respectively.

On the other hand, the values of $f(x)$ may approach a real number c. Two ways in which this may occur are illustrated in Figure 5(c).

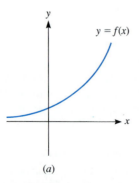

(a)

$f(x)$ approaches positive infinity as x approaches positive infinity.

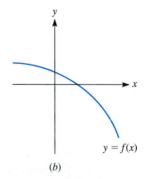

(b)

$f(x)$ approaches negative infinity as x approaches positive infinity.

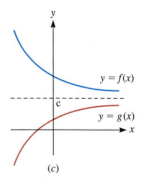

(c)

$f(x)$ and $g(x)$ approach the asymptote $y = c$ as x approaches positive infinity.

Figure 5

Definition 2	Suppose that the values of the rational function $f(x)$ approach a real number
Horizontal Asymptote	c as x approaches either positive or negative infinity. We then say that the line $y = c$ is a **horizontal asymptote** of the graph of $f(x)$.

As x approaches positive infinity, so do all positive powers of x. However, negative powers of x approach 0 (since the reciprocal of a large number is small).

As x approaches negative infinity, negative powers of x still approach 0. However, a positive even power of x approaches positive infinity, and a positive odd power of x approaches negative infinity. (A positive even power of a large negative number is a large positive number; a positive odd power of a large negative number is a large negative number.)

Using these facts, we can determine the behavior of a rational function as x approaches positive infinity or negative infinity. The next example illustrates how this may be done.

➤ **EXAMPLE 3**

Graph with Horizontal Asymptotes

Determine the horizontal asymptotes, if any, of the rational function

$$f(x) = \frac{x^2}{3x^2 - 4x - 1}$$

Solution

Divide the numerator and denominator by the largest power of x in the denominator, namely, x^2, to obtain the following expression for $f(x)$:

$$f(x) = \frac{\dfrac{x^2}{x^2}}{\dfrac{3x^2 - 4x - 1}{x^2}} = \frac{1}{3 - \dfrac{4}{x} - \dfrac{1}{x^2}}$$

As x approaches either positive infinity or negative infinity, the terms

$$\frac{4}{x}, \frac{1}{x^2}$$

approach 0, and the value of $f(x)$ approaches

$$\frac{1}{3 + 0 + 0} = \frac{1}{3}$$

That is, the graph has as horizontal asymptote the line $y = 1/3$. This asymptote is shown along with a sketch of the graph in Figure 6.

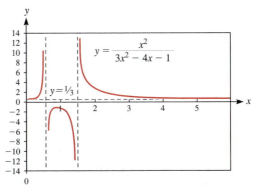

Figure 6

Note that whereas a graph can never cross a vertical asymptote, it is perfectly possible for a graph to cross a horizontal asymptote. In fact, there are graphs that oscillate an infinite number of times around a horizontal asymptote, approaching closer with each oscillation.

Oblique Asymptotes

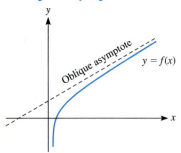

Figure 7

A line is said to be **oblique** if it is nonhorizontal and nonvertical. In our discussion above, we introduced both horizontal and vertical asymptotes. Analogously, we have the concept of an **oblique asymptote**, which is a line that a graph approaches as x approaches either positive infinity or negative infinity. (See Figure 7.)

Here is an example to illustrate how oblique asymptotes arise. Consider the rational function $f(x) = (x^2 + 3x + 1)/x$. By performing the indicated division, we arrive at the following alternate expression for $f(x)$:

$$f(x) = x + 3 + \frac{1}{x}$$

It is clear that as x approaches either positive infinity or negative infinity, the term $1/x$ approaches 0, so the function value approaches the value of the linear function $x + 3$. Geometrically, this means that the graph of $f(x)$ approaches the graph of $x + 3$ as x approaches either positive or negative infinity. That is, the line $y = x + 3$ is an oblique asymptote of the graph. (See Figure 8.)

In a similar fashion, suppose that we are given a rational function $f(x) = p(x)/q(x)$, where the degree of $p(x)$ exceeds the degree of $q(x)$ by 1. Then we may perform division to write the rational function in the form

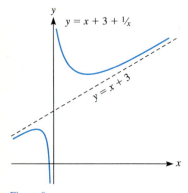

Figure 8

$$f(x) = ax + b + \frac{r(x)}{q(x)}$$

where the degree of $r(x)$ is less than the degree of $q(x)$. In this case, the line $y = ax + b$ is an oblique asymptote of the graph of $f(x)$. (See Figure 9.)

We can summarize our discussion of horizontal and oblique asymptotes with the following result.

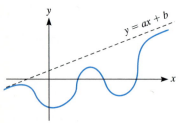

Figure 9

ASYMPTOTES OF A RATIONAL FUNCTION

Let $f(x) = p(x)/q(x)$ be a rational function, where $p(x)$ is a polynomial of degree m and leading coefficient c, and $q(x)$ is a polynomial of degree n and leading coefficient d.

1. If $m < n$, then the x-axis is a horizontal asymptote of the graph of $f(x)$, and $f(x)$ approaches 0 as x approaches either positive infinity or negative infinity.
2. If $m = n$, then the line $y = c/d$ is a horizontal asymptote of the graph of $f(x)$, and $f(x)$ approaches c/d as x approaches either positive infinity or negative infinity.
3. If $m = n + 1$, then we may write $f(x)$ in the form $f(x) = ax + b + r(x)/q(x)$, where $r(x)$ has degree less than n. In this case, $y = ax + b$ is an oblique asymptote of the graph of $f(x)$.
4. If $m > n$, then the graph of $f(x)$ has no horizontal asymptotes. Moreover, as x approaches positive infinity or negative infinity, $f(x)$ approaches either positive infinity or negative infinity. (It is simplest to separately determine which it approaches in each instance using the reasoning of Example 2.)

Some Detailed Examples

Let's now work out some examples that utilize everything we have learned about sketching graphs of rational functions.

➤ **EXAMPLE 4**

Asymptotes of Rational Functions

Describe the behaviors of the following rational functions as x approaches positive infinity and negative infinity.

1. $f(x) = \dfrac{1}{2x - 1}$

2. $f(x) = \dfrac{2x^3 - 1}{(x - 1)(x + 1)}$

3. $f(x) = \dfrac{(x - 1)(3x - 5)}{(2x + 7)(x + 1)}$

Solution

1. In this example, the degree of the numerator is less than the degree of the denominator. By Statement 1 on page 266, the graph has the x-axis as a horizontal asymptote. That is, as x approaches positive or negative infinity, the values of $f(x)$ approach 0.

2. In this case, the degree of the numerator exceeds the degree of the denominator by 1, so the graph has no horizontal asymptotes but has an oblique asymptote. Using the reasoning of Example 3, we may rewrite the expression for $f(x)$ by dividing both numerator and denominator by the highest power of x appearing in the denominator:

$$f(x) = \frac{\dfrac{2x^3 - 1}{x^2}}{\dfrac{x^2 - 1}{x^2}} = \frac{2x - \dfrac{1}{x^2}}{1 - \dfrac{1}{x^2}}$$

As x approaches positive infinity or negative infinity, the expression

$$\frac{1}{x^2}$$

approaches 0, so the expression for $f(x)$ approaches

$$\frac{2x - 0}{1 - 0} = 2x$$

That is, $f(x)$ approaches the value $2x$. When x approaches positive infinity, $f(x)$ along with $2x$ approaches positive infinity. When x approaches negative infinity, $f(x)$ along with $2x$ approaches negative infinity.

3. In this case, the degree of the numerator equals the degree of the denominator. So by Statement 2 on page 266, the graph has a horizontal asymptote. The leading coefficient of the numerator is 3, and the leading coefficient of the denominator is 2, so the horizontal asymptote is the line $y = 3/2$. As x approaches either positive infinity or negative infinity, the value $f(x)$ approaches 3/2.

Sketching Graphs of Rational Functions

Here is a general, organized approach to graphing rational functions.

GRAPHING RATIONAL FUNCTIONS

Suppose that $f(x) = p(x)/q(x)$ is a rational function. To sketch the graph of $f(x)$:

1. Determine any common factor of the numerator and denominator. The zeros of this common factor are points at which the rational function is undefined. Replace the given expression of the rational function by one in lowest terms.

2. Determine the horizontal and oblique asymptotes using the procedure on page 266. If there are none, determine the behavior of $f(x)$ as x approaches both positive infinity and negative infinity.

3. Determine the vertical asymptotes by calculating the real zeros of the denominator. For each zero a, the line $x = a$ is a vertical asymptote.

4. Determine the zeros of $f(x)$ by determining the zeros of the numerator. These are the x-intercepts of the function.

5. Divide the x-axis into intervals determined by the real zeros of $f(x)$ and by the vertical asymptotes. For each interval, choose a test value and determine the sign of $f(x)$ in that interval. Then use the signs to determine whether the graph is above or below the x-axis in each interval.

6. Draw the horizontal and vertical asymptotes and plot the zeros.

7. Draw the portion of the graph corresponding to each interval using the information developed in steps 1–6.

The next example illustrates how to apply this general approach to sketching a graph with vertical asymptotes.

> **EXAMPLE 5**

Graph with Horizontal and Vertical Asymptotes

Determine all asymptotes of the rational function

$$f(x) = \frac{(x - 2)(x - 4)}{(x - 1)(x - 3)}$$

Sketch the graph of the function.

Solution

There is a vertical asymptote corresponding to each zero of the denominator. There are two such zeros, namely, 1 and 3, so the vertical asymptotes are the lines $x = 1$ and $x = 3$. Since the numerator and denominator have the same degree, there is a horizontal asymptote. In finding the horizontal asymptote, use the expanded form of $f(x)$, that is,

$$f(x) = \frac{x^2 - 6x + 8}{x^2 - 4x + 3}$$

The leading coefficients of the numerator and denominator are each 1, so the horizontal asymptote is the line $y = 1/1 = 1$.

To sketch the graph we must first determine the zeros of $f(x)$. These are just the zeros of the numerator. In this case, the numerator is factored for us, so we may read off the zeros directly, namely, 2 and 4. The zeros and the vertical asymptotes determine a division of the x-axis into intervals:

$$(-\infty, 1), (1, 2), (2, 3), (3, 4), (4, +\infty)$$

Interval	Test Point	Sign of $f(x)$
$(-\infty, 1)$	0	+
$(1, 2)$	3/2	−
$(2, 3)$	5/2	+
$(3, 4)$	7/2	−
$(4, +\infty)$	5	+

In each of these intervals, the value of $f(x)$ has a constant sign. Let's choose a test point in each interval and determine the corresponding sign. The results can be summarized in the accompanying table.

We now turn to the graph itself. First we draw asymptotes and plot the zeros. We now plot the section of the graph lying in each interval.

For the leftmost interval, $(-\infty, 1)$, the graph approaches the horizontal asymptote $y = 1$ from above as x approaches negative infinity. On the right side of the interval, the graph must approach the vertical asymptote $x = 1$. The graph must head upward as it does so, since otherwise it would cross the x-axis within the interval and there are no zeros of $f(x)$ within the interval.

In the second interval, $(1, 2)$, the function value is negative. This means that the graph must head downward to the asymptote on the left. Moreover, the graph crosses the x-axis at $x = 2$. In the third interval, $(2, 3)$ the graph is positive. So the graph must increase in its approach to the asymptote $x = 3$ on the right.

In the fourth interval, $(3, 4)$, the function value is negative. This means that the graph must head downward to the asymptote on the left. Moreover, the graph crosses the x-axis at $x = 4$.

In the fifth interval, $(4, +\infty)$, the graph rises from the x-axis on the left and must approach the horizontal asymptote $y = 1$ from below as x approaches positive infinity.

The completed graph is sketched in Figure 10.

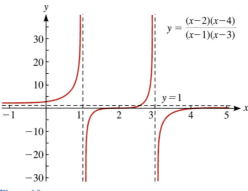

$$y = \frac{(x-2)(x-4)}{(x-1)(x-3)}$$

Figure 10

➤ **EXAMPLE 6**
Inventory Management

The cost $C(x)$ of managing an inventory of x cases of a certain brand of canned beans is determined by a team of financial analysts to be given by the function

$$C(x) = \frac{10,000}{x} + 3x$$

Sketch the graph of $C(x)$.

Solution

Since x represents a number of cases of canned goods, x must be positive. So the domain of the function $C(x)$ is the set of all positive integers. If we write the expression for $C(x)$ in the form

$$\frac{3x^2 + 10,000}{x}$$

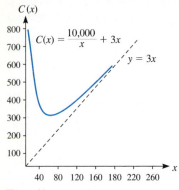

Figure 11

we see that $C(x)$ is a rational function. The only zero of the denominator is 0, so there is a single vertical asymptote, namely, $x = 0$. Moreover, since the degree of the numerator is one more than the degree of the denominator, there is an oblique asymptote and no horizontal asymptotes. If we write

$$C(x) = 3x + \frac{10{,}000}{x}$$

then we see that the line $y = 3x$ is an oblique asymptote.

We now sketch the graph beginning with the single vertical asymptote, $x = 0$. Since $C(x)$ is positive throughout the domain, the graph must rise in approaching the asymptote from the right. We have drawn the oblique asymptote and show the graph approaching this asymptote as x approaches positive infinity. The completed sketch is shown in Figure 11.

In case the numerator and denominator of a rational function have a common factor, the rational function will not be defined at the zeros of the common factor, and the graph will have "holes" in it corresponding to the values of x at which the function is undefined. The next example illustrates this phenomenon.

➤ **EXAMPLE 7**
**Rational Function
with Cancellation**

Sketch the graph of the function

$$f(x) = \frac{x^2 - 25}{x - 5}$$

Solution

Since

$$\frac{x^2 - 25}{x - 5} = \frac{(x + 5)(x - 5)}{x - 5}$$

we see that $f(x)$ has the equivalent expression $f(x) = x + 5$ as long as x is not equal to a zero of the common factor $x - 5$. That is, we have

$$f(x) = x + 5, \quad x \neq 5$$

The graph of this function is shown in Figure 12.

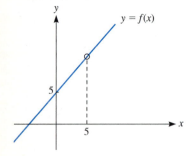

Figure 12

**Graphing
Rational
Functions
Using Graphing
Technology**

If you are using a graphing utility, you have probably seen that even very complicated rational functions can be graphed easily. You may then wonder why it is important to be able to sketch these graphs by hand or to understand the algebra if the technology will do all of the graphing for you. The answer to this is that sometimes a graph on a calculator may not look anything like the graph of the function.

For instance, look at Figure 13. This is a graph of

$$f(x) = \frac{(x - 2)(x - 4)}{(x - 1)(x - 3)}$$

which was sketched in Figure 10. The two graphs do not look similar at all. Why? Figure 13 was graphed only in the range $-1 \leq x \leq 1$. Without a knowledge of asymptotes and rational functions, you might be fooled into thinking that this is what the graph looks like. This is one danger involved in using a graphing utility.

Rational functions provide another example of how you could get a misleading graph from a graphing utility. Figure 14 was produced on a TI-81 graphing calculator. It is the graph of

$$f(x) = \frac{x^2 - 25}{x - 5}$$

Again notice that this is not the same graph that was shown in Figure 12 in Example 7. The reason is that the graphing calculator can only evaluate the function at a certain number of points, and it uses these points to plot the graph. It happened that $x = 5$ was not one of the points evaluated, so the graph does not show the hole that we know is there. Even using a graphing utility's zoom feature does not guarantee that it will show these holes. Figure 15 shows the same function graph in the range $4.95 \le x \le 5.05$. Even this close-up view does not show the hole. Without some understanding of the concepts involved, the technology could mislead us.

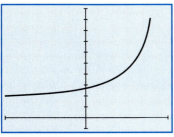

Figure 13

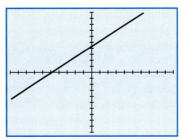

Figure 14

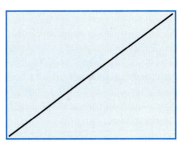

Figure 15   ◀

Exercises 4.6

For each function:

a. Determine all the asymptotes.

b. Sketch the graph of the function.

1. $f(x) = \dfrac{2}{x^2}$

2. $f(x) = \dfrac{1}{x^3}$

3. $f(x) = -\dfrac{1}{x^3}$

4. $f(x) = -\dfrac{3}{x^2}$

5. $f(x) = \dfrac{1}{x + 1}$

6. $f(x) = \dfrac{1}{x - 1}$

7. $f(x) = \dfrac{-2}{x + 2}$

8. $f(x) = \dfrac{-2}{(x - 3)^2}$

9. $f(x) = \dfrac{1}{(x + 3)^2}$

10. $f(x) = \dfrac{x + 1}{2 - x}$

11. $f(x) = \dfrac{3 - x}{x + 2}$

12. $f(x) = \dfrac{2x + 3}{16 - 5x}$

13. $f(x) = \dfrac{8 - 2x}{10 + 3x}$

14. $f(x) = \dfrac{1}{x^2 + 1}$

15. $f(x) = -\dfrac{1}{x^2 + 3}$

16. $f(x) = \dfrac{1}{x^2 - 4}$

17. $f(x) = -\dfrac{1}{x^2 - 9}$

18. $f(x) = \dfrac{2x}{x^2 - x - 6}$

19. $f(x) = \dfrac{x - 3}{2x^2 + 5x - 3}$

20. $f(x) = \dfrac{x^2 - 9}{x + 2}$

21. $f(x) = \dfrac{x^2 - 4}{x - 3}$

22. $f(x) = \dfrac{x^2 - 1}{x^2 + x - 6}$

23. $f(x) = \dfrac{x^2 - 9}{x^2 - x - 2}$

24. $f(x) = \dfrac{x^2 - x - 2}{x + 2}$

25. $f(x) = \dfrac{x^2 - 2x - 8}{x + 4}$

26. $f(x) = x + \dfrac{4}{x}$

27. $f(x) = x + \dfrac{1}{x}$

28. $f(x) = \dfrac{2x^2 - x - 3}{3x^2 - 2x - 8}$

29. $f(x) = \dfrac{x^2 + 2x - 3}{3x^2 - 1}$

30. $f(x) = \dfrac{x^3 + 1}{x}$

31. $f(x) = \dfrac{x^3 - 2}{4x}$

32. $f(x) = \dfrac{x^2 - 1}{x - 1}$

33. $f(x) = \dfrac{x^2 - 9}{x + 3}$

34. $f(x) = \dfrac{(x + 2)(x^2 - 9)}{x + 2}$

35. $f(x) = \dfrac{(x - 1)(3 - x)^2}{x - 1}$

36. $f(x) = \dfrac{x^2 + 2x - 15}{x + 5}$

37. $f(x) = \dfrac{x^2 + x - 2}{x - 1}$
38. $f(x) = \dfrac{x^2 - 4x + 3}{x^2}$

39. $f(x) = \dfrac{4x^2}{x^2 - 1}$
40. $f(x) = \dfrac{x^3 - 2x^2 - 8x}{x^2 - 9}$

41. $f(x) = \dfrac{x^3 + 2x^2 - 3x}{x^2 - 25}$

42. $f(x) = \dfrac{x^3 - 4x^2 + x + 6}{x^2 + x - 2}$

♦ **Applications**

43. **Earned run average**. The earned run average, ERA, of a pitcher is given by ERA $= 9R/I$, where $R =$ the number of earned runs allowed and $I =$ the number of innings pitched.

   a. Suppose $R = 1$. Graph the function over the interval $(0, 20]$.

   b. Suppose $R = 2$. Graph the function over the interval $(0, 20]$.

   c. Suppose $R = 5$. Graph the function over the interval $(0, 20]$.

44. **Distance as a function of time**. The distance from Dallas, Texas, to El Paso, Texas, is 617 miles. A car makes the trip in time t, in hours, at various speeds r, in miles per hour.

   a. Find a rational function for t in terms of r.

   b. Graph the function.

45. **Loudness of sound**. A person is sitting at a distance d from a stereo speaker. (See figure.) The loudness L of the sound varies inversely as the square of the distance d according to the equation

$$L = \dfrac{k}{d^2}$$

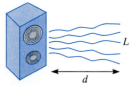

Exercise 45

where k depends on the original intensity and the units used to measure d and L. Graph this function assuming $k = 0.1$.

46. **Inventory management.** The cost $C(x)$ of managing an inventory of x television sets is determined by a team of financial analysts to be given by the function

$$C(x) = 5x + \dfrac{50{,}000}{x} + 22{,}500$$

Sketch the graph of this function over the interval $(0, +\infty)$.

47. **Weight above the earth.** A person's weight W_h at a height h above sea level satisfies the equation

$$W_h = \left(\dfrac{r}{r + h}\right)^2 W_0$$

where W_0 is the person's weight at sea level; h is the height, in miles, of the person above sea level; and r is the radius of the earth, in miles. Assume that $r \approx 4000$ miles.

   a. Suppose a person weighs 200 lbs. at sea level. Find a rational function for the person's weight h miles above the earth.

   b. Graph this function on the interval $[0, 16{,}000]$.

   c. At what height would the person weigh 100 lbs.?

48. **Optics.** Consider the accompanying figure. F is the focal length of the lens in mm, where an object u meters from the lens of a camera has an image v millimeters focused on the film from the lens. A law of optics asserts that $1/F = 1/u + 1/v$. Suppose the focal length is 50 mm.

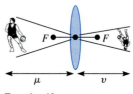

Exercise 48

   a. How far is the lens from the film if the camera is focused on an object that is 6 meters away from the camera?

   b. Let v be a dependent variable. Find a rational function for v in terms of u.

   c. Graph v.

Graph.

49. $f(x) = \left| \dfrac{1}{x - 1} \right|$
50. $f(x) = \dfrac{2x}{\left| x^2 - 2x + 15 \right|}$

51. $f(x) = \dfrac{x^4 - 3x^3 - 21x^2 + 43x + 60}{x^4 - 6x^3 + x^2 + 24x - 20}$

Graph using the same set of axes. Try to discover a pattern for graphing a function and its reciprocal.

52. $f(x) = x^2$ and $g(x) = \dfrac{1}{x^2}$

53. $f(x) = x^2 - 4$ and $g(x) = \dfrac{1}{x^2 - 4}$

54. $f(x) = \dfrac{x^2}{x^2 - 1}$ and $g(x) = \dfrac{x^2 - 1}{x^2}$

55. Sketch the general shape of the graph of $y = 1/x^n$ for any even natural number.

56. Sketch the general shape of the graph of $y = 1/x^n$ for any odd natural number.

➡ **Technology**

Sketch a graph of each of the following. Use computer graphing software, a calculator that sketches graphs, or just a computer or calculator to find function values.

57. $f(x) = \dfrac{1}{x^{26} + 4} - 3$ **58.** $f(x) = (x + 1)^{2/3}$

59. $f(x) = (x^2 - 1)^{2/3}$ **60.** $f(x) = \dfrac{x^{28} - x^2 + 2}{x^{28} - x + 11}$

Match the following functions with graphs (a)–(f).

61. $f(x) = \dfrac{x}{(x - 1)(x + 2)}$ **62.** $f(x) = \dfrac{1}{x + 3}$

63. $f(x) = \dfrac{x^2 + 1}{x}$

64. $f(x) = \dfrac{1}{x(x + 1)(x + 2)}$

65. $f(x) = \dfrac{x^2}{x^2 + 1}$ **66.** $f(x) = \dfrac{1}{x^2 + 1}$

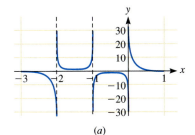

(a)

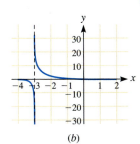

(b)

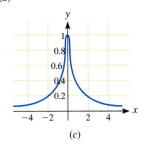

(c)

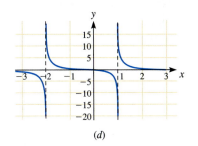

(d)

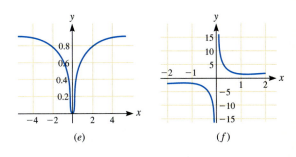

(e) (f)

The graphs of rational functions $f(x)$ and $g(x)$ are shown in graphs (a) and (b). Sketch the graphs of the following functions.

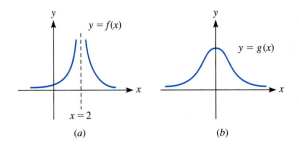

(a) (b)

67. $f(x + 1)$ **68.** $f(x) + 3$

69. $f(x - 2) + 3$ **70.** $g(x + 1) + 1$

71. $g(x - 4)$ **72.** $g(x) - 2$

✍ **In Your Own Words**

73. Write a detailed procedure for determining the vertical asymptotes of a rational function.

74. Write a detailed procedure for determining the horizontal asymptotes of a rational function.

75. Give an applied example of a rational function with a horizontal asymptote. Explain the applied significance of the horizontal asymptote.

76. Give an applied example of a rational function with a vertical asymptote. Explain the applied significance of the vertical asymptote.

4.7 CHAPTER REVIEW

Important Concepts, Properties, and Formulas—Chapter 4

Polynomial function	$f(x) = a_n x^n + a_{n-1} x^{n-1} + \cdots + a_0$	p. 218
Quadratic function	$f(x) = ax^2 + bx + c, \quad a \neq 0$	p. 218
Standard form for a quadratic function	$f(x) = a(x - h)^2 + k$, where $$h = -\frac{b}{2a}, \qquad k = \frac{4ac - b^2}{4a}$$ and where the extreme point (h, k) is maximum if $a < 0$ and minimum if $a > 0$.	p. 220
Descartes' rule of signs	Number of positive real zeros is at most the number of sign changes m of $f(x)$ and differs from m by a multiple of two. The number of negative real zeros is at most the number of sign changes n of $f(-x)$ and differs from n by a multiple of 2.	p. 251
Asymptotes of a rational function	If $$f(x) = \frac{p(x)}{q(x)}$$ where $p(x)$ is of degree m and leading coefficient c, and $q(x)$ is a polynomial of degree n and leading coefficient d, then: 1. If $m < n$, then the x-axis is a horizontal asymptote of the graph of $f(x)$, and $f(x)$ approaches 0 as x approaches either positive infinity or negative infinity. 2. If $m = n$, then the line $y = c/d$ is a horizontal asymptote of the graph of $f(x)$, and $f(x)$ approaches c/d as x approaches either positive infinity or negative infinity. 3. If $m = n + 1$, then we may write $f(x)$ in the form $$f(x) = ax + b + \frac{r(x)}{q(x)}$$ where $r(x)$ has degree less than n. In this case, $y = ax + b$ is an oblique asymptote of the graph of $f(x)$. 4. If $m > n$, then the graph of $f(x)$ has no horizontal asymptotes. Moreover, as x approaches positive infinity or negative infinity, $f(x)$ approaches either positive infinity or negative infinity. (It is simplest to separately determine which it approaches in each instance.)	p. 266

| Graphing rational functions | 1. Determine any common factor of the numerator and denominator. The zeros of this common factor are points at which the rational function is undefined. Replace the given expression of the rational function by one in lowest terms.

2. Determine the horizontal and oblique asymptotes. If there are none, determine the behavior of $f(x)$ as x approaches both positive infinity and negative infinity.

3. Determine the vertical asymptotes by calculating the real zeros of the denominator. For each zero a, the line $x = a$ is a vertical asymptote.

4. Determine the zeros of $f(x)$ by determining the zeros of the numerator. These are the x-intercepts of the function.

5. Divide the x-axis into intervals determined by the real zeros of $f(x)$ and by the vertical asymptotes. For each interval, choose a test value and determine the sign of $f(x)$ in that interval. Then use the signs to determine whether the graph is above or below the x-axis in each interval.

6. Draw the horizontal and vertical asymptotes and plot the zeros.

7. Draw the portion of the graph corresponding to each interval using the information developed in steps 1–6. | p. 268 |

Cumulative Review Exercises—Chapter 4

For each of the following quadratic functions:

a. Determine the standard form.

b. Find the vertex.

c. Determine whether the graph opens upward or downward.

d. Find the maximum or minimum value.

e. Graph the function.

1. $f(x) = x^2 - 3x$

2. $f(x) = 2(x - 1)^2 + 4$

3. $f(x) = 2x^2 + 12x + 2$

4. $f(x) = \frac{1}{2}x^2 + 2x + 5$

5. $f(x) = -\frac{1}{3}x^2 + \frac{2}{5}x - \frac{4}{5}$

Find the maximum or minimum value.

6. $f(x) = 2x^2 + 17x - 35$

7. $f(x) = \frac{1}{4}x^2 - 20x + \frac{2}{3}$

8. $f(x) = 23.45x^2 - 34.67x - 1$

9. $f(x) = -\sqrt{2}x^2 + \sqrt{6}x + \sqrt{8}$

10. Find two numbers whose sum is 31 and whose product is a maximum.

11. The Metro Bus Company charges a person 20 cents to ride the bus. They now carry 8000 passengers per day. They want to raise the fare, but for every 5-cent raise in the fare they will lose 800 passengers. What fare should they charge to maximize passenger income?

12. **Total profit.** A small business is marketing a new tape recorder. It determines that its total revenue from the sale of x tape recorders is given by the function

$$R(x) = 44x - x^2$$

The firm also determines that its total cost of producing x tape recorders is given by

$$C(x) = 700 - 36x$$

Total profit is given by

$$P(x) = R(x) - C(x)$$

a. Find the total profit.

b. How many tape recorders must the company produce and sell to maximize profit?

c. What is the maximum profit?

Sketch graphs of each of the following functions.

13. $f(x) = (x^2 - 1)(x^2 - 4)$

14. $f(x) = 4x^2 - x^3$

15. $f(x) = (x + 4)(x - 1)(x - 3)$

16. $f(x) = -(x - 2)^3 - 3$ 17. $f(x) = (x + 3)^2(x - 2)$

18. $f(x) = 2x^4 + 8$ 19. $f(x) = -\dfrac{2}{x^2}$

20. $f(x) = \dfrac{1}{x + 1}$ 21. $f(x) = \dfrac{8}{x^2 - 4}$

22. $f(x) = \dfrac{x^2 + 2x - 3}{x^2 - x - 2}$ 23. $f(x) = 2x + \dfrac{2}{x}$

24. $f(x) = \dfrac{1}{x^2 + 2}$

25. $f(x) = \dfrac{x^3 + 12x^2 + 4x + 48}{x^3 + 2x^2 - 3x}$

26. $f(x) = \dfrac{x^2 - 5}{x - 4}$

27. $f(x) = \dfrac{(x - 2)(x^2 - 1)}{x - 2}$

28. $f(x) = \dfrac{x^2 - 4}{x - 2}$

29. $f(x) = \dfrac{x^3 + 2}{x}$

30. $f(x) = \dfrac{x^3 + x^2 - 6x}{x^2 + x - 2}$

In Exercises 31 and 32, a polynomial $f(x)$ and a divisor $g(x)$ are given. Calculate the quotient $q(x)$ and the remainder $r(x)$.

31. $f(x) = x^5 + 3x^4 - 2x^2 + 6x - 3$

    $g(x) = x - 1$

32. $f(x) = x^5 + 3x^4 - 2x^2 + 6x - 3$

    $g(x) = x^3 - x^2 + 4$

33. Suppose that $f(x)$ is a quadratic polynomial with zeros $1/2$ and $-2/3$, and $f(-1) = 5$. Determine f.

34. Suppose that $f(x)$ is a quadratic polynomial with zeros $-2 + \sqrt{3}$ and $-2 - \sqrt{3}$, and $f(-1) = 5$. Determine f.

35. Suppose that $f(x)$ is a quadratic polynomial with zeros

$$\dfrac{1 + 3i}{2} \quad \text{and} \quad \dfrac{1 - 3i}{2}$$

and $f(-1) = 5$. Determine f.

36. Suppose that $f(x)$ is a fourth-degree polynomial with zeros $1 + \sqrt{2}, 1 - \sqrt{2}, 3,$ and -2; and $f(0) = 6$. Determine f.

37. Suppose that $f(x)$ is a fifth-degree polynomial with zeros $2i, -2i, 3,$ and -1 with muliplicity 2; and $f(1) = -6$. Determine f.

Use synthetic division to find the quotient and remainder when $f(x)$ is divided by $g(x)$.

38. $f(x) = 4x^5 + 6x + 1$

    $g(x) = x + 2$

39. $f(x) = x^4 - 2x^3 - 5x^2 - x + 1$

    $g(x) = x + 1$

40. $f(x) = \dfrac{1}{3}x^2 + \dfrac{5}{6}x - 3$

    $g(x) = 2x + 9$

41. $f(x) = x^4 - 2x^3 - 2x - 1$

    $g(x) = x + i$

Use synthetic division to determine the given function values.

42. $f(x) = x^4 - 2x^3 - 2x - 1$; find $f(0)$, $f(-1)$, and $f(i)$.

43. $f(x) = x^5 + x^4 + 2x^3 - 2x^3 - 2x^2 + x - 1$; find $f(-1)$, $f(2)$, and $f(-i)$.

Use synthetic division to determine whether each number is a zero of the given polynomial.

44. $2, 3, -2; f(x) = x^3 - 9x^2 + 23x - 15$

45. $-1, i, 2; f(x) = x^4 - x^3 - 7x^2 - 7x - 2$

Factor each polynomial $f(x)$. Then use the result of the factoring to solve the equation $f(x) = 0$.

46. $f(x) = x^3 - 100$

47. $f(x) = x^3 - 6x^2 + 3x + 10$

For each polynomial determine the zeros and their multiplicities.

48. $f(x) = 5x^3(x - 4)^2(2x + 3)$

49. $f(x) = (3x^3 - 6x^2 + 3x)^2$

50. $f(x) = (x^2 + 1)^2(x^2 - 1)^3$

51. How many real zeros can a polynomial of degree 5 have?

52. How many real zeros can a polynomial of degree 6 have?

Use Descartes' Rule of Signs to analyze the zeros of each polynomial.

53. $6x^4 - 30x^3 - 32x^2 - 21$

54. $3x^6 - 5x^4 + 2x^3 - x - 4$

55. $8x^5 + 3x^4 - x^3 + 5x^2 - 3$

56. Find k so that the first polynomial is a factor of the second.

$$x - 2; \qquad 2x^3 + 3x^2 + kx + 10$$

For each polynomial, certain zeros are given. Find the remaining zeros.

57. $f(x) = 2x^4 + 4x^3 + 3x^2 - 6x - 9; -1 - i\sqrt{2}$

58. $f(x) = x^3 - x^2 - 7x + 3;\ 3$

59. $f(x) = x^4 + 2x^3 + x^2 + 60x + 144; -3$ is a root of multiplicity 2.

60. $f(x) = x^4 - 4x^3 + 18x^2 + 8x - 40; 2 + 4i$

61. Find all possible rational zeros of
$$4x^6 - 5x^5 + 6x^4 - 20x^3 - 10x^2 + 6$$

Determine upper and lower bounds for the real zeros of $f(x)$.

62. $f(x) = 6x^4 - 30x^3 - 32x^2 - 23x + 2$

63. $f(x) = 3x^6 - 5x^4 + 2x^3 - x - 4$

64. $f(x) = 8x^5 + 3x^4 - x^3 + 5x^2 - 3$

Find only the rational zeros.

65. $6x^4 - 30x^3 - 32x^2 - 23x + 2$

66. $3x^6 - 5x^4 + 2x^3 - x - 4$

67. $8x^5 + 3x^4 - x^3 + 5x^2 - 3$

68. $x^5 - 2x^4 + x^3 - 1$

Solve. Use any procedures studied in this chapter.

69. $(x - 1)(x^2 + x + 1) = 0$

70. $x^3 + 31x = 10x^2 + 30$

71. $5x^2 = x^3 + 5x + 3$

72. $12x^6 - 52x^5 + 91x^4 - 82x^3 + 40x^2 - 10x + 1 = 0$

73. $x^3 + 2 = 2x^2 + 7x + 2$

74. $x^4 + x^2 + 1 = 0$

75. $2x^4 - 3x^3 + x^2 + 4x - 2 = 0$

Show that each of the following polynomials has a zero in the given interval.

76. $f(x) = x^3 - 2x - 5; (2, 3)$

77. $f(x) = x^4 - x^3 - 2x^2 - 6x - 4; (-1, 0)$

Approximate the real zeros of each of the following polynomial functions with an accuracy of 0.0001 using the method of bisection.

78. $f(x) = x^3 - 2x - 5; (2, 3)$

79. $f(x) = x^4 - x^3 - 2x^2 - 6x - 4; (-1, 0)$

80. $f(x) = x^5 - x + 1; (-2, 0)$

♦ **Applications**

81. An open box of volume 132 cubic inches can be made from a piece of cardboard that is 10 in. by 15 in. by cutting a square from each corner and folding up the sides. What is the length of a side of the squares? Use the method of bisection to find an approximate answer, if appropriate.

82. **Electrical resistance.** Suppose three resistors are connected in parallel, as shown in the figure. The result of

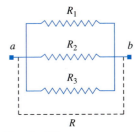

Exercise 82

the three resistances is an equivalent resistance R, given by

$$\frac{1}{R} = \frac{1}{R_1} + \frac{1}{R_2} + \frac{1}{R_3}$$

Suppose $R = 1$, the second resistance is two more than the first, and the third resistance is three more than the first. (See figure.)

a. Find a polynomial equation in R_1 that can be used to find the resistances.

b. Find each of the three resistances. (Resistance is always a nonnegative unit.) Use the method of bisection if appropriate.

83. **Inventory management.** The cost $C(x)$ of managing an inventory of x pool tables is determined by a team of financial analysts to be given by the function

$$C(x) = 10x + \frac{4000}{x} + 1600$$

Sketch the graph of this function over the interval $(0, +\infty)$.

84. There are many rules for determining the dosage of a medication for a child based on an adult dosage. The following are two such rules:

$$\text{Young's rule:} \quad C = \frac{A}{A + 12}D$$

$$\text{Cowling's rule:} \quad C = \frac{A + 1}{24}D$$

where A = the age of the child, in years; D = adult dosage; and C = child's dosage. (*Warning:* Do not apply these formulas without consulting a physician.

These are estimates and may not apply to a particular medication.) An adult dosage of a liquid medication is 5 cc (cubic centimeters).

a. Sketch a graph of Young's rule for the value of A in the interval [0, 18].

b. Sketch a graph of Cowling's rule for values of A in the interval [0, 18].

c. For what ages A are the child's dosages the same?

Sketch a graph of each function and find the x-intercepts. Approximate where appropriate.

85. $f(x) = 3x^3 - 16x^2 + 12x + 16$

86. $f(x) = x^2 + \dfrac{1}{x^2}$

87. $f(x) = \dfrac{1}{x^2 - 1}$

88. $f(x) = \dfrac{2x^2}{x^2 - 16}$

89. $f(x) = \dfrac{x}{x - 3}$

90. Find a such that

$$f(x) = ax^2 - \frac{2}{3}x + 7$$

has a maximum value of 85.

91. Find c such that $f(x) = 8/7x^2 - 3/4x + c$ has a minimum value of 226.

92. Prove that $\sqrt{3}$ is irrational by considering the zeros of the polynomial function $f(x) = x^2 - 3$.

➡ **Technology**

93. Use computer software or a graphing calculator to sketch a graph of the following function. Then approximate the real zeros.

$$f(x) = 3x - (x - 1)^{3/2}, \qquad (x \geq 1)$$

EXPONENTIAL AND LOGARITHMIC FUNCTIONS

*The spiral exhibited in this cross section of a nautilus is a curve called a **logarithmic spiral** and is an illustration of the natural logarithmic function appearing in nature.*

A pplied mathematics uses various functions that describe common situations. In this chapter, we add to our repertoire of functions the exponential and logarithmic functions, which are necessary for describing many situations you may encounter in other fields such as radioactive decay, bacterial growth, hearing, and the transmission of light to ocean depths.

5.1 EXPONENTIAL FUNCTIONS WITH BASE a

Suppose that a is a positive number. We have already defined the power a^x for x a rational number. This suffices to define a^x for x a finite decimal.

In many applications, it is necessary to consider a^x for x an irrational number, say, $x = \pi$ or $x = \sqrt{2}$. Such powers can be defined using a sequence of rational approximations to x. For example, to define the power $3^{\sqrt{2}}$, we approximate $\sqrt{2}$ by the sequence of decimal values

$$1.4, 1.41, 1.414, 1.4142, \ldots$$

Corresponding to this sequence of approximations, we have the following approximations to $3^{\sqrt{2}}$:

$$3^{1.4} = 4.655536\ldots$$
$$3^{1.41} = 4.706965\ldots$$
$$3^{1.414} = 4.727695\ldots$$
$$3^{1.4142} = 4.728733\ldots$$

On the basis of these approximations, we see that $3^{\sqrt{2}}$ is approximately equal to 4.73. In fact, using more precise approximations to the value of the exponent, we get ever closer approximations to the value of

$$3^{\sqrt{2}} = 4.728804\ldots$$

Using similar approximations, we can define the value of a^x for any real number x.

Calculating a^x

To calculate a^x it is most convenient to use a scientific calculator. Such calculators have a key, usually labeled x^y, for calculating powers. To calculate a^x, first enter the value of a. Then press the x^y key. Next enter the value of x. Finally, press the $=$ key. These instructions will work for a calculator using algebraic logic. For example, to calculate

$$2.57^{3.4875}$$

first key in the number 2.57. Then press the x^y key. Next, key in the exponent 3.4875. Finally, press the $=$ key.

Some calculators, most notably Hewlett-Packard calculators, use "reverse Polish logic." For such calculators, a different order of data entry must be used: First enter the value of a and press the **ENTER** key. Next enter the value of x. Finally press the x^y key. No matter what type of calculator you have, the approximate value you get is 26.893.

Laws of Exponents

It is possible to prove that the laws of exponents, introduced in Chapter 1 for rational exponents, continue to hold for real exponents. In fact, we can state the following result.

LAWS OF REAL EXPONENTS

Let a and b be positive real numbers, and let x and y be any real numbers. Then the following laws of exponents hold:

1. $a^x a^y = a^{x+y}$
2. $(a^x)^y = a^{xy}$
3. $(ab)^x = a^x b^x$
4. $\dfrac{a^x}{a^y} = a^{x-y}$

Definition 1
**Exponential Function
with Base *a***

Let a be a positive real number. Then the **exponential function with base a** is the function defined by the formula

$$f(x) = a^x, \quad a \neq 1$$

Note that we excluded the possibility $a = 1$ so that the constant function $f(x) = 1$ would not be considered as an exponential function.

Here are some examples of exponential functions with various bases:

$$f(x) = 2^x$$
$$g(x) = 3^x$$
$$h(x) = \sqrt{5}^x$$
$$k(x) = \left(\frac{1}{2}\right)^x$$

Since a^x is defined for every real number x, the domain of an exponential function is the set **R** of real numbers. Moreover, a^x is positive, so the range of an exponential function is contained in the set

$$\{y : y \text{ is positive}\}$$

Actually, it can be proved that the range of an exponential function is precisely the set of all positive numbers. We will assume this fact in what follows.

➤ **EXAMPLE 1**
Calculating Exponentials

Let $f(x) = 2^x$ be the exponential function with base 2. Calculate the following:

1. $f(1)$ 2. $f(2)$ 3. $f(3)$ 4. $f(-1)$

5. $f(-2)$ 6. $f(0)$ 7. $f\left(\frac{1}{2}\right)$

Solution

1. $f(1) = 2^1 = 2$
2. $f(2) = 2^2 = 4$
3. $f(3) = 2^3 = 8$
4. $f(-1) = 2^{-1} = \dfrac{1}{2^1} = \dfrac{1}{2}$
5. $f(-2) = 2^{-2} = \dfrac{1}{2^2} = \dfrac{1}{4}$

6. $f(0) = 2^0 = 1$

7. $f\left(\dfrac{1}{2}\right) = 2^{1/2} = \sqrt{2}$

Graphs of Exponential Functions

As we have seen in the case of polynomial and rational functions, the graph of a function provides valuable information about the function (where it increases or decreases, its zeros, its asymptotes, etc.). With this in mind, let's now learn to sketch the graphs of exponential functions. The following example discusses the graph of a particular exponential function.

➤ **EXAMPLE 2**
Graphing an Exponential

Sketch the graph of the function $f(x) = 2^x$.

Solution

We may record the results of Example 1 in a table.

x	$f(x)$
-2	$1/4$
-1	$1/2$
0	1
$1/2$	$\sqrt{2}$
1	2
2	4
3	8

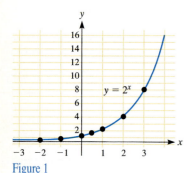

Figure 1

In Figure 1, we plot the points listed in the table and connect them with a smooth curve. Note that as x increases, the graph rises very sharply. As x approaches negative infinity, the graph has the negative x-axis as an asymptote.

For $a > 1$, the graph of $f(x) = a^x$ has the same general shape as the graph in Figure 1. Namely, as x increases, so does the value of $f(x)$. Moreover, as x approaches negative infinity, the graph approaches the negative x-axis as an asymptote. Furthermore, the larger the value of a, the steeper the rate at which the graph rises. In Figure 2, we have drawn the graphs of several exponential functions corresponding to assorted values of a that are larger than 1.

In the next example, we will consider a particular exponential function for which $0 < a < 1$.

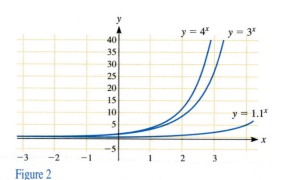

Figure 2

➤ **EXAMPLE 3**

Graphing a Decreasing Exponential

x	$f(x)$
-3	8
-2	4
-1	2
0	1
1	1/2
2	1/4
3	1/8

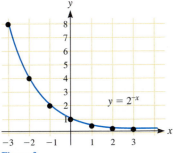

Figure 3

Sketch the graph of the exponential function

$$f(x) = \left(\frac{1}{2}\right)^x$$

Solution

To get a sense of the shape of the graph, we plot points corresponding to some representative values of x. The calculations require the use of the laws of real exponents. For instance, we have

$$f(-1) = \left(\frac{1}{2}\right)^{-1} = \frac{1}{1/2} = 2$$

The accompanying table summarizes various values of the function. You should verify these values as an exercise.

In Figure 3, we have plotted the points corresponding to the table entries and drawn a smooth curve through the points. Note that as x increases, the graph of $f(x)$ decreases. Moreover, as x approaches positive infinity, the graph approaches the positive x-axis as an asymptote. As x decreases and approaches negative infinity, the function values increase without bound and approach infinity.

For $0 < a < 1$, the graph of $f(x) = a^x$ has the same general shape as the graph shown in Figure 3. In Figure 4, we have sketched the graphs of several exponential functions with base values a that are less than 1 and greater than 0. Note that the smaller the value of a, the steeper the graph.

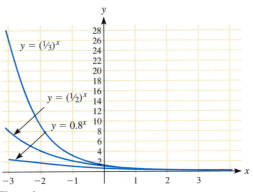

Figure 4

Applications of the Exponential Functions

Exponential functions abound in applications such as radioactive decay, bacteria growth, and the spread of epidemics. Such functions typically arise, multiplied by a constant, in the form

$$f(x) = Aa^x$$

where A is the constant. The following examples illustrate some of the instances in which such functions arise.

➤ **EXAMPLE 4**

Compound Interest

Suppose that P dollars are deposited in a bank account paying annual interest at a rate i and compounded n times per year. After a length of time t, in years, the amount $A(t)$ in the account is given by the formula

$$A(t) = P\left(1 + \frac{i}{n}\right)^{nt}$$

Suppose that the interest rate is 10%, the amount is compounded annually, and the initial deposit is $10,000.

1. Write the function $A(t)$ in terms of an exponential function.
2. Determine the amount in the account after 1, 2, and 10 years.
3. Graph the function $A(t)$.

Solution

1. In this example, we have

$$P = 10,000, \qquad i = 0.1, \qquad n = 1$$

So in this case, the compound interest formula reads

$$A(t) = 10,000\left(1 + \frac{0.1}{1}\right)^{1 \cdot t}$$

$$A(t) = 10,000 \cdot (1.1)^t$$

2. The desired amounts are given, respectively, by the function values $A(1)$, $A(2)$, and $A(10)$. Computing these values, we have

$$A(1) = 10,000 \cdot (1.1)^1 = \$11,000.00$$
$$A(2) = 10,000 \cdot (1.1)^2 = \$12,100.00$$
$$A(10) = 10,000 \cdot (1.1)^{10} = \$25,937.43$$

To calculate these values, we used a calculator.

3. We know the general shape of the graph of the exponential function $f(t) = 1.1^t$ from Figure 2. The graph of $A(t)$ is obtained by scaling the graph of $f(t)$ by a factor of 10,000. From part 2, we have several points that help calibrate the graph. The graph must pass through these points and increase rapidly with increasing values of t. From the context of the application, only nonnegative values of t are relevant, so we limit the domain of the graph to $t > 0$. A sketch of the graph is given in Figure 5.

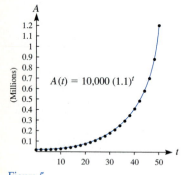

$A(t) = 10,000\,(1.1)^t$

Figure 5

➤ **EXAMPLE 5**

Radioactive Decay

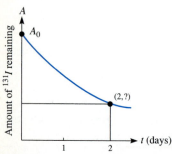

Figure 6

The decay of radioactive materials is described by an exponential function. Suppose that A_0 denotes the amount of a radioactive substance present at time 0, and $Q(t)$ represents the quantity present at time t. As time passes, the amount of radioactive material decays, with the amount approaching 0 as time increases. Suppose H denotes the time it takes for a quantity of the radioactive material to decay to half its original amount. The number H depends on the particular radioactive material and is called the **half-life** of the material. The function $Q(t)$ is then given by the expression

$$Q(t) = A_0 2^{-t/H}$$

Radioactive iodine (^{131}I) is used in various medical diagnostic tests. Its half-life is 8 days.

1. Write the formula for $Q(t)$ for radioactive iodine in terms of an exponential function.
2. What proportion of the original radioactive iodine is present 48 hours after it is originally swallowed? (See Figure 6.)

Solution

1. We are given that $H = 8$ days. Since we are not given the original amount of radioactive iodine, we leave it in the formula as A_0. We then have

$$Q(t) = A_0 2^{-\frac{t}{8}}$$
$$= A_0 2^{-\frac{1}{8} \cdot t}$$
$$= A_0 (2^{-\frac{1}{8}})^t$$
$$= A_0 0.917^t$$

The function $Q(t)$ is a constant, A_0, times an exponential function a^t, where a is equal to 0.917.

2. When t equals 48 hours = 2 days, the amount of radioactive iodine remaining is $Q(2)$, which is equal to

$$Q(2) = A_0 2^{-\frac{1}{8} \cdot 2} = A_0 2^{-\frac{1}{4}} = 0.840896 A_0$$

That is, after 48 hours about 84% of the radioactive iodine remains.

➤ EXAMPLE 6
Exponential Growth

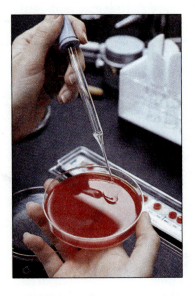

Colonies of certain bacteria exhibit exponential growth. At time 0 hours, suppose that there are N_0 bacteria. Then $N(t)$, the number of bacteria at time t, in hours, is given by

$$N(t) = N_0 2^{\frac{t}{D}}$$

where D denotes the time, in hours, that it takes the colony to double in size.

1. Write $N(t)$ in terms of an exponential function.
2. Suppose that the colony initially contains 5000 bacteria and the doubling time is 4 hours. How many bacteria are there after 10 hours?
3. Consider again the bacteria colony of part 2. How long will it be before the colony contains 160,000 bacteria?

Solution

1. By the laws of exponents, we may write

$$N(t) = N_0 2^{\frac{t}{D}}$$
$$= N_0 2^{\frac{1}{D} \cdot t}$$
$$= N_0 (2^{\frac{1}{D}})^t$$

If we let $a = 2^{\frac{1}{D}}$, then the formula takes the form of a constant, N_0, times an exponential function, a^t:

$$N(t) = N_0 a^t$$

2. We are given that the doubling time $D = 4$ hours and $N_0 = 5000$. So the formula for $N(t)$ is given by

$$N(t) = 5000 \cdot 2^{\frac{t}{4}}$$

The number of bacteria after 10 hours is given by the value $N(10)$, which equals

$$5000 \cdot 2^{\frac{10}{4}} = 5000 \cdot 2^{\frac{5}{2}} = 5000 \cdot \sqrt{2^5} = 28,284$$

Note that the last number was rounded to the nearest integer. The formula for the number of bacteria, like most mathematical descriptions of physical phenomena, is only approximate, and answers must be interpreted to conform to physical reality, in this case the fact that there are no fractional bacteria.

3. We must determine the value of t for which the following equation holds:

$$N(t) = 160,000$$

Using the equation for $N(t)$ derived in part 2, we have

$$5000 \cdot 2^{t/4} = 160,000$$
$$2^{t/4} = 32$$
$$2^{t/4} = 2^5$$

In the last equation, we have two powers of 2 that are equal. For this equation to hold, the exponents must be equal. (Look at the graph of 2^x. It is constantly increasing, so that no two values of x have the same value for y.) Thus, we have the equation

$$\frac{t}{4} = 5$$

$$t = 20$$

That is, after 20 hours, the bacteria colony has 160,000 bacteria.

Exploring the Laws of Real Exponents with a Graphing Utility

Optional

Two nice features of graphing utilities are that they allow you to graph many functions in the time it would take to graph one by hand and that they work easily with functions that would not be convenient to tabulate and graph otherwise. In this discussion, we will take a look at how the power to graph several functions quickly can aid in understanding the Laws of Real Exponents. It is important to recognize that graphing these functions, no matter how many we graph, does not prove the truth of these laws, but it will visually demonstrate their truth for the specific examples we use.

The first law states that $a^x a^y = a^{x+y}$ for a a positive real number and x, y any real numbers. We can visually demonstrate this on a graphing utility by first graphing the functions $f(x) = a^x$ and $g(x) = a^y$. We need a specific example for the purpose of demonstration, so we'll look at $f(x) = 2^x$ and $g(x) = 2^{x/2}$. Let $h(x) = f(x)g(x)$. On a TI-81, these two functions would be entered as Y_1 and Y_2. Next, in Y_3, enter $Y_1 * Y_2$. If the range is set to $-3 \le x \le 3, -1 \le y \le 10$, the resulting graphs can then be seen on the same system as in Figure 7(a). Moving from the bottom to the top of the figure, the functions plotted are $g(x), f(x)$, and $h(x)$.

To demonstrate the law $a^x a^y = a^{x+y}$, we must now graph the function $k(x) = 2^{x+x/2} = 2^{3x/2}$. Clear the other functions from the **Y=** entry screen and enter $k(x)$ in Y_1. The resulting graph is shown in Figure 7(b). A visual comparison shows that this is the same as the graph of $h(x)$ in Figure 7(a). ◀

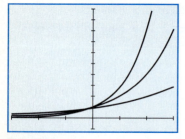

Figure 7(a)

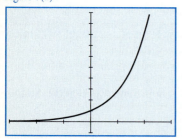

Figure 7(b)

Exercises 5.1

1. Estimate 4^π by making the following calculations:

   a. 4^3 b. $4^{3.1}$ c. $4^{3.14}$ d. $4^{3.141}$

   e. $4^{3.1415}$ f. $4^{3.14159}$ g. $4^{3.141592}$

   h. What seems to be the value of 4^π to two decimal places?

2. Estimate $5^{\sqrt{3}}$ by making the following calculations:

    a. 5^1 *b.* $5^{1.7}$ *c.* $5^{1.73}$ *d.* $5^{1.732}$

    e. $5^{1.73205}$ *f.* $5^{1.7320508}$ *g.* $5^{1.732050808}$

    h. What seems to be the value of $5^{\sqrt{3}}$ to five decimal places?

3. *a.* For $f(x) = 3^x$, find $f(1), f(2), f(3), f(0), f(-1), f(-2), f(-3)$, and $f(1/2)$.

    b. Sketch a graph of $f(x) = (3)^x$.

4. *a.* For $f(x) = (1/3)^x$, find $f(1), f(2), f(3), f(0), f(-1), f(-2), f(-3)$, and $f(1/2)$.

    b. Sketch a graph of $f(x) = (1/3)^x$.

Sketch a graph of each of the following:

5. $f(x) = 4^x$

6. $f(x) = 6^x$

7. $f(x) = \left(\dfrac{1}{4}\right)^x$

8. $f(x) = \left(\dfrac{1}{5}\right)^x$

9. $f(x) = \left(\dfrac{3}{4}\right)^x$

10. $f(x) = \left(\dfrac{2}{3}\right)^x$

11. $f(x) = (2.4)^x$

12. $f(x) = (3.6)^x$

13. $f(x) = (0.38)^x$

14. $f(x) = (0.76)^x$

15. $y = 5^x$

16. $y = 3^x$

17. $y = 10^x$

18. $y = 2.7^x$

19. $f(x) = 2^{-x}$

20. $f(x) = 2^{x-1}$

21. $f(x) = 2^{-(x-3)}$

22. $f(x) = 2^x - 4$

23. $f(x) = 2^{|x|}$

24. $f(x) = 2^x + 2^{-x}$

25. $f(x) = 2^{-x^2}$

26. $f(x) = |2^x - 4|$

27. $f(x) = 2^x - 2^{-x}$

28. $f(x) = 2^{-(x+3)^2}$

29. $f(x) = 1 - 2^{-x^2}$

30. $f(x) = x + 2^{-x^2}$

♦ **Applications**

31. **Compound interest.** Suppose that P dollars are deposited in a bank account paying annual interest at a rate i and compounded n times per year. After a length of time t, the amount $A(t)$ in the account is given by the formula

$$A(t) = P\left(1 + \dfrac{i}{n}\right)^{nt}$$

Suppose that the interest rate is 8%, that the amount is compounded annually, and the initial deposit is $10,000.

    a. Write the function $A(t)$ in terms of an exponential function.

    b. Determine the amount in the account after 1, 2, 3, and 10 years.

    c. Graph the function $A(t)$.

32. **Radioactive decay.** A radioactive decay function is given by

$$Q(t) = A_0 2^{-t/H}$$

where A_0 denotes the amount of a radioactive substance present at time $t = 0$ and H is the half-life. Radioactive cobalt 60 (^{60}Co) has a half-life of 5.3 years. It is used for the chemotherapy treatment of cancer tumors.

    a. Write the formula for $Q(t)$ for radioactive cobalt in terms of an exponential function.

    b. Suppose a chemotherapy device contains 4 grams of ^{60}Co. That is, A_0, is 4 grams. How much will be present after 1 day (1/365 yr.)? 1 year? 2 years? 10.6 years? 20 years?

33. **Radioactive decay.** Plutonium, a common product and ingredient of nuclear reactors, is a great concern to those against the building of such reactors. The half-life of plutonium is 23,105 years.

    a. Write the formula for $Q(t)$ for plutonium in terms of an exponential function.

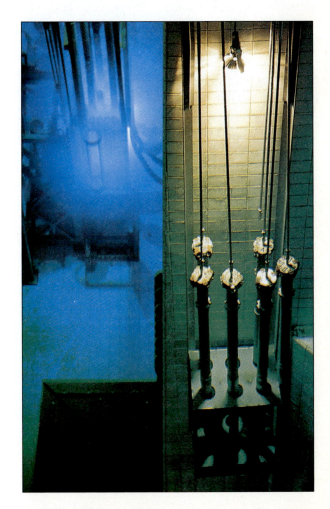

b. A nuclear reactor contains 400 kilograms of plutonium. What proportion of this original amount of plutonium is present after 1 year? 2 years? 200 years?

34. **Exponential growth.** The population of the world was about 5 billion on January 1, 1987, and is doubling every 43 years. Suppose we let $t = 0$ correspond to 1987 and consider the world population N_0 to be 5 billion at $t = 0$. The population after t years is $N(t)$, where

$$N(t) = N_0 2^{t/D}$$

and D is the doubling time.

a. Write $N(t)$ in terms of an exponential function.

b. What will the world population be in 43 years? 86 years? In 1997? In 2000?

35. **Exponential growth.** The population of Texas was 16,370,000 on January 1, 1985, and is doubling every 23 years. Suppose we let $t = 0$ correspond to 1985 and consider the population of Texas N_0 to be 16,370,000 at $t = 0$. The population after t years is $N(t)$, where

$$N(t) = N_0 2^{t/D}$$

and D is the doubling time.

a. Write $N(t)$ in terms of an exponential function.

b. What will the population of Texas be in 23 years? In 46 years? In 1995? In 2000?

36. **Exponential growth.** An investment grows exponentially in such a way that it doubles every 10 years. An initial amount of N_0 dollars at time $t = 0$ will grow to an amount $N(t)$ in t years, where

$$N(t) = N_0 2^{t/D}$$

and D is the doubling time.

a. Write $N(t)$ in terms of an exponential function assuming the initial amount invested is $100,000.

b. What will be the value or amount of the investment after 5 years? After 10 years? After 20 years? After 32 years?

c. How many years will it take for the amount of the investment to reach $1 million?

37. **Light filtration.** A certain type of tinted automobile glass 1 cm thick allows only 75% of the light to pass through.

a. How much light will pass through such glass when its thickness is 2 cm? 3 cm? 10 cm? x cm?

b. Let $P(x)$ represent the percentage of light that passes through glass x cm thick. Find a formula for $P(x)$.

c. Sketch a graph of $P(x)$ for $x \geq 0$.

Match the following functions with their corresponding graphs (a)–(j).

38. $f(x) = 2^x$

39. $f(x) = 2^{-x}$

40. $f(x) = 2^{x+1}$

41. $f(x) = 2^x + 1$

42. $f(x) = 2^{-(x+1)}$

43. $f(x) = 2^{x-3}$

44. $f(x) = 2^x - 3$

45. $f(x) = 3 \cdot 2^x$

46. $f(x) = -2^x$

47. $f(x) = -2^{-x}$

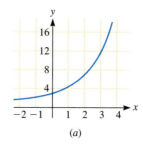

(a)

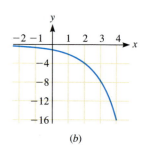

(b)

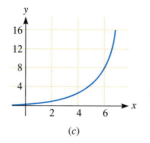

(c)

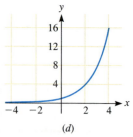

(d)

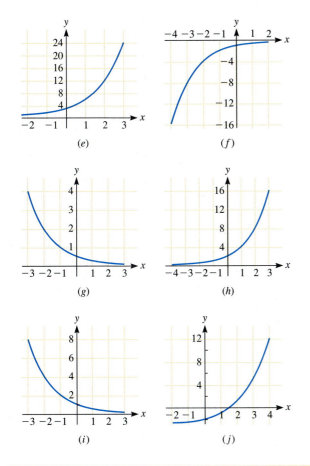

(e)

(f)

(g)

(h)

(i)

(j)

➡ **Technology**

Use a graphing calculator to graph the following functions for $0 \le x \le 5$. Experiment with the y range to include points corresponding to all values of x.

48. $f(x) = 3^x$

49. $f(x) = 2.5 \cdot 2^x$

50. $f(x) = 8.71 \cdot 2^{5x}$

51. $f(x) = -3.1 \cdot 2^{-1.1x}$

52. $f(x) = 2^x - 2^{-x}$

53. $f(x) = 2^{2^x}$

54. Determine all points at which the graphs of $f(x) = 2^x$ and $f(x) = 3^x - 2.5^x$ intersect. Approximate coordinates to two significant digits.

55. Graph the function $f(x) = 1000(1 - 2^{-x})$. By zooming out, determine any asymptotes of the graph.

✍ **In Your Own Words**

56. Describe the sort of physical situation that would best be modeled by an exponential function of the form $f(x) = a^x$, $a > 1$.

57. Describe the sort of physical situation which would best be modeled by an exponential function of the form $f(x) = a^x$, $0 < a < 1$.

58. Visually demonstrate the laws of real exponents using a graphing utility. Look at functions with various integer, irrational, and rational bases and exponents.

59. In the Laws of Real Exponents, explain why the base must be a positive number. Why is this restriction not necessary for the exponents?

▌ 5.2 THE NATURAL EXPONENTIAL FUNCTION

The Natural Number e

In the preceding section, we introduced the exponential function with base a. In this section, we will choose from this infinite family of exponential functions, one which is the "standard" one. This so-called natural exponential function is the most convenient one to use in many modeling problems and in calculus.

Any positive real number $a \ne 1$ can serve as the base for an exponential function. The most important value for a is a particular real number designated e. The number e is an irrational number whose value is given approximately by $2.718281\ldots$. Although this number may seem to come out of nowhere, it arises quite naturally in many applications of mathematics in the biological and physical sciences as well as in business. To introduce the number e, let's consider a problem in compound interest.

Suppose that P dollars are deposited for t years in a bank account yielding interest at an annual rate i and compounded n times per year. As we stated in the preceding section, the value of the account at time t, namely, $A(t)$, is given by the formula

$$A(t) = P\left(1 + \frac{i}{n}\right)^{nt}$$

Suppose that the initial deposit is $1, the length of time is 1 year, and the rate of interest is 100%. In this case, the compound interest reads

$$A(1) = 1 \cdot \left(1 + \frac{1}{n}\right)^{n \cdot 1} = \left(1 + \frac{1}{n}\right)^{n}$$

Let's consider the effect of different periods of compounding on the amount after 1 year. This amounts to computing the value of $A(1)$ for various values of n. We have computed assorted values on a calculator and listed them in the accompanying table.

n	Amount from Compounding n Times per Year
1	2
2	2.25
3	2.370370
4	2.441406
10	2.5937425
50	2.691588
100	2.7048138
1000	2.7169239
10,000	2.7181459
100,000	2.718268
1,000,000	2.7182805
10,000,000	2.7182817

Note that as the frequency of compounding increases without bound, the amount in the account at the end of the year approaches an amount approximately equal to $2.718 \ldots$. This is rather surprising. Your intuition might suggest that as the money is compounded more and more frequently (every day, every hour, every second, every millisecond), the amount in the account would grow without limit. However, this is just not the case. As the frequency of compounding is increased without bound, the amount in the account approaches a specific real number, denoted e, whose decimal expansion is given by $2.718281 \ldots$. Let's record these results for future reference.

$$\left(1 + \frac{1}{n}\right)^{n} \text{ approaches } e \text{ as } n \text{ becomes arbitrarily large}$$
$$e = 2.718281$$

Historical Note: Euler and e

The number e was first defined by the 18th century mathematician Euler (pronounced "Oiler"). Although born in Switzerland, Euler spent most of his life in Russia, as court mathematician to Empress Catherine the Great. One of the greatest mathematicians of the century, Euler made contributions to pure mathematics, fluid dynamics, mechanics, and astronomy. Among his many accomplishments, Euler wrote the first textbook on calculus, *Introductio in Analysin Infinitorum,* in 1740. Amazingly prolific, Euler's work spans more than 60 large volumes. Euler became blind at the end of his life. But he continued his mathematical research by dictating calculations to a servant he trained specially for the purpose. |█

Calculations Involving e

You may calculate with e just as you would with any other decimal number. In fact, expressions involving e may be evaluated using the e^x key found on most scientific calculators. Some calculators don't have an e^x key but use other techniques for calculating powers of e. For such calculators, consult your manual for details.

Algebraic expressions involving e^x can be manipulated using the fact that e^x is just an ordinary power to which the laws of exponents apply. The next example provides some practice in carrying out such algebraic manipulations.

➤ **EXAMPLE 1**

Calculating with e

Simplify the following expressions involving e.

1. $e^3 e^{-1} e^{-5}$ 2. $(e + 1)^2$ 3. $(e^x)^5$

4. $(e^x + e^{-x})^2$ 5. $(e^x + 1)(3e^x - 4)$ 6. $\sqrt{e^x}$

Solution

1. By the laws of exponents, we have
$$e^3 e^{-1} e^{-5} = e^{3 + (-1) + (-5)} = e^{-3}$$

2. Use the formula for the square of a binomial to obtain
$$(e + 1)^2 = e^2 + 2 \cdot e \cdot 1 + 1^2$$
$$= e^2 + 2e + 1$$

3. By the laws of exponents, we have
$$(e^x)^5 = e^{x \cdot 5} = e^{5x}$$

4. Multiply out the binomials and apply the laws of exponents to obtain
$$(e^x + e^{-x})^2 = (e^x)^2 + 2e^x e^{-x} + (e^{-x})^2$$
$$= e^{2x} + 2e^0 + e^{-2x}$$
$$= e^{2x} + e^{-2x} + 2$$

5. Multiply out and apply the laws of exponents:
$$(e^x + 1)(3e^x - 4) = e^x \cdot 3e^x - 4e^x + 3e^x - 4$$
$$= 3e^{2x} - e^x - 4$$

6. Recalling that taking the square root is equivalent to raising to the power 1/2, we see that
$$\sqrt{e^x} = (e^x)^{1/2} = e^{x \cdot 1/2} = e^{x/2}$$

The Natural Exponential Function

The exponential function with base e is the most important exponential function. To indicate its importance, let's assign it a name as follows:

Definition 1
Natural Exponential Function

The **natural exponential function** is the function $f(x)$ defined by
$$f(x) = e^x$$
The natural exponential function is a special case of the more general exponential function a^x. In particular, the domain of the natural exponential function is the set **R** of all real numbers, and the range is the set of all positive real numbers.

In the preceding section, we considered applications that involved exponential functions of the form
$$f(x) = a^x$$

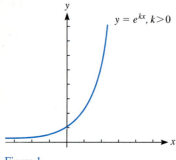

Figure 1

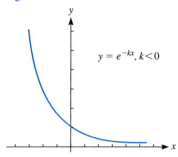

Figure 2

where a is a positive constant. Later in this chapter, we will show that every such function can be written in the alternative form

$$f(x) = e^{Ax}$$

where A is a constant. For this reason, functions of the alternative form are called **exponential functions.**

The case $A = 0$ is uninteresting since it corresponds to $f(x) =$ the constant function 1. The nonzero values of A give rise to two sorts of functions:

$$f(x) = e^{kx}$$
$$f(x) = e^{-kx}$$

where k is a positive constant.

The first sort of function increases very rapidly as x increases. The rapid rate of increase of such functions has passed over into common English usage, where we refer to **exponential growth** as a synonym for extremely rapid growth. In Figure 1, we show the graph of a typical function $f(x) = e^{kx}$ for $k > 0$.

For $k > 0$, the function $f(x) = e^{-kx}$ shows exponential decay and rapidly approaches 0 as x approaches positive infinity. In Figure 2, we show the graph of this sort of function.

Applications of the Natural Exponential Function

The natural exponential function is used in countless applications in fields ranging from biology and medicine to physics and engineering. The following examples provide illustrations of two of these applications.

➤ EXAMPLE 2
Bacteria Growth

Suppose that at time t (in hours), the number $N(t)$ of E. coli bacteria in a culture is given by the formula $N(t) = 5000e^{0.1t}$. How many bacteria are in the culture at time 5 hours?

Solution

At time $t = 5$, the number of bacteria equals $N(5)$, which by the given formula equals

$$5000e^{0.1(5)} = 5000e^{0.5} = 8244$$

We used a calculator to determine the value of the exponential function. Furthermore, we rounded the result to the nearest integer since the answer represents a number of bacteria.

➤ EXAMPLE 3
Skydiving

A skydiver jumps out of an airplane and executes a free-fall for a period of time. The diver's velocity does not increase indefinitely but approaches a limiting velocity, called the **terminal velocity,** and denoted v_T. The value of the terminal velocity depends on the particular skydiver. Let $v(t)$ denote the velocity of a skydiver in free-fall t seconds after jumping from an airplane. The laws of physics can be used to show that $v(t)$ can be described in terms of exponential functions, as follows:

$$v(t) = v_T(1 - e^{-kt})$$

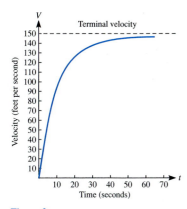

Figure 3

where k is a positive constant. Suppose that for a certain skydiver the terminal velocity is 150 feet per second and k is equal to 0.1. What is the velocity of the skydiver 10 seconds after jumpoff? The graph of the function in this case is shown in Figure 3.

Solution

From the given data, $v_T = 150$ and $k = 0.1$. Inserting these values in the equation for the function, we derive the following formula for $v(t)$:

$$v(t) = 150(1 - e^{-0.1t})$$

The velocity 10 seconds after jumpoff is equal to $v(10)$, which has the value

$$150(1 - e^{-0.1 \cdot 10}) = 150(1 - e^{-1})$$

We used a calculator to evaluate the above expression, which is equal to 94.818. That is, after 10 seconds of free fall, the skydiver is falling with a velocity of 94.818 feet per second.

➤ EXAMPLE 4
Spread of an Epidemic

Suppose that a long-lasting epidemic, against which there is no immunity, spreads through a town with P inhabitants. Let $N(t)$ denote the number of people infected t days after the epidemic breaks out. One model of the spread of the epidemic gives the following formula for $N(t)$:

$$N(t) = \frac{P}{1 + Ae^{-kt}}$$

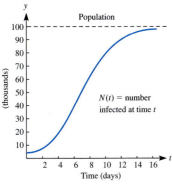

Figure 4

where A and k are constants. The graph of the function $N(t)$ is shown in Figure 4. The graph is called a **logistic curve.** Suppose that the town has 100,000 inhabitants and that statistics show that A has the value 5000 and k the value 1. Note that the line $y = P$ is a horizontal asymptote of the graph. This indicates that after a long period of time, almost everyone is infected with the epidemic. How many people are sick after 5 days? After 10 days?

Solution

According to the given data, the value of P is 100,000, the value of A is 5000, and the value of k is 1. Inserting these values into the formula for $N(t)$, we have

$$N(t) = \frac{100,000}{1 + 5000e^{-t}}$$

The number of people sick after 5 days is

$$N(5) = \frac{100,000}{1 + 5000e^{-5}} = \frac{100,000}{1 + 5000 \cdot 0.006738} = 2883$$

We used a calculator to determine both the value of e^{-5} and the value of the entire expression.

In a similar fashion, the number of people sick after 10 days is

$$N(10) = \frac{100,000}{1 + 5000e^{-10}} = 81,500$$

Note the virulence of the epidemic. After only 10 days more than 81% of the population is sick!

Simplify the following expressions involving e. Where appropriate, approximate values using a calculator with an e^x key.

1. $e^5 e e^{-3}$

2. $e^{-4} e^6 e^0$

3. $e^5 e^{-2} + 2e^3 e^4$

4. $5e^{-2} - 4e^6 e^{-8}$

5. $\dfrac{e^5 - e^{-5}}{e^5 + e^{-5}}$

6. $\dfrac{e^2 + 4e^5}{e^3}$

7. $(e - 1)^2$

8. $(2 + e)^2$

9. $(e^4)^x$

10. $(2e^{-x})^4$

11. $(e^x - e^{-x})^2$

12. $(1 + 2e^x)^2$

13. $(e^x - 1)^3$

14. $(e^x + e^{-x})^3$

15. $(e^{x-2})(4e^x + 5)$

16. $(e^x + e^{-x})(e^x - e^{-x})$

17. $(e^x - 3)(e^x + 3)$

18. $(e^x - 1)(e^{2x} + e^x + 1)$

19. $\sqrt{e^{2x}}$

20. $\sqrt[3]{e^{6x}}$

21. $e^{0.1654} + e^{-2.34}$

22. $2e^{-4.6} - e^{5.71}$

♦ **Applications**

23. **Skydiving.** The velocity of a particular skydiver $v(t)$ can be described in terms of exponential functions as follows:

$$v(t) = v_T(1 - e^{-kt})$$

where k is a positive constant. Suppose that for a certain skydiver the terminal velocity v_T is 160 miles per hour and k is equal to 0.2, and t is in seconds.

a. What is the velocity of the skydiver 1 second after jumpoff? After 2 sec.? After 5 sec.? After 10 sec.? After 20 sec.?

b. Sketch a graph of the function for $t \geq 0$.

24. **Advertising.** A company introduces a new product on a trial run in a city. It advertises the product on TV and gathers data concerning the percentage $P(t)$ of the people in the city who bought the product after it was advertised a certain number of times t. It determines that the **limiting effect** of the advertising is 75%. This means that no matter how many times the company advertises its product, the percentage of people who buy the product will never reach or exceed 75%, although the percentage will get closer and closer to 75%. The company determines that $P(t)$ can be described in terms of exponential functions as follows:

$$P(t) = 0.75(1 - e^{-kt})$$

where $k = 0.05$.

a. What percentage of the city buys the product after 1 advertisement? After 2 ads? After 10 ads? After 20 ads? After 50 ads?

b. Sketch a graph of $P(t)$.

25. **Advertising.** Repeat Exercise 24 assuming the limiting effect is 50% and that $k = 0.08$.

26. **Advertising.** Repeat Exercise 24 assuming the limiting effect is 100% and that $k = 0.06$.

27. **Medicine.** Even after a new medication has been extensively tested and approved by the FDA, it takes time for it to be accepted for usage by doctors. They first have to become aware of it and then try it on their patients. The usage approaches a limiting value of 100%, or 1, as a function of t, in months. The percentage $P(t)$ of doctors who use the medication can be described in terms of exponential functions as follows:

$$P(t) = 100\%(1 - e^{-kt})$$

where $k = 0.3$.

a. What percentage of the doctors have accepted the medication after 1 month? After 2 months? After 5 months? After 10 months? After 1 year?

b. Sketch a graph of $P(t)$ for $t \geq 0$.

28. **Psychology.** The Hullian Model of Learning asserts that the probability $P(t)$ of mastering a certain concept after t learning trials is given by

$$P(t) = 1 - e^{-kt}$$

where k is a constant that depends on the difficulty of the learning. An educational psychologist analyzes the amount of time it takes to learn a concept in mathematics and finds that $k = 0.18$.

a. What is the probability of learning the concept after 1 trial? After 2 trials? After 3 trials? After 5 trials? After 13 trials?

b. Sketch a graph of the function $P(t)$.

29. **Agriculture.** A farmer is growing soybeans in a field. The more fertilizer the farmer spreads, the greater his yield up to some **limiting value** L. The yield per acre $Y(n)$ can be described in terms of exponential functions as

$$Y(n) = L(1 - e^{-kn})$$

where k is a constant that depends on the crop and the fertilizer. (This mathematical model makes an assumption that may not always work in practice, in that most fertilizers will kill a crop if applied to extreme, but past some point it is not cost-effective to keep adding more fertilizer to get just a slight change in yield.) Suppose for a certain crop $k = 0.04$ and the limiting value is 30 bushels per acre.

a. Find $Y(n)$ when $n = 1, 2, 5, 6, 10, 20, 30,$ and 40.

b. Sketch a graph of $Y(n)$ for $n \geq 0$.

30. **Spread of an epidemic.** Suppose that a long-lasting epidemic against which there is no immunity spreads through a town with P inhabitants. Let $N(t)$ denote the number of people infected t days after epidemic breaks out. One model of the spread of the epidemic gives the following formula for $N(t)$:

$$N(t) = \frac{P}{1 + Ae^{-kt}}$$

where A and k are constants. Suppose that the town has 4000 inhabitants and that statistics show that A has the value 200 and k the value 2.

a. How many people are sick after 5 days? After 10 days? After 20 days? After 30 days?

b. Sketch a graph of $N(t)$ for $t \geq 0$.

31. **Limited population growth.** A ship carrying 100 passengers is shipwrecked on a small island and never rescued. The amount of food and fresh water on the island limits the growth of the population to a limiting value of 540, to which the population gets closer and closer but never reaches. The number of people $N(t)$ who make up the population after time t is given by

$$N(t) = \frac{P}{1 + Ae^{-kt}}$$

where A and k are constants and t is time in years. For the passengers on the island, $P = 540$, $A = 4.4$, and $k = 3$.

a. What is the population after 10 years? After 20 years? After 30 years? After 100 years?

b. Sketch a graph of the logistic curve $N(t)$ for $t \geq 0$.

32. **Spread of a rumor.** In a college with a population of 800, a group of students spread the rumor "Study algebra and trigonometry from this book and you will get an A in calculus." The number of people $N(t)$ who have heard the rumor after t minutes is given by

$$N(t) = \frac{P}{1 + Ae^{-kt}}$$

where A and k are constants. In this situation, $P = 800$, $A = 132$, and $k = 0.4$.

a. How many people have heard the rumor after 1 minute? After 2 minutes? After 6 minutes? After 12 minutes? After 30 minutes? After 1 hour?

b. Sketch a graph of the logistic curve $N(t)$ for $t \geq 0$.

33. **Electricity.** (See figure.) If an electromotive force is suddenly connected to a circuit containing a resistor and a capacitor, the charge does not immediately reach its equilibrium value but approaches it exponentially. For a circuit with resistance R ohms, capacitance C farads, and

electromotive force E volts, the charge q (in coulombs) after time t seconds is given by

$$q = CE(1 - e^{-t/RC})$$

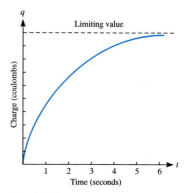

Exercise 33

a. What is the limiting value of the charge?

b. A car battery (12 V) is connected to a circuit containing a 10,000-ohm resistor and a 10-microfarad capacitor. What will the charge be after 0.05 seconds?

Simplify each of the following expressions.

34. $\dfrac{(1 + e^x)(1 + e^{-x}) - (1 + e^x)(1 - e^{-x})}{(1 + e^x)^2}$

35. $\dfrac{(e^x + e^{-x})(e^x + e^{-x}) - (e^x - e^{-x})(e^x - e^{-x})}{(e^x + e^{-x})^2}$

For each of the following functions find and simplify

$$\frac{f(x + h) - f(x)}{h}, \quad h \neq 0$$

36. $f(x) = e^x$

37. $f(x) = \dfrac{e^x - e^{-x}}{2}$

38. $f(x) = \dfrac{e^x + e^{-x}}{2}$

39. $f(x) = e^{-3x}$

➡ **Technology**

Use a graphing calculator to graph the following functions in the range $-4 \leq x \leq 4$. Experiment with the y range so that a point is included for each given x-value.

40. $f(x) = e^x$

41. $f(x) = 5e^{3x}$

42. $f(x) = e^{-0.1x}$

43. $f(x) = \dfrac{e^x - e^{-x}}{e^x + e^{-x}}$

44. $f(x) = \dfrac{1}{e^x + 1}$

45. $f(x) = 5 - 5e^{-x}$

Use a graphing calculator to solve the following exercises:

46. Exercise 23.

47. Exercise 28.

48. Exercise 29.

49. Exercise 30.

5.3 LOGARITHMIC FUNCTIONS

Inverse functions provide us with a useful way of creating new functions from given ones. In this section, we introduce the logarithmic functions by taking the inverse of the exponential function.

Definition of Logarithmic Functions

Suppose that a is a positive real number. We have seen that the exponential function $f(x) = a^x$ for $a \neq 1$ is one-to-one with domain the set $\mathbf{R}$ of all real numbers, and range the set of all positive real numbers. Since the exponential function with base a is one-to-one, it has an inverse function, which depends on the value of a. The inverse function is denoted $\log_a x$ and is called the **logarithm with base** a.

The domain of $\log_a x$ is the range of $f(x)$, that is, the set of all positive real numbers. The range of $\log_a x$ is the domain of a^x, that is, the set $\mathbf{R}$ of all real numbers.

Definition 1
Logarithmic Function with Base a

Let x be a positive real number. Then
$$\log_a x = y \text{ if and only if } a^y = x$$

We may put this definition into words as follows:

> $\log_a x$ is the exponent to which a must be raised to obtain x.

For example, since $3^2 = 9$, we have
$$\log_3 9 = 2$$
That is, 2 (or $\log_3 9$) is the exponent to which 3 must be raised to obtain 9.
Similarly, since
$$2^{10} = 1024$$
we have
$$\log_2 1024 = 10$$
That is, 10 (or $\log_2 1024$) is the exponent to which 2 must be raised to obtain 1024.

A word about notation: In writing logarithmic functions, it is common to omit the parentheses if no confusion will result. For example, it is common to write $\log_a x$ instead of $\log_a (x)$. However, in cases where you are taking the logarithm of a complicated expression, it is best to use the parentheses notation.

Properties of Logarithms

For each fact about raising a number to a power, there is a corresponding fact about logarithms. For example, the property
$$a^1 = a$$
gives us the following property of logarithms:

> $$\log_a a = 1$$

That is, 1 is the power to which a must be raised to obtain a.
Similarly, the property
$$a^0 = 1$$

yields the property of logarithms

$$\log_a 1 = 0$$

That is, 0 is the power to which a must be raised to obtain 1.

From the definition of $\log_a x$, we deduce the following important properties:

FUNDAMENTAL PROPERTIES OF LOGARITHMS

1. $a^{\log_a x} = x, x > 0$
2. $\log_a a^x = x, x$ any real number

Fundamental Property 1 merely restates the fact that $\log_a x$ is the exponent to which a must be raised to obtain x. (In the first property, the logarithm is in the exponent.) The value on the left side of the second property is the exponent to which a must be raised to obtain a^x. And this exponent is clearly x.

Note that these two fundamental properties are just special cases of the two formulas for inverse functions expressed as composites of functions, which we stated in Chapter 3. The two fundamental properties of logarithms are very useful in simplifying expressions involving logarithmic functions, as the following example illustrates.

➤ **EXAMPLE 1**
Calculations Involving
Logarithms

Calculate the values of the following expressions:

1. $\log_{10} 10^{5.781}$ 2. $\log_{10} 1000$

3. $\log_2 \dfrac{1}{8}$ 4. $4^{\log_4 11.1}$

5. $\log_a a^{3x}$ 6. $a^{\log_a x^3}$

Solution

1. By Fundamental Property 2, $\log_a a^x = x$, so the value of the expression is $x = 5.781$.
2. Since $1000 = 10^3$, we see that $\log_{10} 1000 = \log_{10} 10^3 = 3$ (Fundamental Property 2).
3. We have

$$\log_2 \frac{1}{8} = \log_2 2^{-3}$$

$$= -3 \qquad \textit{(Fundamental Property 2)}$$

4. $4^{\log_4 11.1} = 11.1$
5. $\log_a a^{3x} = 3x$
6. $a^{\log_a x^3} = x^3$

The Graph of $\log_a x$

The function $\log_a x$ is the inverse of the function a^x. As we have seen, this relationship implies that the graph of $\log_a x$ can be obtained by reflecting the graph of a^x in the line $y = x$. In Figure 1, we have sketched the graphs of the two functions together with the line $y = x$, in case $a > 1$.

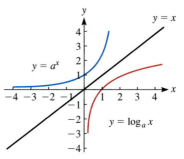

Figure 1

EXAMPLE 2
Graphing Logarithmic Functions

Sketch the graphs of the following functions:

1. $y = \log_2 x$ 2. $y = \log_2 (x - 1)$ 3. $y = \log_2 x + 3$

Solution

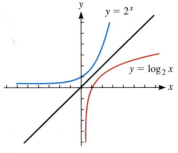

Figure 2

1. The function $y = \log_2 x$ is the inverse of the function $y = 2^x$. Figure 2 shows the graph of $y = 2^x$ in blue. To obtain the graph of the inverse function, we reflect this graph in the line $y = x$. This graph is shown in red in Figure 2. The y-axis is a vertical asymptote of the latter graph, and the graph increases without bound as x approaches positive infinity.

2. According to our discussion on translating graphs of functions, the graph of $y = \log_2(x - 1)$ can be obtained by translating the colored graph of Figure 2 1 unit to the right. The sketch of the graph appears in Figure 3. Note that the vertical asymptote is now the line $x = 1$.

3. According to our discussion on translating graphs of functions, the graph of $y = \log_2 x + 3$ can be obtained by translating the graph $\log_2 x$ three units upward. The sketch is shown in Figure 4.

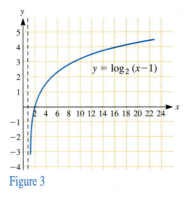

Figure 3

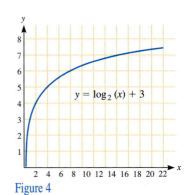

Figure 4

Properties of $\log_a x$

By examining the graph in Figure 5, we can verify the following properties of the function $\log_a x$. (We have already observed and stated several of these, but we record them here for convenient reference.)

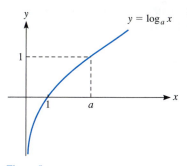

Figure 5

██	**PROPERTIES OF LOGARITHMIC FUNCTIONS**

Suppose that $a > 1$. Then we have the following properties of $\log_a x$:

1. The domain of $\log_a x$ is the set of all positive real numbers. The range is the set **R** of all real numbers.
2. As x increases, the value of $\log_a x$ increases.
3. The y-axis is a vertical asymptote of the graph of $\log_a x$.
4. As x approaches positive infinity, the value of $\log_a x$ approaches positive infinity.
5. The function $\log_a x$ is one-to-one.

The logarithmic function $\log_a x$ has a number of important algebraic properties. The following result is of great help in simplifying expressions involving logarithmic functions.

██	**LAWS OF LOGARITHMS**

The function $\log_a x$ has the following properties for $x, y > 0$:

1. The logarithm of a product is the sum of the logarithms.

$$\log_a(xy) = \log_a x + \log_a y$$

2. To compute the logarithm of x raised to a power, multiply the logarithm of x by the exponent.

$$\log_a x^y = y \log_a x$$

3. The logarithm of a quotient is the difference of the logarithms.

$$\log_a\left(\frac{x}{y}\right) = \log_a x - \log_a y$$

Proof

1. By Fundamental Property 1, we have

$$x = a^{\log_a x}$$
$$y = a^{\log_a y}$$

Multiplying these expressions for x and y gives

$$xy = a^{\log_a x} \cdot a^{\log_a y}$$

By the laws of exponents, the expression can be rewritten as

$$xy = a^{\log_a x + \log_a y}$$

The last equation states that $\log_a x + \log_a y$ is the exponent to which a must be raised to obtain the value xy. By the definition of the logarithmic function, this means that

$$\log_a x + \log_a y = \log_a xy$$

2. Again we use Fundamental Property 1 to obtain $x = a^{\log_a x}$, and we raise both sides of the equation to the power y:

$$x^y = \left(a^{\log_a x}\right)^y = a^{y \log_a x}$$

This equation states that $y \log_a x$ is the exponent to which a must be raised to obtain x^y. From the definition of the logarithmic function, this means that

$$y \log_a x = \log_a x^y$$

3. Write the expression x/y in the form

$$x y^{-1}$$

and apply what we just proved in parts 1 and 2:

$$
\begin{aligned}
\log_a \left(\frac{x}{y} \right) &= \log_a (x y^{-1}) \qquad \textit{Laws of exponents} \\
&= \log_a x + \log_a y^{-1} \qquad \textit{(part 1)} \\
&= \log_a x + (-1) \log_a y \qquad \textit{(part 2)} \\
&= \log_a x - \log_a y
\end{aligned}
$$
◆

Algebraic Expressions Involving Logarithms

The laws of logarithms may be used to simplify expressions involving logarithmic functions, as the following two examples illustrate.

➤ **EXAMPLE 3**
Simplifying Logarithmic Expressions

Write the following expressions in a briefer form by applying the laws of logarithms.

1. $\log_5 [(x - 1)(x - 2)] - \log_5 (x - 2)$
2. $\log_{10} \sqrt{x} + \log_{10} x^{3/2}$

Solution

1. By Law of Logarithms 1, we have

$$
\begin{aligned}
\log_5 [(x - 1)(x - 2)] &- \log_5 (x - 2) \\
&= \left[\log_5 (x - 1) + \log_5 (x - 2) \right] - \log_5 (x - 2) \\
&= \log_5 (x - 1)
\end{aligned}
$$

2. By Law of Logarithms 2, we have

$$
\begin{aligned}
\log_{10} \sqrt{x} + \log_{10} x^{3/2} &= \log_{10} x^{1/2} + \log_{10} x^{3/2} \\
&= \frac{1}{2} \log_{10} x + \frac{3}{2} \log_{10} x \\
&= 2 \log_{10} x
\end{aligned}
$$

➤ **EXAMPLE 4**
Further Practice with the Laws of Logarithms

Write the following expressions using a single logarithm.

1. $3 \log_a x + 4 \log_a y + \dfrac{1}{2} \log_a z$

2. $\log_2 \dfrac{x}{y} - \log_2 \dfrac{y^2}{z}$

Solution

1. Applying Law of Logarithms 2, we may write the given expression in the form

$$\log_a x^3 + \log_a y^4 + \log_a z^{1/2}$$

We may now apply Laws 1 and 3 to combine the three terms to obtain the result

$$\log_a (x^3 y^4 z^{1/2})$$

2.　By the third law of logarithms, we have

$$\log_2 \frac{x}{y} - \log_2 \frac{y^2}{z} = \log_2 \left[\frac{x}{y} \div \frac{y^2}{z} \right]$$

$$= \log_2 \left[\frac{x}{y} \cdot \frac{z}{y^2} \right]$$

$$= \log_2 \left(\frac{xz}{y^3} \right)$$

Equations Involving Logarithms

We may solve equations involving logarithms and exponentials of a variable by using the following approach.

SOLVING EXPONENTIAL AND LOGARITHMIC EQUATIONS

1.　*Solving an equation involving an exponential of a variable*
　　a.　Transform the equation so that it has a single exponential on one side of the equation and a number on the other side.
　　b.　Take logarithms of both sides of the equation.
　　c.　Apply the laws of logarithms to solve for the variable.
2.　*Solving an equation involving a logarithm of a variable*
　　a.　Transform the equation so that it has a single logarithm on one side of the equation and a number on the other side.
　　b.　Exponentiate both sides of the equation.
　　c.　Apply the laws of logarithms to solve for the variable.

The next example illustrates how this approach is carried out in practice.

➤ **EXAMPLE 5**
Solving Logarithmic Equations

Solve the following equations for x:

1.　$\log_5 (3x - 1) = 2$
2.　$7^{2x} = 4$

Solution

1.　For an equation involving logarithms, the best strategy is to use the properties of logarithms to eliminate them from the equation. In this case, we can raise 5 to the quantity on both sides of the equation to obtain

$$5^{\log_5(3x-1)} = 5^2$$

By Fundamental Property 1, the left side of the equation equals $3x - 1$ and the equation reads

$$3x - 1 = 5^2$$
$$3x - 1 = 25$$
$$x = \frac{26}{3}$$

2. Apply the function $\log_7$ to both sides of the equation:

$$\log_7 7^{2x} = \log_7 4$$
$$2x \log_7 7 = \log_7 4$$
$$2x = \log_7 4$$
$$x = \frac{\log_7 4}{2}$$

➤ **EXAMPLE 6**

More Solving Logarithmic Equations

Solve the following equation:

$$\log_2 (x + 1) + \log_2 (x - 1) = 3$$

Solution

We may combine the two terms on the left using the first law of logarithms:

$$\log_2[(x + 1)(x - 1)] = 3$$
$$\log_2 (x^2 - 1) = 3$$
$$2^{\log_2 (x^2 - 1)} = 2^3$$
$$x^2 - 1 = 8$$
$$x^2 = 9$$
$$x = \pm 3$$

We now test these potential solutions to the equation to determine whether they satisfy the equation. For $x = 3$,

$$\log_2(x + 1) + \log_2(x - 1) = \log_2(3 + 1) + \log_2(3 - 1)$$
$$= \log_2 4 + \log_2 2$$
$$= 2 + 1$$
$$= 3 \qquad\qquad 3$$

So $x = 3$ is a solution to the equation. Note, however, that for $x = -3$, the expression $\log_2 (x + 1)$ is not even defined, so $x = -3$ is not a solution.

➤ **EXAMPLE 7**

Binary Representation of Numbers

A **binary number** is a number composed of a sequence of digits 0 and 1, such as 1001 or 111101000111. It is possible to represent any decimal integer as a binary number. For example, 189 can be written as the binary number 10111101. (At the moment it is of no concern to us how to determine this binary representation.) The electronic circuits of a computer work with numbers in binary form, which accounts for the great significance of binary representation of integers. Suppose that x is a positive integer. It can be shown that the number of digits $B(x)$ in its binary representation is given by the formula

$$B(x) = \text{INT}(\log_2 x) + 1$$

where INT denotes the greatest integer function, that is, $\text{INT}(x)$ is equal to the greatest integer less than or equal to x. How many binary digits are required to represent the integer 32,767? The integer 32,768?

Solution

To solve this problem, we must first obtain some feeling for the values assumed by the function $\log_2 x$. We begin by tabulating the various values of the powers of 2 and the corresponding values of $\log_2$:

x	$\log_2 x$	x	$\log_2 x$
1	0	1,024	10
2	1	2,048	11
4	2	4,096	12
8	3	8,192	13
16	4	16,384	14
32	5	32,768	15
64	6	65,536	16
128	7	131,072	17
256	8	262,144	18
512	9	524,288	19

Since the function $y = \log_2 x$ is the inverse of the function $y = 2^x$, we can reverse the two columns in the table to obtain a table of values for the function $\log_2 x$. According to the resulting table, the value of $\log_2 32{,}767$ is greater than 14 and less than 15. Therefore, the value of INT($\log_2 32{,}767$) is 14, and the value of $B(32{,}767)$ is equal to $14 + 1 = 15$. That is, it takes 15 binary digits to represent the decimal number 32,767.

Also according to the table, the value of $\log_2 32{,}768$ is exactly 15, so the value of INT($\log_2 32{,}767$) is 15, and the value of $B(32{,}768)$ is $15 + 1 = 16$. That is, it takes 16 binary digits to represent the decimal number 32,768.

The next example shows how the properties of logarithms may be applied in dealing with the exponential functions that arise in describing radioactive decay.

➤ EXAMPLE 8
Radioactive Decay

A physicist measures the decay of a radioactive isotope. She finds that after 20 days, only 40% of the original amount of the isotope remains. Determine the half-life of the isotope.

Solution

Let A_0 denote the original quantity of the radioactive isotope, let $Q(t)$ denote the quantity left after time t, and let H denote the half-life of the isotope. Earlier in the chapter, we stated the result from physics that

$$Q(t) = A_0 2^{-t/H}$$

In this example, we are given that

$$Q(20) = 0.4A_0$$

Inserting the value 20 for t in the first formula and substituting the value $0.4A_0$ for $Q(20)$, we have

$$0.4A_0 = A_0 2^{-20/H}$$
$$0.4 = 2^{-20/H}$$

To solve this equation, we take logarithms with base 2 of both sides to obtain

$$\log_2 0.4 = \log_2(2^{-20/H})$$

$$\log_2 0.4 = -\frac{20}{H}$$

$$(\log_2 0.4)H = -20$$

$$H = -\frac{20}{\log_2 0.4}$$

This gives an expression from which a numerical value for the half-life H may be computed. We will show how to compute the numerical value for H in the next section.

Exercises 5.3

Determine the equivalent exponential equation.

1. $\log_2 8 = 3$
2. $\log_5 3125 = 5$
3. $\log_3 243 = 5$
4. $\log_9 729 = 3$
5. $\log_2 \frac{1}{8} = -3$
6. $\log_5 \frac{1}{25} = -2$
7. $\log_x 3 = -12$
8. $\log_4 Q = t$

Determine the equivalent logarithmic equation.

9. $4^2 = x$
10. $20^0 = 1$
11. $10^{-3} = 0.001$
12. $27^{2/3} = 9$
13. $\sqrt{4} = 2$
14. $\sqrt[3]{27} = 3$
15. $x^{-3} = 0.01$
16. $a^b = c$

Simplify the following expressions.

17. $\log_x x$
18. $\log_e e$
19. $\log_{10} 10$
20. $\log_{23} 23$
21. $\log_m 1$
22. $\log_{10} 1$
23. $\log_e 1$
24. $\log_k 1$
25. $25^{\log_{25} 4}$
26. $9^{\log_9 3x}$
27. $\log_{10} 10^{-3}$
28. $\log_8 8^k$
29. $e^{\log_e 0.8241}$
30. $10^{\log_{10} 25}$
31. $\log_2 8$
32. $\log_3 9^2$
33. $y^{\log_7 (2x-1)}$
34. $\log_5 5^{(3x+5)}$
35. $\log_m m^{x^4}$
36. $t^{\log_t (x^2+3)}$

Sketch graphs of the following functions.

37. a. $f(x) = \log_3 (x)$
    b. $f(x) = \log_3 (x + 1)$
    c. $f(x) = \log_3 (x) - 2$
38. a. $f(x) = \log_4 (x)$
    b. $f(x) = \log_4 (x - 1)$
    c. $f(x) = \log_4 (x) + 3$
39. $f(x) = \log_5 (x)$
40. $f(x) = \log_{10} (x)$
41. $f(x) = \log_2 (x - 3)$
42. $f(x) = \log_2 (x + 3)$
43. $f(x) = \log_2 |x|$
44. $f(x) = |\log_2 (x)|$
45. $f(x) = \log_2 (x^2)$
46. $f(x) = \log_2 (\sqrt{x})$
47. $f(x) = \log_2 \left(\frac{1}{x}\right)$
48. $f(x) = \frac{1}{\log_2 x}$

Write the following expressions using a single logarithm.

49. $\frac{1}{3}(2 \log_{10} 8 - 6 \log_{10} 3) - 2 \log_{10} 2 + \log_{10} 3$
50. $\frac{1}{2}(\log_{10} 6 + \log_{10} 4 - \log_{10} 12 + \log_{10} 2) + 2 \log_{10} 5$
51. $\frac{1}{2} \log_a 36 - \frac{1}{3} \log_a 8$
52. $2 \log_a 25x - \log_a 125x$
53. $2 \log_b \sqrt{bx} - \log_b x$
54. $\frac{1}{2} \log_a 8x^2 - \frac{1}{6} \log_a 8x^3$
55. $\log_a (x^2 - 9) - \log_a (x - 3)$
56. $\log_b (x^2 - 1) - [\log_b (x + 1) + \log_b (x - 1)]$
57. $\log_3 (2x - 3) - \log_3 (2x^2 - x - 3) + \log_3 (3x + 3)$
58. $\log_a (3x^2 - 5x - 2) - \log_a (x^2 - 4) - \log_a (3x + 1)$
59. $\log_a (x^3 - y^3) - \log_a (x^2 + xy + y^2)$
60. $\log_a (m + n) + \log_a (m^2 - mn + n^2)$

In Exercises 61–70, express the given logarithm in terms of $\log x$, $\log y$ and $\log z$.

61. $\log_b \left(\frac{x \sqrt{y}}{z^2}\right)$
62. $\log_c \left(x^3 \sqrt[4]{\frac{y}{z}}\right)$
63. $\log_2 (x^2 y^2 z^4)$
64. $\log_5 x^3 (yz)^4$
65. $\log_3 \frac{x^4 y}{z^2}$
66. $\log_2 \frac{1}{x^2 y z^3}$
67. $\log_{10} \frac{x^2 y^{2/3}}{z^4}$
68. $\log_a \sqrt{\frac{x}{y^2 z^3}}$
69. $\log_{10} \sqrt{x^3 \sqrt{yz^5}}$
70. $\log_6 \frac{x^3 y^2 z^{-1}}{x y^4 z^{-2}}$

In Exercises 71–76, write the given expression in terms of logarithms of polynomials of degree at most 2.

71. $\log_a (m^3 - n^3)$
72. $\log_t (P^3 - 8Q^3)$

73. $\log_a \sqrt[3]{x^2 - 6x + 5}$

74. $\log_a \sqrt{\dfrac{x^2 - y^2}{x + y}}$

75. $\log_b (x^3 + 8)^2$

76. $\log_b (6x^2 + 7xy + 2y^2)$

77. Graph $f(x) = 2^x$ and $f^{-1}(x) = \log_2 x$ using the same set of axes.

78. Graph $f(x) = \log_3 x$ and $f^{-1}(x) = 3^x$ using the same set of axes.

Given that
$$\log_7 2 = 0.3562, \quad \log_7 3 = 0.5646, \quad \log_7 5 = 0.8271$$
find each of the following by applying laws of logarithms.

79. $\log_7 4$

80. $\log_7 6$

81. $\log_7 \dfrac{2}{3}$

82. $\log_7 \dfrac{3}{5}$

83. $\log_7 \dfrac{7}{3}$

84. $\log_7 \dfrac{3}{7}$

85. $\log_7 \dfrac{1}{7}$

86. $\log_7 \dfrac{1}{49}$

87. $\log_7 50$ (*Hint:* $50 = 2 \cdot 5^2$)

88. $\log_7 90$

89. $\log_7 \sqrt{5}$

90. $\log_7 \sqrt[3]{2}$

91. $\log_7 \sqrt[5]{24}$

92. $\log_7 \sqrt{\dfrac{3}{5}}$

93. $\log_7 0.9^{2/3}$

94. $\log_7 \left(\dfrac{8}{9}\right)^{2.3}$

Binary representation of numbers. Use the function
$$B(x) = \text{INT}[\log_2 (x) + 1]$$
to determine how many digits are required to represent the following integers in binary representation.

95. 78,723

96. 123,876

97. 235,669

98. 380,000

♦ **Applications**

99. A physicist measures the decay of a radioisotope. She finds that after 33 days, only 40% of the original amount remains. Determine the half-life of the isotope.

100. Xenon-133 is a radioactive chemical. It is known that 1000 grams of this chemical will decay to 246.6 grams in 10 days. What is its half-life?

101. The population of Mexico City was 14.5 million in 1980 and 21.5 million in 1989. Find the doubling time of the population.

102. The bacteria *Escherichia coli* is a common cause of urinary tract infections. A population of 10,000,000 will grow to 10,352,650 in 1 minute. Find the doubling time of the population.

Solve each of the following equations.

103. $\log_x 125 = 3$

104. $\log_x 3 = 2$

105. $3.4^x = 1$

106. $56^x = 0$

107. $\log_3 81 = x$

108. $\log_7 x = -2$

109. $\log_{13} x = 3$

110. $\log_x 256 = 4$

111. $\log_x \dfrac{1}{36} = -2$

112. $\log_5 \dfrac{1}{25} = x$

113. $\log_x 5 = 3$

114. $\log_5 1 = x$

115. $5^{x^2 - 5} = 14$

116. $2^{5x+7} = 16$

117. $5^{3x-1} = 125^{2x}$

118. $4^{x^2} = 8^{2/3} \cdot 2^{-3x}$

119. $b^x = 5$

120. $\dfrac{1}{8} = 4^{x-3}$

121. $5^x = 7$

122. $4^{3x+5} = 13$

123. $16^{-x} = 1024$

124. $8^{2x-5} = 2^{x+3}$

125. $4^{2x-1} + 8 = 40$

126. $2^{2-3x} + 2 = 18$

127. $5^{2x}(25^{x^2}) = 125$

128. $\left(\dfrac{1}{25}\right)^{x+3} = 0.2^{x+9}$

129. $24^{\log_{24} (x^2 - 8)} = 1$

130. $5^{-2 \cdot \log_5 x} = 4$

True or false.

131. $\log_a \dfrac{xy}{wz} = \log_a x + \log_a y - \log_a w + \log_a z$

132. $\log_2 12 - (\log_2 2 - \log_2 5) = \log_2 30$

133. $\log_a (P + Q) = \log_a P + \log_a Q$

134. $(\log_a P)(\log_a Q) = \log_a PQ$

135. $\dfrac{\log_a P}{\log_a Q} = \log_a \dfrac{P}{Q}$

136. $\log_a (ax)^n = n + n \log_a x$

137. $\log_a x^2 = 2 \log_a x$

138. $\log_a 5x = 5 \log_a x$

Solve each of the following inequalities by examining graphs of functions.

139. $\log_3 x \geq 0$

140. $\log_3 x < 0$

141. $\log_2 (x + 3) < 0$

142. $\log_2 (x - 4) \geq 0$

143. $2^x < 1$

144. $3^x \geq 1$

145. $2^{x-2} < 1$

146. $3^{x+2} \geq 1$

Simplify.

147. $\log_a (x^3 + 3x^2 y + 3xy^2 + y^3)$

148. $\log_b (a^4 - 4a^3 b + 6a^2 b^2 - 4ab^3 + b^4)$

Prove each of the following.

149. $\log_b \left(\dfrac{1}{P}\right) = -\log_b P$

150. $\log_a \left(\dfrac{x + \sqrt{x^2 - 1}}{x - \sqrt{x^2 - 1}}\right) = 2 \log_a (x + \sqrt{x^2 - 1})$

151. Suppose that $f(x) = 3^x$. Determine $f^{-1}(x)$.

152. Suppose that $f(x) = \log_2 x$. Determine $f^{-1}(x)$.

153. Suppose that $f(x) = 2^x + 2^{-x}$. Prove that $f(x)$ is one-to-one. Determine $f^{-1}(x)$.

154. Suppose that $f(x) = 2^{x^2}$, $x \geq 0$. Prove that $f(x)$ is one-to-one. Determine $f^{-1}(x)$.

155. Visually demonstrate the laws of logarithms using a graphing utility. Look at functions with various integer, irrational, and rational bases.

5.4 COMMON LOGARITHMS AND NATURAL LOGARITHMS

Rather than a single function, $\log_a x$ is really an infinite family of functions, one for each value of a. It is very inconvenient to have calculations stated in terms of many different logarithmic functions. For instance, calculations involving the strength of earthquakes and intensity of light are most conveniently phrased in terms of the function $\log_{10} x$, whereas calculations involving information theory are most conveniently phrased in terms of the function $\log_e x$. In this section, we examine more closely the two logarithmic functions that are most commonly used, namely, $\log_{10} x$ and $\log_e x$.

Common Logarithms

The decimal number system is based on the number 10. Place values are determined by powers of 10. Therefore, it should come as no surprise that logarithms to base 10 play a special role in applications.

Definition 1
Common Logarithms

Logarithms to base 10 are called **common (or Briggsian) logarithms,** and the function $\log_{10} x$ (also written $\log x$), is called the **common logarithmic func-tion.**

Here are some special values of $\log_{10} x$:
$$\log_{10} 10 = 1$$
$$\log_{10} 100 = \log_{10} 10^2 = 2$$
$$\log_{10} 1000 = \log_{10} 10^3 = 3$$
$$\log_{10} 0.01 = \log_{10} 10^{-2} = -2$$

Before the common availability of scientific calculators, values of common logarithms were determined from tables. However, the need for such tables to-day is virtually nil. Scientific calculators have a key for calculating values of the common logarithm function. This key is usually labeled **log** or **log₁₀**. On most calculators, to compute the value of **log₁₀x**, press the **log** key, and then enter the value of x and press **=** .

➤ **EXAMPLE 1**
Calculators and Logarithms

Calculate the following with the use of a calculator.

1. $\log_{10} 50.39$

2. $\log_{10} \dfrac{5000}{4.1}$

Solution

We assume the use of a calculator with algebraic logic.

1. Here are the keystrokes needed: **log₁₀ 50.39 =**. The answer is 1.7023.

2. Here are the keystrokes needed: $\log_{10}$ **(5000 ÷ 4.1) =**. The answer is 3.08619.

Historical Note: Common Logarithms and Calculations

Before the invention of computers, common logarithms were used to perform arithmetic calculations. By the properties of logarithms, the logarithm of a product is the sum of the logarithms. So to multiply two numbers, one could first determine their logarithms, add them, and then determine the number whose logarithm is the sum. Values of common logarithms were determined from tables, and it was commonplace for most textbooks in the sciences to include a table of common logarithms to aid the computations in the text. Book-length tables of logarithms, carried out to many decimal places, were considered a standard tool for anyone who needed to perform lengthy scientific calculations, especially those involving powers.

Common logarithms were the invention of the English mathematician Henry Briggs in the 17th century. In some old books, common logarithms were called Briggsian logarithms. Although common logarithms are an anachronism for computational purposes today, it is impossible to overemphasize the great advance in calculation that they afforded to scientists of the 17th through the 19th centuries. Important calculations in astronomy, physics, and chemistry became possible only after tables of logarithms became available.

So important were tables of logarithms to calculation that when the Works Progress Administration (WPA) was looking for jobs for unemployed scientists and mathematicians during the Great Depression, they commissioned a new set of logarithm tables, carried out to 14 decimal places.   ❙❚

➤ **EXAMPLE 2**
Loudness of Sound

The loudness D of a sound, as measured in decibels, may be calculated in terms of the sound's intensity I using the formula

$$D(I) = 10 \log\left(\frac{I}{I_0}\right)$$

where I is measured in watts per square centimeter and I_0 is the minimum perceptible intensity, set by international standard at 10^{-16} watts per square centimeter.

1. Determine the number of decibels in a human conversation if the intensity of the sound of a human voice is 10^{-10} watts per square centimeter.
2. An antinoise ordinance bans noise of more than 80 decibels after midnight. What is the intensity of an 80-decibel sound?

Solution

1. The number of decibels in a human conversation is equal to
$$D(10^{-10})$$

By the given formula, this quantity is equal to

$$10 \log\left(\frac{10^{-10}}{10^{-16}}\right) = 10 \log(10^{-10-(-16)})$$

$$= 10 \log(10^6)$$

$$= 10 \cdot 6$$

$$= 60 \text{ decibels}$$

2. The intensity I of an 80-decibel sound satisfies the equation

$$80 = 10 \log\left(\frac{I}{I_0}\right)$$

$$8 = \log\left(\frac{I}{I_0}\right)$$

$$10^8 = 10^{\log(I/I_0)}$$

$$10^8 = \frac{I}{I_0}$$

$$I = 10^8 I_0$$

$$I = 10^8 10^{-16}$$

$$= 10^{-8} \text{ watts per square centimeter.}$$

Natural Logarithms

The number e assumes its importance because mathematical formulas (particularly in calculus) involving exponential and logarithmic functions assume an especially simple form when expressed in terms of base e. We have already explored such exponential functions. Let's now discuss the corresponding logarithmic functions.

Definition 2 Natural Logarithmic Function	Logarithms to base e are called **natural logarithms**. The function $\log_e(x)$ is called the **natural logarithm function** and is denoted $\ln x$.

Here the notation "ln" is shorthand for the Latin expression for "natural logarithm." This notation has been in use for several hundred years and has become traditional.

As with the common logarithmic function, it has become traditional to omit the parentheses, where possible, in writing the natural logarithm function. That is, we write $\ln x$ rather than $\ln(x)$, and $\ln e^x$ rather than $\ln(e^x)$. We will use this notation so long as it does not result in confusion. However, in taking the natural logarithm of complicated expressions, we will always use the parentheses for clarity.

To calculate values of the natural logarithmic function, we can use a scientific calculator, which has a key labeled ln for this purpose. Calculate $\ln x$ by pressing the ln key, keying in the value of x, and then pressing the = key.

Applying to the natural logarithmic function the properties previously established for logarithmic functions in general allows us to state the following facts:

PROPERTIES OF NATURAL LOGARITHMS

1. $\ln 1 = 0$
2. $\ln e = 1$
3. $\ln e^x = x$
4. $e^{\ln x} = x, \quad x > 0$

➤ **EXAMPLE 3**
Using Laws of Logarithms

Determine the values of the following expressions.

1. $\ln e^5$
2. $(\ln e)^5$
3. $e^{5\ln(2e)}$

Solution

1. By Property 3 of natural logarithms, $\ln e^x = x$, so the value of the expression is $x = 5$.

2. The value of $\ln e$ is 1, so the value of the given expression equals
$$(1)^5 = 1$$

3. By Law of Logarithms 2 in Section 5.3, we have
$$e^{5\ln(2e)} = e^{\ln(2e)^5}$$

    By Property 4 of natural logarithms, $e^{\ln 2e} = 2e$, so
$$e^{\ln(2e)^5} = (2e)^5 = 32e^5$$

The laws of logarithms proved in Section 5.3 are also valid in the special case of the natural logarithmic function.

LAWS OF NATURAL LOGARITHMS

Let $x, y > 0$. Then the following properties of the natural logarithmic function hold.

1. $\ln xy = \ln x + \ln y$
2. $\ln x^a = a \ln x$
3. $\ln \dfrac{x}{y} = \ln x - \ln y$

➤ **EXAMPLE 4**
Ecology

A reservoir has become polluted from an industrial waste spill. The pollution has caused a buildup of algae. The number of algae $N(t)$ present per 1000 gallons of water t days after the spill is given by the formula
$$N(t) = 100e^{2.1t}$$
How long will it take before the algae count reaches 10,000?

Solution

We are asked to determine the value of t satisfying the equation
$$N(t) = 10{,}000$$
Inserting this value for $N(t)$ into the given formula, we obtain
$$10{,}000 = 100e^{2.1t}$$
$$e^{2.1t} = 100$$

To solve the last equation, we take natural logarithms of each side and apply the properties of natural logarithms to simplify the resulting equation:

$$\ln(e^{2.1t}) = \ln 100$$

$$2.1t = \ln 100$$

$$t = \frac{\ln 100}{2.1} = \frac{4.6052}{2.1} = 2.1929$$

To compute the final numerical value, we used a scientific calculator. Thus, the number of algae per 1000 gallons of water will reach 10,000 after 2.1929 days. That is, the number of algae will equal 10,000 in a little more than 2 days.

➤ **EXAMPLE 5**
Archaeology

If $Q(t)$ denotes the amount of radioactive carbon-14 present t years after an organism dies, then $Q(t)$ is given by the formula

$$Q(t) = A_0 e^{-0.00012t}$$

where A_0 denotes the amount of carbon-14 present at the time the organism died. Suppose that archaeologists discover a papyrus scroll in which they measure carbon-14 in an amount equal to 30% of that present in freshly manufactured papyrus. How old is the papyrus?

Solution

We are asked to determine the value of t for which

$$Q(t) = 0.3A_0$$

Using this value for $Q(t)$, we derive the equation

$$0.3A_0 = A_0 e^{-0.00012t}$$

Dividing both sides by A_0, we remove any dependence of the equation on that number. We obtain

$$e^{-0.00012t} = 0.3$$

To solve this equation, we take natural logarithms of both sides and apply the properties of natural logarithms:

$$\ln(e^{-0.00012t}) = \ln 0.3$$

$$-0.00012t = \ln 0.3$$

$$t = -\frac{\ln 0.3}{0.00012}$$

To obtain a numerical value for t, we use a scientific calculator to evaluate the quantity on the right side:

$$-\frac{\ln 0.3}{0.00012} = -\frac{-1.204}{0.00012} = 10,033$$

That is, the piece of papyrus is 10,033 years old.

**Logarithms
to Various Bases**

In the preceding discussion, we saw that the values of common and natural logarithms can be calculated using scientific calculators. However, these are the only special cases of the function $\log_a x$ that can be calculated directly in this way. But for any a, we can compute $\log_a x$ using the following formula.

CHANGE OF BASE FORMULA

Let a and b be positive numbers. Then

$$\log_a x = \frac{\log_b x}{\log_b a}$$

Proof   Begin with Fundamental Property 1 of logarithms:

$$x = a^{\log_a x}$$

Then

$$\log_b x = \log_b a^{\log_a x} \qquad \textit{Substituting for x}$$
$$= \log_a x \cdot \log_b a \qquad \textit{Law of Logarithms 2}$$

Hence

$$\log_a x = \frac{\log_b x}{\log_b a} \qquad \blacklozenge$$

The Change of Base formula asserts that any logarithmic function is a constant multiple of any other given logarithmic function. In particular, we can calculate logarithms with base a in terms of either natural logarithms or common logarithms using the formulas

$$\log_a x = \frac{\ln x}{\ln a}$$

$$\log_a x = \frac{\log x}{\log a}$$

By setting a equal to 10 in the first formula, we have the following relationship between common and natural logarithms:

$$\log x = \frac{\ln x}{\ln 10}$$

➤ **EXAMPLE 6**
Using Change
of Base Formula

Calculate $\log_{100} 10$ without using a calculator.

Solution

By the Change of Base formula, we have

$$\log_{100} 10 = \frac{\log_{10} 10}{\log_{10} 100}$$
$$= \frac{1}{2}$$

since $\log_{10} 10 = 1$ and $\log_{10} 100 = 2$.

Exercises 5.4

Find each common logarithm without using a calculator.

1. $\log_{10} 10$
2. $\log_{10} 100$
3. $\log_{10} 1000$
4. $\log_{10} 10{,}000$
5. $\log_{10} 0.1$
6. $\log_{10} 0.01$
7. $\log_{10} 0.001$
8. $\log_{10} 0.0001$

Find each of the following using a calculator.

9. $\log_{10} 83.47$
10. $\log_{10} 124.7$

11. $\log_{10} 3.456$

12. $\log_{10} 9.87$

13. $\log_{10} 0.458$

14. $\log_{10} 0.00778$

15. $\log_{10} (6.732 \times 10^6)$

16. $\log_{10} (4.012 \times 10^8)$

17. $\log_{10} (9.152 \times 10^{49})$

18. $\log_{10} (1.302 \times 10^{28})$

19. $\log_{10} (8.6022 \times 10^{-4})$

20. $\log_{10} (5.554432 \times 10^{-7})$

21. $\log_{10} (1.21223 \times 10^{-23})$

22. $\log_{10} (9.902 \times 10^{-19})$

♦ Applications

23. **Loudness of sound.** The loudness D of a sound, as measured in decibels, may be calculated in terms of the sound's intensity I using the formula

$$D(I) = 10 \log \left(\frac{I}{I_0} \right)$$

where I is measured in watts per square centimeter and I_0 is the minimum perceptible intensity, set by international standard at 10^{-16} watts per square centimeter. Determine the number of decibels from the sound of an elevated train if the intensity is 10^{-3} watts per square meter.

24. Determine the number of decibels from the sound of a riveter if the intensity is 3.2×10^{-3} watts per square meter.

25. The noise of rustling leaves is 10 decibels. What is the intensity of a 10-decibel sound?

26. Sound actually hurts the ear at levels of 120 decibels or higher. What is the intensity of a 120-decibel sound?

27. Two sounds have intensities I_1 and I_2, respectively. Show that the difference in their sound levels can be expressed as

$$D(I_1) - D(I_2) = 10 \log \frac{I_1}{I_2}$$

28. Find a formula for the sum of the sound levels of two sounds with intensities I_1 and I_2.

29. Solve the sound level formula for the intensity I.

30. **Earthquake magnitude.** The magnitude R of an earthquake, as measured on what is called the Richter scale, may be calculated in terms of an intensity I, where

$$R(I) = \log \frac{I}{I_0}$$

and I_0 is a minimum perceptible intensity. What is the value of $R(I_0)$?

31. Find the magnitudes of earthquakes whose intensities are

a. $10 I_0$ b. $100 I_0$ c. $1000 I_0$ d. $10,000 I_0$

e. $100,000 I_0$ f. $100,000,000 I_0$

32. One earthquake has a magnitude 10 times as intense as another. How much higher is its magnitude on the Richter scale?

33. One earthquake has a magnitude 100 times as intense as another. How much higher is its magnitude on the Richter scale?

34. One of the worst earthquakes ever recorded was near Lebu, Chile, on May 21–23, 1960. It had an intensity of $10^{8.9} \cdot I_0$. What was its magnitude on the Richter scale?

35. Two earthquakes have intensities I_1 and I_2, respectively. Show that the difference in their magnitudes can be expressed as

$$R(I_1) - R(I_2) = \log \frac{I_1}{I_2}$$

36. Find a formula for the sum of the magnitudes of two earthquakes with intensities I_1 and I_2.

37. Solve the Richter scale magnitude formula for I.

38. **pH in chemistry.** In chemistry, the pH of a substance is defined by

$$pH = -\log_{10} H^+$$

where H^+ represents the concentration of hydrogen ions, in moles per liter. Find the pH concentration of each of the following:

a. Normal blood, $H^+ = 3.4 \times 10^{-8}$

b. Ammonia, $H^+ = 1.6 \times 10^{-11}$

c. Wine, $H^+ = 4.0 \times 10^{-4}$

39. Given the following pH values, find the corresponding hydrogen ion concentrations, H^+.

a. Eggs, pH $= 7.8$

b. Tomatoes, pH $= 4.2$

40. Solve the pH formula for H^+.

41. A substance is said to be an **acid** if its pH is greater than 7. For what hydrogen ion concentrations is a substance an acid?

42. A substance is said to be a **base** if its pH is less than 7. For what hydrogen ion concentrations is a substance a base?

Determine the values of the following expressions.

43. $\ln 1$ 44. $\ln e^2$ 45. $\ln e^{-7}$ 46. $(\ln e)^{-7}$

47. $(\ln e)^2$ 48. $e^{\ln t}$ 49. $e^{\ln 7k}$ 50. $e^{3 \ln(4e)}$

51. $e^{-4 \ln(5e^2)}$ 52. $\ln \sqrt{e^8}$ 53. $\ln \sqrt{e^6}$

54. $\ln e^2 + 4^3$ 55. $\ln e^3 + 5^3$

Simplify the following expressions.

56. $\dfrac{1}{2 \ln 10} (10)^{2x}$ 57. $\dfrac{\ln x^2}{x}, x > 0$ 58. $\dfrac{\ln n}{\ln n^2}$

59. $\dfrac{1}{n \ln(n^3)}$ 60. $e^{2\ln x}$ 61. $e^{\ln x - 2\ln y}$

62. $\dfrac{1}{2}(\ln 4 - \ln 1)$ 63. $\ln 8 - \ln 2$

64. $\dfrac{1}{2}e^{2(\ln 6)+1} - \dfrac{1}{2}e^{2(\ln 2)+1}$

65. $2^x(\ln 3)3^{2x} \cdot 2 + 3^{2x}(\ln 2)2^x$

66. $\ln(4^2 + 4 - 1) - \ln(1^2 + 1 - 1)$

67. $-\dfrac{1}{3}\ln\dfrac{8}{4} + \dfrac{1}{3}\ln(3 + \sqrt{10})$

68. $\dfrac{x(\ln 4)4^{1/x}\left(-\dfrac{1}{x^2}\right) - 4^{1/x}}{x^2}$

♦ **Applications**

69. **Ecology**. A reservoir has become polluted due to an industrial waste spill. The pollution has caused a buildup of algae. The number of algae $N(t)$ present per 1000 gallons of water t days after the spill is given by the formula

$$N(t) = 100e^{2.1t}$$

How long will it take before the algae count reaches 20,000?

70. **Archaeology**. Use the formula of Example 5. Suppose archaeologists find a Chinese artifact that has lost 40% of its carbon-14. How old is the artifact?

71. **The population of the United States**. (See figure.) The population of the United States was 240 million on January 1, 1986. Think of that date as $t = 0$. The population $P(t)$ t years later is given by

$$P(t) = P_0 e^{kt}$$

where $P_0 = 240$ million and $k = 0.009$.

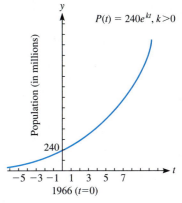

Exercise 71

a. Estimate the population of the United States in 1995 ($t = 9$). Estimate the U.S. population in 2001.

b. After what period of time will the population be double that in 1986?

72. Exponential population growth is often modeled by the function

$$P(t) = P_0 e^{kt}$$

where P_0 is the initial population at time $t = 0$ and k is a positive constant, called the *exponential growth rate*.

a. Show that k and the doubling time T are related by the formula

$$k = \frac{\ln 2}{T}$$

b. The growth rate of the U.S. population is 0.9%. Find the doubling time.

c. The doubling time of the population of Texas is 23 years. Find the exponential growth rate.

73. **Population growth of California.** The population of California was 23,669,000 in 1980 and was estimated to be 27,526,000 in 1990. Assuming this was exponential growth:

a. Find the exponential growth rate k.

b. Find the equation of exponential growth.

c. Estimate the population in 2000.

74. **Interest compounded continuously**. A method of compounding interest that used to be quite prevalent was *compounding continuously*. It is used less now because the public did not understand it. A "population" of money, P_0, invested at interest rate k, compounded continuously, grows to an amount $P(t)$ in t years, where

$$P(t) = P_0 e^{kt}$$

Suppose $10,000 is invested in a bank account at 8% interest, compounded continuously.

a. How much is in the account after 1 year? After 2 years? After 5 years? After 10 years?

b. How much would be in the account after the time periods in part (*a*) if interest had been compounded annually?

c. What is the doubling time of this population of money?

d. Suppose the money doubles in value every 9.9 years. What is the interest rate?

75. A new father deposits $250 in a bank account paying 10 1/4%, compounded continuously. How much will be in the account in 18 years?

76. If the father in Exercise 75 wanted to have $30,000 in the bank account after 18 years, how much should he now invest?

77. What would the interest rate have to be if the father wanted the $250 to grow to $40,000 in 18 years?

78. **Atmospheric pressure**. At altitude *a*, under standard conditions of temperature, the atmospheric pressure $P(a)$ is given by

$$P(a) = P_0 e^{-0.0000385a}$$

At sea level $P_0 = 1013$ millibars.

a. What is the pressure at 1000 ft.? At 2000 ft.? At 10,000 ft.? At 30,000 ft.?

b. A device for measuring atmospheric pressure changes from 1013 millibars on the ground floor of the Empire State Building to 965.4038 millibars at the top of the building. How tall is the Empire State Building?

79. **Magnitude of a star**. In astronomy, the **magnitude** *M* of a star is a measure of the brightness of the star. (Stars with the largest magnitude are the faintest.) The magnitude of stars made visible through a telescope depends

on the aperture *a* of the telescope, as given by the formula

$$M(a) = 9 + \log_{10} a$$

Calculate the magnitude of the faintest star made visible by a telescope with aperture:

a. 3 inches.

b. 30 inches.

80. **Beer-Lambert law**. A beam of light enters a medium, such as water or smoky air, with initial intensity I_0. Its intensity is decreased depending on the thickness (or concentration) of the medium. The intensity *I* at a depth (or concentration) of *x* units is given by

$$I = I_0 e^{-\mu x}$$

The constant μ (pronounced "mu"), called the **coefficient of absorption**, varies with the medium. Light through smog has $\mu = 1.4$.

a. What percentage of I_0 remains at a depth of sea water that is 1 m? 2 m? 3 m?

b. Plant life cannot exist below 10 meters. What percentage of I_0 remains at 10 meters?

81. **Salvage value**. A business estimates that the salvage value *V* of a piece of machinery after *t* years is given by $V(t) = \$80,000e^{-t}$.

a. What did the machinery cost initially?

b. What is the salvage value after 4 years?

82. **Length of a parabolic arc**. (See figure.) When an object is thrown into the air at an angle, it follows a parabolic path. If the height *H* to which the object rises and the horizontal distance *D* are known, the actual distance traveled *L*, or the length of the arc, is given by

$$L = 2D\left\{ \sqrt{N^2 + \frac{1}{16}} + \frac{1}{16}N\left[\ln 4\left(N + \sqrt{N^2 + \frac{1}{16}} \right) \right] \right\}$$

where $N = H/D$.

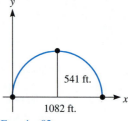

541 ft.

1082 ft.

Exercise 82

A cannonball shot at an angle of 45° with an initial velocity of 180 mph will rise to a height of 541 ft. and will travel a horizontal distance of 1082 ft. What is the actual distance (the length of the parabola) the ball travels?

83. A soccer ball kicked at an angle of 37° with an initial speed of 50 ft./sec. will rise to a height of 14 ft. and will travel a horizontal distance of 75 ft. What is the actual distance the ball travels?

84. **Newton's law of cooling.** An object hotter than its surroundings will cool down to its surroundings and a cool object will warm up to its surroundings exponentially. If a body of temperature T_1 is placed in surroundings of temperature T_0, it will cool or warm to temperature $T(t)$ after time t, in minutes, where

$$T(t) = T_0 + |T_1 - T_0|e^{-kt}$$

a. An object of temperature 98.4°C is placed in a 26°C room. After 12 seconds (1/5 min.), the temperature is 66.4°C. What will its temperature be 5 minutes after placement in a room?

b. Another object, at water's freezing temperature (0°C), is placed in the same room. After 30 seconds, it has warmed up to 10°C. What will its temperature be in 10 minutes?

Use the Change of Base formula to find the following logarithms.

85. $\log_7 50$

86. $\log_{12} 265$

87. $\log_{23} 0.0089$

88. $\log_8 2.347$

Convert each of the following to natural logarithms.

89. $\log 4578$

90. $\log 2.347$

91. $\log x^2$

92. $\log \sqrt{x}$

Convert each of the following to common logarithms.

93. $\ln 0.987$

94. $\ln 78.56$

95. $\ln \sqrt{y}$

96. $\ln t^4$

Simplify.

97. $(e^{e^x})^2$

98. $\dfrac{x^2}{2} \ln [x(x^2 + 1)] + \dfrac{1}{2} \ln (x^2 + 1) - \dfrac{3}{4} x^2$

Simplify by rationalizing the numerator of the fractional expression involving z.

99. $\dfrac{1}{\sqrt{3}} \ln \left| \dfrac{z + 2 - \sqrt{3}}{z + 2 - \sqrt{3}} \right|$

➡ **Technology**

100. **Population growth of Hawaii.** The population of Hawaii in 1980 was 964,691. The total area of Hawaii is 6425 square miles. The growth rate was 1.8% per year. Assuming this was exponential growth, after how many years will the population be such that there will be one person for every square yard of land?

Use a graphing calculator to approximate the solutions to the following equations to two significant digits.

101. $2^x - 3^x = 40$

102. $5^x = 5^{-x} - 4$

103. $3^x + x^2 = 50$

104. $2^x - 3^x + e^x = x^2 + x^3$

Match the following functions with the graphs (a)–(d).

105. $f(x) = \ln x$ 106. $f(x) = \ln x + 1$

107. $f(x) = \ln x - 1$ 108. $f(x) = -\ln x$

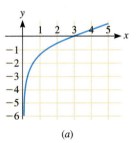

(a)

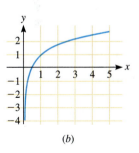

(b)

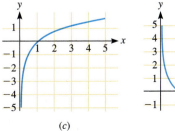

(c)

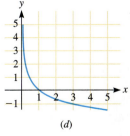

(d)

5.5 EXPONENTIAL AND LOGARITHMIC EQUATIONS

In Section 5.4, we saw how equations involving logarithms and exponential functions arise in applications. Actually, the discussion of that section was only the tip of the iceberg. Equations involving logarithms and exponential functions are common throughout mathematics, and it is important to become proficient in solving them. In this section, we will solve a number of typical equations of this sort as a first step in developing that proficiency.

➤ **EXAMPLE 1**
Exponential Equation

Solve the following equations:

1. $5^{x^2} = 3$
2. $e^{8x+2} = 3^x$

Solution

1. The presence of the variable in the exponent suggests that we take logarithms of both sides. But logarithms to what base? Actually, it doesn't really matter, since logarithms to all bases are related to one another by the Change of Base formula. But it is fairly standard to perform all calculations in terms of the natural logarithmic function. Taking natural logarithms of both sides yields

$$\ln 5^{x^2} = \ln 3$$

Therefore, we obtain

$$x^2 \cdot \ln 5 = \ln 3$$
$$x^2 = \frac{\ln 3}{\ln 5}$$
$$x = \pm \sqrt{\frac{\ln 3}{\ln 5}} \approx \pm 0.8262$$

2. Again we take natural logarithms of the expressions on both sides to obtain

$$\ln e^{8x+2} = \ln 3^x$$

Therefore, we have

$$8x + 2 = x \ln 3$$

In spite of the complex appearance of this equation, it is just a linear equation in x, since $\ln 3$ is a constant. Collecting terms involving x on the left side of the equation yields

$$(8 - \ln 3)x = -2$$
$$x = -\frac{2}{8 - \ln 3}$$
$$\approx -0.2898$$

➤ **EXAMPLE 2**
Logarithmic Equation

Solve the equation

$$\ln \frac{x-1}{2x} = 4$$

Solution

To eliminate the natural logarithmic function on the left side of the equation, we express both sides as a logarithm with base e:

$$e^{\ln[(x-1)/2x]} = e^4$$

$$\frac{x-1}{2x} = e^4$$

$$x - 1 = 2e^4 x$$

$$x - 2e^4 x = 1$$

$$(1 - 2e^4)x = 1$$

$$x = \frac{1}{1 - 2e^4}$$

$$\approx -0.0092425$$

➤ **EXAMPLE 3**
Hyperbolic Function

The function $\sinh(x)$, called the **hyperbolic sine** of x, is defined by the equation

$$\sinh(x) = \frac{e^x - e^{-x}}{2}$$

Determine all values of x for which $\sinh(x)$ is equal to 1/4.

Solution

The equation $\sinh(x) = 1/4$, is equivalent to

$$\frac{e^x - e^{-x}}{2} = \frac{1}{4}$$

$$2e^x - 2e^{-x} = 1$$

To eliminate the negative exponent, we multiply both sides by e^x to obtain

$$2e^x e^x - 2e^{-x} e^x = e^x$$

$$2e^{2x} - 2e^0 - e^x = 0$$

$$2e^{2x} - e^x - 2 = 0$$

To solve this last equation, we write it in the form

$$2(e^x)^2 - e^x - 2 = 0$$

This equation is a quadratic equation in the variable e^x. Applying the quadratic formula gives us the solutions

$$e^x = \frac{1 \pm \sqrt{(-1)^2 - 4(2)(-2)}}{4} = \frac{1 \pm \sqrt{17}}{4}$$

To determine the value of x, we now take natural logarithms of both sides. Recall, however, that the domain of the natural logarithmic function consists of the positive real numbers. In particular, the natural logarithmic function is undefined at the value of the right side corresponding to the minus sign, since this gives a negative value for the right side. The only possible value of x corresponds to

$$\ln e^x = x = \ln\left(\frac{1 + \sqrt{17}}{4}\right) \approx 0.2475$$

➤ **EXAMPLE 4**
A More Complex
Logarithmic Equation

Solve the equation

$$\ln\left(x^2 + 2x\right) = 0$$

Solution

Express both sides of the equation as a logarithm with base e:

$$e^{\ln\left(x^2+2x\right)} = e^0$$

$$x^2 + 2x = 1$$

$$x^2 + 2x - 1 = 0$$

$$x = \frac{-2 \pm \sqrt{2^2 - 4(1)(-1)}}{2} = \frac{-2 \pm \sqrt{8}}{2}$$

$$= \frac{-2 \pm 2\sqrt{2}}{2} = -1 \pm \sqrt{2}$$

Exercises 5.5

Solve the following equations.

1. $7^{x^2} = 4$

2. $8^{x^2} = 5$

3. $e^{5x-2} = 4^x$

4. $e^{7x+4} = 5^{2x}$

5. $\ln\left(\dfrac{x+1}{3x}\right) = 6$

6. $\ln\left(\dfrac{2x-3}{5x+2}\right) = 7$

7. $e^t = 100$

8. $e^t = 60$

9. $e^{-t} = 0.01$

10. $e^{-0.02t} = 0.06$

11. $184.50 = 100e^{10k}$

12. $52 = 25e^{9k}$

13. $90 = 25e^{-10k} + 75$

14. $25e^{-0.05t} + 75 = 80$

15. $\ln\left(x^2 + 3x\right) = 0$

16. $\ln\left(x^2 - 5x\right) = 0$

17. $\log_3 (6x + 5) = 2$

18. $\log_4 (8 - 3x) = 3$

19. $\log x + \log(x + 9) = 1$

20. $\log(x - 9) + \log x = 1$

21. $\log(x + 3) + \log x = 1$

22. $\log(x - 3) + \log x = 1$

23. $\log_4 (x + 5) + \log_4 (x - 5) = 2$

24. $\log_5 (x - 5) + \log_5 (x + 2) = 2$

25. $\log_2 x - \log_2 (x - 3) = 5$

26. $\log_3 (2x - 5) - \log_3 (3x + 1) = 4$

27. $(\log_2 x)^2 + \log_2 x - 6 = 0$

28. $(\log x)^2 - \log x = 6$

29. $\dfrac{\log(x + 1)}{\log x} = 2$

30. $\dfrac{\log(x + 2)}{\log x} = 2$

31. $2 = 5\log x + 12(\log x)^2$

32. $\log(x - 1) + \log(x + 2) = \log_7 7$

33. $\log_5 x^3 + \log_5 x = 4$

34. $3\log x = \log 125$

35. $\log_3 (\log_2 x) = 1$

36. $\log_5 (\log_6 x) = 0$

37. $e^{2x} + 5e^x - 14 = 0$

38. $e^{4x} - e^{2x} = 12$

39. $e^x + e^{-x} - 1 = 0$

40. $5e^x - 3e^{-x} = 8$

41. The function $\sinh(x)$, called the hyperbolic sine of x, is defined by

$$\sinh(x) = \frac{e^x - e^{-x}}{2}$$

Determine all values of x for which $\sinh(x)$ equals 8.

42. The function $\cosh(x)$, called the **hyperbolic cosine** of x, is defined by

$$\cosh(x) = \frac{e^x + e^{-x}}{2}$$

Determine all values of x for which $\cosh(x)$ equals 3.

43. The function $\tanh(x)$, called the **hyperbolic tangent** of x, is defined by

$$\tanh(x) = \frac{e^x - e^{-x}}{e^x + e^{-x}}$$

Determine all values of x for which $\tanh(x)$ equals 1/2.

44. Determine all values of x for which $\tanh(x)$ equals $-1/4$.

Solve the following equations.

45. $|\log_3 x| = 2$

46. $\log_2 |x| = 5$

47. $\log_2 \sqrt[3]{x - 12} = 2 - \log_2 \sqrt[3]{x}$

48. $\log \sqrt{\dfrac{x + 5}{x - 4}} = 2$

49. $\log x^{\log x} = 9$

50. $x^{\log x} = 1000x$

51. $3^{\log_2 x} = 3x$

52. $\sqrt{\dfrac{\left(e^{3x} \cdot e^{-7x}\right)^{-6}}{e^{-x}/e^x}} = e^{10}$

Solve each of the following equations for t.

53. $P(t) = P_0 e^{kt}$

54. $N(t) = N_0 e^{-kt}$

55. $N = \dfrac{P}{1 + Ae^{-kt}}$

56. $T = T_0 + |T_1 - T_0| e^{-kt}$

57. $\dfrac{e^t + e^{-t}}{2} = x$

58. $\dfrac{e^t - e^{-t}}{2} = x$

59. $\dfrac{e^t - e^{-t}}{e^t + e^{-t}} = x$

60. $v = v_T(1 - e^{-kt})$

61. $I = \dfrac{E}{R}(1 - e^{-Rt/L})$

62. $N = GB^t$

Prove each of the following.

63. $\tanh(x) = \dfrac{\sinh(x)}{\cosh(x)}$

64. $\left[\cosh(x)\right]^2 - \left[\sinh(x)\right]^2 = 1$

65. $\sinh(a + b) = \sinh(a)\cosh(b) + \cosh(a)\sinh(b)$

66. $\sinh(x)$ is an odd function.

67. $\sinh(2x) = 2\sinh(x)\cosh(x)$

Determine a function of the form $f(x) = Ce^{kx}$ which has the properties shown in the graphs.

68.

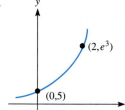

69.

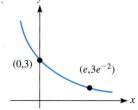

70.

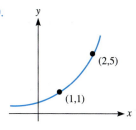

71.

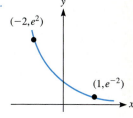

72.

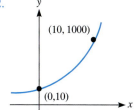

73.

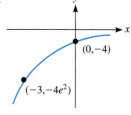

➡ **Technology**

Use a graphing calculator to approximate numerically the solutions to the following equations.

74. $\log(x + 1) + \log(x - 3) = 4.1$

75. $(\log x)^2 - 3x = 1$

76. $\log\log x = 0.1x$

77. $e^x = 3\log x$

78. $\log x + 10\sqrt{x} = 4$

79. $e^{3.7x} - e^{1.5x} = 8$

▌ 5.6 CHAPTER REVIEW

Important Concepts, Properties, and Formulas—Chapter 5

Laws of exponents	$a^x a^y = a^{x+y}$ $(a^x)^y = a^{xy}$ $(ab)^x = a^x b^x$	p. 281
Compound interest formula	$A(t) = P\left(1 + \dfrac{i}{n}\right)^{nt}$	p. 283
Radioactive decay	$Q(t) = A_0 2^{-t/H} \qquad Q(t) = Q_0 e^{-kt}$	p. 284
Exponential growth	$N(t) = N_0 2^{t/D} \qquad P(t) = P_0 e^{kt}$	p. 285
Other exponential formulas	$v(t) = v_T(1 - e^{-kt})$ $N(t) = \dfrac{P}{1 + Ae^{-kt}}$	pp. 292–293

Properties and laws of logarithms	$\log_a x = y$ equivalent to $a^y = x$ $\log_a a = 1; \log_a 1 = 0$ $a^{\log_a y} = y; \log_a a^x = x$ $\log_a(xy) = \log_a x + \log_a y$ $\log_a\left(\dfrac{x}{y}\right) = \log_a x - \log_a y$ $\log_a x^y = y \log_a x$ $\log x = \log_{10} x, \quad \ln x = \log_e x$ $\ln xy = \ln x + \ln y, \quad \ln \dfrac{x}{y} = \ln x - \ln y$ $\log_a x = \dfrac{\log_b x}{\log_b a}, \quad \ln x^k = k \ln x$	p. 299

Cumulative Review Exercises—Chapter 5

Determine the equivalent exponential equation.

1. $\log_e 2 = 0.6931$

2. $\log_{10} 2 = 0.3010$

3. $\log_{0.5} 0.25 = 2$

4. $\log_{0.25} 0.5 = \dfrac{1}{2}$

5. $\ln p = k$

6. $\log_3 \sqrt{27} = \dfrac{3}{2}$

Determine the equivalent logarithmic equation.

7. $2^1 = 2$

8. $2^{-1} = \dfrac{1}{2}$

9. $e^2 = 7.3891$

10. $16^x = 4.015$

♦ **Applications**

11. **Exponential growth.** The population of Kenya was 20 million on January 1, 1985, and is doubling every 23 years. Suppose we let $t = 0$ correspond to 1985. The population after t years is $N(t)$ where

$$N(t) = N_0 2^{t/D}$$

and D is the doubling time.

a. Write $N(t)$ in terms of an exponential function.

b. What will the population of Kenya be in 23 years? In 46 years? In 1995? In 2000?

c. How many years will it take for the population of Kenya to reach 25 million?

12. **Radioisotope power supply.** A space satellite has a radioisotope power supply. The power output $W(t)$ is given by the equation

$$W(t) = W_0 e^{-t/250}$$

where t is the time in days, $W(t)$ is measured in watts, and $W_0 = 50$ watts.

a. How much power is available at the beginning of the flight?

b. How much power will be available after 30 days? After 1 year?

c. What is the half-life of the power supply?

d. The equipment aboard the satellite requires 10 watts of power to operate properly. How long can the satellite operate properly?

Sketch a graph of each of the following functions.

13. $f(x) = 2^x$

14. $f(x) = \log_2 x$

15. $f(x) = 2^{-(x-1)^2}$

16. $f(x) = |3^x - 4|$

17. $f(x) = \log_2 (x + 3) - 5$

18. $f(x) = \log_3 |x|$

19. $f(x) = |\log_3 x|$

20. $f(x) = \log_3\left(\dfrac{3}{x}\right)$

Solve each of the following equations.

21. $\log_9 81 = x$

22. $\log_x \dfrac{9}{16} = 2$

23. $\log_2 x = -\dfrac{1}{2}$

24. $\ln x = 4$

25. $\log_8 x = -\dfrac{2}{3}$

26. $\log x = -3$

27. $\log_2 (x^2 - 1) - \log_2 (x + 1) + \log_2 (x - 1) = 0$

28. $2 \log_6 \sqrt{6x} - \log_6 x = 1$

29. $a^{3 \log_a x} x = 16$

30. $\log_3 x = -2$

31. $27^{2x+3} - 9 = 72$

32. $\log_4 (x - 2) + \log_4 (x - 8) = 2$

33. $e^{-2x} = 0.01$

34. $\log_2 \left[\log_4 \left(\log_{10} x \right) \right] = -1$

35. $\log_{10} \left[\log_2 \left(\log_x 25 \right) \right] = 0$

36. $3e^x + 2 = 7e^{-x}$

37. $3(\log_8 x)^2 + \log_8 x - 2 = 0$

38. $\dfrac{e^x + e^{-x}}{2} = \dfrac{1}{3}$

39. $\log x^{\log x} = 16$

40. $21 = 60(1 - e^{-10k})$

41. $|\log_4 x| = 2$

42. $10^x = 3.6308$

Simplify the following expressions involving e. Where appropriate, approximate values using a calculator with an e^x key.

43. $\sqrt{e^{4x}}$

44. $2700(e^{0.008 \cdot 10} - e^{0.008 \cdot 0})$

45. $(e^x + e^y)(e^x - e^y)$

46. $(e^x + 2)(e^{2x} - 2e^x + 4)$

◆ **Applications**

47. **Sales of a firm.** A firm's sales S increase toward a limiting value of \$20 million and satisfy the following function

$$S(t) = \$20(1 - e^{-kt})$$

where $k = 0.11$ and time t is in years.

  a. Find $S(t)$ for $t = 0, 1, 2, 5,$ and 10 years.

  b. Sketch a graph of $S(t)$ for $t \geq 0$.

  c. After what time will the sales be \$10 million?

48. **Spread of a rumor**. In a college with a population of 6400, a group of students spread the rumor "People who study mathematics are people you can count on!" The number of people $N(t)$ who have heard the rumor after t minutes is given by

$$N(t) = \dfrac{P}{1 + Ae^{-kt}}$$

where A and k are constants. In this situation, $P = 6400$, $A = 8$, and $k = 0.23$.

  a. How many people start the rumor at $t = 0$?

  b. How many people have heard the rumor after 1 minute? After 2 min.? After 6 min.? After 12 min.? After 30 min.? After 1 hr.?

  c. Sketch a graph of the **logistic curve** $N(t)$ for $t \geq 0$.

  d. How much time passes before half the students have heard the rumor?

Simplify each of the following expressions.

49. $6^{\log_6 (5x - 2)}$

50. $\log_5 25^{(3x - 1)}$

51. $\log_b b$

52. $\ln 1$

53. $\log 10^3$

54. $\log_4 4x$

55. $\log_x x^4$

56. $4^{2 \log_4 x + (1/2) \log_4 9}$

57. $2 \log_a x - 3 \log_a y + \log_a (x + y)$

58. $\log_a x^3 - \log_a \sqrt{x}$

59. $\dfrac{1}{3} \left[\log_a x - \dfrac{1}{2} \log_a w \right] + 3 \log_a v$

60. $\log (1 - t^3) - \log (1 + t + t^2)$

Express each of the following as a combination of logarithms.

61. $\log_3 \dfrac{x^3(y + z)}{x^2}$

62. $\log_a \left(\dfrac{x^2 y}{\sqrt{xz}} \right)$

Given that

$$\log_b 3 = 1.099, \qquad \log_b 5 = 1.609,$$

Find each of the following by applying laws of logarithms.

63. $\log_b b^2$

64. $\log_b 3b$

65. $\log_b b^b$

66. $\log_b \dfrac{b}{b + 1}$

67. $\log_b \sqrt[3]{b}$

68. $\log_b \dfrac{1}{5}$

69. $\log_b \sqrt{b}$

70. $\log_b (b^{4.7 \sqrt{b}})$

Binary representation of numbers. Use the function

$$B(x) = \text{INT}[\log_2(x) + 1]$$

to determine how many digits are required to represent the following integers in binary representation.

71. 178,723

72. 443,876

◆ **Applications**

73. A physicist measures the decay of a radioisotope. She finds that after 13 days, 80% of the original amount remains. Determine the half-life of the isotope.

74. The population of Maryland was 4,216,000 in 1980 and is estimated to be 4,491,000 in 1990. Find the doubling time of the population.

75. Calculate the pH of a substance whose hydrogen ion concentration is 8.2×10^{-12} mol/l.

76. Determine the number of decibels in busy street traffic if the intensity of the sound is 10^{-5} watts per square meter.

77. **Medicine.** A patient takes a dosage A_0 of a medication. The medication leaves the circulatory system through the liver, the urinary system, the sweat glands, and by being consumed by those organs that use it. The amount

of the medication present after time t, in hours, is given by

$$A(t) = A_0 e^{-kt}$$

where k is a constant that depends on the metabolism of the patient and the medication. A patient takes a 2 mg tablet for which $k = 0.3$.

a. What amount of the medication will be in the patient's system after 1 hr.? 2 hr.? 4 hr.? 6 hr.?

b. After what amount of time will the amount of medication in the patient's system be 1 mg?

78. For a certain medication and patient, an initial dosage of 5 mg will decrease to 3 mg after 1 hr.

a. Find the decay constant k.

b. Find an equation of exponential decay.

c. After what amount of time will the amount of medication in the patient's system be 2 mg?

79. The growth rate of Kuwait is 5.9% per year. What is the doubling time?

80. The growth rate of a population of rabbits is 12% per day. What is the doubling time?

81. **Population growth of Arizona.** The population of Arizona was 2,286,000 in 1980 and was estimated to be 2,580,000 in 1990. Assuming this was exponential growth:

a. Find the exponential growth rate k.

b. Find the equation of exponential growth.

c. Estimate the population in 2010.

82. Refer to the previous exercise. When will the population of Arizona increase by 50% over its level in 1990?

Solve each of the following equations for t.

83. $\dfrac{e^t + e^{-t}}{e^t - e^{-t}} = y$

84. $N = e^{e^t}$

85. $\log_a y = 1 + n \log_a t$

86. $y = an^t$

87. $A = P\left(1 + \dfrac{i}{n}\right)^{nt}$

88. $0.49 = 0.5(1 - e^{-0.07/t})$

89. $400 = \dfrac{4800}{6 + 794e^{-800(12.2)t}}$

For each of the following functions, find and simplify

$$\dfrac{f(x + h) - f(x)}{h}, \quad h \neq 0$$

90. $f(x) = e^{-x}$

91. $f(x) = \ln x$

Prove each of the following.

92. $(\log_x a)(\log_a b) = \log_x b$

93. $\log_a\left(\dfrac{x - \sqrt{x^2 - h}}{h}\right) = -\log_a (x + \sqrt{x^2 - h})$

94. $\log_x b = \dfrac{1}{\log_b x}$

95. If $\log_a y = a + \dfrac{1}{2}\log_a x$, then $y = a^a \cdot \sqrt{x}$.

LINEAR AND NONLINEAR SYSTEMS

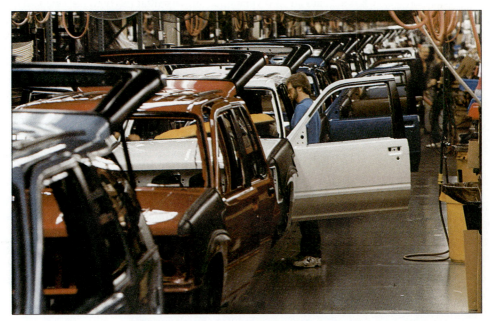

*Applied mathematicians often must determine how to maximize
production of industrial goods in situations where a set of restrictions
on production are described by a system of inequalities.*

I n preceding chapters, it has often been necessary to solve equations and inequalities. Our attention has been concentrated, for the most part, on expressions in a single variable. In this chapter, we will turn to equations and inequalities involving several variables.

6.1 SYSTEMS OF EQUATIONS

In our applications thus far, the problems we have considered boiled down to solving a single equation in one unknown. However, in many applied problems, it is necessary to simultaneously satisfy several equations in several unknowns. For instance, consider the following problem. Suppose that a pension fund seeks to invest $1 million. In order to satisfy current and future obligations of the fund, it must earn $90,000 per year. Suppose that the fund has two sorts of investments open to it: stocks paying 8% per year and corporate bonds paying 12% per year. How much should the pension fund invest in each in order to achieve the desired income?

To solve the problem, let $x =$ the amount invested in stocks and $y =$ the amount invested in bonds. Denote money amounts in multiples of $1 million. So, for instance, the required annual income from the portfolio is $0.09 million. The requirement that $1 million be invested may be expressed through the equation

$$x + y = 1$$

The requirement that the income of the portfolio be 0.09 may be expressed as

$$0.08x + 0.12y = 0.09$$

To solve the problem, we must determine x and y so that the two equations are *simultaneously* satisfied. The pair of equations is an example of a **system of equations** in the variables x and y.

In general, a system of equations is a list of equations in two or more variables. In this section, we will restrict ourselves to systems of two equations in two variables. For such systems, let us make the following definition.

Definition 1 **Solution of a System of Equations**	A **solution of a system in two variables** x and y is an ordered pair of numbers (a, b) such that all equations of the system are satisfied for $x = a$ and $y = b$.

For example, consider the system

$$\begin{cases} x - y = -1 \\ 5x - 4y = 3 \end{cases}$$

Then $(x, y) = (7, 8)$ is a solution to the system, since when we substitute $x = 7$ and $y = 8$, both equations are satisfied:

$$\begin{cases} 7 - 8 = -1 \\ 5(7) - 4(8) = 3 \end{cases}$$

As we shall soon see, this is the only solution of the system. However, as we will illustrate in this chapter, some systems have a single solution, some a finite number of solutions, others an infinite number of solutions, and still others no solutions at all.

Our goal in this chapter is to develop techniques to determine the solutions of systems of equations (if there are any). Actually, this is quite an ambitious goal, since as the number of equations and number of variables increase, it becomes much more difficult to determine the solutions. However, we will go part way toward achieving our goal in developing techniques that can be used to solve a wide variety of systems.

The Method of Substitution

Very often, it is possible to solve one of the equations of a system for one variable in terms of another. When this is possible, we can often determine the solutions of the system by using the **method of substitution**, described as follows:

THE METHOD OF SUBSTITUTION (FOR TWO EQUATIONS IN TWO VARIABLES)

1. Solve an equation for one of the variables in terms of the other.
2. Substitute the expression found in step 1 into the other equation.
3. Solve the resulting equation for the possible values of one variable.
4. For each value found in step 3, substitute into the expression of step 1 to determine the corresponding value of the other variable.

The method of substitution is illustrated in the following example.

➤ **EXAMPLE 1**
Applying Substitution to a System of Two Linear Equations in Two Variables

Determine all solutions of the following system:

$$\begin{cases} x - y = 1 \\ 3x + 2y = 7 \end{cases}$$

Solution

Let's begin by solving the first equation for y in terms of x:

$$x - y = 1$$
$$y = x - 1 \qquad (1)$$

Next we substitute the expression for y just obtained into the second equation:

$$3x + 2y = 7$$
$$3x + 2(x - 1) = 7$$
$$3x + 2x - 2 = 7$$
$$5x = 9$$
$$x = \frac{9}{5}$$

To obtain the value of y, we substitute the value for x into equation (1), which gives y in terms of x.

$$y = x - 1$$
$$= \frac{9}{5} - 1$$
$$= \frac{4}{5}$$

Thus, $x = 9/5$ and $y = 4/5$, so the system has a single solution, namely, $(9/5, 4/5)$. We may check this solution by substituting the values for x and y into each equation:

$$
\begin{array}{c|c}
x - y = 1 & 3x + 2y = 7 \\
\dfrac{9}{5} - \dfrac{4}{5} = 1 & 3\left(\dfrac{9}{5}\right) + 2\left(\dfrac{4}{5}\right) = 7 \\
\dfrac{5}{5} = 1 & \dfrac{27}{5} + \dfrac{8}{5} = 7 \\
1 = 1 & \dfrac{35}{5} = 7 \\
& 7 = 7
\end{array}
$$

Let's now apply the method of substitution to solve the pension fund problem stated at the beginning of this section.

➤ **EXAMPLE 2**

Solving the Pension Fund Problem

Determine all solutions of the system

$$
\begin{cases}
x + y = 1 \\
0.08x + 0.12y = 0.09
\end{cases}
$$

Solution

Recall the pension fund problem posed at the beginning of the section. We set $x =$ the amount of money invested in stocks (in millions) and $y =$ the amount invested in bonds (in millions). We wished to determine x and y, which are a solution to the given system. Solve the first equation for y in terms of x:

$$y = 1 - x$$

Now substitute this expression into the second equation:

$$0.08x + 0.12(1 - x) = 0.09$$
$$-0.04x + 0.12 = 0.09$$
$$0.04x = 0.03$$
$$x = 0.75$$

Now substitute the value of x into the expression for y:

$$y = 1 - x = 1 - 0.75 = 0.25$$

That is, the pension fund should invest \$750,000 in stocks and \$250,000 in bonds.

Examples 1 and 2 applied the method of substitution to systems whose equations were linear. Here is an example of the method applied to solve a nonlinear system.

➤ **EXAMPLE 3**

Applying Substitution to a Nonlinear System

Solve the following system of equations in two variables:

$$
\begin{cases}
x^2 + 3y^2 + 3x + y = 1 \\
x - 2y = -2
\end{cases}
$$

Solution

We apply the method of substitution. The second equation may be simply solved for x in terms of y:

$$x - 2y = -2$$
$$x = 2y - 2$$

Substitute the last expression for x into the first equation:

$$x^2 + 3y^2 + 3x + y = 1$$
$$(2y - 2)^2 + 3y^2 + 3(2y - 2) + y = 1$$
$$\left(4y^2 - 8y + 4\right) + 3y^2 + (6y - 6) + y = 1$$
$$7y^2 - y - 3 = 0$$

Applying the quadratic formula to this last equation, we see that

$$y = \frac{-(-1) \pm \sqrt{(-1)^2 - 4(7)(-3)}}{2 \cdot 7}$$
$$= \frac{1 \pm \sqrt{85}}{14}$$

In this example, there are two values of y. For each of these values, we go back to the equation expressing x in terms of y and obtain a corresponding value of x:

$$x = 2y - 2$$
$$= 2 \cdot \frac{1 \pm \sqrt{85}}{14} - 2$$
$$= \frac{2 \pm 2\sqrt{85}}{14} - 2$$
$$= \frac{-26 \pm 2\sqrt{85}}{14}$$
$$= \frac{-13 \pm \sqrt{85}}{7}$$

There are two pairs which are solutions of the system:

$$\left(\frac{-13 + \sqrt{85}}{7}, \frac{1 + \sqrt{85}}{14}\right), \left(\frac{-13 - \sqrt{85}}{7}, \frac{1 - \sqrt{85}}{14}\right)$$

We may check the solution by substituting each of the solutions into the original equations of the system. We leave the calculations as an exercise.

➤ **EXAMPLE 4**
A System with No Solutions

Solve the following system of linear equations:

$$\begin{cases} 3x - y = 1 \\ -6x + 2y = 4 \end{cases}$$

Solution

Solve the first equation for y in terms of x:

$$y = 3x - 1$$

Substitute this expression into the second equation:

$$-6x + 2y = 4$$
$$-6x + 2(3x - 1) = 4$$
$$-6x + 6x - 2 = 4$$
$$-2 = 4$$

The last equation is inconsistent and so has no solutions. Therefore, the system has no solutions.

Graphical Analysis of Systems in Two Variables

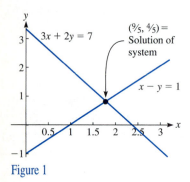

Figure 1

As we have seen in preceding chapters, an equation in two variables has a graph that is a curve in the x-y plane. The points on the curve correspond to the ordered pairs (x, y) that satisfy the equation.

We may graph a system of equations by graphing separately each equation of the system. The solutions of the system then correspond to the points where the graphs intersect. For example, Figure 1 shows the graph of the system of Example 1. Each of the equations has a straight line as a graph. The intersection point of the two lines has coordinates that correspond to the solution of the system. Figure 2 shows the graph of the system in the pension fund problem. Again the graph of each equation is a straight line, and the intersection of the lines corresponds to the solution of the system. Figure 3 shows the graph of the nonlinear system of Example 3. In this case, there are two solutions, one corresponding to each of the intersection points of the two graphs. Finally, Figure 4 shows the graph of the system of Example 4. The fact that the two lines don't intersect corresponds to the result of Example 4 that the system has no solutions.

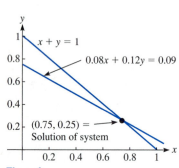

Figure 2

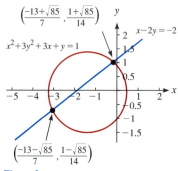

Figure 3

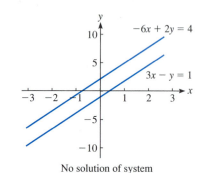

Figure 4

Sometimes it is difficult to solve a system using algebraic methods. In such cases, the solutions may usually be approximated numerically using a graphing calculator, as shown in the following example.

➤ **EXAMPLE 5**
Graphical Solution
of a System

Solve the following system graphically:

$$\begin{cases} y = x^2 \\ y = \sin x \end{cases}$$

Solution

To solve this system by substitution would require that we solve an equation of the form

$$\sin x = x^2$$

which is impossible to do exactly. However, we can approximate the solutions of the system numerically by drawing the graph, as shown in Figure 5. It is clear from the graph that the system has two solutions. The first is $(0, 0)$, and the second corresponds to a value of x between 0 and 1. By enlarging the portion of the graph corresponding to $0.7 \le x \le 1.0$ (Figure 6), we see that with single-digit accuracy, we have $(x, y) = (0.8, 0.7)$. By further enlarging portions of the graph, we could read off the second solution with greater accuracy.

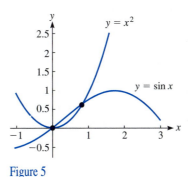

Figure 5

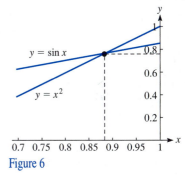

Figure 6

Applications of Systems in Two Variables

Here are some applied problems that give rise to systems of two equations in two variables.

➤ **EXAMPLE 6**

Farming

A rectangular cow pen is fenced on three sides, with the fourth side a barn. The fence opposite the barn cost \$4 per foot. The other sides cost \$3 per foot. The amount of fencing needed is 400 feet, and the cost of the fencing is \$1400. (See Figure 7.) What are the dimensions of the pen?

Solution

In the figure, we have labeled the side opposite the barn as x and the other sides as length y. The amount of fencing required is $x + 2y$, so we have the equation

$$x + 2y = 400$$

The cost of the side opposite the barn is $4x$, and the cost of the other sides is $3y + 3y = 6y$. Therefore, the total cost of the fencing is $4x + 6y$, so that we have the equation

$$4x + 6y = 1400$$

Thus, we must solve the system of linear equations

$$\begin{cases} x + 2y = 400 \\ 4x + 6y = 1400 \end{cases}$$

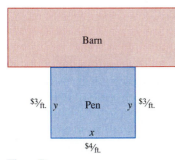

Figure 7

To do so, we use the method of substitution:

$$x = 400 - 2y$$
$$4x + 6y = 1400$$
$$4(400 - 2y) + 6y = 1400 \qquad \textit{(Substitution in second equation)}$$
$$-2y = -200$$
$$y = 100$$
$$x = 400 - 2 \cdot 100$$
$$= 200$$

That is, the side opposite the barn is 200 ft., and the other two sides are each 100 ft. A simple check shows that these lengths satisfy the conditions set forth.

➢ **EXAMPLE 7**
Architectural Design

A cylindrical water cistern is to contain 10,000 cubic feet of water, and its sides are to be constructed of 5000 square feet of aluminum. What are the dimensions of the cistern?

Solution

Let r denote the radius of the cistern and h its height. (See Figure 8.) By the formula for the surface area of a cylinder, the area of the side of the cistern is equal to $2\pi r h$. Therefore, we have the equation

$$2\pi r h = 5000$$

By the formula for the volume of a cylinder, the volume of the cistern is equal to $\pi r^2 h$. Therefore, we have the second equation

$$\pi r^2 h = 10,000$$

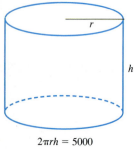

$2\pi rh = 5000$

Figure 8

So to solve the problem, we must solve the system of equations

$$\begin{cases} 2\pi r h = 5000 \\ \pi r^2 h = 10,000 \end{cases}$$

We may solve the second equation for h in terms of r:

$$h = \frac{10,000}{\pi r^2} \qquad (2)$$

Substituting this expression for h into the first equation gives us

$$2\pi r h = 5000$$
$$2\pi r \cdot \frac{10,000}{\pi r^2} = 5000$$
$$\frac{20,000}{r} = 5000$$
$$r = 4 \text{ feet}$$

Substituting this value into equation (2), we have

$$h = \frac{10,000}{\pi r^2}$$
$$= \frac{10,000}{16\pi}$$
$$= 198.94 \text{ ft.}$$

Exercises 6.1

Determine all solutions of the following systems.

1. $\begin{cases} 3x - 6y = -5 \\ 2x + 9y = 1 \end{cases}$

2. $\begin{cases} 3x + 8y = 4 \\ 12y + 6x = -1 \end{cases}$

3. $\begin{cases} x - 3y = 17 \\ 2x + y = 6 \end{cases}$

4. $\begin{cases} x - 3y = 10 \\ 3x + 2y = 2 \end{cases}$

5. $\begin{cases} 3x + 2y = 4 \\ 5x + 3y = 2 \end{cases}$

6. $\begin{cases} \dfrac{3}{2}x = 2 + \dfrac{5}{4}y \\ \dfrac{1}{2}x = \dfrac{3}{2} - \dfrac{5}{3}y \end{cases}$

7. $\begin{cases} x^2 + y^2 = 16 \\ y - 2x = 3 \end{cases}$

8. $\begin{cases} x + y = 1 \\ x^2 - y^2 = 4 \end{cases}$

9. $\begin{cases} x + 3 = y \\ 4y^2 - 16 = x^2 \end{cases}$

10. $\begin{cases} x + 2y = 3 \\ y^2 = 9 - x^2 \end{cases}$

11. $\begin{cases} 4y = 1 + 3x \\ 2x^2 - 8y^2 + 3x + 2y = 2 \end{cases}$

12. $\begin{cases} 5x - 2y = 6 \\ 4x^2 + 4x - y^2 - 4y = 12 \end{cases}$

13. $\begin{cases} xy + x^2 + 2y = 3 \\ 3x - 2y + 2 = 0 \end{cases}$

14. $\begin{cases} 5xy = 4x + y \\ x = 5 - 4y \end{cases}$

15. $\begin{cases} 3x^2 - 2y^2 = 9 \\ x^2 - y^2 = -1 \end{cases}$

16. $\begin{cases} x^2 + y^2 = 4 \\ x^2 - 2y^2 = 1 \end{cases}$

17. $\begin{cases} xy = 8 \\ y = -\dfrac{1}{2}x^2 + 2x + 2 \end{cases}$

18. $\begin{cases} xy = 8 \\ \dfrac{1}{x} + \dfrac{1}{y} = \dfrac{1}{4} \end{cases}$

♦ Applications

19. Tickets for a pre-Broadway showing of a play sold at $40.00 for the main floor and $27.50 for the balcony. If the receipts from the sale of 1600 tickets were $55,250.00, how many tickets were sold at each price?

20. Tickets to a class play were $2.50 and $5.00. In all, 275 tickets were sold, and the total receipts were $1187.50. Find the number of tickets sold at each price.

21. The atomic number of antimony is 1 less than four times the atomic number of aluminum. If twice the atomic number of aluminum is added to that of antimony, the result is 77. What are the atomic numbers of aluminum and antimony?

22. Two temperature scales are established. In the x scale, water freezes at 10° and boils at 300°, and in the y scale,

water freezes at 30° and boils at 1430°. If x and y are linearly related, find an expression for any temperature y in terms of a temperature x.

23. You have $5000 to invest for one year. Part of your investment went into a risky stock that paid 12% annual interest, and the rest went into a savings account that paid 6% annual interest. If your annual return was $540 from both investments, how much was invested at each rate?

24. Suppose you invested a certain sum of money at 8.5% and another sum at 9.5%. If your total investment was $17,000 and your total interest for one year of both investments was $1535, how much did you invest at each rate?

25. Find two numbers whose sum is 10 and the sum of whose reciprocals is 5/12.

26. Find two numbers whose difference is 6 and the difference of whose reciprocals is 2/9.

27. Find the value of C for which the graph of $y = 2x + C$ is tangent to the graph of $y = x^2 + 1$. (*Hint:* There is only one point of intersection.)

28. Find the value of C for which the graph of $y = 2x + C$ is tangent to $x^2 + y^2 = 25$.

29. The sum of the tens digit and the ones digit of a three-digit number is 5. The product of the three digits is 24, and the hundreds digit is 2 less than twice the ones digit. Find the number.

30. The sum of the digits of a three-digit number is 8, the product of the digits is 10, and if the hundreds digit is subtracted from the ones digit, the result is 4. Find the number.

31. Marvin wants to make a window box from a rectangular sheet of metal whose area is 680 cm². He cuts 6 cm squares from each corner and folds the sides to make a box with base area of 176 cm². Find the dimensions of the original metal sheet.

32. An art dealer bought some paintings for $360. When a thief stole two of them, he sold the rest for $380, thus making a profit of $8 on the original cost of each picture. How many pictures did he buy?

33. Find two numbers such that the square of their sum is 20 more than the square of their difference, and the difference of their squares is 24.

34. Dave made a round trip between two cities 270 miles apart. On his return trip, heavy snow reduced his average

rate by 5 miles per hour and increased his traveling time by 3/4 hours. What was his original rate?

35. A number of people shared the purchase price of a skiing chalet equally. If two more people would have shared the chalet, each would have paid $12 less. If there had been three fewer persons, each would have paid $24 more. How much did each pay?

36. **East-west air travel.** Airline schedules allow 3 hrs., 31 min. (terminal to terminal) to fly from Chicago to Phoenix, and 3 hrs., 9 min. to fly from Phoenix to Chicago. Of this, about 1/2 hr. might be used in getting off the ground and getting back down, so the actual flying times at cruising speed might be 3 hrs. in

one direction and 2.7 hrs. in the other. The difference is in the jet stream, a wind that usually blows west to east.

a. If there were no wind, or jet stream, how much time would you expect a flight to take?

b. The airline distance between Chicago and Phoenix is 1453 mi. Estimate the average speed of the jet stream and the average speed of a plane.

37. **Catering costs.** A caterer finds that the expenses for a breakfast consist of fixed costs and costs per guest. A meal function for 40 guests costs $150 and for 100 guests is $225. Find the fixed costs and the cost per guest.

6.2 LINEAR SYSTEMS IN TWO VARIABLES

In this section, we will concentrate on solving systems of two linear equations in two unknowns.

The Idea of Elimination

Linear equations in one variable are simple to solve. One method of solving a system of two linear equations in two variables is to eliminate one of the variables to obtain a single equation in one unknown. This procedure is called **elimination** and is accomplished by multiplying the system equations by suitable numbers to obtain equal coefficients for one of the variables. Subtraction then gives the desired equation in one unknown. This procedure is illustrated in the following example.

➤ **EXAMPLE 1**
Solving Using Elimination

Use elimination to solve the following system:

$$\begin{cases} 11x - 7y = 11 \\ 5x + 3y = -4 \end{cases}$$

Solution

Let's eliminate x by creating equations with equal x-coefficients and subtracting:

$$\begin{cases} 55x - 35y = 55 \quad \text{\textit{Multiply first equation by 5}} \\ 55x + 33y = -44 \quad \text{\textit{Multiply second equation by 11}} \end{cases}$$

$$-68y = 99 \qquad \text{\textit{Subtract second equation from first}}$$

$$y = -\frac{99}{68} \qquad \text{\textit{Solve for y}}$$

$$5x + 3 \cdot \left(-\frac{99}{68}\right) = -4 \qquad \text{\textit{Substitute into second equation}}$$

$$x = \frac{5}{68} \qquad \text{\textit{Solve for x}}$$

We may summarize the process of elimination as follows:

> ### THE METHOD OF ELIMINATION
> ### (TWO LINEAR EQUATIONS IN TWO VARIABLES)
>
> 1. Multiply one or both equations of the system by suitable numbers to obtain a new system in which the coefficients of one of the variables are equal.
> 2. Subtract the equations to obtain a linear equation in one variable.
> 3. Solve the equation obtained in step 2.
> 4. Substitute the value obtained into either of the original system equations and solve for the second variable.

Graphical Analysis of the Solutions

We may graph the equations in a linear system in two variables. The graph of each equation is a straight line. The solutions of the system correspond to points where the lines intersect. There are three possible configurations:

1. The graphs of the equations correspond to two distinct lines. In this case, the system has a single solution, corresponding to the point of intersection of the lines. (See Figure 1.)
2. The two straight lines coincide. In this case, the system has infinitely many solutions, corresponding to all points on the line. (See Figure 2.)
3. The two straight lines are different but parallel. In this case, the system has no solutions. (See Figure 3.)

Example 1 provides an illustration of a system with a single solution. The graphs of the two equations and the solution (intersection point) are shown in Figure 4. The next example illustrates the second case.

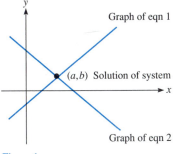

Figure 1

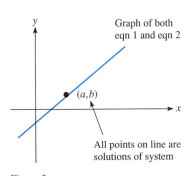

Figure 2

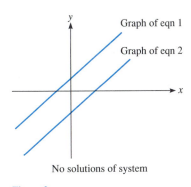

Figure 3

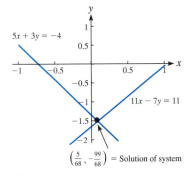

Figure 4

> ### ➤ EXAMPLE 2
> A System in Two Variables with Infinitely Many Solutions

Solve the following system of linear equations:

$$\begin{cases} 3x + 8y = 12 \\ \dfrac{3}{4}x + 2y = 3 \end{cases}$$

Solution

Eliminate x:

$$\begin{cases} 3x + 8y = 12 \\ 3x + 8y = 12 \end{cases} \quad \textit{Multiply equation 2 by 4}$$

$$0 = 0 \quad \textit{Subtract equation 2 from equation 1}$$

This equation is satisfied by any value of y. For a particular value of y, the corresponding value of x is obtained by solving one of the equations for x in terms of y:

$$3x = 12 - 8y$$

$$x = 4 - \frac{8}{3}y$$

Figure 5 shows the graphs of the two equations of the system. Both equations have the same graph, since one is four times the other. Any point on the line shown is a solution to the system.

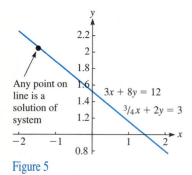

Any point on line is a solution of system

$3x + 8y = 12$

$3/4x + 2y = 3$

Figure 5

> ## EXAMPLE 3

A System in Two Variables with No Solutions

Determine all solutions of the system

$$\begin{cases} 2x - 10y = 1 \\ 3x - 15y = 7 \end{cases}$$

Solution

Eliminate x:

$$\begin{cases} 6x - 30y = 3 & \textit{Multiply equation 1 by 3} \\ 6x - 30y = 14 & \textit{Multiply equation 2 by 2} \end{cases}$$

$$0 = -11 \quad \textit{Subtract equation 2 from equation 1}$$

Clearly, the last equation has no solutions, so the same can be said of the system. In this case, the graphs of the two equations are parallel, as shown in Figure 6.

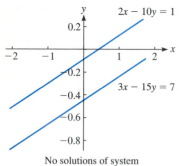

$2x - 10y = 1$

$3x - 15y = 7$

No solutions of system

Figure 6

Application

Let's now apply the method of elimination to solve a linear system that arises in an application.

> ## EXAMPLE 4

Aviation

A plane can fly against the wind and cover a 500-mile distance in 2 hours. The same plane can fly with the wind and cover the same route in the opposite direction in 1.5 hours. What is the speed of the plane in still air, and what is the speed of the wind?

Solution

Let v denote the speed of the plane in still air, and let w denote the speed of the wind. Against the wind, the speed of the plane is $v - w$. And since the plane covers 500 miles in 2 hours, we have

$$\text{Distance} = \text{Rate} \cdot \text{Time}$$

$$500 = (v - w) \cdot 2$$

$$2v - 2w = 500$$

With the wind, the speed of the plane is $v + w$. And since it takes 1.5 hours to cover the same 500 miles in this case, we have

$$500 = 1.5(v + w)$$
$$1.5v + 1.5w = 500$$

To solve the problem, we therefore must solve the following system:

$$\begin{cases} 2v - 2w = 500 \\ 1.5v + 1.5w = 500 \end{cases}$$

$$\begin{cases} 6v - 6w = 1{,}500 & \textit{Multiply equation 1 by 3} \\ 6v + 6w = 2{,}000 & \textit{Multiply equation 2 by 4} \end{cases}$$

$$12v = 3{,}500 \qquad \textit{Add equations}$$
$$v = 291.67 \text{ mph} \qquad \textit{Solve}$$
$$2 \cdot 291.67 - 2w = 500 \qquad \textit{Substitute in equation 1}$$
$$w = 41.67 \text{ mph}$$

Thus, the airplane goes 291.67 mph in still air, and the wind speed is 41.67 mph.

Exercises 6.2

Use elimination to determine all solutions of the following systems of equations.

1. $\begin{cases} 2x - y = 0 \\ 3x + y = 5 \end{cases}$

2. $\begin{cases} 3x + y = 7 \\ x - y = 1 \end{cases}$

3. $\begin{cases} 3x + 4y = -3 \\ 2x + 2y = 2 \end{cases}$

4. $\begin{cases} 3x - y = 3 \\ 2x + 2y = -6 \end{cases}$

5. $\begin{cases} 2x + 4y = 0 \\ 3x + 6y = 6 \end{cases}$

6. $\begin{cases} 3x - 6y = -5 \\ 3x + 9y = 1 \end{cases}$

7. $\begin{cases} 3x + 8y = 4 \\ 12y + 6x = -1 \end{cases}$

8. $\begin{cases} x - 3y = 17 \\ 2x + y = 6 \end{cases}$

9. $\begin{cases} x - 3y = 10 \\ 3x + 2y = 2 \end{cases}$

10. $\begin{cases} 3x + 2y = 4 \\ 5x + 3y = 2 \end{cases}$

11. $\begin{cases} \dfrac{3}{2}x = 2 + \dfrac{5}{4}y \\ \dfrac{1}{2}x = \dfrac{3}{2} - \dfrac{5}{3}y \end{cases}$

12. $\begin{cases} \dfrac{9}{2}x - 4y = -3 \\ \dfrac{4}{3}x - \dfrac{1}{2}y = \dfrac{7}{6} \end{cases}$

♦ Applications

13. The perimeter of a rectangular pool is 84 meters. The width of the pool is 3/4 of the length. What are the length and the width of the pool?

14. The perimeter of a rectangular pool is 70 meters. The length is 4/3 the width of the pool. What are the width and the length of the pool?

15. A bank contains $37 in quarters and dimes. The number of dimes is 10 more than twice the number of quarters. How many dimes and how many quarters are there in the bank?

16. Tickets to a high school play cost $4.50 and $6.00. If 450 tickets were sold and the receipts were $2550, how many of each ticket were sold?

17. A lawn service company has two types of mowers, a new one and an old one. With both mowers working together, a certain job is done in 2 hours and 24 minutes. On another job of the same kind, the old mower worked alone for 3 hours and then was joined by the new mower, and the two mowers finished the job in an additional 1 hour and 12 minutes. How long would it take each mower operating alone to do the second job?

18. One bricklayer can build the wall to a standard basement in 6 hours less than a second bricklayer. If the faster bricklayer works for 2 hours alone on a standard basement and is then joined by the second bricklayer, they complete the basement in an additional 1 3/7 hours working together. How long would it take each bricklayer working alone to build a standard basement?

19. Traveling downstream, the riverboat *New Baltimore* can travel 12 miles in 2 hours. Going up the same river, it can travel 2/3 of this distance in twice the time. What is the rate of the river?

20. Flying with the wind, a SAC bomber can cover a distance of 3360 miles to its target in 5 hours and 15 minutes. On its return against the same wind, the trip would require 6 hours. Find the rate of the wind.

Solve the following systems for x and y in terms of a and b.

21. $\begin{cases} x + y = 2a \\ 2x + y = 3a + b \end{cases}$

22. $\begin{cases} 2x + y = \dfrac{a}{b} \\ x + y = 0 \end{cases}$

6.3 LINEAR SYSTEMS IN ANY NUMBER OF VARIABLES

In the preceding section, we learned to solve systems of two linear equations in two variables. Let's now consider systems of linear equations in any number of variables. We will develop a method called **Gauss elimination** for determining all solutions of such a system.

Systems in Triangular Form

Let's begin by introducing a type of linear system that is extremely easy to solve. Consider a linear system in the variables $x, y, z, w, \ldots$. We say that the system is in **triangular form** (also called **echelon form**) provided that:

1. The second equation does not involve x, the third equation does not involve x and y, the fourth equation does not involve x, y, and z, and so forth.
2. The first nonzero coefficient of each equation is 1.

Here are two systems in triangular form:

$$\begin{cases} x + y + z = -1 \\ \quad\ \ y - 2z = 4 \\ \quad\qquad\ z = 4 \end{cases}$$

$$\begin{cases} x - 5y + z = -1 \\ \qquad\qquad z = 6 \end{cases}$$

In each system, the second equation does not involve x, and the third equation does not involve x and y. Note that in the second system, the second equation has all 0 coefficients and is omitted. However, in terms of triangular form, we consider it as part of the system.

A triangular system is simple to solve. Just start from the last equation and determine the value of the last variable, then use the next-to-last equation to determine the value of the next-to-last variable, and so forth. The next example illustrates how this is done.

> **EXAMPLE 1**
> Solving a Triangular System

Solve the following system of equations in triangular form:

$$\begin{cases} x + 2y - 5z = -1 \\ \quad\ \ y - 4z = 7 \\ \qquad\qquad z = -2 \end{cases}$$

Solution

Start from the bottom equation and work upward. The bottom equation gives the value of z, namely, $z = -2$. Proceed upward to the next equation and substitute the value of z just obtained:

$$y - 4z = 7$$
$$y - 4(-2) = 7$$
$$y = -1$$

Now we have determined the values of y and z. Move up to the first equation and substitute these values:

$$x + 2y - 5z = -1$$
$$x + 2(-1) - 5(-2) = -1$$
$$x = -9$$

Thus, the system has the single solution

$$(x, y, z) = (-9, -1, -2)$$

The method just illustrated may be used to solve any system in triangular form and is called **back-substitution**.

> ➤ **EXAMPLE 2**
> Triangular System with Infinitely Many Solutions

Solve the triangular system:

$$\begin{cases} x - y + z = 5 \\ y - z = 4 \end{cases}$$

Solution

The equation that would allow you to determine the value of z is missing. So let z be any real number. In terms of this choice for z, we may find y from the second equation:

$$y = 4 + z$$

And we may find x from the first equation:

$$\begin{aligned} x &= y - z + 5 \\ &= (4 + z) - z + 5 \\ &= 9 \end{aligned}$$

The solution of the system is

$$z = \text{any value}, \qquad y = 4 + z, \qquad x = 9$$

This system has infinitely many solutions.

> ➤ **EXAMPLE 3**
> Another System with Infinitely Many Solutions

Determine all solutions of the triangular system.

$$\begin{cases} x + 3y - 2z = 1 \\ z = 4 \end{cases}$$

Solution

The equation used to determine the value of y is missing. So let y have any value. In terms of this value, we obtain the solution

$$\begin{aligned} z &= 4 \\ x &= -3y + 2z + 1 \\ &= -3y + 2 \cdot 4 + 1 \\ &= -3y + 9 \end{aligned}$$

So the solutions of the system are

$$y = \text{any value}, \qquad x = -3y + 9, \qquad z = 4$$

Elementary Operations on a Linear System

We can transform a linear system into triangular form using a sequence of the following operations, which don't change the solutions of the system.

ELEMENTARY OPERATIONS ON LINEAR SYSTEMS

1. Interchange two equations.
2. Multiply an equation by a nonzero real number.
3. Add to an equation a multiple of another equation.

The next example illustrates how these operations can be used to transform a system into triangular form.

➤ **EXAMPLE 4**

Reducing a System in Two Variables to Triangular Form

Consider the following system of linear equations:

$$\begin{cases} 3x - y = -5 \\ x + y = 2 \end{cases}$$

1. Use a sequence of elementary row operations to transform the system into echelon form.
2. Determine all solutions of the system.

Solution

1. Start by interchanging the first equation with the second (to avoid fractions where possible):

$$\begin{cases} x + y = 2 \\ 3x - y = -5 \end{cases}$$

Multiply the first equation by -3 and add it to the second equation:

$$\begin{cases} x + y = 2 \\ -4y = -11 \end{cases}$$

Multiply the second equation by $-1/4$:

$$\begin{cases} x + y = 2 \\ y = \dfrac{11}{4} \end{cases}$$

2. Back-substituting $y = 11/4$ into the first equation, we have

$$x + \frac{11}{4} = 2$$

$$x = -\frac{3}{4}$$

Therefore, the system has the single solution

$$\left(-\frac{3}{4}, \frac{11}{4} \right)$$

The main advantage of using the echelon form in solving a linear system is that the method works for any number of equations in any number of variables, and the method, although tedious, is simple to describe in a computer program. This is the preferred method for solving large systems. In the next section, we will discuss in detail how it works for solving linear systems in three or more variables.

➤ **EXAMPLE 5**

Reducing a System with Three Variables to Triangular Form

Use a sequence of elementary operations to transform the following system into triangular form. Determine all solutions of the system.

$$\begin{cases} 3y + 4z = 15 \\ x - y - z = -4 \\ 2x + 5y + 9z = 39 \end{cases}$$

Solution

Since the first equation has no x term, we interchange the first and second equations to obtain the system

$$\begin{cases} x - y - z = -4 \\ 3y + 4z = 15 \\ 2x + 5y + 9z = 39 \end{cases}$$

Next, multiply the second equation by 1/3 to obtain

$$\begin{cases} x - y - z = -4 \\ y + \dfrac{4}{3}z = 5 \\ 2x + 5y + 9z = 39 \end{cases}$$

Next, add (-2) times the first equation to the third:

$$\begin{cases} x - y - z = -4 \\ y + \dfrac{4}{3}z = 5 \\ 7y + 11z = 47 \end{cases}$$

Next, add (-7) times the second equation to the third equation:

$$\begin{cases} x - y - z = -4 \\ y + \dfrac{4}{3}z = 5 \\ \dfrac{5}{3}z = 12 \end{cases}$$

Finally, multiply the third equation by 3/5 to obtain the triangular form

$$\begin{cases} x - y - z = -4 \\ y + \dfrac{4}{3}z = 5 \\ z = \dfrac{36}{5} \end{cases}$$

We now use back-substitution to solve the triangular system:

$$z = \frac{36}{5}$$

$$y = 5 - \frac{4}{3}z = 5 - \frac{48}{5} = -\frac{23}{5}$$

$$x = y + z - 4 = -\frac{23}{5} + \frac{36}{5} - 4 = -\frac{7}{5}$$

Matrices and Gauss Elimination

The elementary row operations on a system of linear equations involve only the numbers that appear on both sides of the equations; they don't involve the variable names in any way. Therefore, as a shorthand, it is convenient to write the system

$$\begin{cases} 3x - 2y = 4 \\ x + 5y = 0 \end{cases}$$

as the rectangular array of numbers:

$$\begin{bmatrix} 3 & -2 & 4 \\ 1 & 5 & 0 \end{bmatrix}$$

Actually, rectangular arrays such as this one occur often in mathematics and are called **matrices**. An extensive theory of matrices has been developed over the last century, and matrices have been used to describe diverse phenomena, including elementary particles in physics and the interaction of various segments of an economy.

Definition 1
Matrix, Entry of a Matrix

A **matrix** is a rectangular array

$$\begin{bmatrix} a_{11} & a_{12} & \cdots & a_{1n} \\ a_{21} & a_{22} & \cdots & a_{2n} \\ \vdots & \vdots & \vdots & \\ a_{m1} & a_{m2} & \cdots & a_{mn} \end{bmatrix}$$

where $a_{11}, a_{21}, \ldots$ are algebraic expressions, called the **elements** or **entries** of the matrix.

The size of a matrix is measured by the numbers of rows and columns it contains. The matrix provided as an example contains m rows and n columns and is called an $m \times n$ (read: m by n) matrix. Note that in specifying the size of a matrix, the number of rows always comes first. The rows and columns are numbered beginning from 1, with the top row and the left-hand column each numbered 1.

The elements of a matrix are often denoted, as in the example, using subscript notation a_{ij}. In this notation, the first subscript, i, refers to the row number, and the second subscript, j, refers to the column number. Thus, for example, a_{32} refers to the element in row 3, column 2. The elements $a_{11}, a_{22}, \ldots$, whose row and column numbers are equal, are said to form the **main diagonal** of the matrix. A matrix with an equal number of rows and columns is called a **square matrix**.

Suppose that we are given a system of linear equations:

$$\begin{cases} a_{11}x_1 + a_{12}x_2 + \cdots + a_{1n}x_n = b_1 \\ a_{21}x_1 + a_{22}x_2 + \cdots + a_{2n}x_n = b_2 \\ \qquad\qquad\qquad \vdots \\ a_{m1}x_1 + a_{m2}x_2 + \cdots + a_{mn}x_n = b_m \end{cases}$$

The matrix

$$\begin{bmatrix} a_{11} & a_{12} & \cdots & a_{1n} \\ a_{21} & a_{22} & \cdots & a_{2n} \\ \vdots & \vdots & \vdots & \vdots \\ a_{m1} & a_{m2} & \cdots & a_{mn} \end{bmatrix}$$

is called the **coefficient matrix** of the system. The matrix

$$\begin{bmatrix} a_{11} & a_{12} & \cdots & a_{1n} & b_1 \\ a_{21} & a_{22} & \cdots & a_{2n} & b_2 \\ \vdots & \vdots & \vdots & \vdots & \vdots \\ a_{m1} & a_{m2} & \cdots & a_{mn} & b_m \end{bmatrix}$$

is called the **augmented matrix** of the system. We may perform elementary row operations on this matrix rather than on the system in order to solve the corresponding system of linear equations. Here is a summary of the elementary row operations in matrix form:

ELEMENTARY ROW OPERATIONS ON MATRICES

E_{ij}:   Interchange rows i and j.

kE_i: Multiply row i by a nonzero real number k.

$kE_i + E_j$:   Add k times row i to row j.

The notations introduced in this list are a compact way of describing elementary row operations without using long phrases. The next example provides some practice in using this notation.

➤ **EXAMPLE 6**
Elementary Row Operations

Consider the matrix

$$\begin{bmatrix} 3 & 0 & 1 \\ -1 & 5 & 8 \\ 0 & -4 & 7 \end{bmatrix}$$

Perform the following elementary row operations:

1. E_{13}

2. $\dfrac{1}{2}E_2$

3. $3E_1 + E_2$

Solution

1. The elementary row operation specified requires us to interchange rows 1 and 3. The result is

$$\begin{bmatrix} 0 & -4 & 7 \\ -1 & 5 & 8 \\ 3 & 0 & 1 \end{bmatrix}$$

2. The elementary row operation specified requires us to multiply row 2 by 1/2. The result is

$$\begin{bmatrix} 3 & 0 & 1 \\ -\frac{1}{2} & \frac{5}{2} & 4 \\ 0 & -4 & 7 \end{bmatrix}$$

3. The elementary row operation specified requires us to add three times row 1 to row 2. The result is

$$\begin{bmatrix} 3 & 0 & 1 \\ 8 & 5 & 11 \\ 0 & -4 & 7 \end{bmatrix}$$

Corresponding to the echelon form of a linear system, we make the following definition:

Let A be a matrix. We say that A is in **echelon form** provided that:

1. The first nonzero element of each nonzero row is a 1.
2. The position of the first 1 in a row is to the right of the first 1 in the row above it.
3. Any all-zero rows are at the bottom of the matrix.

Here are some matrices in echelon form:

$$\begin{bmatrix} 1 & 1 & 2 & -5 \\ 0 & 1 & 3 & 2 \\ 0 & 0 & 1 & 4 \end{bmatrix}$$

$$\begin{bmatrix} 1 & -1 & 5 & 4 \\ 0 & 0 & 1 & 9 \\ 0 & 0 & 0 & 0 \end{bmatrix}$$

By using a sequence of elementary operations, we may transform a matrix into echelon form. The next example illustrates a typical calculation.

➤ **EXAMPLE 7**

Transforming a Matrix Into Echelon Form

Use a sequence of elementary operations to transform the following matrix into echelon form:

$$\begin{bmatrix} 4 & 1 & 0 & -1 \\ 1 & 0 & 1 & 0 \\ 2 & -1 & 3 & 0 \end{bmatrix}$$

Solution

We transform a column at a time, proceeding from the top of the column down, taking subsequent columns in left-to-right order:

$$\begin{bmatrix} 4 & 1 & 0 & -1 \\ 1 & 0 & 1 & 0 \\ 2 & -1 & 3 & 0 \end{bmatrix} \xrightarrow{\frac{1}{4}E_1} \begin{bmatrix} 1 & \frac{1}{4} & 0 & -\frac{1}{4} \\ 1 & 0 & 1 & 0 \\ 2 & -1 & 3 & 0 \end{bmatrix}$$

$$\xrightarrow{(-1)E_1 + E_2} \begin{bmatrix} 1 & \frac{1}{4} & 0 & -\frac{1}{4} \\ 0 & -\frac{1}{4} & 1 & \frac{1}{4} \\ 2 & -1 & 3 & 0 \end{bmatrix}$$

$$\xrightarrow{(-2)E_1 + E_3} \begin{bmatrix} 1 & \frac{1}{4} & 0 & -\frac{1}{4} \\ 0 & -\frac{1}{4} & 1 & \frac{1}{4} \\ 0 & -\frac{3}{2} & 3 & \frac{1}{2} \end{bmatrix}$$

$$\xrightarrow{(-4)E_2} \begin{bmatrix} 1 & \frac{1}{4} & 0 & -\frac{1}{4} \\ 0 & 1 & -4 & -1 \\ 0 & -\frac{3}{2} & 3 & \frac{1}{2} \end{bmatrix}$$

$$\left(\frac{3}{2}\right)E_2 + E_3 \xrightarrow{} \begin{bmatrix} 1 & \frac{1}{4} & 0 & -\frac{1}{4} \\ 0 & 1 & -4 & -1 \\ 0 & 0 & -3 & -1 \end{bmatrix}$$

$$\left(-\frac{1}{3}\right)E_3 \xrightarrow{} \begin{bmatrix} 1 & \frac{1}{4} & 0 & -\frac{1}{4} \\ 0 & 1 & -4 & -1 \\ 0 & 0 & 1 & \frac{1}{3} \end{bmatrix}$$

The last matrix is the echelon form for the given matrix.

Once we have the echelon form of the matrix of a linear system, we may determine the solutions of the system using back-substitution. This method of solving a linear system is called **Gauss elimination**, named for the 19th-century mathematician Karl Friedrich Gauss.

THE METHOD OF GAUSS ELIMINATION

To solve a system of linear equations in any number of variables:

1. Form the augmented matrix of the system.
2. Use elementary row operations to write the augmented matrix in echelon form.
3. Write the linear system corresponding to the echelon form. This system is triangular.
4. Use back-substitution to solve the triangular system.

The following examples illustrate the computational procedures of Gauss elimination.

➤ **EXAMPLE 8**

Solving a System
in Three Variables
Using Gauss Elimination

Solve the following system of linear equations:

$$\begin{cases} 4x + y & = -1 \\ x + & z = 0 \\ 2x - y + 3z = 0 \end{cases}$$

Solution

The matrix of the system is precisely the matrix considered in the preceding example. We found that the echelon form of the matrix is

$$\begin{bmatrix} 1 & \frac{1}{4} & 0 & -\frac{1}{4} \\ 0 & 1 & -4 & -1 \\ 0 & 0 & 1 & \frac{1}{3} \end{bmatrix}$$

The corresponding linear system is

$$\begin{cases} x + \dfrac{1}{4}y & = -\dfrac{1}{4} \\ y - 4z = -1 \\ z = \dfrac{1}{3} \end{cases}$$

Using back-substitution, we find that the solution is $\left(-\frac{1}{3}, \frac{1}{3}, \frac{1}{3}\right)$.

Nonsquare Systems

In the preceding examples, we considered only square linear systems, that is, systems in which the number of equations equals the number of variables. However, Gauss elimination works just as well on nonsquare systems, as the next two examples show.

➤ **EXAMPLE 9**

A System with Infinitely Many Solutions

Solve the following system of linear equations:

$$\begin{cases} 2x - 3y + z - w = 4 \\ 2x - 9z - 3w = -2 \end{cases}$$

Solution

The matrix of the system is

$$\begin{bmatrix} 2 & -3 & 1 & -1 & 4 \\ 2 & 0 & -9 & -3 & -2 \end{bmatrix}$$

We transform the first column of the matrix, proceeding as in the preceding example:

$$\begin{bmatrix} 2 & -3 & 1 & -1 & 4 \\ 2 & 0 & -9 & -3 & -2 \end{bmatrix} \xrightarrow{\frac{1}{2}E_1} \begin{bmatrix} 1 & -\frac{3}{2} & \frac{1}{2} & -\frac{1}{2} & 2 \\ 2 & 0 & -9 & -3 & -2 \end{bmatrix}$$

$$\xrightarrow{(-2)E_1 + E_2} \begin{bmatrix} 1 & -\frac{3}{2} & \frac{1}{2} & -\frac{1}{2} & 2 \\ 0 & 3 & -10 & -2 & -6 \end{bmatrix}$$

$$\xrightarrow{\frac{1}{3}E_2} \begin{bmatrix} 1 & -\frac{3}{2} & \frac{1}{2} & -\frac{1}{2} & 2 \\ 0 & 1 & -\frac{10}{3} & -\frac{2}{3} & -2 \end{bmatrix}$$

The two rows are both in the proper form, so the last matrix is the echelon form. The corresponding system is

$$\begin{cases} x - \dfrac{3}{2}y + \dfrac{1}{2}z - \dfrac{1}{2}w = 2 \\ y - \dfrac{10}{3}z - \dfrac{2}{3}w = -2 \end{cases}$$

The equations from which the values of z and w would be determined are not present in this echelon form. Accordingly, the values of z and w may be any real numbers. The echelon form may be used to determine the values of x and y in terms of the values chosen for z and w. From the second equation, we have

$$y = \frac{10z}{3} + \frac{2}{3}w - 2$$

Therefore, from the first equation, we have

$$x = \frac{3}{2}y - \frac{1}{2}z + \frac{1}{2}w + 2$$

$$= \frac{3}{2}\left(\frac{10z}{3} + \frac{2}{3}w - 2\right) - \frac{1}{2}z + \frac{1}{2}w + 2$$

$$= \frac{9}{2}z + \frac{3}{2}w - 1$$

Therefore, the solution of the system is

$$z = \text{any value}, \qquad w = \text{any value}$$

$$x = \frac{9}{2}z + \frac{3}{2}w - 1, \qquad y = \frac{10}{3}z + \frac{2}{3}w - 2$$

In particular, we see that the system has an infinite number of solutions. Here are three particular solutions of the system

$$\left(\frac{7}{2}, \frac{4}{3}, 1, 0\right), \left(\frac{1}{2}, -\frac{4}{3}, 0, 1\right), (-1, -2, 0, 0)$$

➢ **EXAMPLE 10**
A System with No Solution

Determine all solutions of the following system of linear equations:

$$\begin{cases} x + y - 2z = 1 \\ \quad\quad y - 3z = 4 \\ 2x + 3y - 7z = 5 \end{cases}$$

Solution

The first two rows are already in the proper echelon form. We use row operations on the system matrix to transform the third row, proceeding from left to right:

$$\begin{bmatrix} 1 & 1 & -2 & 1 \\ 0 & 1 & -3 & 4 \\ 2 & 3 & -7 & 5 \end{bmatrix} \xrightarrow{(-2)E_1 + E_3} \begin{bmatrix} 1 & 1 & -2 & 1 \\ 0 & 1 & -3 & 4 \\ 0 & 1 & -3 & 3 \end{bmatrix}$$

$$\xrightarrow{(-1)E_2 + E_3} \begin{bmatrix} 1 & 1 & -2 & 1 \\ 0 & 1 & -3 & 4 \\ 0 & 0 & 0 & -1 \end{bmatrix}$$

Note that the last row of the final matrix corresponds to the equation

$$0 = -1$$

This equation is inconsistent. Therefore, the system has no solutions. Whenever the echelon form of a system contains a row consisting of zeros in all columns except the rightmost one, the system has no solutions. Such systems are called **inconsistent**.

Exercises 6.3

Solve the following systems of equations in echelon form:

1. $\begin{cases} x + 2y - z = 4 \\ \quad 5y - 2z = 8 \\ \quad\quad z = 1 \end{cases}$

2. $\begin{cases} 2x + y + 2z = 12 \\ \quad 4y - 3z = 17 \\ \quad\quad z = 15 \end{cases}$

3. $\begin{cases} 3x + 2y - z = 15 \\ \quad y - 3z = 5 \\ \quad\quad z = -2 \end{cases}$

4. $\begin{cases} x + 3y - z = 11 \\ \quad 4y - 7z = 14 \\ \quad\quad z = 2 \end{cases}$

5. $\begin{cases} x + y + 5z = 13 \\ \quad y - 4z = -10 \\ \quad\quad z = 3 \end{cases}$

6. $\begin{cases} 3x - 5y - z = 7 \\ \quad 6y - 2z = 8 \\ \quad\quad z = -4 \end{cases}$

Consider the matrix

$$\begin{bmatrix} 4 & 1 & 0 \\ -1 & 2 & 3 \\ 0 & -1 & 5 \end{bmatrix}$$

Perform the following elementary row operations:

7. E_{12} 8. E_{13} 9. $\frac{1}{2}E_2$ 10. $-2E_1$

11. $2E_1 - E_2$ 12. $4E_3 + 2E_2$

13. $-1E_1 + 2E_2 + E_3$ 14. $E_2 - E_3 + 2E_1$

Transform the following matrices into echelon form.

15. $\begin{bmatrix} 1 & 2 & 0 & 3 \\ 4 & 5 & 1 & 0 \\ -2 & 0 & 1 & -1 \end{bmatrix}$

16. $\begin{bmatrix} 1 & -1 & 5 & -6 \\ 3 & 3 & -1 & 10 \\ 1 & 3 & 2 & 5 \end{bmatrix}$

17. $\begin{bmatrix} 1 & 1 & 2 \\ 2 & 2 & 5 \end{bmatrix}$

18. $\begin{bmatrix} 2 & 6 & 18 \\ 4 & -3 & -19 \end{bmatrix}$

19. $\begin{bmatrix} 1 & 3 & 2 & 1 \\ 2 & 1 & -1 & 2 \\ 1 & 1 & 1 & 2 \end{bmatrix}$

20. $\begin{bmatrix} 2 & 1 & 1 & 3 \\ 3 & -4 & 2 & -7 \\ 1 & 1 & 1 & 2 \end{bmatrix}$

21. $\begin{bmatrix} 1 & 5 & 6 \\ 0 & 1 & 1 \end{bmatrix}$

22. $\begin{bmatrix} 2 & 7 & 1 \\ 5 & 0 & -15 \end{bmatrix}$

23. $\begin{bmatrix} 3 & 2 & 1 & 1 \\ 0 & 2 & 4 & 22 \\ -1 & -2 & 3 & 15 \end{bmatrix}$

24. $\begin{bmatrix} 4 & 2 & 6 & 24 \\ 4 & -3 & 0 & 10 \\ 5 & 0 & -4 & -11 \end{bmatrix}$

25. $\begin{bmatrix} 1 & 2 & 3 & 0 \\ 1 & 0 & -2 & 3 \\ 0 & 1 & 1 & 1 \end{bmatrix}$

26. $\begin{bmatrix} 2 & 1 & 2 & 1 \\ 1 & 2 & -3 & 4 \\ 3 & -1 & 1 & 0 \end{bmatrix}$

Solve the following systems of linear equations.

27. $\begin{cases} x + 2y + 3z = 0 \\ x \quad\quad - 2z = 3 \\ \quad\quad y + z = 1 \end{cases}$

28. $\begin{cases} x + y + 4z = 1 \\ -2x - y + z = 2 \\ 3x - 2y + 3z = 1 \end{cases}$

29. $\begin{cases} 3x + 2y - z = 4 \\ 2x - y + 3z = 5 \\ x + 3y + 2z = -1 \end{cases}$

30. $\begin{cases} 4x - 2y + 3z = 4 \\ 5x - y + 4z = 7 \\ 3x + 5y + z = 7 \end{cases}$

31. $\begin{cases} 2x + 3y + z = 1 \\ x - y + 2z = 3 \end{cases}$

32. $\begin{cases} -5x + 3y - z = 0 \\ 2x + y - z = 4 \end{cases}$

33. $\begin{cases} 2x \quad\quad - z = 0 \\ \quad y + 2z = -2 \\ x + y \quad\quad = 1 \end{cases}$

34. $\begin{cases} x \quad\quad + z = 1 \\ x + y \quad\quad = -1 \\ \quad y + z = 4 \end{cases}$

35. $\begin{cases} x + y + z + w = -2 \\ -2x + 3y + 2z - 3w = 10 \\ 3x + 2y - z + 2w = -12 \\ 4x - y + z + 2w = 1 \end{cases}$

36. $\begin{cases} x - y - 3z - 2w = 2 \\ 4x + y + z + 2w = 2 \\ x + 3y - 2z - w = 9 \\ 3x + y - z + w = 5 \end{cases}$

37. $\begin{cases} x + 2y + 3z = 0 \\ x \quad\quad - 2z = 3 \\ \quad\quad y + z = 1 \end{cases}$

38. $\begin{cases} 3x - 5y = 6 \\ -6x + 10y = 7 \end{cases}$

39. $\begin{cases} 3x - 6y = 12 \\ -2x + 4y = -8 \end{cases}$

40. $\begin{cases} x - y + z = 0 \\ 5x + y + 3z = 1 \end{cases}$

41. $\begin{cases} -5x + 3y - z = 1 \\ 2x + y - z = 4 \end{cases}$

42. $\begin{cases} x - y \quad\quad\quad = 5 \\ \quad y - z \quad\quad = 6 \\ \quad\quad z - w = 7 \\ x \quad\quad\quad + w = 8 \end{cases}$

43. $\begin{cases} x + y - z + w = 4 \\ 2x + 3y + z - w = -1 \\ -x - 3y + 2z + 3w = 3 \\ 3x + 2y - 3z + w = 8 \end{cases}$

44. $\begin{cases} x + 2y + 3z - w = 7 \\ x - 4y + z \quad\quad = 3 \\ \quad 2y - 3z + w = 4 \end{cases}$

45. $\begin{cases} x + 2y = 5 \\ 2x - 5y = -8 \end{cases}$

◆ Applications

46. Flying with the wind, an airplane can travel 1080 miles in six hours, but flying against the same wind it goes only 1/3 this distance in half the time. Find the speed of the plane in still air and the wind speed.

47. To start a record company, Barry G. borrowed $50,000 from three different banks. He borrowed some money at 8%, and he borrowed $3000 more than that amount at 6.9%. The rest was borrowed at 12%. If his annual interest totals $4755 from all three loans, how much was borrowed at each rate?

48. Barry G.'s record company makes two types of records, 45s and 33 1/3s. Both require time on two machines: 45s, 1 hour on the recorder and 2 hours on the cutter; 33 1/3s, 3 hours on the recorder and 1 hour on the cutter. Both the recorder and the cutter operate 15 hours a day. How many of each record can be produced in a 5-day work week under these conditions?

49. A famous national appliance store sells both Sony and Sanyo stereos. Sony sells for $280, and Sanyo sells for $315. During a one-day sale, a total of 85 Sonys and Sanyos were sold for a total of $23,975. How many of each brand were sold during this one-day sale?

▌ 6.4 SYSTEMS OF INEQUALITIES

In Chapter 1, we introduced inequalities involving a single variable and showed how to graph such inequalities on the number line. Let's now turn our attention to inequalities in two variables x and y. Here are some typical inequalities of this sort:

$$3x + 5y < 4$$

$$x^2 + 3xy \le 10$$

$$x + \frac{1}{y} > 1$$

Suppose that we are given an inequality in two variables. An ordered pair (a, b) is said to **satisfy the inequality** (or **is a solution of the inequality**) provided that the inequality is true when x is replaced by a and y is replaced by b. For instance, the ordered pair $(-1, 2)$ is not a solution of the first inequality shown since when

x is replaced by -1 and y is replaced by 2, we have

$$3x + 5y = 3(-1) + 5(2)$$
$$= -3 + 10$$
$$= 7$$

Since 7 is not less than 4, we see that $(-1, 2)$ is not a solution. However, $(-4, 1)$ is a solution of the inequality, since when x is replaced by -4 and y is replaced by 1, we have

$$3x + 5y = 3(-4) + 5(1)$$
$$= -7$$
$$< 4 \qquad \textit{(True)}$$

The set of ordered pairs that are solutions of an inequality is called the **solution set** of the inequality. The solution set may be represented graphically as a region in the x-y plane. The next few examples determine the solution sets of some typical inequalities in two variables and the corresponding plane regions.

➤ **EXAMPLE 1**
Determining Solution Set

Graph the inequality $x^2 + y^2 < 4$.

Solution

Note that $x^2 + y^2$ equals the square of the distance of the point (x, y) from the origin. The inequality specifies that the square of the distance of (x, y) be less than 4; that is, the distance is less than 2. These are the points in the interior of the circle shown in Figure 1. The dotted line indicates that the points on the circle are excluded from the shaded region.

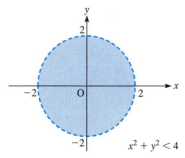

Figure 1

In Figure 2, we have drawn the graph of an equation of the form $y = f(x)$, where $f(x)$ is a function of the variable x. A point (x, y) for which $y = f(x)$ lies on the graph. A point (x, y) for which $y > f(x)$ lies above the graph. A point (x, y) for which $y < f(x)$ lies below the graph. That is, we have the following result.

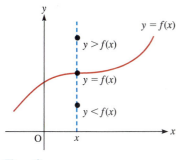

Figure 2

GRAPHING SOLUTION SETS

The graph of the inequality $y > f(x)$ consists of the region above the graph of the equation $y = f(x)$. (See Figure 3(a).) The graph of the inequality $y < f(x)$ consists of the region below the graph of the equation $y = f(x)$. (See Figure 3(b).)

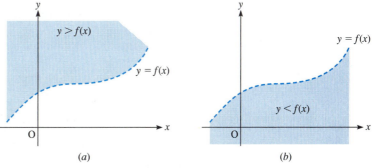

(a)   (b)

Figure 3

The next example gives an application of this result.

> **EXAMPLE 2**
Graphing a Solution Set

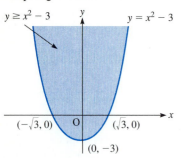

Figure 4

Graph the inequality $y \geq x^2 - 3$.

Solution

In Figure 4, we have drawn the graph of the equation $y = x^2 - 3$. The graph of the given inequality is the region above and on the graph. This is the shaded region in Figure 4.

In Chapter 1, we discussed the rules that allow us to add, subtract, multiply, and divide inequalities. It is often possible to apply these rules to transform an inequality into one of the forms $y < f(x)$ or $y > f(x)$, to which we may apply the rules for graphing solution sets to determine the graph. The next two examples illustrate how this may be done.

> **EXAMPLE 3**
Solution Set
of a Linear Inequality

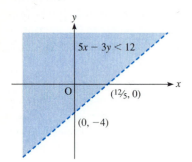

Figure 5

Graph the inequality $5x - 3y < 12$.

Solution

Let's transform the inequality so that y is isolated on the left side. First subtract $5x$ from both sides to obtain

$$-3y < 12 - 5x$$

Next, multiply by $-1/3$. Since this quantity is negative, we must reverse the sign of the inequality:

$$y > -\frac{1}{3}(12 - 5y)$$

$$y > -4 + \frac{5}{3}x$$

We now graph the equation $y = -4 + \frac{5}{3}x$, as shown in Figure 5. The graph of the given inequality is the region above the line in the figure.

> **EXAMPLE 4**
Solution Set
of a Nonlinear Inequality

Graph the solution set of the inequality

$$xy \geq 1$$

Solution

As in the preceding example, we apply the laws of inequalities to isolate y on the left side. Basically, we wish to multiply the inequality by $1/x$. However, this quantity is positive if $x > 0$ and negative if $x < 0$. In the first case, the inequality is equivalent to

$$y \geq \frac{1}{x}, \quad x > 0$$

In the second case, the inequality is equivalent to

$$y \leq \frac{1}{x}, \quad x < 0$$

For the first case, draw the graph $y = 1/x$, $x > 0$, as shown in Figure 6(a). The points corresponding to solutions of the given inequality are those lying on or above the graph. These are the shaded points in the figure. For the second case, draw the graph $y = 1/x$, $x < 0$, as shown in Figure 6(b). The point corresponding to solutions of the given inequality are those lying on or below the graph. These are the shaded points in the figure. The solution set of the given inequality consists of the points belonging to the solution sets for either of the cases. This set is the shaded region shown in Figure 6(c).

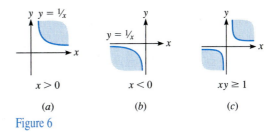

$$x > 0 \qquad\qquad x < 0 \qquad\qquad xy \geq 1$$

$$(a) \qquad\qquad\qquad (b) \qquad\qquad\qquad (c)$$

Figure 6

   In the preceding examples we used the laws of inequalities to transform the inequality so that the rule for graphing solution sets could be applied to determine the graph. This is not really necessary. Another method of graphing the inequality is to replace the inequality sign by equal signs and graph the resulting equation. The graph of the inequality consists of all points lying to one side of the graph. To determine which side, choose a point (a, b) not on the graph of the equation and substitute this point into the inequality. If the result is a true inequality, then the point lies on the same side as the graph of the inequality. Otherwise, the point lies on the opposite side from the graph of the inequality. The point (a, b) is called a **test point**.

   For example, consider the inequality $x^2 + y^2 < 4$ of Example 1 and the test point $(2, 2)$. The equation $x^2 + y^2 = 4$ has a circle as its graph, and the test point lies outside the circle. At the test point, we have

$$x^2 + y^2 = 2^2 + 2^2$$
$$= 8$$

That is, the inequality is false at the test point because $8 < 4$ is false. This means that the region containing the test point, the exterior of the circle, does not belong to the graph of the inequality. The graph consists of the points within the circle. And since the sign of the inequality is $<$ rather than $\leq$, the circle itself is not included in the graph.

   In the previous sections of this chapter, we studied systems of equations and their solutions. By analogy, let's now turn to systems of *inequalities* and their solutions. A **system of inequalities** is a collection of one or more inequalities. We will concentrate on systems of inequalities in two variables. A **solution** of a system of inequalities is an ordered pair that is a solution for each of the inequalities of the system. The set of solutions of a system of inequalities is called the **feasible set** (or **graph**) of the system. One method for determining the feasible set is as follows:

1. Solve separately each inequality of the system.
2. Graph on the same coordinate system the solution set of each of the inequalities. Shade each solution set with a different sort of shading (hatching).
3. The solution set of the system consists of the region that is shaded with all possible shading types.

The following examples illustrate how to solve a simple system of inequalities.

➤ **EXAMPLE 5**
Graph a Feasible Set

Determine the feasible set of the following system of inequalities:

$$\begin{cases} y \geq x^2 - 3 \\ 5x + 3y < 12 \end{cases}$$

Solution

In Figure 7(a), we have sketched the solution of the first inequality of the system. In Figure 7(b), we have sketched the solution set of the second inequality. Figure 7(c) shows the two solution sets superimposed on a single coordinate system. The solution set of the system consists of the region that is shaded twice. This is the shaded region in Figure 7(d).

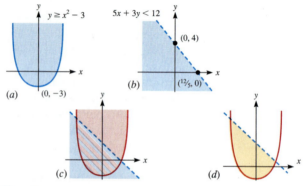

Figure 7

➤ **EXAMPLE 6**
Feasible Set
of a Linear System

Sketch the graph of the following system of linear inequalities:

$$\begin{cases} x \leq 8 \\ y \leq 3x - 7 \\ y > x + 4 \end{cases}$$

Solution

In this case, there are three solution sets to determine, one for each inequality. Figure 8(a) shows the three solution sets, each shaded with a different color. The region that is shaded three times is the solution set of the system. (See Figure 8(b).)

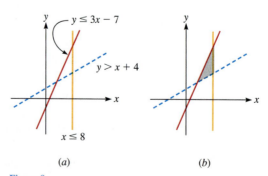

Figure 8

➤ **EXAMPLE 7**

Manufacturing

A pharmaceutical company manufactures two drugs using the same ingredients. Each case of drug 1 requires 300 grams of ingredient A and 600 grams of ingredient B. Each case of drug 2 requires 500 grams of ingredient A and 100 grams of ingredient B. Market conditions dictate that at most 100,000 grams of ingredient A and 50,000 grams of ingredient B are available each week. Determine a system of inequalities describing the production limitations for the two drugs and determine the feasible set of the system of inequalities.

Solution

Let

$$x = \text{number of cases of drug 1}$$
$$y = \text{number of cases of drug 2}$$

First, all the values of x and y must be nonnegative. That is,

$$x \geq 0, \quad y \geq 0$$

The condition of ingredient A is

$$\text{total ingredient A} \leq 100,000$$
$$\text{ingredient A in drug 1} + \text{ingredient A in drug 2} \leq 100,000$$
$$300x + 500y \leq 100,000$$

Similarly, the condition on ingredient B is

$$\text{total of ingredient B} \leq 50,000$$
$$\text{ingredient B in drug 1} + \text{ingredient B in drug 2} \leq 50,000$$
$$600x + 100y \leq 50,000$$

Thus, the production limitations are described by the following system of four inequalities:

$$\begin{cases} x \geq 0, \quad y \geq 0 \\ 3x + 5y \leq 1000 \\ 6x + 1y \leq 500 \end{cases}$$

In Figure 9(*a*), we have graphed the system of inequalities. The region that is shaded four times is the feasible set for the system. This is the shaded region in Figure 9(*b*).

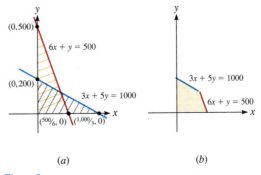

(*a*)   (*b*)

Figure 9

Exercises 6.4

Graph the inequality.

1. $x + 2y \leq 4$ 2. $3x - y > 10$
3. $x^2 + y^2 < 4$ 4. $x^2 + 9y^2 \leq 36$
5. $x < -y^2$ 6. $x + y^2 \geq -4$

7. $xy > 4$ 8. $x \leq \dfrac{16}{y}$

9. $x^2 + 4y^2 - 10x + 24y + 57 \leq 0$
10. $x^2 - y^2 + 2x - 2y \geq 0$
11. $3x - 2y > 6$ 12. $x < y$
13. $y \leq x^2 - 4x + 2$ 14. $y \geq -1 - x^2$

Determine the feasible set of each of the following systems of inequalities.

15. $\begin{cases} x + 2y > 6 \\ x^2 > 2y \end{cases}$ 16. $\begin{cases} x^2 + y^2 \leq 9 \\ \dfrac{x^2}{16} + \dfrac{y^2}{4} \geq 1 \end{cases}$

17. $\begin{cases} 4x + 3y < 12 \\ 4x + y > -4 \end{cases}$ 18. $\begin{cases} x^2 + y^2 \leq 36 \\ -4 \leq x \leq 4 \end{cases}$

19. $\begin{cases} 9x^2 - y^2 \geq 0 \\ x^2 + y \leq 25 \end{cases}$ 20. $\begin{cases} x^2 + 4y^2 < 16 \\ x^2 - 2xy \geq 0 \end{cases}$

21. $\begin{cases} y \geq (x - 2)^2 + 3 \\ 3x - 4y \leq -16 \end{cases}$ 22. $\begin{cases} y \leq -(x - 3)^2 + 2 \\ 3x + 5y \geq -20 \end{cases}$

23. $\begin{cases} |x| \geq 6 \\ |y| \leq 1 \end{cases}$ 24. $\begin{cases} y > |x + 2| \\ 3 \geq |x| \end{cases}$

Sketch the graphs of the following systems of linear inequalities.

25. $\begin{cases} x + y > 6 \\ 2x - y \leq 0 \\ y \geq -2 \end{cases}$ 26. $\begin{cases} 2x - y \leq 6 \\ x + y > 6 \\ x \geq -1 \end{cases}$

27. $\begin{cases} x - y \leq 2 \\ x + 2y \geq 8 \\ y \leq 4 \end{cases}$ 28. $\begin{cases} 2x - y \geq -1 \\ 2x + y \geq 1 \\ x \leq 2 \end{cases}$

29. $\begin{cases} x \geq 0 \\ y \geq 0 \\ 2x + 3y \leq 12 \\ 3x + y \leq 6 \end{cases}$ 30. $\begin{cases} x \geq 0 \\ y \geq 0 \\ 2x + y \leq 4 \\ 2x - 3y \leq 6 \end{cases}$

31. $\begin{cases} x \geq 2 \\ y \geq 5 \\ 3x - y \geq 12 \\ x + y \leq 15 \end{cases}$ 32. $\begin{cases} x \geq 10 \\ y \geq 20 \\ 2x + 3y \leq 100 \\ 5x + 4y \leq 200 \end{cases}$

♦ Applications

For Exercises 33–39, (a) determine a system of inequalities describing the given problems, and (b) determine the feasible set of the systems of inequalities.

33. A manufacturer of skis can produce as many as 60 pairs of recreational skis and as many as 45 pairs of racing

skis per day. It takes 3 hours of labor to produce a pair of recreational skis and 4 hours of labor to produce a pair of racing skis. The company has up to 240 hours of labor available for ski production each day.

34. General Mills wants to make a cereal from a mixture of oats and wheat so that it will contain at least 88 g of protein and at least 36 mg of iron. Given that wheat contains 80 g of protein and 40 mg of iron per kg, and that oats contain 100 g of protein and 30 mg of iron per kg, write the system of inequalities.

35. You are taking a test in which items of type x are worth 10 points and items of type y are worth 15 points. It takes 3 minutes for each item of type x and 6 minutes for each item of type y. Total time allowed is 60 minutes and you may not answer more than 16 questions.

36. Every day Marvin B. needs a dietary supplement of 4 mg of vitamin A, 11 mg of vitamin B, and 100 mg of vitamin C. Either of two brands of vitamin pills can be used: brand x, at 6 cents a pill, or brand y, at 8 cents a pill. Brand x supplies 2 mg of vitamin A, 3 mg of vitamin B, and 25 mg of vitamin C. Likewise a brand y pill supplies 1, 4, and 50 mg, respectively, of vitamins A, B, and C.

37. **Hockey.** A hockey team figures that it needs at least 60 points for the season to make the playoffs. A win (w) is worth 2 points in the standings, and a tie (t) is worth 1 point.

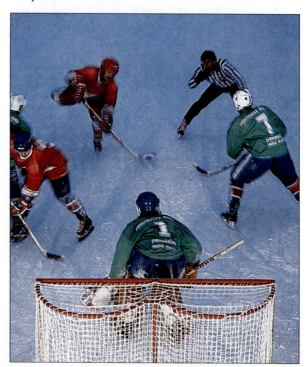

38. **Elevators.** Many elevators have a capacity of 1 metric ton (1000 kg). Suppose c children, each weighing 35 kg and a adults, each weighing 75 kg, are on an elevator.

39. **Width of a basketball floor.** Sizes of basketball floors vary considerably due to building dimensions and other constraints, such as cost. The length L is to be at most 74 ft., and the width w is to be at most 50 ft.

6.5 LINEAR PROGRAMMING

Linear programming is an important application involving systems of linear inequalities. To introduce this important subject, let's return to the pharmaceutical company example at the end of the preceding section. Suppose that the company earns \$50 for each case of drug 1 and \$75 for each case of drug 2. How many cases of each drug should it produce in order to maximize its profit?

Let's state the problem in algebraic form: Suppose that the drug company produces x cases of drug 1 and y cases of drug 2. The profit P it earns is given by

$$P = 50x + 75y$$

At first, you might ask: Why not just manufacture a huge number of each type of drug and thereby make an equally huge profit? The answer is that production has a number of constraints imposed on it, as stated in the example of the preceding section. Namely, production is subject to the inequalities

$$300x + 500y \leq 100,000$$
$$600x + 100y \leq 50,000$$
$$x \geq 0, \qquad y \geq 0$$

The four inequalities are called **constraints**, since they constrain the choice of x and y. As we saw, the set of ordered pairs (x, y) that satisfy the system of inequalities is represented by the points in the shaded region in Figure 1. Our problem is somehow to choose the point (x, y) of the feasible set at which the value of P is a maximum.

A problem of this sort is called a **linear programming problem**. More precisely, suppose that we are given a function

$$P = ax + by$$

of two variables x and y and a system of linear inequalities in the same variables.

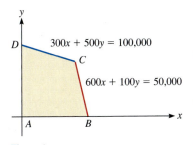

Figure 1

LINEAR PROGRAMMING PROBLEM

Determine the ordered pair (x, y) that is a solution of the system of inequalities and for which the value of P is maximized (respectively minimized).

The function P is called the **objective function**, and the inequalities are called **constraints**. The corners of the feasible set are called **vertices**. Then we have the following result.

FUNDAMENTAL THEOREM OF LINEAR PROGRAMMING

The objective function assumes its maximum (respectively minimum) value at a vertex of the feasible set.

We will omit the proof of this theorem.

➤ **EXAMPLE 1**
Manufacturing

Determine the number of cases of each type of drug that the pharmaceutical company must produce in order to maximize its profit.

Solution

According to the Fundamental Theorem of Linear Programming, the maximum profit occurs at a vertex of the feasible set. Let's determine the vertices. In Figure 1, we have drawn the feasible set and labeled each of the vertices. Also, each of the bounding lines of the feasible set is labeled with its equation. Vertex A is the origin $(0,0)$. Vertex B is the intersection of the lines with equations

$$y = 0$$
$$600x + 100y = 50{,}000$$

The solution of this system is obtained by substituting $y = 0$ into the second equation:

$$600x + 100(0) = 50{,}000$$
$$x = \frac{250}{3}$$
$$B = \left(\frac{250}{3}, 0\right)$$

Vertex C is the intersection of the lines with equations

$$3x + 5y = 1000$$
$$6x + 1y = 500$$

To solve this system, multiply the second equation by -5 and add it to the first to obtain:

$$\begin{array}{r} 3x + 5y = 1000 \\ +\quad -30x - 5y = -2500 \\ \hline -27x = -1500 \end{array}$$

$$x = \frac{1500}{27} = \frac{500}{9}$$
$$y = 500 - 6x$$
$$= 500 - 6\left(\frac{500}{9}\right)$$
$$= \frac{1500}{9}$$
$$= \frac{500}{3}$$
$$C = \left(\frac{500}{9}, \frac{500}{3}\right)$$

The final vertex, D, is the intersection of the lines with equations

$$x = 0$$
$$300x + 500y = 100{,}000$$
$$300 \cdot 0 + 500y = 100{,}000$$
$$y = \frac{1000}{5} = 200$$
$$D = (0, 200)$$

Let's now make a table of the vertices and the value of the function P at each of them:

Vertex	$P = 50x + 75y$
$A : (0, 0)$	0
$B : \left(\dfrac{250}{3}, 0\right)$	$\dfrac{12{,}500}{3} = 4166.67$
$C : \left(\dfrac{500}{9}, \dfrac{500}{3}\right)$	$15{,}277.78 \leftarrow$ Maximum
$D : (0, 200)$	$15{,}000$

From the table, we see that the maximum profit occurs for vertex C. By the Fundamental Theorem of Linear Programming, the maximum value of the function P for (x, y) in the feasible set is 15,277.78 and this value occurs for $x = 500/9$, $y = 500/3$. That is, the company should produce 500/9 cases of drug 1 and 500/3 cases of drug 2 per week.

➤ **EXAMPLE 2**

Minimization
of a Linear Function

Determine the minimum value of the function $P = 3x - 5y$ if x and y are constrained to lie in the triangle with vertices $(1, 1)$, $(3, 2)$, and $(1, 5)$. (See Figure 2.)

Solution

By the Fundamental Theorem of Linear Programming, the minimum point must occur at one of the vertices of the triangle. So let's tabulate the value of the function at each of the vertices:

Vertex	$P = 3x - 5y$
$(1, 1)$	-2
$(3, 2)$	-1
$(1, 5)$	-22

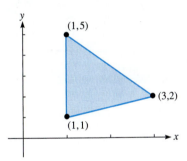

Figure 2

The minimum value of $3x - 5y$ is -22, which occurs for $x = 1$, $y = 5$.

Exercises 6.5

Determine the maximum and minimum values of the function and the values of x and y where they occur, given a feasible set with vertices as follows:

1. $f(x, y) = 3x + 5y; (4, 8), (2, 4), (1, 1), (5, 2)$

2. $f(x, y) = x + 4y; (0, 7), (0, 0), (6, 2), (5, 4)$

3. $Q = x + 3y$ subject to

$$\begin{cases} x \le 0 \\ y \ge 0 \\ 5x + 2y \ge 20 \\ 2y \ge x \end{cases}$$

4. $F = 2x - 3y$ subject to

$$\begin{cases} x \ge 0 \\ y \ge 0 \\ 3x + 5y \le 15 \\ 5x + 3y \ge 15 \end{cases}$$

5. $f(x, y) = 10x + 12y$ subject to

$$\begin{cases} 2x + 5y \ge 22 \\ 4x + 3y \ge 28 \\ 2x + 2y < 17 \\ x \ge 0 \\ y \ge 0 \end{cases}$$

6. $f(x, y) = -3x + y$ subject to

$$\begin{cases} x + 2y \ge 10 \\ 2x + y \le 12 \\ x - y \le 8 \\ x \ge 0 \\ y \ge 0 \end{cases}$$

Find the maximum and minimum values of the given function in the indicated region.

7. $f(x, y) = 2x - 3y$

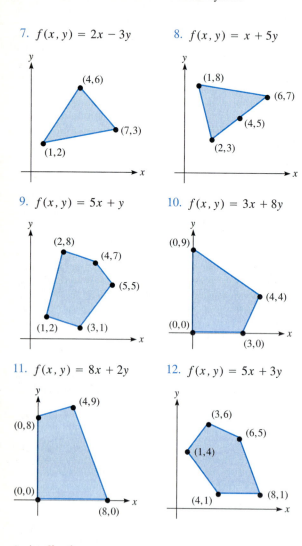

8. $f(x, y) = x + 5y$

9. $f(x, y) = 5x + y$

10. $f(x, y) = 3x + 8y$

11. $f(x, y) = 8x + 2y$

12. $f(x, y) = 5x + 3y$

◆ Applications

13. RCA Manufacturing Company in Fort Wayne, Indiana, makes a $60 profit on each 19″ TV it produces and a $40 profit on each 13″ TV. A 19″ TV requires 1 hour on machine X, 1 hour on machine Y, and 4 hours on machine Z. The 13″ TV requires 2 hours on X, 1 hour on Y, and 1 hour on Z. In a given day, machines X, Y, and Z can work a maximum of 16, 9, and 24 hours, respectively. How many 19″ TVs and how many 13″ TVs should be produced per day to maximize the profit?

14. The Buick manufacturing plant in Flint, Michigan, must fill orders for Park Avenues from two dealers. The first dealer, Farmington Buick, has ordered 20 Park Avenues, and the second dealer, Brighton Buick, has ordered 30. The manufacturer has the cars stored in two different areas: southeast Flint and southwest Flint. There are 40 cars in southeast Flint and only 15 cars in southwest Flint. The shipping costs, per car, are: $15 from southeast Flint to Brighton; $13 from southeast Flint to Farm-

ington; $14 from southwest Flint to Brighton; $16 from southwest Flint to Farmington. With these conditions, find the number of cars to be shipped from each area to each dealer if the total shipping cost is to be a minimum. What is the cost?

15. Kellogg's of Battle Creek, Michigan, is going to produce a new cereal from a mixture of bran and rice so that it contains at least 88 g of protein and at least 36 mg of iron. Knowing that bran contains 80 g of protein and 40 mg of iron per kilogram, and that rice contains 100 g of protein and 30 mg of iron per kilogram, find the minimum cost of producing this new cereal, "Rice Bran," if bran costs 50 cents per kilogram and rice costs 40 cents per kilogram.

16. Sanyo makes stereo receivers. It produces a 30-watt receiver that it sells for a $100 profit and a 50-watt receiver that it sells for a $150 profit. In manufacturing, the 30-watt receiver requires 3 hours, and the 50-watt receiver takes 5 hours. The cabinet shop spends 1 hour on a 30-watt receiver and 3 hours on a 50-watt receiver. Packing takes 2 hours for both types of receivers. Per week, Sanyo has available 3900 work hours for manufacturing, 2100 hours for cabinet making, and 2200 hours for packing. How many receivers of each type should Sanyo produce per week to maximize its profit, and what is the maximum profit per week?

17. Almosttexas makes two types of calculators: Deluxe sells for $12 and Top of the Line sells for $10. It costs Almosttexas $9 to produce a Deluxe and $8 to produce a Top of the Line calculator. In one week, Almosttexas can produce between 200 and 300, inclusive, Deluxe calculators, and between 100 and 250, inclusive, Top of the Line calculators, but no more than 500 total calculators. How many of each type should be produced per week to maximize the profits for Almosttexas, Inc.?

18. You have $40,000 to invest in stocks and bonds. The least you are allowed to invest in stocks is $6000, and you cannot invest more than $22,000 in stocks. You may also invest no more than $30,000 in bonds. The interest on stocks is 8% tax-free, and the interest on bonds is

7 1/2% tax-free. How much should you invest in each type to maximize your income? What is your income from the $40,000 invested?

19. The Glodfelty's 312-acre farm grows corn and beans. The work of growing and picking the corn takes 35 labor-hours per acre, and the work of growing and picking the beans takes 27 labor-hours per acre. There are only 9500 labor-hours available for these crops. The profit per acre on corn is $173, and the profit per acre on beans is $152. With this in mind, how many acres should be planted in corn to maximize the Glodfelty's profits? What are the maximum profits?

20. Fenton Cement Company has been contracted to haul 360 tons of cement per day for a highway construction job on U.S. 23. Fenton has seven 6-ton trucks, four 10-ton trucks, and nine drivers to haul cement per day. The 6-ton trucks can make eight trips per day and the 10-ton trucks can make only six trips per day, with the costs being $15 and $24, respectively, per day. Using all 9 drivers, how many trucks of each type should be used to minimize the cost per day? What is that cost?

21. The North Manchester building code requires that the area of the windows be at least 1/8 the area of the walls and roof in any new building. The annual heating cost of a new building is $3 per square foot of windows and $1 per square foot of walls and roof. To the nearest square foot, what is the largest surface area a new building can have if its annual heating cost cannot exceed $1000?

22. Gary Smelting Company receives a monthly order for at least 40 tons of iron, 60 tons of copper, and 40 tons of lead. It can fill this order by smelting either alloy A or alloy B. Each railroad car of A will produce 1 ton of iron, 3 tons of copper, and 4 tons of lead after smelting. Each railroad car of B will produce 2 tons of iron, 2 tons of copper, and 1 ton of lead after smelting. If the cost of smelting the alloy in railroad car A is $350 and the cost

of smelting the alloy in railroad car B is $200, then how many carloads of each should be used to fill the order at a minimum cost to Gary Smelting? What is that minimum cost?

23. A lumber company can convert logs into either lumber or plywood. In a given week, the mill can turn out 400 units of production, of which 100 units of lumber and 150 units of plywood are required by regular customers. The profit on a unit of lumber is $20, and on a unit of plywood it is $30. How many units of each type should the mill produce per week in order to maximize profit?

24. A farm consists of 240 acres of cropland. The farmer wishes to plant part or all of the acreage in corn and/or oats. Profit per acre in corn production is $40, and in oats it is $30. An additional restriction is that the total hours of labor during the production be no more than 320. Each acre of land in corn production uses 2 hours of labor during the production period; production of oats requires only 1 hour per acre. How many acres of land should be planted in corn and how many in oats in order to maximize profit?

6.6 CHAPTER REVIEW

Important Concepts, Properties, and Formulas—Chapter 6

| Method of substitution for systems of two equations in two variables | 1. Solve an equation for one of the variables in terms of the other.
2. Substitute the expression found in step 1 into the other equation.
3. Solve the resulting equation for the possible values of one variable.
4. For each value found in step 3, substitute into the expression of step 1 to determine the corresponding value of the other variable. | p. 325 |

Method of elimination for systems of two linear equations in two variables	1. Multiply one or both equations of the system by suitable numbers to obtain a new system in which the coefficients of one of the variables are equal. 2. Subtract the equations to obtain a linear equation in one variable. 3. Solve the equation obtained in step 2. 4. Substitute the value obtained into either of the original system equations and solve for the second variable.	p. 333
Graphical analysis of solutions of two linear equations in two variables.	1. The graphs of the equations correspond to two distinct lines. In this case, the system has a single solution, corresponding to the point of intersection of the lines. 2. The two straight lines coincide. In this case, the system has infinitely many solutions, corresponding to all points on the line. 3. The two straight lines are different but parallel. In this case, the system has no solutions.	p. 333
Elementary operations on linear systems	1. Interchange two equations. 2. Multiply an equation by a nonzero real number. 3. Add to an equation a multiple of another equation.	p. 334
Determining the feasible set of a system of inequalities	1. Solve separately each inequality of the system. 2. Graph on the same coordinate system the solutions sets of each of the inequalities. Shade each solution set with a different sort of shading (hatching). 3. The solution set of the system consists of the region which is shaded with all possible shading types.	p. 346
Fundamental theorem of linear programming	The objective function assumes its maximum (respectively minimum) value at a vertex of the feasible set.	p. 350

Cumulative Review Exercises—Chapter 6

Determine all solutions of the following systems of equations.

1. $\begin{cases} 3x + y = 11 \\ 2x - 5y = 13 \end{cases}$

2. $\begin{cases} 4x + 3y = 7 \\ 2x + 4y = 16 \end{cases}$

3. $\begin{cases} x + y = 1 \\ x^2 + y^2 = 25 \end{cases}$

4. $\begin{cases} x = 3 - 2y \\ x^2 = 9 - y^2 \end{cases}$

5. $\begin{cases} x + y + z = 6 \\ xyz = 6 \\ 2x - y = 0 \end{cases}$

6. $\begin{cases} x - y - z = 0 \\ xyz = 6 \\ 3x + 4y = 5 \end{cases}$

◆ **Applications**

7. I have a total of 30 coins in my pocket. Some are dimes, and the rest are quarters. If I have a total of $4.50, then how many of the coins are dimes?

8. Tickets to the Fisher Theater cost $20 for the main floor and $16 for the balcony. If the receipts from the sale of 1420 tickets were $26,060, how many tickets were sold at each price?

9. Find two numbers whose sum is 14 and the sum of whose reciprocals is 14/33.

10. The sum of the digits of a three-digit number is 11, the product of the digits is 40, and if the hundreds digit is subtracted from the ones digit the result is 1. Find the number.

Solve the following in echelon form.

11. $\begin{cases} x + y + z = 9 \\ 3y + 2z = 7 \\ z = -1 \end{cases}$ 12. $\begin{cases} x - y + 5z = 13 \\ y - 4z = -10 \\ z = 3 \end{cases}$

13. $\begin{cases} 3x - 2y = 1 \\ 4x + y = 5 \end{cases}$ 14. $\begin{cases} x + 2y = 3 \\ 2x + 3y = 4 \end{cases}$

15. $\begin{cases} 2x - 3y = -12 \\ -10x + 15y = 60 \\ 6x - 9y = -36 \end{cases}$ 16. $\begin{cases} 3x - 4y = -1 \\ x + 6y = -4 \\ -2x + 8y = -2 \end{cases}$

♦ Applications

17. Suppose you invested a certain sum of money at 12.5% and another sum at 14%. If your total investment was $60,000 and your total yearly interest for one year on both investments was $8190, how much did you invest at each rate?

18. The perimeter of a rectangular pool is 68 meters. If the width of the pool is 6 m less than the length, what are the dimensions of the pool?

Graph the inequality.

19. $x + 3y \geq 12$ 20. $x - y < 4$

21. $x^2 + y^2 < 25$ 22. $x^2 + 4y^2 \leq 36$

23. $xy > 16$ 24. $x + y^2 < -4$

25. $y > x^2 + 16$ 26. $xy \leq 25$

Determine the feasible set for each of the following systems of inequalities.

27. $\begin{cases} x \geq 0 \\ y \geq 0 \\ x + y \leq 8 \end{cases}$ 28. $\begin{cases} x \geq 1 \\ y < 4 \\ x + y > 8 \end{cases}$

29. $\begin{cases} x^2 + y^2 \leq 36 \\ -2 \leq x \leq 2 \end{cases}$ 30. $\begin{cases} 9x^2 - y^2 \geq 0 \\ x^2 + y^2 \leq 25 \end{cases}$

Sketch the graph of the following systems of linear inequalities.

31. $\begin{cases} 4x + 7y \leq 28 \\ 2x - 3y \geq -6 \\ y \geq -2 \end{cases}$ 32. $\begin{cases} 3x + 9y \leq 18 \\ 6x - 4y \geq 8 \\ 3y \geq 0 \end{cases}$

33. $\begin{cases} x + y \leq 10 \\ x - 3y \geq -18 \\ x \geq 0 \\ y \geq 0 \end{cases}$ 34. $\begin{cases} 2x + 3y \geq -120 \\ 6x - 3y \geq -150 \\ x \leq 0 \\ y \leq 0 \end{cases}$

Determine the maximum and minimum values of the functions and the values of x and y where they occur, given a feasible set with vertices as follows:

35. $f(x, y) = 2x + 3y; (3, 1), (5, 2), (1, 4), (8, 1)$

36. $f(x, y) = 4x + y; (2, 1), (1, 9), (3, 5), (2, 7)$

37. $Q = 2x + 3y$ subject to

$$\begin{cases} x \geq 0 \\ y \geq 0 \\ x + 2y \leq 7 \\ 5x - 8y \leq -3 \end{cases}$$

38. $F = \dfrac{3}{2}x + y$ subject to

$$\begin{cases} x \leq 6 \\ y \geq 0 \\ x + 2y \geq 4 \\ x - 2y \geq 2 \end{cases}$$

Find the maximum and the minimum values of the given function in the indicated region.

39. $f(x, y) = x + 4y$ 40. $f(x, y) = 3x + y$

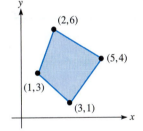
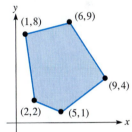

♦ Applications

For each of the following linear programming problems, (a) determine a system of inequalities describing the given problem, (b) determine the feasible set of the systems of inequalities, and (c) solve the problem.

41. A cabinet company makes two types of cabinet drawers, one plain and one fancy. Each plain drawer takes 2 hours of work to assemble and 1 hour of sanding. Each fancy drawer takes 1 hour of work to assemble and 4 hours of sanding. The four assembly workers and six sanding workers are each used 12 hours per day. Each plain drawer nets a $3 profit, and each fancy drawer nets a $5 profit. If the company can sell all the drawers it makes, how many of each kind should be produced each day in order to maximize profit?

42. Seiberling Tire Company of Akron, Ohio, has 1000 units of raw rubber to use in producing radial tires for cars and for tractor tires. Each car tire requires 5 units of rubber, and each tractor tire requires 20 units of rubber. Labor costs are $8 for a car tire and $12 for a tractor tire. If Seiberling does not want to pay more than $1500 in labor costs and wishes to make a profit of $10 per car tire and $25 per tractor tire, how many of each type of tire should be made in order to maximize profits?

APPLICATIONS OF
LINEAR SYSTEMS

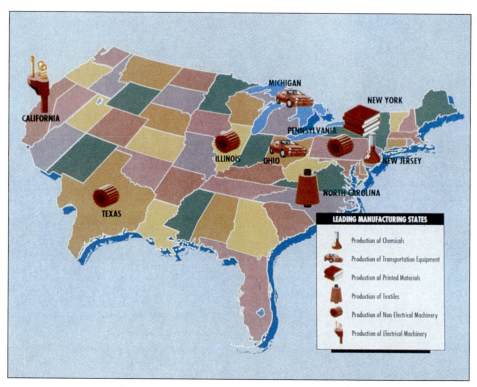

*The various segments of an economy are interrelated in the sense that they use the outputs of other segments. These interrelationships may be described using **input-output analysis**, a mathematical technique using matrices. Using input-output analysis, the effect of a change in one segment of the economy on the other segments may be calculated.*

I n the preceding chapter, we developed the mathematics of systems of linear equations and inequalities. In this chapter, we apply these subjects to the theory of matrices, determinants, and partial fractions.

7.1 PARTIAL FRACTIONS

We have learned to add rational expressions in a single variable. We found that the sum of rational expressions is another rational expression. Moreover, the denominator of the sum is the product of the denominators of the summands (at least before any reduction is carried out). For example, we have

$$\frac{1}{x - 2} + \frac{2}{x - 1} = \frac{3x - 5}{(x - 2)(x - 1)}$$

In many contexts (most notably in performing integration in calculus), an expression such as the one on the left is far easier to work with than one like the expression on the right. The method of **partial fractions** is a procedure for writing rational expressions (such as the one on the right) in terms of simple rational expressions (such as the ones on the left).

Suppose that you are given a rational expression $f(x)/g(x)$, where $f(x)$ and $g(x)$ are polynomials with real coefficients. By the division algorithm for polynomials, we can divide $f(x)$ by $g(x)$ to obtain

$$\frac{f(x)}{g(x)} = q(x) + \frac{h(x)}{g(x)}$$

where $q(x)$ is the quotient and $h(x)$ is the remainder and has degree less than the degree of $g(x)$. We then write the rational expression $h(x)/g(x)$ as a sum of simple expressions as follows.

As we discussed in Chapter 4, the polynomial $g(x)$ may be factored into a product of linear and irreducible quadratic factors, all with real coefficients. A typical linear factor is of the form $(ax + b)^n$. Corresponding to each such factor, the expression for $h(x)/g(x)$ has a sum of simple expressions of the form

$$\frac{A_1}{ax + b} + \frac{A_2}{(ax + b)^2} + \cdots + \frac{A_n}{(ax + b)^n}$$

for suitable real numbers $A_1, A_2, \ldots, A_n$. A typical quadratic factor is of the form $(cx^2 + dx + e)^n$. Corresponding to each such factor, the expression for $h(x)/g(x)$ has a sum of simple expressions of the form

$$\frac{A_1 + B_1 x}{cx^2 + dx + e} + \frac{A_2 + B_2 x}{(cx^2 + dx + e)^2} + \cdots + \frac{A_n + B_n x}{(cx^2 + dx + e)^n}$$

where $A_1, B_1, A_2, B_2, \ldots, A_n, B_n$ are suitable real numbers.

The expression of $f(x)/g(x)$ as the sum of polynomials and rational expressions of the preceding form is called a **partial fraction expansion**. We may summarize the discussion as follows:

DETERMINING THE PARTIAL FRACTION EXPANSION OF $f(x)/g(x)$

1. Use long division to write $f(x)/g(x)$ in the form

$$\frac{f(x)}{g(x)} = q(x) + \frac{h(x)}{g(x)}, \quad \deg(h) < \deg(g)$$

2. Factor $g(x)$ into a product of distinct real factors, some linear and some quadratic.
3. For a linear factor $(ax + b)^n$, the partial fraction expansion has a component

$$\frac{A_1}{ax + b} + \frac{A_2}{(ax + b)^2} + \cdots + \frac{A_n}{(ax + b)^n}$$

4. For an irreducible quadratic factor $(cx^2 + dx + e)^n$, the partial fraction expansion has a component

$$\frac{A_1 + B_1 x}{cx^2 + dx + e} + \frac{A_2 + B_2 x}{(cx^2 + dx + e)^2} + \cdots + \frac{A_n + B_n x}{(cx^2 + dx + e)^n}$$

5. The partial fraction expansion is the sum of $q(x)$, from step 1, and each of the components from steps 3 and 4.

The principal difficulty in determining the partial fraction expansion of a given rational expression is in determining the values of the various coefficients. The following example shows how this may be done.

➤ **EXAMPLE 1**
Simple Partial Fraction Expansion

Determine the partial fraction expansion of the rational expression

$$\frac{3x - 5}{(x - 2)(x - 1)}$$

Solution

Since the degree of the numerator is less than the degree of the denominator, it is unnecessary to carry out the division indicated in the previous discussion. (The result would indicate that $q(x)$ is the zero polynomial and $h(x)$ equals $f(x)$.) The denominator is already factored. According to the previous discussion, there is a component corresponding to each factor in the denominator. The component corresponding to the factor $x - 1$ is of the form

$$\frac{A}{x - 1}$$

The component corresponding to the factor $x - 2$ is of the form

$$\frac{B}{x - 2}$$

That is, we must have

$$\frac{3x - 5}{(x - 2)(x - 1)} = \frac{A}{x - 1} + \frac{B}{x - 2}$$

Multiply both sides by $(x - 2)(x - 1)$ to obtain

$$3x - 5 = A(x - 2) + B(x - 1)$$

This last equation must hold for all values of x. Let's choose two particular values for x to use in determining the values of A and B. The most convenient values are, respectively, $x = 1$ and $x = 2$. Substituting these values into the last equation, we see that when $x = 1$,

$$3(1) - 5 = A(1 - 2) + B(0)$$
$$-A = -2$$
$$A = 2$$

When $x = 2$,

$$3(2) - 5 = A(0) + B(2 - 1)$$
$$B = 1$$

Thus, the partial fraction expansion of the given rational expression is

$$\frac{3x - 5}{(x - 2)(x - 1)} = \frac{2}{x - 1} + \frac{1}{x - 2}$$

The calculation just performed reversed the addition of rational functions we performed at the beginning of the section.

➤ **EXAMPLE 2**

Partial Fractions
with Three Factors

Determine the partial fraction expansion of the following rational expression:

$$\frac{2x^4 - 1}{x^3 - x}$$

Solution

Begin by dividing the numerator by the denominator. The quotient is $q(x) = 2x$, and the remainder is $r(x) = 2x^2 - 1$. That is, we have

$$\frac{2x^4 - 1}{x^3 - x} = 2x + \frac{2x^2 - 1}{x^3 - x}$$

Next, factor the denominator:

$$x^3 - x = x(x^2 - 1)$$
$$= x(x - 1)(x + 1)$$

The partial fraction expansion has the form

$$\frac{2x^2 - 1}{x^3 - x} = \frac{A}{x} + \frac{B}{x - 1} + \frac{C}{x + 1}$$

Multiply both sides by $x(x - 1)(x + 1)$ to obtain the identity

$$2x^2 - 1 = A(x - 1)(x + 1) + Bx(x + 1) + Cx(x - 1)$$

To obtain the values of A, B, C, we substitute three different values for x into the above identity. The three most convenient values are $x = 0, x = 1,$ and $x = -1$, since these values result in various terms being equal to 0. When $x = 0$, we have

$$2(0)^2 - 1 = A(0 - 1)(0 + 1) + B \cdot 0(0 + 1) + C \cdot 0(0 - 1)$$
$$-1 = -A$$
$$A = 1$$

When $x = 1$, we have

$$2(1)^2 - 1 = A(1-1)(1+1) + B \cdot 1(1+1) + C \cdot 1(1-1)$$

$$1 = 2B$$

$$B = \frac{1}{2}$$

When $x = -1$, we have

$$2(-1)^2 - 1 = A(-1-1)(-1+1) + B \cdot (-1)(-1+1) + C \cdot (-1)(-1-1)$$

$$1 = 2C$$

$$C = \frac{1}{2}$$

Thus, the partial fraction expansion is given by

$$\frac{2x^4 - 1}{x^3 - x} = 2x + \frac{1}{x} + \frac{1/2}{x-1} + \frac{1/2}{x+1}$$

$$= 2x + \frac{1}{x} + \frac{1}{2(x-1)} + \frac{1}{2(x+1)}$$

In the preceding two examples, we were able to determine the unknown co-efficients in the partial fraction expansion one at a time. However, in some instances, it is necessary to determine the coefficients by solving a system of linear equations. This phenomenon is illustrated in the next example.

➤ **EXAMPLE 3**
Partial Fractions
with a Multiple Factor

Determine the partial fraction expansion of the following rational expression:

$$\frac{3x^3 - 5x + 1}{x^4 + x^2}$$

Solution

The degree of the numerator is less than the degree of the denominator, so the initial division step is unnecessary. Factoring the denominator gives us

$$x^4 + x^2 = x^2(x^2 + 1)$$

Note that the quadratic factor

$$x^2 + 1$$

cannot be factored further since the factorization is assumed to be factored into polynomials with real coefficients. The general form of the partial fraction expansion is then

$$\frac{3x^3 - 5x + 1}{x^4 + x^2} = \frac{A}{x} + \frac{B}{x^2} + \frac{Cx + D}{x^2 + 1}$$

Multiplying both sides by $x^4 + x^2$ yields the identity

$$3x^3 - 5x + 1 = Ax(x^2 + 1) + B(x^2 + 1) + (Cx + D)x^2$$

Substituting $x = 0$ yields the value of B

$$1 = 0 + B(0^2 + 1) + 0$$

$$B = 1$$

Inserting this value back into the identity gives us

$$3x^3 - 5x + 1 = Ax(x^2 + 1) + (x^2 + 1) + (Cx + D)x^2$$
$$= (Ax^3 + Ax) + (x^2 + 1) + (Cx^3 + Dx^2)$$
$$= (A + C)x^3 + (D + 1)x^2 + Ax + 1$$

We now have an identity between two polynomials in x. For them to be equal, the coefficients of their respective powers of x must be equal. That is, we have the system of equations

$$A + C = 3 \qquad \text{Coefficient of } x^3$$
$$D + 1 = 0 \qquad \text{Coefficient of } x^2$$
$$A = -5 \qquad \text{Coefficient of } x$$

Solving this system gives us

$$D = -1$$
$$A = -5$$
$$-5 + C = 3$$
$$C = 8$$

Therefore, the desired partial fraction decomposition is

$$\frac{3x^3 - 5x + 1}{x^4 + x^2} = \frac{-5}{x} + \frac{1}{x^2} + \frac{8x - 1}{x^2 + 1}$$

Note that in our solution we compared like coefficients on both sides of the identity rather than substituting assorted values of x to arrive at a system of equations. Both approaches are valid. For instance, we could just as well have arrived at a (different) system by substituting for x the values $x = 0, 1, 2, 3$. Why not carry out the calculations and verify that the resulting system leads to the same partial fraction expansion?

Exercises 7.1

Determine the partial fractional expansion of the given rational expressions.

1. $\dfrac{6x - 2}{(x + 1)(x - 1)}$

2. $\dfrac{4x + 1}{(x - 2)(x + 1)}$

3. $\dfrac{4x + 2}{x(x + 1)(x + 2)}$

4. $\dfrac{2x^2 - 6x - 2}{x(x + 2)(x - 1)}$

5. $\dfrac{7x - 10}{(2x - 1)(x - 2)}$

6. $\dfrac{2x + 8}{(x - 3)(3x - 2)}$

7. $\dfrac{x^2 - 17x + 35}{(x^2 + 1)(x - 4)}$

8. $\dfrac{7x^2 - 29x + 24}{(2x - 1)(x - 2)^2}$

9. $\dfrac{13x + 5}{3x^2 - 7x - 6}$

10. $\dfrac{4x - 13}{2x^2 + x - 6}$

11. $\dfrac{3x^3 + 4x^2 - x}{x^2 + x - 2}$

12. $\dfrac{2x^3 - x^2 - 13x + 6}{x^2 - x - 6}$

13. $\dfrac{3x^3 - 8x^2 + 9x - 6}{x^2 - 3x + 2}$

14. $\dfrac{4x^3 - 16x^2 + 9x - 15}{2x^2 - 9x + 4}$

15. $\dfrac{5x + 5}{x^2 - x - 6}$

16. $\dfrac{11x + 2}{2x^2 + x - 1}$

17. $\dfrac{-2x^2 + 7x + 2}{x^3 - 2x^2 + x}$

18. $\dfrac{-3x^2 - 10x - 4}{x(x + 1)^2}$

19. $\dfrac{-x^2 + 26x + 6}{(2x - 1)(x + 2)^2}$

20. $\dfrac{5x^2 + 9x - 56}{(x - 4)(x - 2)(x + 1)}$

21. $\dfrac{6x^3 + 29x^2 - 6x - 5}{2x^2 + 9x - 5}$

22. $\dfrac{-9x^2 + 7x - 4}{x^3 - 3x^2 - 4x}$

23. $\dfrac{x^3 + x^2 + 2}{(x^2 + 2)^2}$

24. $\dfrac{2x^3 + 8x^2 + 2x + 4}{(x + 1)^2(x^2 + 3)}$

25. $\dfrac{6x^3 + 5x^2 - 7}{3x^2 - 2x - 1}$

26. $\dfrac{3x^2 - 11x - 26}{(x^2 - 4)(x + 1)}$

27. $\dfrac{3x^2 + 10x + 9}{(x + 2)^3}$

28. $\dfrac{x^2 - x - 4}{(x - 2)^3}$

29. $\dfrac{5}{(x + 2)^2 (x + 1)}$

30. $\dfrac{-8x + 23}{2x^2 + 5x - 12}$

Decompose into partial fractions.

31. $\dfrac{x}{x^4 - a^4}$

32. $\dfrac{2x^2 + 2ax + 3x + 2a^2 - 3a}{x^3 - a^3}$

Decompose into partial fractions and graph by addition of ordinates.

33. $y = \dfrac{x - 11}{x^2 - 1}$

34. $y = \dfrac{5x - 1}{x^2 - x - 2}$

35. $y = \dfrac{2x^2 + 4x - 1}{x^2 + x}$

36. $y = \dfrac{9}{x^2 + x - 2}$

Find constants x, y, and z so that the following statements are true.

37. $\dfrac{2a + 1}{(a + 1)(a^2 - 3a + 5)} = \dfrac{x}{(a + 1)} + \dfrac{ya + z}{a^2 - 3a + 5}$

38. $\dfrac{a + 2}{a^3 + 2a^2 + a} = \dfrac{x}{a} + \dfrac{ya + z}{a^2 + 2a + 1}$

7.2 MATRICES

In Chapter 9, we introduced the notion of a matrix and showed how elementary row operations on a matrix can be used to solve systems of linear equations. Actually, matrices are an important mathematical notion in their own right, and their applications extend throughout pure and applied mathematics. Matrices are used by physicists extensively, for instance, in describing mechanics problems and in representing elementary particles. Economists use matrices to describe the interactions of various sectors of an economy. Matrices are even used by computer scientists in describing computer graphics calculations. Mathematicians have developed an extensive algebraic theory of matrices that parallels (and draws its motivation from) the elementary algebraic notions for real numbers. In this section, we will provide an introduction to the algebra of matrices.

Recall that a matrix is a rectangular array of real numbers. Matrices are usually denoted with capital letters, such as A, B, C, X, Y, Z. The entries of a matrix are typically denoted using the lowercase version of the matrix name, with subscripts to indicate position. For instance, the entries of the matrix A are denoted a_{ij}, where i denotes the row number and j denotes the column number. We write

$$A = \left[a_{ij} \right]$$

to indicate that a typical entry of A is a_{ij}.

Suppose that $A = \left[a_{ij} \right]$ and $B = \left[b_{ij} \right]$ are matrices. We say that the two matrices are equal, denoted $A = B$, provided that the two matrices are of the same size and their corresponding entries are equal, that is, $a_{ij} = b_{ij}$ for all possible i and j.

A matrix whose entries are all 0 is called a **zero matrix** and is denoted **0**. Actually, there is a zero matrix corresponding to each possible matrix size. It is customary to determine the size of a zero matrix from context.

Definition 1
Sum of Two Matrices,
Product of a Matrix
and a Real Number

Suppose that A and B are matrices of the same size. The **sum of A and B**, denoted $A + B$, is the matrix obtained by adding corresponding entries of A and B. Suppose that k is a real number. Then the product kA is the matrix whose entries are the entries of A each multiplied by k.

> **EXAMPLE 1**
Matrix Arithmetic

Let

$$A = \begin{bmatrix} 5 & 0 \\ 3 & -1 \end{bmatrix}, \qquad B = \begin{bmatrix} -2 & 4 \\ 1 & 3 \end{bmatrix}$$

Calculate the following:

    1. $A + B$ 2. $(-2)A$ 3. $3A + 4B$

Solution

1. $A + B = \begin{bmatrix} 5 & 0 \\ 3 & -1 \end{bmatrix} + \begin{bmatrix} -2 & 4 \\ 1 & 3 \end{bmatrix}$

$$= \begin{bmatrix} 5-2 & 0+4 \\ 3+1 & -1+3 \end{bmatrix}$$

$$= \begin{bmatrix} 3 & 4 \\ 4 & 2 \end{bmatrix}$$

2. $(-2)A = (-2)\begin{bmatrix} 5 & 0 \\ 3 & -1 \end{bmatrix}$

$$= \begin{bmatrix} (-2)\cdot 5 & (-2)\cdot 0 \\ (-2)\cdot 3 & (-2)\cdot(-1) \end{bmatrix}$$

$$= \begin{bmatrix} -10 & 0 \\ -6 & 2 \end{bmatrix}$$

3. $3A + 4B = 3\begin{bmatrix} 5 & 0 \\ 3 & -1 \end{bmatrix} + 4\begin{bmatrix} -2 & 4 \\ 1 & 3 \end{bmatrix}$

$$= \begin{bmatrix} 3\cdot 5 + 4\cdot(-2) & 3\cdot 0 + 4\cdot 4 \\ 3\cdot 3 + 4\cdot 1 & 3\cdot(-1) + 4\cdot 3 \end{bmatrix}$$

$$= \begin{bmatrix} 7 & 16 \\ 13 & 9 \end{bmatrix}$$

Addition of matrices obeys both the commutative and associative laws. That is, we know the following formulas hold for matrices of the same size:

$$A + B = B + A \qquad \textit{Commutative law}$$
$$A + (B + C) = (A + B) + C \qquad \textit{Associative law}$$

Moreover, the zero matrix (of appropriate size) acts as the identity for the operation of addition. That is, we have

$$A + 0 = 0 + A = A \qquad \textit{Identity}$$

The **additive inverse of the matrix** A, denoted $-A$, is the matrix whose entries are the negatives of the corresponding entries of A. Then the additive inverse has the property

$$A + (-A) = (-A) + A = 0$$

The difference $A - B$ of two matrices is defined via the formula

$$A - B = A + (-B)$$

All of these formulas should remind you of the corresponding properties of real numbers with respect to the operations of addition and subtraction. In fact, from the previously mentioned definitions, we can deduce analogues of many of the elementary algebraic properties of real numbers. The exercises contain a number of these for you to deduce.

A matrix consisting of a single row is called a **row matrix**, and a matrix consisting of a single column is called a **column matrix.** Suppose that we are given a row matrix and a column matrix of the same size. The product of the row times the column is the 1×1 matrix formed by multiplying corresponding entries of the two matrices and adding. That is, we have

$$[a_1 \quad a_2 \quad \cdots \quad a_n] \begin{bmatrix} b_1 \\ b_2 \\ \vdots \\ b_n \end{bmatrix} = [a_1 b_1 + a_2 b_2 + \cdots + a_n b_n]$$

Thus, for example, we have

$$[2 \quad 1 \quad -3] \begin{bmatrix} 5 \\ 2 \\ -4 \end{bmatrix} = [2 \cdot 5 + 1 \cdot 2 + (-3) \cdot (-4)]$$

$$= [24]$$

Using this way of multiplying a row matrix by a column matrix, we may define multiplication of matrices as follows:

Definition 2 **Matrix Multiplication**	Suppose that A is an $m \times n$ matrix and B is an $n \times k$ matrix. Then the product AB is the $m \times k$ matrix with entries as follows: The entry in the ith row and jth column of AB is obtained by multiplying the ith row of A by the jth column of B.

Suppose that the ith row of A is equal to

$$[a_{i1} \quad a_{i2} \quad \ldots \quad a_{in}]$$

and that the jth column of B is equal to

$$\begin{bmatrix} b_{1j} \\ b_{2j} \\ \vdots \\ b_{nj} \end{bmatrix}$$

as shown by the following product:

$$\begin{bmatrix} a_{11} & a_{12} & \cdots & a_{1n} \\ \vdots & \vdots & \vdots & \vdots \\ a_{i1} & a_{i2} & \cdots & a_{in} \\ \vdots & \vdots & \vdots & \vdots \\ a_{m1} & a_{m2} & \cdots & a_{mn} \end{bmatrix} \begin{bmatrix} b_{11} & \cdots & b_{1j} & \cdots & b_{1k} \\ b_{21} & \cdots & b_{2j} & \cdots & b_{2k} \\ \vdots & \vdots & \vdots & \vdots & \vdots \\ b_{n1} & \cdots & b_{nj} & \cdots & b_{nk} \end{bmatrix}$$

Then the entry c_{ij} in the ith row and jth column of the product is given by

$$c_{ij} = a_{i1} b_{1j} + a_{i2} b_{2j} + \cdots + a_{in} b_{nj}$$

The next example shows how this calculation works in practice.

➤ **EXAMPLE 2**
Product of 2×2 Matrices

Compute the following matrix product:

$$\begin{bmatrix} 5 & -1 \\ 2 & 0 \end{bmatrix} \begin{bmatrix} 3 & 1 \\ 0 & 9 \end{bmatrix}$$

Solution

The number of columns of the left matrix equals the number of rows of the right matrix, so the product is defined. Moreover, the number of rows in the product equals the number of rows (2) in the left matrix, and the number of columns in the product equals the number of columns (2) in the right matrix. That is, the product is a 2×2 matrix. Let's now work out the entries in the product. The entry in row 1, column 1 is obtained by multiplying the first row on the left by the first column on the right. The product is

$$5 \cdot 3 + (-1) \cdot 0 = 15$$

This gives us

$$\begin{bmatrix} 5 & -1 \\ 2 & 0 \end{bmatrix} \begin{bmatrix} 3 & 1 \\ 0 & 9 \end{bmatrix} = \begin{bmatrix} 15 & * \\ * & * \end{bmatrix}$$

The entry in row 1, column 2 of the product is obtained as the product of the first row on the left times the second column on the right. This product is

$$5 \cdot 1 + (-1) \cdot 9 = -4$$

That is,

$$\begin{bmatrix} 5 & -1 \\ 2 & 0 \end{bmatrix} \begin{bmatrix} 3 & 1 \\ 0 & 9 \end{bmatrix} = \begin{bmatrix} 15 & -4 \\ * & * \end{bmatrix}$$

Next, we compute the entry in row 2, column 1 as the product of the second row times the first column:

$$\begin{bmatrix} 5 & -1 \\ 2 & 0 \end{bmatrix} \begin{bmatrix} 3 & 1 \\ 0 & 9 \end{bmatrix} = \begin{bmatrix} 15 & -4 \\ 6 & * \end{bmatrix}$$

Finally, we compute the entry in row 2, column 2 as the product of the second row times the second column:

$$\begin{bmatrix} 5 & -1 \\ 2 & 0 \end{bmatrix} \begin{bmatrix} 3 & 1 \\ 0 & 9 \end{bmatrix} = \begin{bmatrix} 15 & -4 \\ 6 & 2 \end{bmatrix}$$

Note that the definition of matrix multiplication is based on the fact that we can form the product of rows of the first matrix times columns of the second matrix. In order for this to happen, the number of columns of the first matrix must equal the number of rows of the second. If this is not the case, the product is not defined.

➤ **EXAMPLE 3**
Which Products are Defined?

Determine which of the following products are defined and the size of those products.

1. A is 2×5; B is 5×7.
2. A is 3×4; B is 2×4.
3. A is 4×4; B is 4×3.

Solution

1. The product AB is defined since the inner dimensions of A and B match. That is, the number of columns of A (5) equals the number of rows of B (5). The size of the product is obtained from the outer dimensions and is 2×7.

2. The product is undefined since the number of columns of A (4) is unequal to the number of rows of B (2).

3. The product AB is defined and is of size 4×3.

Matrix multiplication obeys the distributive and associative laws. That is, the following formulas hold provided the products are defined:

$$A(BC) = (AB)C$$
$$A(B + C) = AB + BC$$

We will omit the proofs of these facts.

Unlike multiplication of real numbers, matrix multiplication does not necessarily obey the commutative law, as the following example illustrates.

➤ **EXAMPLE 4**

Matrix Multiplication Is Noncommutative

Suppose that

$$A = \begin{bmatrix} 5 & 1 \\ 2 & 3 \end{bmatrix}, \qquad B = \begin{bmatrix} 1 & 4 \\ 3 & 0 \end{bmatrix}$$

Show that $AB \neq BA$.

Solution

We may determine the values of AB and BA by direct calculation:

$$AB = \begin{bmatrix} 5 & 1 \\ 2 & 3 \end{bmatrix} \cdot \begin{bmatrix} 1 & 4 \\ 3 & 0 \end{bmatrix}$$
$$= \begin{bmatrix} 8 & 20 \\ 11 & 8 \end{bmatrix}$$
$$BA = \begin{bmatrix} 1 & 4 \\ 3 & 0 \end{bmatrix} \cdot \begin{bmatrix} 5 & 1 \\ 2 & 3 \end{bmatrix}$$
$$= \begin{bmatrix} 13 & 13 \\ 15 & 3 \end{bmatrix}$$

Two matrices are equal only when their corresponding entries are the same. By examining the entries here, they are certainly not the same, so $AB \neq BA$.

For each positive integer n, let I_n denote the $n \times n$ matrix with 1s down the main diagonal and 0s everywhere else. For instance,

$$I_1 = [1] = 1$$
$$I_2 = \begin{bmatrix} 1 & 0 \\ 0 & 1 \end{bmatrix}$$
$$I_3 = \begin{bmatrix} 1 & 0 & 0 \\ 0 & 1 & 0 \\ 0 & 0 & 1 \end{bmatrix}$$

I_n is called the **identity matrix of size n**. Indeed, I_n acts as the identity for multiplication with square matrices of size $n \times n$. That is, if A is an $n \times n$ matrix, then one can prove that

$$I_n A = A I_n = A \qquad (1)$$

In case $n = 2$, this statement follows from the equations:

$$I_2 \cdot \begin{bmatrix} a & b \\ c & d \end{bmatrix} = \begin{bmatrix} 1 & 0 \\ 0 & 1 \end{bmatrix}\begin{bmatrix} a & b \\ c & d \end{bmatrix}$$

$$= \begin{bmatrix} 1 \cdot a + 0 \cdot c & 1 \cdot b + 0 \cdot d \\ 0 \cdot a + 1 \cdot c & 0 \cdot b + 1 \cdot d \end{bmatrix}$$

$$= \begin{bmatrix} a & b \\ c & d \end{bmatrix}$$

Similarly,

$$\begin{bmatrix} a & b \\ c & d \end{bmatrix} \cdot I_2 = \begin{bmatrix} a & b \\ c & d \end{bmatrix}$$

The proof of (1) for general n will be omitted.

Exercises 7.2

Given

$$A = \begin{bmatrix} 2 & 0 \\ 1 & 3 \end{bmatrix}, \qquad B = \begin{bmatrix} -1 & 3 \\ 0 & -2 \end{bmatrix}, \qquad C = \begin{bmatrix} 4 & 2 \\ -1 & 0 \end{bmatrix},$$

$$D = \begin{bmatrix} 1 & 0 \\ 0 & 1 \end{bmatrix}, \qquad E = \begin{bmatrix} 2 & -1 \\ 0 & 0 \end{bmatrix}, \qquad F = \begin{bmatrix} 1 & 3 \\ 2 & 6 \end{bmatrix},$$

calculate the following.

1. $A + B$
2. $A + E$
3. $2F$
4. $3C$
5. $B - C$
6. $C - F$
7. $2A + 3D$
8. $4F - 2B$
9. $B + D$
10. $E + D$
11. $F - 2A$
12. $E + 2B$
13. $-5B$
14. $-3E$
15. AB
16. EF
17. CD
18. BC
19. $(AB)C$
20. $A(BF)$

Given

$$A = \begin{bmatrix} 1 & 2 \\ 3 & 4 \end{bmatrix}, \quad B = \begin{bmatrix} -1 \\ 7 \end{bmatrix}, \quad C = \begin{bmatrix} 3 & -4 & 1 \\ 5 & 0 & 2 \end{bmatrix},$$

$$D = \begin{bmatrix} -2 & -3 & -4 \\ 2 & -1 & 0 \\ 4 & -2 & 3 \end{bmatrix}, \quad E = \begin{bmatrix} -1 & 2 \\ 3 & -2 \end{bmatrix}, \quad F = \begin{bmatrix} 5 \\ -2 \end{bmatrix},$$

$$G = \begin{bmatrix} -1 & 2 & 3 \\ 0 & 1 & 0 \end{bmatrix}, \quad H = \begin{bmatrix} 0 & 1 & 4 \\ 1 & 2 & -1 \\ 3 & 2 & -2 \end{bmatrix},$$

$$I = \begin{bmatrix} 4 & 0 & -1 \end{bmatrix}$$

find each of the following matrices, whenever possible.

21. AB
22. DH
23. CD
24. EF
25. FG
26. GH
27. CH
28. BI
29. DG
30. EA
31. IH
32. IF
33. FE
34. AF

Determine which of the following products are defined and the size of those products.

35. A is 3×2; B is 2×5.
36. A is 4×5; B is 5×2.
37. A is 2×2; B is 1×2.
38. A is 5×7; B is 5×7.
39. A is 1×4; B is 4×3.
40. A is 5×1; B is 1×3.

Find the values of w, x, y, and z in the following matrix equations.

41. $\begin{bmatrix} 3 & 1 \\ 4 & 5 \end{bmatrix} = \begin{bmatrix} x & y \\ z & 5 \end{bmatrix}$

42. $\begin{bmatrix} 3 & 5 & x \\ 2 & y & 3 \end{bmatrix} = \begin{bmatrix} z & 5 & 2 \\ 2 & 7 & w \end{bmatrix}$

43. $\begin{bmatrix} x - 7 & 4y & 8z \\ 6w & 2 & 5 \end{bmatrix} + \begin{bmatrix} -9 & 8y & 3 \\ 2 & 5 & 4 \end{bmatrix} = \begin{bmatrix} 2 & 36 & 27 \\ 20 & 7 & 9 \end{bmatrix}$

44. $\begin{bmatrix} x + 2 & 3y + 1 & 5w \\ 4z & 0 & 18 \end{bmatrix}$
$+ \begin{bmatrix} 3x & 2y & 5w \\ 2z & 7 & -6 \end{bmatrix} = \begin{bmatrix} 10 & -14 & 80 \\ 10 & 7 & 12 \end{bmatrix}$

Prove the following for

$$A = \begin{bmatrix} a_{11} & a_{12} \\ a_{21} & a_{22} \end{bmatrix}, \quad B = \begin{bmatrix} b_{11} & b_{12} \\ b_{21} & b_{22} \end{bmatrix},$$

$$C = \begin{bmatrix} c_{11} & c_{12} \\ c_{21} & c_{22} \end{bmatrix}, \quad I = \begin{bmatrix} 1 & 0 \\ 0 & 1 \end{bmatrix}$$

45. $IA = AI = A$
46. $A + B = B + A$
47. $k(A + B) = kA + kB$
48. $(A + B) + C = A + (B + C)$

7.3 DETERMINANTS

Let A be a square matrix. Associated to A is a real number, its determinant. The determinant plays a significant role in matrix theory, as we will see in the next section, where we discuss Cramer's rule for calculating the inverse of a matrix. Note that the concept of a determinant is defined only for square matrices. Throughout this section, all matrices will be assumed to be square.

The determinant of a square matrix A is denoted $|A|$ and is a real number. The concept of a determinant is defined for a matrix of a given size in terms of determinants of matrices of smaller sizes.

For a 1×1 matrix, the determinant is defined as follows: Suppose that $A = [a]$ is a 1×1 matrix. Then $|A|$ is defined as a.

Definition 1
Determinant of a 2 × 2 Matrix

Suppose that A is the 2×2 matrix

$$A = \begin{bmatrix} a & b \\ c & d \end{bmatrix}$$

Then $|A|$ is defined by the formula:

$$|A| = ad - bc$$

That is, $|A|$ is computed by forming the product of the elements along the diagonal from top left to bottom right, minus the product of the elements along the diagonal from bottom left to top right. (See Figure 1.)

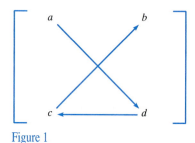

Figure 1

➤ **EXAMPLE 1**
Determinant of a 2×2 Matrix

Calculate $|A|$, where A is the matrix

$$A = \begin{bmatrix} 2 & -3 \\ 4 & 1 \end{bmatrix}$$

Solution

From the definition just given, the value of the determinant equals

$$|A| = 2 \cdot 1 - 4 \cdot (-3)$$
$$= 14$$

Definition 2
Minor of a 3×3 Matrix

By eliminating a row and a column of the matrix, we can arrive at a 2×2 matrix. The determinant of such a matrix is called a **minor** of A corresponding to the eliminated row and column.

For instance, suppose that

$$A = \begin{bmatrix} -1 & 0 & 3 \\ 2 & 1 & 0 \\ 3 & 4 & -1 \end{bmatrix}$$

The minor obtained by eliminating the second row and third column is denoted M_{23}, given by

$$M_{23} = \begin{vmatrix} -1 & 0 \\ 3 & 4 \end{vmatrix} = (-1) \cdot 4 - 3 \cdot 0 = -4$$

The subscript in the notation M_{23} indicates that the minor was formed from A by eliminating the second row and the third column. Note that the row comes first, then the column. More generally, we denote by M_{ij} the minor obtained by eliminating the ith row and jth column.

Definition 3	Suppose that A is a 3×3 matrix. Then the **cofactor** A_{ij} is defined through the
Cofactor of a Minor M_{ij}	formula

$$A_{ij} = (-1)^{i+j} M_{ij}$$

For instance, let A be the matrix defined above, then the cofactor A_{23} is equal to

$$A_{23} = (-1)^{2+3} M_{23}$$
$$= -(-4)$$
$$= 4$$

The determinant of a 3×3 matrix may be defined in terms of cofactors as follows:

Definition 4	Let A be a 3×3 matrix. Then its determinant $|A|$ is defined by the formula
Determinant of a 3×3 Matrix	

$$|A| = \begin{vmatrix} a_{11} & a_{12} & a_{13} \\ a_{21} & a_{22} & a_{23} \\ a_{31} & a_{32} & a_{33} \end{vmatrix} = a_{11}A_{11} + a_{12}A_{12} + a_{13}A_{13}$$

That is, to calculate the determinant of a 3×3 matrix, form the sum of products of each element of the first row times its corresponding cofactor, that is, the cofactor formed by eliminating the row and column containing the element.

We may derive an alternative formula for the determinant of a 3×3 matrix by using the definitions of the cofactors. We have

$$A_{11} = (-1)^{1+1} M_{11}$$
$$= \begin{vmatrix} a_{22} & a_{23} \\ a_{32} & a_{33} \end{vmatrix}$$
$$= a_{22}a_{33} - a_{23}a_{32}$$
$$A_{12} = (-1)^{1+2} M_{12}$$
$$= (-1) \begin{vmatrix} a_{21} & a_{23} \\ a_{31} & a_{33} \end{vmatrix}$$
$$= -(a_{21}a_{33} - a_{31}a_{23})$$

$$A_{13} = (-1)^{1+3} M_{13}$$

$$= \begin{vmatrix} a_{21} & a_{22} \\ a_{31} & a_{32} \end{vmatrix}$$

$$= a_{21} a_{32} - a_{31} a_{22}$$

Therefore, by the definition of a determinant for a 3×3 matrix, this gives us

$$|A| = a_{11} A_{11} + a_{12} A_{12} + a_{13} A_{13}$$
$$= a_{11} (a_{22} a_{33} - a_{32} a_{23}) - a_{12} (a_{21} a_{33} - a_{31} a_{23})$$
$$\quad + a_{13} (a_{21} a_{32} - a_{31} a_{22})$$
$$= a_{11} a_{22} a_{33} - a_{11} a_{32} a_{23} - a_{12} a_{21} a_{33} + a_{12} a_{31} a_{23}$$
$$\quad + a_{13} a_{21} a_{32} - a_{13} a_{31} a_{22}$$

When calculating determinants, it is usually most convenient to go back to the definition rather than use this last formula.

➤ **EXAMPLE 2**
Determinant of a 3×3 Matrix

Calculate the determinant of the matrix

$$A = \begin{bmatrix} 4 & -1 & 5 \\ 0 & -1 & 1 \\ 1 & -1 & 1 \end{bmatrix}$$

Solution

We form the sum of products of elements of the first row times the corresponding cofactor:

$$|A| = 4 \cdot (-1)^{1+1} \begin{vmatrix} -1 & 1 \\ -1 & 1 \end{vmatrix} + (-1) \cdot (-1)^{1+2} \begin{vmatrix} 0 & 1 \\ 1 & 1 \end{vmatrix}$$
$$\quad + 5 \cdot (-1)^{1+3} \begin{vmatrix} 0 & -1 \\ 1 & -1 \end{vmatrix}$$
$$= 4 \cdot [(-1) \cdot 1 - (-1) \cdot 1] + 1 \cdot (0 \cdot 1 - 1 \cdot 1)$$
$$\quad + 5 \cdot [0 \cdot (-1) - 1 \cdot (-1)]$$
$$= 4 \cdot 0 + 1 \cdot (-1) + 5 \cdot 1$$
$$= 4$$

This definition of a determinant involves products of elements of the first row times the corresponding cofactor. Actually, the determinant can also be computed in analogous fashion using any row or column. We describe this computation as **expanding the determinant about a given row or column.** For example, here is the calculation of the determinant of Example 2 expanding about the second column:

$$|A| = (-1) \cdot (-1)^{1+2} \begin{vmatrix} 0 & 1 \\ 1 & 1 \end{vmatrix} + (-1) \cdot (-1)^{2+2} \begin{vmatrix} 4 & 5 \\ 1 & 1 \end{vmatrix}$$
$$\quad + (-1) \cdot (-1)^{3+2} \begin{vmatrix} 4 & 5 \\ 0 & 1 \end{vmatrix}$$
$$= 1 \cdot (0 \cdot 1 - 1 \cdot 1) - 1 \cdot (4 \cdot 1 - 1 \cdot 5) + 1 \cdot (4 \cdot 1 - 0 \cdot 5)$$
$$= -1 + 1 + 4$$
$$= 4$$

Note that this agrees with the result derived in Example 2. You will find that we arrive at the same result in this case no matter what row or column we use. This observation is a special case of the following result.

CALCULATING A DETERMINANT BY COFACTORS

Let A be a 3×3 matrix. Then the determinant of A can be computed as the sum of the products of the elements in a particular row or column times the corresponding cofactors.

We will omit the proof of this result, which properly belongs to a course in linear algebra.

In our discussion so far, we have defined the determinants for square matrices of size 1, 2, and 3. Let's now define determinants for square matrices of any size. The general idea is to make the definition by **recursion**. That is, determinants for matrices of a given size are defined in terms of determinants of matrices of smaller size.

Definition 5
Determinant
of an $n \times n$ Matrix

Let A be an $n \times n$ matrix. Then the determinant $|A|$ is defined by the formula
$$|A| = a_{11}A_{11} + a_{12}A_{12} + \cdots + a_{1n}A_{1n}$$
where A_{ij} denotes the cofactor determined by eliminating the ith row and jth column.

Note that the size of the determinants involved in the cofactors is $n - 1$. This means, for example, that this formula defines the determinant of a 4×4 matrix in terms of determinants of 3×3 matrices, the determinant of a 5×5 matrix in terms of determinants of 4×4 matrices, and so forth.

As with 3×3 matrices, the definition given involves expansion about the first row. However, an analogue of this definition holds for determinants of any size. That is, the determinant may be calculated by expanding about any row or column. The next example illustrates how this definition may be applied to calculate the determinant of a 4×4 matrix.

➤ **EXAMPLE 3**
Determinant of a 4×4 Matrix

Calculate the determinant of the matrix
$$A = \begin{bmatrix} 2 & 0 & 1 & 0 \\ 0 & 4 & 2 & 0 \\ 0 & 1 & 0 & 1 \\ 0 & -1 & 1 & 0 \end{bmatrix}$$

Solution

Note that the first column has a single nonzero element. So the computation will be simplified if we expand the determinant about the first column, since all but the first product are 0. We have
$$|A| = 2 \cdot \begin{vmatrix} 4 & 2 & 0 \\ 1 & 0 & 1 \\ -1 & 1 & 0 \end{vmatrix}$$

The simplest calculation for expanding the determinant on the right is about the third column, since two of the resulting products are 0. The result is

$$|A| = 2 \cdot 1 \cdot (-1)^{2+3} \begin{vmatrix} 4 & 2 \\ -1 & 1 \end{vmatrix}$$

$$= -2 \cdot [4 \cdot 1 - (-1) \cdot 2] = -2(4 + 2) = -2(6)$$

$$= -12$$

Cramer's Rule

Cramer's rule gives a general formula for the solution of a system of linear equations in terms of determinants formed from the coefficients of the system. To motivate the general statement of Cramer's rule, let's consider the specialized case of a linear system in two variables:

$$\begin{cases} ax + by = S \\ cx + dy = T \end{cases}$$

Use the addition and subtraction method. Multiply the first equation by c and the second equation by a and subtract:

$$cax + cby = cS$$
$$-\ \underline{cax + ady = aT}$$
$$cby - ady = cS - aT$$
$$(cb - ad)y = cS - aT$$
$$y = \frac{aT - cS}{ad - bc}, \quad ad - bc \neq 0$$

Substituting this value for y into the first equation of the system, we have

$$ax + by = S$$
$$ax + b\left(\frac{aT - cS}{ad - bc}\right) = S$$
$$ax = S - \frac{baT - bcS}{ad - bc}$$
$$= \frac{adS - baT}{ad - bc}$$
$$x = \frac{Sd - Tb}{ad - bc}, \quad ad - bc \neq 0$$

The numerator and denominator of x and y may each be expressed as a determinant, namely,

$$x = \frac{\begin{vmatrix} S & b \\ T & d \end{vmatrix}}{\begin{vmatrix} a & b \\ c & d \end{vmatrix}}, \qquad y = \frac{\begin{vmatrix} a & S \\ c & T \end{vmatrix}}{\begin{vmatrix} a & b \\ c & d \end{vmatrix}}$$

We may describe these formulas as follows:

1. The denominator in each case is the determinant of the matrix of coefficients A of the system. We have assumed that this determinant is nonzero and shown that with this assumption the system has a unique solution.

2. In the formula for x, the column of coefficients $\frac{a}{b}$ of x is replaced by the column of numbers $\frac{S}{T}$ on the left side of the system. The matrix appearing in the numerator is denoted A_1, so that

$$x = \frac{|A_1|}{|A|}$$

3. In the formula for y, the column of coefficients $\frac{c}{d}$ of y is replaced by the column of numbers $\frac{S}{T}$ on the left side of the system. The matrix appearing in the numerator is denoted A_2, so that

$$y = \frac{|A_2|}{|A|}$$

The formulas

$$x = \frac{|A_1|}{|A|}, \qquad y = \frac{|A_2|}{|A|}$$

are a special case of **Cramer's rule**.

➤ **EXAMPLE 4**

Application of Cramer's Rule

Use Cramer's rule to solve the system

$$\begin{cases} 5x - 2y = 8 \\ 3x + 4y = -1 \end{cases}$$

Solution

The determinant of the coefficients is

$$|A| = \begin{vmatrix} 5 & -2 \\ 3 & 4 \end{vmatrix} = 5 \cdot 4 - 3 \cdot (-2) = 26$$

This value is nonzero, so the system has a unique solution. Applying Cramer's rule, we have

$$x = \frac{|A_1|}{|A|} = \frac{\begin{vmatrix} 8 & -2 \\ -1 & 4 \end{vmatrix}}{26} = \frac{30}{26} = \frac{15}{13}$$

$$y = \frac{|A_2|}{|A|} = \frac{\begin{vmatrix} 5 & 8 \\ 3 & -1 \end{vmatrix}}{26} = \frac{-29}{26} = -\frac{29}{26}$$

So the system has the solution $(x, y) = (15/13, -29/26)$.

Cramer's rule may be generalized to systems of linear equations in n variables. Suppose that we are given a system with coefficient matrix A. For the nth variable, let A_n denote the matrix obtained by replacing the nth column of A with the column of numbers on the right side of the system. Then we have the following result.

CRAMER'S RULE FOR $n \times n$ LINEAR SYSTEMS

If a linear system of n equations in n variables $x, y, z, \ldots$ has a nonzero determinant $|A|$, then the system has a unique solution given by the formulas

$$x = \frac{|A_1|}{|A|}, \qquad y = \frac{|A_2|}{|A|}, \qquad \ldots$$

Note that Cramer's rule applies only to systems in which the number of equations equals the number of variables. Contrast this with Gauss elimination, which can be used to solve any linear system. The proof of Cramer's rule is beyond the scope of this book and will be omitted. With the advent of computers, Cramer's rule is rarely used to solve systems of linear equations, since it is far less efficient than Gauss elimination. However, Cramer's rule is still useful as a theoretical tool for proving facts about solutions of linear systems.

Properties of Determinants

Here are some properties of determinants that are useful to know, but whose proofs will be omitted. They are useful in shortening the task of evaluating determinants using cofactors which, as we have seen, can be a tedious task.

ROW OR COLUMN INTERCHANGE

If two rows or two columns of a matrix are interchanged, then the determinant changes sign.

For example, the determinant of the matrix

$$\begin{bmatrix} 2 & 3 \\ -1 & 8 \end{bmatrix}$$

is 19. Interchanging the first and second columns gives the matrix

$$\begin{bmatrix} 3 & 2 \\ 8 & -1 \end{bmatrix}$$

which has determinant -19.

MULTIPLYING A ROW BY A CONSTANT

If all elements of a row (column) are multiplied by k, then the determinant is multiplied by k.

For instance, if we multiply each element of the second row of the matrix

$$\begin{bmatrix} 2 & 3 \\ -1 & 8 \end{bmatrix}$$

by 3, we get the matrix

$$\begin{bmatrix} 2 & 3 \\ -3 & 24 \end{bmatrix}$$

which has determinant 57, which is $3 \cdot 19$.

ADDING ONE ROW TO ANOTHER

The determinant is unchanged if a row (column) is replaced by adding to it or subtracting from it any multiple of another row (column).

For instance, if we add twice the first row to the second of the matrix

$$\begin{bmatrix} 2 & 3 \\ -1 & 8 \end{bmatrix}$$

we get the matrix

$$\begin{bmatrix} 2 & 3 \\ -1 + 2 \cdot 2 & 8 + 2 \cdot 3 \end{bmatrix} = \begin{bmatrix} 2 & 3 \\ 3 & 14 \end{bmatrix}$$

which has determinant 19, the same as the original matrix.

Equations of Lines and Areas of Triangles

Determinants can be used to write many of the formulas in analytic geometry and physics in a simple and elegant form. For example, consider the equation of a line passing through the points (x_1, y_1) and (x_2, y_2). We have seen that this equation may be written in the form

$$(y - y_1) = \frac{y_2 - y_1}{x_2 - x_1}(x - x_1)$$

Clear the fraction and write the equation in the form

$$(y - y_1)(x_2 - x_1) - (y_2 - y_1)(x - x_1) = 0$$
$$x_2 y - x_1 y - x_2 y_1 + x_1 y_1 - x y_2 + x_1 y_2 + x y_1 - x_1 y_1 = 0$$
$$(x y_1 - x_1 y) - (x y_2 - x_2 y) + (x_1 y_2 - x_2 y_1) = 0$$

Each of the differences is a determinant:

$$\begin{vmatrix} x & y \\ x_1 & y_1 \end{vmatrix} - \begin{vmatrix} x & y \\ x_2 & y_2 \end{vmatrix} + \begin{vmatrix} x_1 & y_1 \\ x_2 & y_2 \end{vmatrix} = 0$$

It is now easy to recognize this expression as the expansion of the 3×3 determinant

$$\begin{vmatrix} x & y & 1 \\ x_1 & y_1 & 1 \\ x_2 & y_2 & 1 \end{vmatrix} = 0$$

in minors about the third column. Thus, we have proven the following result:

DETERMINANT FORM OF THE EQUATION OF A LINE

The equation of a line passing through the points (x_1, y_1) and (x_2, y_2) is given in determinant form by

$$\begin{vmatrix} x & y & 1 \\ x_1 & y_1 & 1 \\ x_2 & y_2 & 1 \end{vmatrix} = 0$$

Suppose we are given three points (x_1, y_1), (x_2, y_2), and (x_3, y_3). They lie on a common line, provided that the third point lies on the line determined by the first two. Therefore, by the preceding result, we have:

CONDITION FOR COLLINEARITY

Three points (x_1, y_1), (x_2, y_2), and (x_3, y_3) are collinear provided that

$$\begin{vmatrix} x_1 & y_1 & 1 \\ x_2 & y_2 & 1 \\ x_3 & y_3 & 1 \end{vmatrix} = 0$$

It is possible to calculate the area of a triangle in terms of a determinant according to the following result:

DETERMINANT FORMULA FOR AREA OF A TRIANGLE

The area of the triangle with vertices (x_1, y_1), (x_2, y_2), and (x_3, y_3) is given by the formula

$$\text{Area} = \pm\frac{1}{2}\begin{vmatrix} x_1 & y_1 & 1 \\ x_2 & y_2 & 1 \\ x_3 & y_3 & 1 \end{vmatrix}$$

➤ **EXAMPLE 5**
Area of a Triangle

Verify the last result in case the triangle is a right triangle with the right angle at the origin.

Solution

In this case, we may set $(x_1, y_1) = (0, 0)$, $(x_2, y_2) = (x_2, 0)$, and $(x_3, y_3) = (0, y_3)$. The determinant formula then gives

$$\text{Area} = \pm\frac{1}{2}\begin{vmatrix} 0 & 0 & 1 \\ x_2 & 0 & 1 \\ 0 & y_3 & 1 \end{vmatrix}$$

$$= \pm\frac{1}{2}\left[0 \cdot \begin{vmatrix} 0 & 1 \\ y_3 & 1 \end{vmatrix} - 0 \cdot \begin{vmatrix} x_2 & 1 \\ 0 & 1 \end{vmatrix} + 1 \cdot \begin{vmatrix} x_2 & 0 \\ 0 & y_3 \end{vmatrix} \right]$$

$$= \pm\frac{1}{2}x_2 y_3$$

$$= \pm\frac{1}{2} \cdot \text{Base} \cdot \text{Altitude}$$

Exercises 7.3

Calculate $|A|$, where A is the matrix given.

1. $\begin{bmatrix} 1 & 2 \\ 3 & -1 \end{bmatrix}$ 2. $\begin{bmatrix} 4 & -2 \\ 1 & 3 \end{bmatrix}$ 3. $\begin{bmatrix} -5 & 0 \\ -2 & 1 \end{bmatrix}$

4. $\begin{bmatrix} 0 & -2 \\ 1 & 3 \end{bmatrix}$ 5. $\begin{bmatrix} 3 & -1 \\ -2 & -4 \end{bmatrix}$ 6. $\begin{bmatrix} -1 & -3 \\ -2 & -5 \end{bmatrix}$

Use

$$A = \begin{bmatrix} 4 & -1 & -6 \\ 3 & 0 & 7 \\ 1 & 2 & -1 \end{bmatrix}, \qquad B = \begin{bmatrix} 2 & -1 & 0 & 3 \\ -1 & 4 & 1 & 3 \\ 0 & 1 & 0 & 2 \\ -1 & 3 & -2 & 1 \end{bmatrix}$$

for the following exercises.

7. Find A_{21}.   8. Find A_{12}.
9. Find B_{43}.   10. Find B_{31}.
11. Find A_{11}.   12. Find A_{23}.
13. Find B_{22}.   14. Find B_{44}.

Calculate $|A|$, where A is the given matrix.

15. $\begin{bmatrix} 1 & 2 & -1 \\ 2 & -1 & 1 \\ 4 & 0 & 2 \end{bmatrix}$ 16. $\begin{bmatrix} 1 & 2 & -3 \\ 4 & 5 & -9 \\ 0 & 0 & 1 \end{bmatrix}$

17. $\begin{bmatrix} 3 & 3 & -1 \\ 2 & 6 & 0 \\ -6 & -6 & 2 \end{bmatrix}$ 18. $\begin{bmatrix} 1 & 2 & -1 \\ 3 & 2 & 1 \\ 1 & 0 & -2 \end{bmatrix}$

19. $\begin{bmatrix} -2 & 0 & 4 & 2 \\ 3 & 6 & 0 & 4 \\ 0 & 0 & 0 & 3 \\ 9 & 0 & 2 & -1 \end{bmatrix}$ 20. $\begin{bmatrix} 7 & -8 & 1 & 2 \\ 21 & 4 & 3 & -1 \\ -35 & 8 & 3 & -2 \\ 14 & 16 & 0 & 1 \end{bmatrix}$

21. $\begin{bmatrix} 4 & 2 & 1 & 0 \\ -2 & 4 & -1 & 7 \\ -5 & 2 & 3 & 1 \\ 6 & 4 & -3 & 2 \end{bmatrix}$ 22. $\begin{bmatrix} 1 & -1 & 0 & 2 \\ 0 & 1 & -1 & 0 \\ 2 & 1 & 0 & -1 \\ -2 & 2 & 1 & 1 \end{bmatrix}$

23. $\begin{bmatrix} 11 & -15 & 20 \\ 16 & 24 & -8 \\ 6 & 9 & 15 \end{bmatrix}$ 24. $\begin{bmatrix} 2 & 1 & 1 \\ 2 & -3 & -1 \\ -4 & 5 & 2 \end{bmatrix}$

25. $\begin{bmatrix} -3 & 0 & 2 & 6 \\ 2 & 4 & 0 & -1 \\ -1 & 0 & -5 & 2 \\ 0 & -1 & -2 & -3 \end{bmatrix}$ 26. $\begin{bmatrix} 4 & 0 & 0 & 2 \\ -1 & 0 & 3 & 0 \\ 2 & 4 & 0 & 1 \\ 0 & 0 & 1 & 2 \end{bmatrix}$

Use Cramer's rule to solve the following linear systems.

27. $\begin{cases} 4x - y = 10 \\ 3x + 5y = 19 \end{cases}$

28. $\begin{cases} 2x + 5y = 18 \\ 3x + 4y = 7 \end{cases}$

29. $\begin{cases} 2x + 3y = -9 \\ 3x + 5y = -13 \end{cases}$

30. $\begin{cases} 4x - y = 3 \\ 2x - 3y = -1 \end{cases}$

31. $\begin{cases} x + 2y = 5 \\ 2x - 5y = -8 \end{cases}$

32. $\begin{cases} 2x + 3y + z = 1 \\ x - y + 2z = 3 \end{cases}$

33. $\begin{cases} 3x - 5y = 6 \\ -6x + 10y = 7 \end{cases}$

34. $\begin{cases} 3x - 6y = 12 \\ -2x + 4y = -8 \end{cases}$

35. $\begin{cases} x + 2y + 3z = 0 \\ x - 2z = 3 \\ y + z = 1 \end{cases}$

36. $\begin{cases} x + y + 4z = 1 \\ -2x - y + z = 2 \\ 3x - 2y + 3z = 5 \end{cases}$

37. $\begin{cases} 3x + 2y - z = 4 \\ 2x - y + 3z = 5 \\ x + 3y + 2z = -1 \end{cases}$

38. $\begin{cases} 4x - 2y + 3z = 4 \\ 5x - y + 4z = 7 \\ 3x + 5y + z = 7 \end{cases}$

39. $\begin{cases} x + 2y + 3z - w = 7 \\ x - 4y + z = 3 \\ 2y - 3z + w = 4 \end{cases}$

40. $\begin{cases} x + y + z + w = -2 \\ -2x + 3y + 2z - 3w = 10 \\ 3x + 2y - z + 2w = -12 \\ 4x - y + z + 2w = 1 \end{cases}$

41. $\begin{cases} x - y - 3z - 2w = 2 \\ 4x + y + z + 2w = 2 \\ x + 3y - 2z - w = 9 \\ 3x + y - z + w = 5 \end{cases}$

42. $\begin{cases} x - y = 5 \\ y - z = 6 \\ z - w = 7 \\ x + w = 8 \end{cases}$

Determine the determinant form of the equation of the line passing through the following points.

43. $(3, 4), (4, 8)$ 44. $(1, 1), (0, -1)$

45. $(0, 0), (5, 2)$ 46. $(10, 0), (0, 8)$

Determine if the following points are collinear.

47. $(1, 1), (4, 4), (8, 11)$

48. $(0, 0), (10, -3), (30, -9)$

49. $(3, 2), (2, 1), (1, -1)$

50. $(0, 0), (1, 0), (0, 1)$

Calculate the area of the triangle with the following vertices.

51. $(5, 3), (3, 1), (1, -2)$

52. $(10, 0), (1, 1), (4, 8)$

53. $(-5, 8), (4, 1), (0, 0)$

54. Prove the determinant formula for the area of a triangle.

Use the appropriate results from this section to solve the following.

55. $\begin{vmatrix} -20 & 2 & -3 \\ 6 & -4 & 1 \\ -1 & -1 & 2 \end{vmatrix} = 12x$

56. $\begin{vmatrix} 2 & -20 & -3 \\ 1 & 6 & 1 \\ 4 & -1 & 2 \end{vmatrix} = 2x + 1$

57. $\begin{vmatrix} a-4 & 0 & 0 \\ 0 & a+4 & 0 \\ 0 & 0 & a+1 \end{vmatrix} = 0$

58. $\begin{vmatrix} 1 & m & m^2 \\ 1 & 1 & 1 \\ 4 & 5 & 0 \end{vmatrix} = 0$

Calculate $|A|$, where A is the given matrix and x, y, and z are real numbers.

59. $\begin{bmatrix} x & y & z \\ 0 & -4 & 2 \\ -1 & 3 & 1 \end{bmatrix}$ 60. $\begin{bmatrix} x & -1 & 4x \\ y & 2 & 4y \\ z & -3 & 4z \end{bmatrix}$

Show that

61. $\begin{vmatrix} x_{11} & x_{12} & x_{13} & x_{14} \\ x_{21} & x_{22} & x_{23} & x_{24} \\ 0 & 0 & x_{33} & x_{34} \\ 0 & 0 & x_{43} & x_{44} \end{vmatrix} = \begin{vmatrix} x_{11} & x_{12} \\ x_{21} & x_{22} \end{vmatrix} \cdot \begin{vmatrix} x_{33} & x_{34} \\ x_{43} & x_{44} \end{vmatrix}$

62. $\begin{vmatrix} x_1 & y_1 & z_1 \\ x_2 & y_2 & z_2 \\ x_3 & y_3 & z_3 \end{vmatrix} = - \begin{vmatrix} x_2 & y_2 & z_2 \\ x_1 & y_1 & z_1 \\ x_3 & y_3 & z_3 \end{vmatrix}$

7.4 INVERSE OF A SQUARE MATRIX

In this section, we will introduce the inverse of a matrix and provide a technique for calculating the inverse. As motivation for the inverse of a matrix, consider a real number a. We say that a^{-1} is the inverse of a under multiplication provided that a^{-1} satisfies

$$aa^{-1} = a^{-1}a = 1$$

Also, recall that the number 1 is the identity for multiplication.

Definition 1
Inverse of a Square Matrix

Suppose that A is an $n \times n$ matrix. By analogy with the situation for real numbers, we say that A^{-1} is an **inverse** of A, provided that A^{-1} satisfies the equations

$$AA^{-1} = A^{-1}A = I_n$$

Here I_n is the identity matrix of size n, which is the identity for multiplication for square matrices of size $n \times n$.

For instance, suppose that

$$A = \begin{bmatrix} 5 & 2 \\ 1 & 3 \end{bmatrix}$$

Then an inverse of A is given by

$$A^{-1} = \begin{bmatrix} 3/13 & -2/13 \\ -1/13 & 5/13 \end{bmatrix}$$

Indeed, a simple computation shows that

$$\begin{bmatrix} 5 & 2 \\ 1 & 3 \end{bmatrix} \begin{bmatrix} 3/13 & -2/13 \\ -1/13 & 5/13 \end{bmatrix} = \begin{bmatrix} 1 & 0 \\ 0 & 1 \end{bmatrix}$$
$$AA^{-1} = I_2$$

and similarly that

$$A^{-1}A = I_2$$

This shows that A^{-1} is an inverse of A.

The inverse of a 2×2 matrix may be determined from the following result.

INVERSE OF A 2 × 2 MATRIX

Let

$$A = \begin{bmatrix} a & b \\ c & d \end{bmatrix}$$

be a 2×2 matrix for which $ad - bc \neq 0$. Then A has an inverse given by the formula

$$A^{-1} = \begin{bmatrix} \dfrac{d}{\Delta} & -\dfrac{b}{\Delta} \\ -\dfrac{c}{\Delta} & \dfrac{a}{\Delta} \end{bmatrix}$$

where $\Delta = |A| = ad - bc$.

Proof Let's calculate the product AA^{-1} and show that it equals I_2:

$$AA^{-1} = \begin{bmatrix} a & b \\ c & d \end{bmatrix} \begin{bmatrix} \dfrac{d}{\Delta} & -\dfrac{b}{\Delta} \\ -\dfrac{c}{\Delta} & \dfrac{a}{\Delta} \end{bmatrix}$$

$$= \begin{bmatrix} a \cdot \dfrac{d}{\Delta} + b \cdot \dfrac{-c}{\Delta} & a \cdot -\dfrac{b}{\Delta} + b \cdot \dfrac{a}{\Delta} \\ c \cdot \dfrac{d}{\Delta} + d \cdot -\dfrac{c}{\Delta} & c \cdot -\dfrac{b}{\Delta} + d \cdot \dfrac{a}{\Delta} \end{bmatrix}$$

$$= \begin{bmatrix} \dfrac{ad - bc}{\Delta} & 0 \\ 0 & \dfrac{ad - bc}{\Delta} \end{bmatrix}$$

$$= \begin{bmatrix} 1 & 0 \\ 0 & 1 \end{bmatrix}$$

$$= I_2 \quad \text{since } ad - bc = \Delta$$

Similarly, $A^{-1}A = I_2$.   ◆

It is possible to prove that a matrix has at most one inverse. So instead of referring to *an* inverse, we may refer to *the* inverse—provided, of course, that the matrix has an inverse.

➤ **EXAMPLE 1**
Matrix Inverse

Calculate the inverse of the matrix

$$A = \begin{bmatrix} 5 & -1 \\ 4 & 3 \end{bmatrix}$$

Solution

In this case, we have

$$\Delta = |A| = 5 \cdot 3 - 4 \cdot (-1) = 19$$

therefore, by the previous result, we have

$$A^{-1} = \begin{bmatrix} \dfrac{3}{19} & -\left(-\dfrac{1}{19}\right) \\ -\dfrac{4}{19} & \dfrac{5}{19} \end{bmatrix} = \begin{bmatrix} \dfrac{3}{19} & \dfrac{1}{19} \\ -\dfrac{4}{19} & \dfrac{5}{19} \end{bmatrix}$$

The previous result provides a formula for the inverse of a 2×2 matrix in case Δ is nonzero. It is possible to show that if $\Delta = 0$, then the matrix has no inverse. We will prove this in particular cases in the exercises.

There is a straightforward computational procedure for determining the inverse of any square matrix A (provided that the inverse exists). The procedure relies on performing elementary row operations and is described as follows:

1. Form the matrix

$$\left[\begin{array}{cccc|cccc} a_{11} & a_{12} & \cdots & a_{1n} & 1 & 0 & \cdots & 0 \\ a_{21} & a_{22} & \cdots & a_{2n} & 0 & 1 & \cdots & 0 \\ \vdots & \vdots & & \vdots & \vdots & \vdots & & \vdots \\ a_{n1} & a_{n2} & \cdots & a_{nn} & 0 & 0 & \cdots & 1 \end{array} \right]$$

where the right half is the identity matrix I_n.

2. If possible, perform elementary row operations on this matrix to transform the left half into the identity matrix I_n.

3. After the transformation, the matrix in the right half is A^{-1}.

The proof that this method works belongs to a more advanced book and won't be included here. The next example illustrates this technique for calculating the inverse.

➤ **EXAMPLE 2**

Inverse of a 3 × 3 Matrix

Determine the inverse of the matrix

$$\begin{bmatrix} 4 & 8 & 0 \\ 3 & 1 & 4 \\ 0 & 0 & 2 \end{bmatrix}$$

Solution

We first form the matrix

$$\left[\begin{array}{ccc|ccc} 4 & 8 & 0 & 1 & 0 & 0 \\ 3 & 1 & 4 & 0 & 1 & 0 \\ 0 & 0 & 2 & 0 & 0 & 1 \end{array}\right]$$

Now we perform a sequence of elementary row operations to this matrix to transform the left half into the identity matrix I_3:

$$\left[\begin{array}{ccc|ccc} 4 & 8 & 0 & 1 & 0 & 0 \\ 3 & 1 & 4 & 0 & 1 & 0 \\ 0 & 0 & 2 & 0 & 0 & 1 \end{array}\right] \xrightarrow{-2E_3 + E_2} \left[\begin{array}{ccc|ccc} 4 & 8 & 0 & 1 & 0 & 0 \\ 3 & 1 & 0 & 0 & 1 & -2 \\ 0 & 0 & 2 & 0 & 0 & 1 \end{array}\right]$$

$$\xrightarrow{-8E_2 + E_1} \left[\begin{array}{ccc|ccc} -20 & 0 & 0 & 1 & -8 & 16 \\ 3 & 1 & 0 & 0 & 1 & -2 \\ 0 & 0 & 2 & 0 & 0 & 1 \end{array}\right]$$

$$\xrightarrow{-\frac{1}{20}E_1} \left[\begin{array}{ccc|ccc} 1 & 0 & 0 & -\frac{1}{20} & \frac{8}{20} & -\frac{16}{20} \\ 3 & 1 & 0 & 0 & 1 & -2 \\ 0 & 0 & 2 & 0 & 0 & 1 \end{array}\right]$$

$$\xrightarrow{-3E_1 + E_2} \left[\begin{array}{ccc|ccc} 1 & 0 & 0 & -\frac{1}{20} & \frac{8}{20} & -\frac{16}{20} \\ 0 & 1 & 0 & \frac{3}{20} & -\frac{4}{20} & \frac{8}{20} \\ 0 & 0 & 2 & 0 & 0 & 1 \end{array}\right]$$

$$\xrightarrow{\frac{1}{2}E_3} \left[\begin{array}{ccc|ccc} 1 & 0 & 0 & -\frac{1}{20} & \frac{8}{20} & -\frac{16}{20} \\ 0 & 1 & 0 & \frac{3}{20} & -\frac{4}{20} & \frac{8}{20} \\ 0 & 0 & 1 & 0 & 0 & \frac{1}{2} \end{array}\right]$$

The left side of the matrix is now the identity matrix I_3. The inverse may then be read off the right side:

$$A^{-1} = \begin{bmatrix} -\frac{1}{20} & \frac{2}{5} & -\frac{4}{5} \\ \frac{3}{20} & -\frac{1}{5} & \frac{2}{5} \\ 0 & 0 & \frac{1}{2} \end{bmatrix}$$

This result may be checked by verifying that A^{-1} satisfies the equations

$$AA^{-1} = A^{-1}A = I_3$$

We leave the calculations as an exercise.

Our original motivation for introducing matrices was to have a convenient way of performing row operations on a system of linear equations. Actually, the connection between matrices and systems of linear equations is much more fundamental than our previous discussion indicates. Let's now explore this connection in greater depth. Suppose that we are given the system of linear equations

$$\begin{cases} a_{11}x_1 + a_{12}x_2 + \cdots + a_{1n}x_n = b_1 \\ a_{21}x_1 + a_{22}x_2 + \cdots + a_{2n}x_n = b_2 \\ \qquad\qquad\vdots \\ a_{m1}x_1 + a_{m2}x_2 + \cdots + a_{mn}x_n = b_m \end{cases}$$

Define the matrices

$$A = \begin{bmatrix} a_{11} & a_{12} & \cdots & a_{1n} \\ a_{21} & a_{22} & \cdots & a_{2n} \\ \vdots & \vdots & & \vdots \\ a_{m1} & a_{m2} & \cdots & a_{mn} \end{bmatrix}, \qquad B = \begin{bmatrix} b_1 \\ b_2 \\ \vdots \\ b_m \end{bmatrix}, \qquad X = \begin{bmatrix} x_1 \\ x_2 \\ \vdots \\ x_n \end{bmatrix}$$

From the definition of matrix multiplication, we may write the system of linear equations in the compact form

$$AX = B$$

Indeed, by multiplying out the matrices on the left and equating the resulting entries to those of the matrix on the right, we arrive precisely at the given system of linear equations.

At first, the previous equation might seem like just a notational convenience, a shorthand way of writing the equations of the system. However, it can be used to arrive at a solution to the system in case A has an inverse. Indeed, multiplying both sides of the equation by A^{-1}, we have

$$AX = B$$
$$A^{-1}(AX) = A^{-1}B$$
$$(A^{-1}A)X = A^{-1}B \qquad \textit{(Associative law)}$$
$$I_n X = A^{-1}B \qquad \textit{(Definition of inverse)}$$
$$X = A^{-1}B \qquad \textit{(I_n is the identity for multiplication)}$$

Thus, we have the following equation:

$$X = A^{-1}B$$

The matrix X is just the column matrix of variables. The right side of the equation is a matrix whose value may be calculated in terms of the constants of the system. By equating corresponding entries on both sides of the equation, we arrive at a solution of the system. We have now derived a new method for solving a system of linear equations:

1. Calculate A^{-1}.
2. Form the product $A^{-1}B$.
3. Use the equation $X = A^{-1}B$ to determine the values of the variables.

The next example illustrates how to solve a system using this method.

➤ **EXAMPLE 3**
Matrix Solution
of a Linear System

Solve the following system using inverse matrices:

$$\begin{cases} 3x - 8y = 7 \\ -x + 4y = 6 \end{cases}$$

Solution

Let

$$A = \begin{bmatrix} 3 & -8 \\ -1 & 4 \end{bmatrix}, B = \begin{bmatrix} 7 \\ 6 \end{bmatrix}, X = \begin{bmatrix} x \\ y \end{bmatrix}$$

Then the system may be written in the matrix form

$$AX = B$$
$$X = A^{-1}B$$

Using the rule for determining the inverse of a 2×2 matrix, we can calculate A^{-1}:

$$\Delta = 3 \cdot 4 - (-1) \cdot (-8) = 4$$

$$A^{-1} = \begin{bmatrix} \frac{4}{4} & -(-\frac{8}{4}) \\ -(-\frac{1}{4}) & \frac{3}{4} \end{bmatrix}$$

$$= \begin{bmatrix} 1 & 2 \\ \frac{1}{4} & \frac{3}{4} \end{bmatrix}$$

Thus,

$$X = A^{-1}B$$

$$= \begin{bmatrix} 1 & 2 \\ \frac{1}{4} & \frac{3}{4} \end{bmatrix}\begin{bmatrix} 7 \\ 6 \end{bmatrix}$$

$$= \begin{bmatrix} 1 \cdot 7 + 2 \cdot 6 \\ \frac{1}{4} \cdot 7 + \frac{3}{4} \cdot 6 \end{bmatrix}$$

$$= \begin{bmatrix} 19 \\ \frac{25}{4} \end{bmatrix}$$

Thus, the solution of the system is $x = 19, y = \frac{25}{4}$.

Note that the method employed in the preceding example makes use of the inverse matrix. We should emphasize that not every matrix has an inverse; therefore, the previous method is not applicable to solving all systems of linear equations, just those for which the matrix of coefficients has an inverse. It can be proved that these are precisely the systems with a single solution. As we have previously seen, there are, in addition, systems that have no solutions and systems with an infinite number of solutions. For such systems, we can't use the method of the preceding example. Instead, we must rely on the method of Gauss elimination discussed earlier.

Exercises 7.4

Calculate the inverse of the given matrix, if it exists.

1. $\begin{bmatrix} 3 & 5 \\ -2 & -14 \end{bmatrix}$
2. $\begin{bmatrix} 3 & 8 \\ 2 & 5 \end{bmatrix}$
3. $\begin{bmatrix} -3 & 4 \\ 1 & -2 \end{bmatrix}$
4. $\begin{bmatrix} 2 & 4 \\ 1 & -1 \end{bmatrix}$
5. $\begin{bmatrix} 2 & -4 \\ 1 & -2 \end{bmatrix}$
6. $\begin{bmatrix} 2 & -3 \\ -3 & 5 \end{bmatrix}$

7. $\begin{bmatrix} 2 & -1 & 1 \\ 1 & -2 & 3 \\ 4 & 1 & 2 \end{bmatrix}$
8. $\begin{bmatrix} 1 & -4 & 8 \\ 1 & -3 & 2 \\ 2 & -7 & 10 \end{bmatrix}$

9. $\begin{bmatrix} -1 & -1 & -1 \\ 4 & 5 & 0 \\ 0 & 1 & -3 \end{bmatrix}$
10. $\begin{bmatrix} 2 & 4 & 6 \\ -1 & -4 & -3 \\ 0 & 1 & -1 \end{bmatrix}$

11. $\begin{bmatrix} 3 & 1 & 0 \\ 1 & 1 & 1 \\ 1 & -1 & 2 \end{bmatrix}$
12. $\begin{bmatrix} 2 & -1 & 0 \\ 3 & 0 & 1 \\ -2 & 4 & 0 \end{bmatrix}$

13. $\begin{bmatrix} 1 & -2 & 3 & 0 \\ 0 & 1 & -1 & 1 \\ -2 & 2 & -2 & 4 \\ 0 & 2 & -3 & 1 \end{bmatrix}$
14. $\begin{bmatrix} 1 & 2 & 3 & 4 \\ 0 & 1 & 3 & -5 \\ 0 & 0 & 1 & -2 \\ 0 & 0 & 0 & 1 \end{bmatrix}$

Determine if the given matrices are inverses of each other.

15. $\begin{bmatrix} 2 & -1 & 1 \\ 1 & -2 & 3 \\ 4 & 1 & 2 \end{bmatrix}$ and $\begin{bmatrix} \frac{7}{15} & -\frac{1}{5} & \frac{1}{15} \\ -\frac{2}{3} & 0 & \frac{1}{3} \\ -\frac{3}{5} & \frac{2}{5} & \frac{1}{5} \end{bmatrix}$

16. $\begin{bmatrix} 1 & -1 & 2 \\ 0 & 1 & 3 \\ 2 & 1 & -2 \end{bmatrix}$ and $\begin{bmatrix} \frac{1}{3} & 0 & \frac{1}{3} \\ -\frac{2}{5} & \frac{2}{5} & \frac{1}{5} \\ \frac{2}{15} & \frac{1}{5} & -\frac{1}{15} \end{bmatrix}$

17. $\begin{bmatrix} 11 & 3 \\ 7 & 2 \end{bmatrix}$ and $\begin{bmatrix} 2 & -3 \\ -7 & 11 \end{bmatrix}$

18. $\begin{bmatrix} 3 & 2 \\ 5 & 3 \end{bmatrix}$ and $\begin{bmatrix} -3 & -2 \\ 5 & -3 \end{bmatrix}$

19. $\begin{bmatrix} 1 & 3 & 3 \\ 1 & 4 & 3 \\ 1 & 3 & 4 \end{bmatrix}$ and $\begin{bmatrix} 7 & -3 & -3 \\ -1 & 1 & 0 \\ -1 & 0 & 1 \end{bmatrix}$

20. $\begin{bmatrix} -1 & 0 & 2 \\ 3 & 1 & 0 \\ 0 & 2 & -3 \end{bmatrix}$ and $\begin{bmatrix} -\frac{1}{5} & \frac{4}{15} & -\frac{2}{15} \\ \frac{3}{5} & \frac{1}{5} & \frac{2}{5} \\ \frac{2}{5} & \frac{2}{15} & -\frac{1}{15} \end{bmatrix}$

Solve the following systems of equations using inverse matrices.

21. $\begin{cases} x + 2y = 6 \\ 2x + y = 9 \end{cases}$
22. $\begin{cases} x + 3y = 8 \\ 3x + 4y = 9 \end{cases}$

23. $\begin{cases} 3x - 2y = 9 \\ 2x + 5y = 8 \end{cases}$
24. $\begin{cases} 2x + 4y = 6 \\ x - 3y = 3 \end{cases}$

25. $\begin{cases} 4x - y = 3 \\ 6x + 4y = -1 \end{cases}$
26. $\begin{cases} 3x - 6y = 1 \\ -5x + 9y = -1 \end{cases}$

27. $\begin{cases} 4x - 3y = -23 \\ -3x + 2y = 16 \end{cases}$
28. $\begin{cases} 2x + 3y = 13 \\ x + 2y = 8 \end{cases}$

29. $\begin{cases} 4x + z = 1 \\ 2x + 2y = 3 \\ x - y + z = 4 \end{cases}$
30. $\begin{cases} x + 2y + 3z = -1 \\ 2x - 3y + 4z = 2 \\ -3x + 5y - 6z = 4 \end{cases}$

31. $\begin{cases} x + y - 3z = 4 \\ 2x + 4y - 4z = 8 \\ -x + y + 4z = -3 \end{cases}$
32. $\begin{cases} 2x - 2y = 5 \\ 4y + 8z = 7 \\ kx + 2z = 1 \end{cases}$

In each of the following, state the conditions under which A^{-1} exists, and then find a formula for A^{-1}.

33. $A = \begin{bmatrix} a & 0 \\ 0 & b \end{bmatrix}$

34. $A = \begin{bmatrix} x & 0 & 0 \\ 0 & y & 0 \\ 0 & 0 & z \end{bmatrix}$

35. If $A = \begin{bmatrix} 1 & 2 \\ 3 & 4 \end{bmatrix}$ and $B = \begin{bmatrix} -1 & -1 \\ 2 & 3 \end{bmatrix}$, show that $(AB)^{-1} = B^{-1}A^{-1}$.

36. Suppose A is a 2×2 matrix. Prove that if $\Delta = 0$, then the inverse of A does not exist.

37. Suppose A is a 3×3 matrix. Prove that if $|A| = 0$, then the inverse of A does not exist.

7.5 CHAPTER REVIEW

Important Concepts, Properties, and Formulas—Chapter 7

Determining a partial fraction expansion	1. Use long division to write $\dfrac{f(x)}{g(x)}$ in the form $$\frac{f(x)}{g(x)} = q(x) + \frac{h(x)}{g(x)}, \quad \deg(h) < \deg(g)$$ 2. Factor $g(x)$ into a product of distinct real factors, some linear and some quadratic. 3. For a linear factor $(ax+b)^n$, the partial fraction expansion has a component: $$\frac{A_1}{ax + b} + \frac{A_2}{(ax + b)^2} + \cdots + \frac{A_n}{(ax + b)^n}$$	p. 362

Determining a partial fraction expansion (*continued*)	4. For a linear factor $(cx^2 + dx + e)^n$, the partial fraction expansion has a component: $$\frac{A_1 + B_1 x}{cx^2 + dx + e} + \frac{A_2 + B_2 x}{(cx^2 + dx + e)^2}$$ $$+ \cdots + \frac{A_n + B_n x}{(cx^2 + dx + e)^n}$$ 5. The partial fraction expansion is the sum of $q(x)$ in step 1 and each of the components in steps 3 and 4.	p. 362
Matrices	Subscript notation for matrix entries. Zero matrix—Matrix of all zeros. Sum of two matrices—Add corresponding entries. Row matrix—only one row. Column matrix—only one column. Matrix multiplication—Suppose that A is an $m \times n$ matrix and B is an $n \times k$ matrix. Then the product AB is the $m \times k$ matrix with entries as follows: The entry in the ith row and jth column of AB is obtained by multiplying the ith row of A by the jth column of B. Identity matrix I_n of size n.	pp. 366, 368, 371
Determinant of a square matrix	Let A be an $n \times n$ matrix. Then the determinant $\|A\|$ is defined by the formula: $$\|A\| = a_{11}A_{11} + a_{12}A_{12} + \cdots + a_{1n}A_{1n},$$ where A_{ij} denotes the cofactor determined by eliminating the ith row and jth column.	p. 375
Cramer's rule	If a linear system of n equations in n variables $x, y, z, \ldots$ has a nonzero determinant $\|A\|$, then the system has a unique solution given by the formulas: $$x = \frac{\|A_1\|}{\|A\|}, y = \frac{\|A_2\|}{\|A\|}, \ldots$$	p. 377
Determinant form of the equation of a line	The equation of a line passing through the points (x_1, y_1) and (x_2, y_2) is given in determinant form: $$\begin{vmatrix} x & y & 1 \\ x_1 & y_1 & 1 \\ x_2 & y_2 & 1 \end{vmatrix} = 0$$	p. 379
Determinant condition for collinearity	Three points (x_1, y_1), (x_2, y_2), and (x_3, y_3) are collinear provided that: $$\begin{vmatrix} x_1 & y_1 & 1 \\ x_2 & y_2 & 1 \\ x_3 & y_3 & 1 \end{vmatrix} = 0$$	p. 379

Determinant formula for the area of a triangle	The area of the triangle with the vertices: (x_1, y_1), (x_2, y_2), and (x_3, y_3) is given by the formula: $$\text{Area} = \pm\frac{1}{2}\begin{vmatrix} x_1 & y_1 & 1 \\ x_2 & y_2 & 1 \\ x_3 & y_3 & 1 \end{vmatrix}$$	p. 380
Inverse of a square matrix	A^{-1} is an inverse of A provided that A^{-1} satisfies the equations: $$AA^{-1} = A^{-1}A = I_n$$	p. 382
Computing the inverse of a square matrix	1. Form the matrix $$\left[\begin{array}{cccc|cccc} a_{11} & a_{12} & \cdots & a_{1n} & 1 & 0 & \cdots & 0 \\ a_{21} & a_{22} & \cdots & a_{2n} & 0 & 1 & \cdots & 0 \\ \vdots & & & & \vdots & & & \\ a_{n1} & a_{n2} & \cdots & a_{nn} & 0 & 0 & \cdots & 1 \end{array}\right]$$ where the matrix on the right is the identity matrix I_n. 2. If possible, perform elementary row operations on this matrix to transform the left half into the identity matrix I_n. 3. After the transformation, the matrix in the right half is A^{-1}.	p. 383
Matrix form of a linear system	The solution of the matrix equation $AX = B$ is $X = A^{-1}B$.	p. 385

Cumulative Review Exercises—Chapter 7

Determine the partial fraction expansion of the given rational expressions.

1. $\dfrac{-3a^2 - 10a - 4}{a(a+1)^2}$

2. $\dfrac{13y + 5}{2y^2 - 7y + 6}$

3. $\dfrac{7x - 1}{6x^2 - 5x + 1}$

4. $\dfrac{5x^2 + 9x - 56}{(x-4)(x-2)(x+1)}$

Given

$A = \begin{bmatrix} 4 & 2 \\ 1 & -3 \end{bmatrix}$, $B = \begin{bmatrix} 0 \\ 3 \end{bmatrix}$, $C = \begin{bmatrix} 2 & -1 & 4 \\ 3 & 0 & 5 \end{bmatrix}$,

$D = \begin{bmatrix} -2 & 4 & 5 \\ 1 & -1 & 2 \\ 3 & 2 & -3 \end{bmatrix}$, $E = \begin{bmatrix} 3 & 4 \\ -1 & 5 \end{bmatrix}$, $F = \begin{bmatrix} 5 \\ -2 \end{bmatrix}$,

$G = \begin{bmatrix} -1 & 3 & 4 \\ 0 & 2 & -1 \end{bmatrix}$, $H = \begin{bmatrix} 0 & 4 & -2 \\ 2 & -1 & 3 \\ 3 & 5 & 2 \end{bmatrix}$,

$I = \begin{bmatrix} 4 & 1 & 0 \end{bmatrix}$

find each of the following matrices, whenever possible.

5. DH 6. EF 7. $4A + 3E$ 8. $2E - 3A$

9. GH 10. CH 11. AB 12. CD

13. BI 14. FG

Determine which of the following products are defined and the size of those products.

15. A is 3×5; B is 4×5.   16. A is 3×4; B is 4×3.

17. A is 1×3; B is 3×1.   18. A is 2×2; B is 4×2.

Find the values of $w, x, y,$ and z in the following matrix equations.

19. $\begin{bmatrix} x & 5 & 2 \\ 2 & 7 & y \end{bmatrix} = \begin{bmatrix} 3 & 5 & z \\ 2 & w & 3 \end{bmatrix}$

20. $\begin{bmatrix} 4 & 1 \\ 2 & 4 \end{bmatrix} = \begin{bmatrix} x & w \\ y & 4 \end{bmatrix}$

Calculate $|A|$, where A is the given matrix.

21. $\begin{bmatrix} 1 & -1 \\ 2 & 3 \end{bmatrix}$

22. $\begin{bmatrix} 4 & 3 \\ -1 & -2 \end{bmatrix}$

23. $\begin{bmatrix} -1 & -2 \\ -4 & 5 \end{bmatrix}$

24. $\begin{bmatrix} 0 & 3 \\ 2 & -5 \end{bmatrix}$

Use

$$A = \begin{bmatrix} 1 & 3 & -1 \\ 4 & 5 & 1 \\ 3 & -1 & 2 \end{bmatrix}, \quad B = \begin{bmatrix} 7 & 2 & 0 & 4 \\ 2 & -1 & 2 & 2 \\ 1 & 4 & 3 & -1 \\ -3 & 5 & 1 & 5 \end{bmatrix}$$

for the following exercises.

25. Find A_{12}.

26. Find A_{11}.

27. Find B_{44}.

28. Find B_{22}.

Calculate $|A|$, where A is the given matrix.

29. $\begin{bmatrix} 1 & 3 & -1 \\ 2 & 4 & 5 \\ -1 & 2 & 0 \end{bmatrix}$

30. $\begin{bmatrix} 0 & 3 & 1 \\ 4 & -2 & -1 \\ -1 & 4 & 2 \end{bmatrix}$

31. $\begin{bmatrix} 2 & 0 & 1 & 4 \\ 1 & 2 & 3 & 0 \\ 4 & -1 & 0 & 2 \\ 0 & -2 & 3 & 1 \end{bmatrix}$

32. $\begin{bmatrix} -3 & 0 & 2 & 1 \\ 2 & 4 & 0 & -1 \\ 3 & 0 & 1 & 2 \\ 0 & 2 & -1 & 4 \end{bmatrix}$

Solve the following for x.

33. $\begin{bmatrix} 3 & -1 \\ 2 & 5 \end{bmatrix} = 2x - 1$

34. $\begin{bmatrix} 2 & -1 & 1 \\ 3 & 4 & 0 \\ 1 & 0 & 2 \end{bmatrix} = x - 5$

Calculate the inverse of the given matrix, if it exists.

35. $\begin{bmatrix} 3 & 5 \\ 1 & 2 \end{bmatrix}$

36. $\begin{bmatrix} 4 & -3 \\ 1 & 2 \end{bmatrix}$

37. $\begin{bmatrix} 1 & -4 & 8 \\ 1 & -3 & 2 \\ 2 & -7 & 10 \end{bmatrix}$

38. $\begin{bmatrix} 1 & -1 & 2 \\ 0 & 1 & 3 \\ 2 & 1 & -2 \end{bmatrix}$

39. $\begin{bmatrix} 1 & 2 & 3 & 4 \\ 0 & 1 & 3 & -5 \\ 0 & 0 & 1 & -2 \\ 0 & 0 & 0 & -1 \end{bmatrix}$

40. $\begin{bmatrix} -2 & -3 & 4 & 1 \\ 0 & 1 & 1 & 0 \\ 0 & 4 & -6 & 1 \\ -2 & -2 & 5 & 1 \end{bmatrix}$

Determine if the given matrices are inverses of each other.

41. $\begin{bmatrix} 1 & 2 & -1 \\ 3 & 4 & 2 \\ 1 & 1 & 0 \end{bmatrix}$ and $\begin{bmatrix} 1 & 1 & 0 \\ 3 & 4 & 2 \\ 1 & 2 & -1 \end{bmatrix}$

42. $\begin{bmatrix} 2 & 3 \\ 1 & 5 \end{bmatrix}$ and $\begin{bmatrix} \frac{1}{2} & \frac{1}{3} \\ 1 & \frac{1}{5} \end{bmatrix}$

Solve the following systems of equations using inverse matrices.

43. $\begin{cases} 2x + y = 7 \\ 3x + 2y = 9 \end{cases}$

44. $\begin{cases} 6x + y = 0 \\ 3x + 5y = 9 \end{cases}$

45. $\begin{cases} x + y = 4 \\ y + z = 2 \\ 2x - z = 1 \end{cases}$

46. $\begin{cases} x + z = 1 \\ 2x + y = 3 \\ x - y + z = 4 \end{cases}$

CHAPTER
8

SEQUENCES, SERIES, AND PROBABILITY

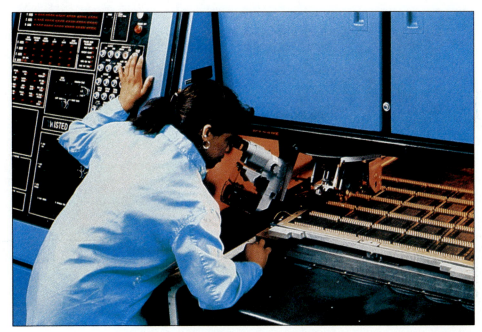

Counting and probability concepts form the groundwork for statistical quality control. Controlling quality and preventing defects is a key concern in manufacturing environments, particularly those with sensitive electronic parts, such as these computer chips at IBM.

I n this chapter, we discuss three topics that are very important for more advanced courses, especially calculus. Our first main topic is the **binomial theorem**, which states how to raise a binomial expression to any positive integer power. Second, we discuss **mathematical induction**, a valuable technique for proving many theorems, including the binomial theorem. Finally, we introduce **sequences** and **series**, which play an important role in calculus, and apply them to permutations, combinations, and probability.

8.1 THE BINOMIAL THEOREM

A **binomial** is an algebraic expression that is a sum of two terms. Here are some examples:

$$a + b, \qquad 2x^2 - y, \qquad \frac{5}{t} - 2t, \qquad \sqrt{x} + 3y$$

Note that we have been working with such expressions throughout the book. In many applications, it is necessary to raise a binomial to a nonnegative integer power. Here are some examples of such powers:

$$(a + b)^2$$
$$(a + b)^3$$
$$(2x^2 - y)^5$$
$$(4\sqrt{x} + 3y)^{18}$$

It is possible to compute such powers by repeated multiplication. For example, we did just that in the case of the first two powers above and came up with the following formulas in Chapter 1:

$$(a + b)^2 = a^2 + 2ab + b^2$$
$$(a + b)^3 = a^3 + 3a^2b + 3ab^2 + b^3$$

For higher powers, such results are tedious to obtain by direct multiplication. However, the **binomial theorem** provides a shortcut.

The binomial theorem allows you to write out the terms of the power $(a+b)^n$, where n is a nonnegative integer. To motivate the binomial theorem, let's examine some particular expansions:

$$(a + b)^0 = 1$$
$$(a + b)^1 = a + b$$
$$(a + b)^2 = a^2 + 2ab + b^2$$
$$(a + b)^3 = a^3 + 3a^2b + 3ab^2 + b^3$$
$$(a + b)^4 = a^4 + 4a^3b + 6a^2b^2 + 4ab^3 + b^4$$
$$(a + b)^5 = a^5 + 5a^4b + 10a^3b^2 + 10a^2b^3 + 5ab^4 + b^5$$

Based on these expansions, we can guess at the general pattern for $(a + b)^n$.

1. The first term is
$$a^n b^0 = a^n \cdot 1 = a^n$$

2. Each successive term is formed by decreasing the exponent of a by 1 and increasing the exponent of b by 1.

3. The sum of the exponents of a and b in any term is n.

4. The last term is
$$a^0 b^n = 1 \cdot b^n = b^n$$

5. Each term has a numerical coefficient that must be determined.

If we denote the coefficients by $c_0, c_1, \ldots, c_n$, then we have the following result:

BINOMIAL THEOREM (FIRST FORM)

The expansion of $(a + b)^n$ can be written in the form

$$(a + b)^n = c_0 a^n + c_1 a^{n-1} b + c_2 a^{n-2} b^2 + \cdots + c_{n-1} a b^{n-1} + c_n b^n$$

where the coefficients on the right are calculated as follows:

1. The first coefficient, c_0, equals 1.
2. Multiply the preceding coefficient by the exponent of a in the preceding term.
3. Divide the result by the exponent of b in the present term.

We will prove the binomial theorem in the next section, once we have developed the method of mathematical induction.

➤ **EXAMPLE 1**
Cube of a Binomial

Use the binomial theorem to verify the cube of a binomial formula cited earlier.

Solution

We calculate the coefficients beginning with that for the term 0 coefficient:

$$c_0 = 1$$

Term 1 coefficient:

$$c_1 = c_0 \cdot \frac{3}{1} = 1 \cdot \frac{3}{1} = 3$$

Term 2 coefficient: The preceding term has exponent 2 on a. This term has exponent 2 on b.

$$c_2 = c_1 \cdot \frac{2}{2} = 3 \cdot \frac{2}{2} = 3$$

Term 3 coefficient: The preceding term has exponent 1 on a. This term has exponent 3 on b.

$$c_3 = c_2 \cdot \frac{1}{3} = 3 \cdot \frac{1}{3} = 1$$

These coefficients agree with those cited in the expansion given at the beginning of the section, namely,

$$(a + b)^3 = a^3 + 3a^2 b + 3ab^2 + b^3$$

➤ **EXAMPLE 2**
Sixth Power of a Binomial

1. Calculate the coefficients in the expansion of $(a + b)^6$.
2. Write out the expansion of $(a + b)^6$.

Solution

1. We follow the calculational procedure described earlier:

$$c_0 = 1$$

$$c_1 = c_0 \cdot \frac{6}{1} = 6$$

$$c_2 = c_1 \cdot \frac{5}{2} = 6 \cdot \frac{5}{2} = 15$$

$$c_3 = c_2 \cdot \frac{4}{3} = 15 \cdot \frac{4}{3} = 20$$

$$c_4 = c_3 \cdot \frac{3}{4} = 20 \cdot \frac{3}{4} = 15$$

$$c_5 = c_4 \cdot \frac{2}{5} = 15 \cdot \frac{2}{5} = 6$$

$$c_6 = c_5 \cdot \frac{1}{6} = 6 \cdot \frac{1}{6} = 1$$

2. From the previous calculation of the coefficients, the expansion of $(a + b)^6$ is equal to

$$a^6 + 6a^5b + 15a^4b^2 + 20a^3b^3 + 15a^2b^4 + 6ab^5 + b^6$$

Factorials, Binomial Coefficients, and Applications

The method just described allows us to calculate the coefficients of any binomial expansion. However, the method suffers from a serious difficulty: it requires that we compute the coefficients in order. To determine the 10th coefficient, say, we must compute the first 9. In many applications, such as probability and statistics, it is necessary to calculate single terms in binomial expansions. For such applications, the method we have given is very cumbersome. Let's now present another method that can be used to compute any specified term of a binomial expansion.

Definition 1 **Factorial**	Let n be a positive integer. Then n **factorial**, denoted $n!$, is the product of all positive integers from 1 to n. That is, $$n! = n(n - 1)(n - 2) \cdots \cdot 2 \cdot 1$$

Here are some calculations of factorials:

$$1! = 1$$
$$2! = 2 \cdot 1 = 2$$
$$3! = 3 \cdot 2 \cdot 1 = 6$$
$$4! = 4 \cdot 3 \cdot 2 \cdot 1 = 24$$
$$5! = 5 \cdot 4 \cdot 3 \cdot 2 \cdot 1 = 120$$

> **EXAMPLE 3**
Factorials

Calculate the value of $11!/6!5!$.

Solution

This type of quotient arises in calculating the coefficients in binomial expansions (as we shall see later). In computing such a quotient, it is best to write out the definitions of all the factorials and do whatever simplification is possible:

$$\frac{11!}{6!5!} = \frac{11 \cdot 10 \cdot 9 \cdot 8 \cdot 7 \cdot 6 \cdot 5 \cdot 4 \cdot 3 \cdot 2 \cdot 1}{(6 \cdot 5 \cdot 4 \cdot 3 \cdot 2 \cdot 1)(5 \cdot 4 \cdot 3 \cdot 2 \cdot 1)}$$

$$= \frac{11 \cdot 10 \cdot 9 \cdot 8 \cdot 7}{5 \cdot 4 \cdot 3 \cdot 2 \cdot 1}$$

$$= 11 \cdot 3 \cdot 2 \cdot 7$$

$$= 462$$

The preceding definition of factorials defines $n!$ for a positive integer n. For use in connection with the binomial theorem, it is convenient to define $0!$ according to the formula

$$0! = 1$$

Let's now introduce a new notation. The coefficient of the $(k + 1)$th term in the binomial expansion of $(a + b)^n$ is customarily denoted with the symbol

$$\binom{n}{k}$$

This symbol is called a **binomial coefficient** and is read "binomial n over k." (Note, however, that "n over k" does not mean the quotient of n divided by k. Rather, the term is just a convenient way of reminding us of the notation for the binomial coefficient.)

From the binomial expansion for $(a + b)^3$, for example, we can read off the values of the following binomial coefficients:

$$\binom{3}{0} = 1, \qquad \binom{3}{1} = 3, \qquad \binom{3}{2} = 3, \qquad \binom{3}{3} = 1$$

We can calculate binomial coefficients using the following result.

FACTORIAL FORMULA FOR BINOMIAL COEFFICIENTS

The binomial coefficients may be computed using the formula

$$\binom{n}{k} = \frac{n!}{k!(n - k)!}$$

We will delay proving this result until the next section. However, in the meantime, we may use it to evaluate binomial coefficients, as illustrated in the next example.

➤ **EXAMPLE 4**
Calculating Binomial Coefficients

Use the factorial formula to compute the following binomial coefficients:

1. $\binom{3}{2}$

2. $\binom{4}{3}$

3. $\binom{7}{2}$

Solution

1. $\binom{3}{2} = \dfrac{3!}{2!(3 - 2)!} = \dfrac{3!}{2! \cdot 1!} = \dfrac{3 \cdot 2 \cdot 1}{2 \cdot 1} = 3$

Note that this result agrees with the value of the coefficient stated earlier.

2. $\binom{4}{3} = \dfrac{4!}{3!(4 - 3)!} = \dfrac{4!}{3! \cdot 1!} = \dfrac{4 \cdot 3 \cdot 2 \cdot 1}{3 \cdot 2 \cdot 1 \cdot 1} = 4$

3. $\dbinom{7}{2} = \dfrac{7!}{2!(7-2)!}$

$$= \dfrac{7!}{2! \cdot 5!}$$

$$= \dfrac{7 \cdot 6 \cdot 5 \cdot 4 \cdot 3 \cdot 2 \cdot 1}{(2 \cdot 1)(5 \cdot 4 \cdot 3 \cdot 2 \cdot 1)}$$

$$= \dfrac{7 \cdot 6}{2 \cdot 1}$$

$$= 7 \cdot 3 = 21$$

Using the binomial coefficient notation, we may compute the $(k + 1)$th term in the expansion of the binomial $(a + b)^n$ according to the formula:

$$(k + 1)\text{th term} = \binom{n}{k} a^{n-k} b^k$$

Using this formula, we can restate the expansion of a binomial in a slightly different form.

BINOMIAL THEOREM (SECOND FORM)

Let n be a nonnegative integer. Then

$$(a + b)^n = \binom{n}{0} a^n + \binom{n}{1} a^{n-1} b + \binom{n}{2} a^{n-2} b^2 + \cdots + \binom{n}{n} b^n$$

Using the factorial formula for the binomial coefficients, we may easily calculate any particular term of a binomial expansion, as illustrated in the following example.

➤ **EXAMPLE 5**
Calculating a Term
in a Binomial Expansion

Determine the fifth term in the expansion of $(x + y)^9$.

Solution

We have

$$n = 9$$
$$k + 1 = 5$$
$$k = 4$$

The fifth term of the expansion is

$$\binom{9}{4} x^{9-4} y^4 = \dfrac{9!}{4!(9-4)!} x^5 y^4$$

$$= \dfrac{9!}{4!5!} x^5 y^4$$

$$= \dfrac{9 \cdot 8 \cdot 7 \cdot 6}{4 \cdot 3 \cdot 2 \cdot 1} x^5 y^4$$

Simplifying the fractional coefficient gives the equivalent expression

$$3 \cdot 7 \cdot 6 x^5 y^4 = 126 x^5 y^4$$

➤ **EXAMPLE 6**

Medicine

The recovery rate for a certain serious viral infection is 80%. A group of 100 persons are infected with the virus. The probability that exactly k will recover is given by the $(k + 1)$th term of binomial expansion $(0.8 + 0.2)^{100}$. Determine the probability that exactly three persons will recover from the infection.

Solution

The described probability is given by the fourth term in the expansion, which is equal to

$$\binom{100}{3}(0.8)^{97}(0.2)^3 = \frac{100!}{3!(100-3)!}(0.8)^{97}(0.2)^3$$

$$= \frac{100!}{3!97!}(0.8)^{97}(0.2)^3$$

$$= \frac{100 \cdot 99 \cdot 98}{3 \cdot 2 \cdot 1}(0.8)^{97}(0.2)^3$$

To compute a numerical value of this last expression, you should use a calculator with a power function. (Computing a 97th power by hand is not fun!) You will find that the value of the above expression is approximately $5.1467 \times 10^{-7} = 0.00000051467$.

Expansions Using the Binomial Theorem

The binomial theorem gives the expansion of a binomial $(a + b)^n$. By replacing a and b with algebraic expressions, we may use the binomial theorem to compute the expansion of any binomial to a nonnegative integer power. The following examples illustrate some typical expansions.

➤ **EXAMPLE 7**

Expansion of a More Complex Binomial

Determine the expansion of the following binomial:
$$(2x^2 - y)^3$$

Solution

We apply the binomial theorem with a replaced by $2x^2$ and b replaced by $-y$. The result is

$$(2x^2 - y)^3 = \binom{3}{0}(2x^2)^3(-y)^0 + \binom{3}{1}(2x^2)^2(-y)^1$$
$$+ \binom{3}{2}(2x^2)^1(-y)^2 + \binom{3}{3}(2x^2)^0(-y)^3$$

Now we replace the binomial coefficients by their respective values and simplify the various terms to obtain the expansion

$$(2x^2)^3 + 3(2x^2)^2(-y)^1 + 3(2x^2)^1(-y)^2 + (-y)^3 = 8x^6 - 12x^4y + 6x^2y^2 - y^3$$

➤ **EXAMPLE 8**

A Binomial Involving a Rational Expression

Determine the expansion of the following binomial:
$$\left(\frac{5}{t} - 2t\right)^4$$

Solution

By the binomial theorem, with a replaced by $5/t$ and b replaced by $-2t$, we have

$$\left(\frac{5}{t} - 2t\right)^4 = \binom{4}{0}\left(\frac{5}{t}\right)^4 + \binom{4}{1}\left(\frac{5}{t}\right)^3(-2t) + \binom{4}{2}\left(\frac{5}{t}\right)^2(-2t)^2$$

$$+ \binom{4}{3}\left(\frac{5}{t}\right)^1(-2t)^3 + \binom{4}{4}(-2t)^4$$

Inserting the values of the binomial coefficients, we have the following expansion:

$$\left(\frac{5}{t}\right)^4 + 4\left(\frac{5}{t}\right)^3(-2t) + 6\left(\frac{5}{t}\right)^2(-2t)^2 + 4\left(\frac{5}{t}\right)(-2t)^3 + (-2t)^4$$

Simplifying each of the terms, we obtain the following expansion:

$$\frac{625}{t^4} - \frac{1000}{t^2} + 600 - 160t^2 + 16t^4$$

➤ **EXAMPLE 9**

A Binomial Involving a Radical

Determine term 5 in the expansion of the following binomial:

$$(\sqrt{x} + 3y)^{18}$$

Solution

The desired term is given by the expression:

$$\binom{18}{4}(\sqrt{x})^{18-4}(3y)^4 = \frac{18!}{4!(18-4)!}(\sqrt{x})^{14}(3y)^4$$

$$= \frac{18 \cdot 17 \cdot 16 \cdot 15}{4 \cdot 3 \cdot 2 \cdot 1}x^7 \cdot 81y^4$$

$$= 247{,}860x^7y^4$$

Pascal's Triangle

Pascal

The 17th-century mathematician and philosopher Blaise Pascal discovered a simple method for computing binomial coefficients using only addition (as opposed to the methods just presented, which involve multiplying numbers that tend to be rather large).

To understand this method, list the binomial coefficients corresponding to $n = 1, 2, 3, \ldots$, with the coefficients corresponding to a given value of n written across a row, as follows:

$$1$$
$$1 \quad 1$$
$$1 \quad 2 \quad 1$$
$$1 \quad 3 \quad 3 \quad 1$$
$$1 \quad 4 \quad 6 \quad 4 \quad 1$$
$$1 \quad 5 \quad 10 \quad 10 \quad 5 \quad 1$$

Each row of the table may be computed from the preceding row as follows: The first element of the row is always 1. Each subsequent element is obtained by adding the two elements immediately above. For example, consider row 3. The first element is 1. The next element, 2, is obtained by adding the two elements immediately above in the preceding row, namely, $1 + 1$. The next element, 1, is obtained by adding a blank and a 1.

Consider now row 4. The first element is 1. The next is obtained by adding $1 + 2$ from the row above. The next element is obtained by adding $2 + 1$ from the row above. The final element is obtained by adding a blank and 1 from the row above.

In this manner, we may calculate the rows of the table one after another. Each row gives the set of binomial coefficients corresponding to one particular value of n. This method of computing the binomial coefficients is called **Pascal's triangle**.

The advantage of Pascal's triangle is that it allows you to calculate the binomial coefficients using addition rather than multiplication. The disadvantage, however, is that to apply Pascal's triangle, you must calculate all rows preceding the one you are interested in. If you need a row corresponding to a large value of n, then a considerable amount of calculation will be involved.

Exercises 8.1

Evaluate.

1. $0!$
2. $1!$
3. $8!$
4. $5!$
5. $\dfrac{6!}{4!2!}$
6. $\dfrac{8!}{5!3!}$
7. $\dfrac{7!}{2!5!}$
8. $\dfrac{10!}{6!4!}$
9. $\dbinom{6}{1}$
10. $\dbinom{8}{0}$
11. $\dbinom{9}{5}$
12. $\dbinom{12}{3}$

Determine the indicated term of the expansion.

13. Fourth, $(x + y)^7$
14. Sixth, $(a + b)^8$
15. 11th, $(t - 3)^{15}$
16. 10th, $(y - 4)^{13}$
17. Fourth, $(\sqrt{x} + 4t^3)^9$
18. Fifth, $\left(\dfrac{a}{5} - \dfrac{1}{a^2}\right)^{10}$
19. Middle two, $(\sqrt{t} - \sqrt{2})^9$
20. Middle, $(5a^2 + 7b^3)^8$
21. Last, $(a - b^2)^{100}$

Expand.

22. $(a + b)^4$
23. $(p - q)^5$
24. $(4a - b)^6$
25. $(a^2 + 3b)^5$
26. $(\sqrt{2} + 2\sqrt{3})^4$
27. $\left(t - \dfrac{t}{2}\right)^8$
28. $\left(\dfrac{a^2}{b} - b^3\right)^6$
29. $\left(2 + \dfrac{t}{2}\right)^8$
30. $(3x^2 + 2xy^3)^5$
31. $\left(\dfrac{3a^2}{2b} - \dfrac{2b^2}{3a}\right)^4$
32. $\left(\dfrac{1}{\sqrt{x}} + \sqrt{x}\right)^6$
33. $(a^3 - a^{-3})^5$
34. $(2 - 2)^n$
35. $(1 + 1)^n$
36. $(1 + \sqrt{2})^5 - (1 + \sqrt{2})^5$
37. $(1 - \sqrt{3})^4 + (1 - \sqrt{3})^4$

38. $(a + \sqrt{5})^6$
39. $(a - \sqrt{3})^4$
40. $(a^{2/3} + b^{1/3})^3$
41. $(2a - b)^{10}$, to four terms.
42. $(2p + 3q)^9$, to four terms.
43. $\left(\dfrac{1}{\sqrt{x}} - 3x^2\right)^8$, to three terms.
44. $(\sqrt{a} + \sqrt{b})^{10}$, to five terms.
45. $\left(3x - \dfrac{1}{y}\right)^{12}$, last four terms.
46. $\left(\dfrac{1}{a} + \dfrac{1}{3}a\sqrt{b}\right)^{10}$, last three terms.
47. Find the term that does not contain a in the expansion of $\left(\dfrac{3a^{-2}}{2} + \dfrac{1}{3}a\right)^{12}$.
48. Find the term containing $\dfrac{1}{t^3}$ in $\left(4 - \dfrac{5}{2}t\right)^9$.

◆ Applications

49. Before the advent of modern calculators and computers, mathematicians were always looking for ways to approximate the results of difficult or lengthy computations. For example, consider 1.01^{17}. An approximation might have been found by considering this as $1.01^{17} = (1.00 + 0.01)^{17}$. Use the sum of the first four terms of this binomial expansion to approximate 1.01^{17} to six decimal places. Approximate 1.01^{17} to six decimal places on your calculator. Compare the two approximations.

50. About 9% of the female population will contract breast cancer. There are 100 women at a sales meeting. The probability that exactly 4 of them will contract breast cancer in their lifetimes is given by the fifth term of the binomial expansion of $(0.91 + 0.09)^{100}$. Find this term and use your calculator to approximate the probability.

51. An electrical company manufactures computer chips. About 2% of the chips are defective. A quality control inspector pulls 50 chips at random off an assembly line. The probability that exactly 10 of them are defective is given by the 11th term of the binomial expansion of $(0.98 + 0.02)^{50}$. Find this term and use your calculator to approximate the probability.

52. At one point in a recent season, Eric Davis of the Los Angeles Dodgers was batting .320. Suppose he came to bat five times in one game. The probability of his getting at most four hits in the game is found by adding the first five terms of the binomial expansion of $(0.680 + 0.320)^5$. Find these terms and use your calculator to approximate the probability.

53. Use Pascal's triangle to find the coefficients of the binomial expansion of $(a + b)^n$ for:

   a. $n = 7$

   b. $n = 8$

   c. $n = 9$

   d. $n = 10$

   Then find each expansion.

54. Divide the fourth term of $(x^2 + 3yx^{1/3})^5$ by the third term and simplify.

55. Expand and simplify: $\dfrac{(x + h)^5 - x^5}{h}$

56. Expand and simplify: $\dfrac{(x + h)^n - x^n}{h}$

57. Find a formula for $(a + b + c)^n$.

58. Expand: $(a^2 + b^2 + c^2)^4$.

In calculus it can be shown that the formula for the binomial expansion can be altered in such a way that includes powers other than nonnegative integers. The formula is as follows:

$$(1 + x)^n = 1 + nx + \frac{n(n-1)}{2!}x^2 + \frac{n(n-1)(n-2)}{3!}x^3 + \cdots$$

where n is any real number, x is any real number such that $|x| < 1$, and the dots $\cdots$ mean that the expansion continues on indefinitely. Find each of the following expansions to eight terms, assuming $|x| < 1$:

59. $(1 + x)^{-1}$

60. $(1 - x)^{-2}$

61. $(1 + x)^{1/2}$

62. $(1 + x)^{1/3}$

Simplify.

63. $\dfrac{\dfrac{(n+2)^2 2^{n+1}}{(n+1)!}}{\dfrac{n^2 2^n}{n!}}$

64. $\dfrac{\dfrac{(3x+6)^{n+1}}{(n+1)!}}{\dfrac{(3x+6)^n}{n!}}$

8.2 MATHEMATICAL INDUCTION

Let n be a positive integer. Consider the following formula, which we will prove later in the section:

$$1 + 2 + \cdots + n = \frac{n(n+1)}{2}$$

It can be viewed as a series of statements, one corresponding to each value of n. Let's denote by P_n the preceding statement. Then P_1 is the statement

$$1 = \frac{1 \cdot (1+1)}{2}$$

P_2 is the statement

$$1 + 2 = \frac{2 \cdot (2+1)}{2}$$

and so forth. The following result can be used to prove such a series of statements.

> ### PRINCIPLE OF MATHEMATICAL INDUCTION
>
> Suppose that for each positive integer k there is given a statement P_k. Furthermore, suppose that the following two conditions are fulfilled:
>
> 1. P_1 is true.
> 2. By assuming that P_k is true, we can deduce the truth of P_{k+1}. (Here k is any positive integer.)
>
> Then P_k is true for every positive integer k.

The operation of the principle of mathematical induction is analogous to a row of dominoes. The truth of statement P_n is represented by the falling of the nth domino. Condition 1 is analogous to the first domino falling. Condition 2 is analogous to domino n knocking down domino $n + 1$. Conditions 1 and 2 together are analogous to all the dominoes falling. That is, the statement P_n is true for every value of n.

In more advanced courses, the principle of mathematical induction (in a slightly different form) is used as the fundamental tool for proving many of the most important properties of the natural numbers. In this text, we will assume its truth. The next three examples show how the principle of mathematical induction may be used as a technique of proof, called **proof by mathematical induction**.

➤ **EXAMPLE 1**
Sum of the First n Integers

Let n be a positive integer. Use mathematical induction to prove the formula

$$1 + 2 + \cdots + n = \frac{n(n + 1)}{2}$$

Solution

According to the principle of mathematical induction, we may prove the formula by verifying conditions 1 and 2. Condition 1 is simple, since P_1 is the statement

$$1 = \frac{1 \cdot (1 + 1)}{2}$$

and this is a true statement. As for condition 2, let's assume that statement P_n is true. That is, we assume that

$$1 + 2 + \cdots + n = \frac{n(n + 1)}{2}$$

Let's deduce from this statement the truth of statement P_{n+1}. To do so, start from the last equation and add $n + 1$ to each side to obtain

$$1 + 2 + \cdots + n + (n + 1) = \frac{n(n + 1)}{2} + (n + 1)$$

Let's simplify the right side by combining the two expressions found there:

$$1 + 2 + \cdots + n + (n + 1) = \frac{n(n + 1) + 2(n + 1)}{2}$$

Factoring $n + 1$ from each term in the numerator, we have the equation

$$1 + 2 + \cdots + n + (n + 1) = \frac{(n + 1)(n + 2)}{2}$$

But this is precisely the statement P_{n+1}. Indeed, if we take the statement P_n and replace n with $n+1$ everywhere, we arrive at the last equation. Thus, starting from the truth of P_n, we have deduced the truth of P_{n+1}. So condition 2 is verified. Since conditions 1 and 2 hold, the principle of mathematical induction says that P_n holds for all n. That is, the given formula holds for every positive integer n. This completes the proof.

Condition 2 requires that you prove that P_n implies P_{n+1}. This part of a proof by induction is often called the **induction step**.

The next example provides further practice in constructing proofs by induction.

➤ **EXAMPLE 2**
Sum of First n Odd Integers

Use mathematical induction to prove that for each positive integer n the following formula holds:

$$1 + 3 + 5 + \cdots + (2n - 1) = n^2$$

Solution

In this example, the statement P_1 is the formula

$$1 = 1^2$$

which is clearly true. So condition 1 holds. To prove the inductive step, assume that P_n holds. That is,

$$1 + 3 + 5 + \cdots + (2n - 1) = n^2$$

We must deduce P_{n+1} from this formula. That is, we must deduce from the last formula the result

$$1 + 3 + 5 + \cdots + [2(n + 1) - 1] = (n + 1)^2$$

To see how to proceed, let's rewrite what we wish to deduce. Explicitly write out the next-to-last term to obtain

$$1 + 3 + 5 + \cdots + (2n - 1) + [2(n + 1) - 1] = (n + 1)^2$$

Note that all the terms but the last form the expression on the left side of the statement P_n. This suggests that we start from P_n and add $2(n + 1) - 1 = 2n + 1$ to both sides of the equation to obtain

$$1 + 3 + \cdots + (2n - 1) + (2[n + 1] - 1) = n^2 + [2n + 1]$$

Factoring the right side yields

$$1 + 3 + \cdots + (2[n + 1] - 1) = (n + 1)^2$$

This is the statement P_{n+1} That is, we have deduced P_{n+1} from P_n. This is the inductive step and completes the proof of the formula.

Recall that in the preceding section, we stated and applied the binomial theorem, but did not supply a proof. Let's now remedy that omission.

➤ **EXAMPLE 3**
Binomial Theorem

Let n be a positive integer. Use mathematical induction to prove the binomial theorem:

$$(a + b)^n = a^n + \binom{n}{1} a^{n-1}b + \binom{n}{2} a^{n-2}b^2 + \cdots + \binom{n}{n-1} ab^{n-1} + b^n$$

Solution

Here P_1 is the statement

$$(a + b)^1 = a^1 + b^1$$

This is clearly true, so let's prove the inductive step. Assume that P_n is true. Namely, assume that

$$(a + b)^n = a^n + \binom{n}{1}a^{n-1}b + \binom{n}{2}a^{n-2}b^2 + \cdots + \binom{n}{n-1}ab^{n-1} + b^n$$

Let's multiply both sides of the equation by $a + b$. On the left, we obtain $(a + b)^{n+1}$, so that the equation reads

$$(a+b)^{n+1} = (a+b)\left[a^n + \binom{n}{1}a^{n-1}b + \binom{n}{2}a^{n-2}b^2 + \cdots + \binom{n}{n-1}ab^{n-1} + b^n\right]$$

Apply the distributive law to the right side to obtain

$$(a + b)^{n+1} =$$

$$\left[\binom{n}{0}a^{n+1} + \binom{n}{1}a^n b + \binom{n}{2}a^{n-1}b^2 + \cdots + \binom{n}{n-1}a^2 b^{n-1} + \binom{n}{n}ab^n\right]$$

$$+ \left[\binom{n}{0}a^n b + \binom{n}{1}a^{n-1}b^2 + \binom{n}{2}a^{n-2}b^3 + \cdots + \binom{n}{n-1}ab^n + \binom{n}{n}b^{n+1}\right]$$

Collecting like terms, we then have

$$(a + b)^{n+1} = \binom{n}{0}a^{n+1} + \left[\binom{n}{0} + \binom{n}{1}\right]a^n b + \left[\binom{n}{1} + \binom{n}{2}\right]a^{n-1}b^2$$

$$+ \left[\binom{n}{2} + \binom{n}{3}\right]a^{n-2}b^3 + \cdots + \left[\binom{n}{n-1} + \binom{n}{n}\right]ab^n + \binom{n}{n}b^{n+1}$$

$$= a^{n+1} + \left[\binom{n}{0} + \binom{n}{1}\right]a^n b + \left[\binom{n}{1} + \binom{n}{2}\right]a^{n-1}b^2$$

$$+ \left[\binom{n}{2} + \binom{n}{3}\right]a^{n-2}b^3 + \cdots + \left[\binom{n}{n-1} + \binom{n}{n}\right]ab^n + b^{n+1}$$

To complete simplification of the right side, let's obtain an alternate formula for the sums of binomial coefficients that appear. Let j be one of the integers $0, 1, \ldots, n - 1$. Then we have

$$\binom{n}{j} + \binom{n}{j+1} = \frac{n!}{j!(n-j)!} + \frac{n!}{(j+1)![n-(j+1)]!}$$

$$= n!\left[\frac{1}{j!(n-j)!} + \frac{1}{j! \cdot (j+1) \cdot (n-j-1)!}\right]$$

$$= \frac{n!}{j!}\left[\frac{1}{(n-j-1)! \cdot (n-j)} + \frac{1}{(j+1) \cdot (n-j-1)!}\right]$$

$$= \frac{n!}{j!(n-j-1)!}\left(\frac{1}{n-j} + \frac{1}{j+1}\right)$$

$$= \frac{n!}{j!(n-j-1)!} \cdot \frac{n+1}{(n-j)(j+1)}$$

$$= \frac{(n+1)!}{(j+1)!(n-j)!}$$

$$= \binom{n+1}{j+1}$$

Using this formula, we replace the sums of the binomial coefficients in the preceding equation to obtain

$$(a + b)^{n+1} = a^{n+1} + \binom{n+1}{1}a^n b + \binom{n+1}{2}a^{n-1}b^2 + \binom{n+1}{3}a^{n-2}b^3$$
$$+ \cdots + \binom{n+1}{n}ab^n + b^{n+1}$$

This equation is precisely statement P_{n+1}. This completes the inductive step and, with it, the proof of the binomial theorem.

Note that in the first two examples, we added a term to each side of the formula to obtain the inductive step. In the third example, the inductive step was more complicated and involved first multiplying both sides of the equation by $a + b$, then performing considerable simplification, and finally proving an identity about binomial coefficients. This is fairly typical of the ingenuity that must sometimes be employed to prove the inductive step.

All the preceding examples used induction to prove formulas. Induction may also be used to prove inequalities, as the next example shows.

➤ **EXAMPLE 4**

Inductive Proof of an Inequality

Let n be a positive integer. Prove that $2n + 1 \leq 3^n$.

Solution

In this case, the statement P_1 is the true inequality

$$2 \cdot 1 + 1 \leq 3^1$$
$$3 \leq 3$$

Let's now prove the inductive step. We assume that P_n holds. That is, we assume that

$$2n + 1 \leq 3^n$$

We must prove (somehow) that

$$2(n + 1) + 1 \leq 3^{n+1}$$

To prove this, let's start from the assumed inequality and add 2 to both sides to obtain

$$(2n + 1) + 2 \leq 3^n + 2$$
$$2(n + 1) + 1 \leq 3^n + 2$$

Since n is a positive integer, we have

$$3^n \geq 3 > 2$$

Therefore, applying this fact to the preceding inequality, we have

$$2(n + 1) + 1 \leq 3^n + 2$$
$$< 3^n + 3^n \qquad \textit{since } 3^n > 2$$
$$= 2 \cdot 3^n$$
$$\leq 3 \cdot 3^n = 3^{n+1}$$
$$2(n + 1) + 1 \leq 3^{n+1}$$

This last inequality is the assertion for $n + 1$. This proves the inductive step and completes the proof of the inequality.

Exercises 8.2

Use mathematical induction to prove that the following formulas hold for all positive integers n.

1. $1 + 5 + 9 + \cdots + (4n - 3) = 2n^2 - n$

2. $1 + 4 + 7 + \cdots + (3n - 2) = \dfrac{n(3n - 1)}{2}$

3. $1 \cdot 2 + 3 \cdot 4 + 5 \cdot 6 + \cdots + (2n - 1)2n = \dfrac{n(n + 1)(4n - 1)}{3}$

4. $1^2 + 3^2 + 5^2 + \cdots + (2n - 1)^2 = \dfrac{n(4n^2 - 1)}{3}$

5. $\dfrac{1}{2} + \dfrac{1}{4} + \dfrac{1}{8} + \cdots + \dfrac{1}{2^n} = \dfrac{2^n - 1}{2^n}$

6. $3 + 3^2 + 3^3 + \cdots + 3^n = \dfrac{3^{n+1} - 3}{2}$

Prove by mathematical induction that the following are true for all positive integers n.

7. $1^3 + 2^3 + 3^3 + \cdots + n^3 = \left[\dfrac{n(n + 1)}{2}\right]^2$

8. $\dfrac{1}{1 \cdot 2} + \dfrac{1}{2 \cdot 3} + \dfrac{1}{3 \cdot 4} + \cdots + \dfrac{1}{n(n + 1)} = \dfrac{n}{n + 1}$

9. $1^2 + 2^2 + 3^2 + \cdots + n^2 = \dfrac{n(n + 1)(2n + 1)}{6}$

10. $2 + 2^2 + 2^3 + \cdots + 2^n = 2(2^n - 1)$

11. $2 + 4 + 6 + \cdots + 2n = n^2 + n$

12. $1 + 3 + 5 + \cdots + (2n - 1) = n^2$

13. $1 + 2 + 3 + \cdots + n = \dfrac{n(n + 1)}{2}$

14. $1 + 2 + 3 + \cdots + n < \dfrac{(2n + 1)^2}{8}$

15. $n(n^2 + 5)$ is a multiple of 6.

16. $3^{2n} - 1$ is a multiple of 8.

17. 3 is a factor of $n^3 - n + 3$.

18. 4 is a factor of $5^n - 1$.

19. $n^2 + n$ is divisible by 2.   20. $n^3 - n$ is divisible by 6.

21. 7 is a factor of $11^n - 4^n$.   22. 3 is a factor of $n^3 + 2n$.

23. $4^n > n^4$ for $n \geq 5$. (*Hint:* Start by proving P_5 true.)

24. $3^n < 3^{n+1}$

25. $n^2 < 2^n$ for $n > 4$.   26. $2n < 2^n$ for $n \geq 3$.

For the following $f(n)$: (*a*) complete the table; (*b*) on the basis of the results in the table, what would you believe to be the value of $f(7)$? Compute $f(7)$; (*c*) make a conjecture about the value of $f(n)$.

27. $f(n) = 1 + 2 + 4 + \cdots + 2^{n-1}$

n	1	2	3	4	5	6
$f(n)$						

28. $f(n) = 3 + 7 + 11 + \cdots + (4n - 1)$

n	1	2	3	4	5	6
$f(n)$						

29. $f(n) = 2 + 9 + 16 + \cdots + (7n - 5)$

n	1	2	3	4	5	6
$f(n)$						

30. $f(n) = 1 + 3 + 5 + \cdots + (2n - 1)$

n	1	2	3	4	5	6
$f(n)$						

31. Prove that it is possible to pay any debt of \$4, \$5, \$6, \$7, ..., \$n by using only \$2 and \$5 bills.

32. Prove that it is possible to pay any debt of \$8, \$9, \$10, ..., \$n by using only \$3 and \$5 bills.

33. Prove that the number of diagonals of an n-sided convex polygon is $n(n - 3)/2$ for $n \geq 3$.

34. Prove that the sum of the measures of the angles in an n-sided convex polygon is $180°(n - 2)$ for $n \geq 3$.

35. If, in a room with n people ($n \geq 2$), every person shakes hands once with every other person, prove that there will be $n(n - 1)/2$ handshakes.

36. Use mathematical induction to prove $[r(\cos\theta + i\sin\theta)]^n = r^n(\cos n\theta + i\sin n\theta)$ (De Moivre's theorem) for n a positive integer.

37. Prove: $\sin(\theta + n\pi) = (-1)^n \sin\theta$ for all positive integers n.

38. Prove: $\cos(\theta + n\pi) = (-1)^n \cos\theta$ for all positive integers n.

Find the smallest positive integer n for which the given statement is true. Then prove that the statement is true for all integers greater than or equal to that smallest value.

39. $n + 5 < 2^n$ 40. $\log n < n$

41. $3^n > 2^n + 20$

42. $(1 + x)^n \geq 1 + nx$, if $x \geq -1$

For the following problems, tell what you can conclude from the given information about the sequence of statements. (*Example:* Given P_4 is true and that P_k implies P_{k+1} for any k, then you may conclude that P_n is true for every integer $n \geq 4$.)

43. P_{18} is true, and P_k implies P_{k+1}.

44. P_{18} is not true, and P_k implies P_{k+1}.

45. P_1 is true, but P_k does not imply P_{k+1}.

46. P_1 is true, and P_k implies P_{k+2}.

47. P_1 and P_2 are true, and P_k and P_{k+1} together imply P_{k+2}.

48. P_1 is true, and P_k implies P_{4k}.

8.3 SEQUENCES AND SERIES

Many applications involve quantities that are reported at discrete intervals. For instance, the balance on a mortgage is computed monthly, the unemployment rate is reported monthly, and the temperature is reported hourly by the weather bureau. To record and manipulate such data, mathematicians use sequences.

Definition 1
Sequence, Terms

A **sequence** is a collection of real numbers organized in a particular order. The numbers contained in a sequence are called **terms**.

Here are some examples of sequences:

$$1, \frac{1}{2}, \frac{1}{4}, \frac{1}{8}, \dots$$

$$1, -1, 1, -1, \dots$$

$$5, 4, 3, 2, 1$$

The first two sequences above contain infinitely many terms and accordingly are called **infinite sequences**. The third sequence contains only finitely many terms and is therefore called a **finite sequence**. The terms of a sequence are customarily labeled using positive integer subscripts to indicate the term's position. For example, here is how a typical sequence is written using subscript notation:

$$a_1, a_2, a_3, \dots, a_n, \dots$$

That is, a_1 denotes the first term of the sequence, a_2 the second term, and a_n the nth term.

One method for defining a sequence (finite or infinite) is to specify a formula for computing a_n in terms of n. In the next example, we calculate terms of sequences specified in this fashion.

➢ **EXAMPLE 1**
Determining Terms of Sequences

Determine terms 1, 2, 3, and 7 of the sequences defined by

1. $a_n = \dfrac{1}{n}$

2. $a_n = 1 + \dfrac{(-1)^n}{n}$

3. $a_n = \dfrac{n^n}{n!}$

Solution

In each case, we calculate a_n by substituting the value of n in the given formula for a_n.

1. $a_1 = \dfrac{1}{1} = 1$

$a_2 = \dfrac{1}{2}$

$a_3 = \dfrac{1}{3}$

$a_7 = \dfrac{1}{7}$

2. $a_1 = 1 + \dfrac{(-1)^1}{1} = 1 + \dfrac{-1}{1} = 0$

$a_2 = 1 + \dfrac{(-1)^2}{2} = 1 + \dfrac{1}{2} = \dfrac{3}{2}$

$a_3 = 1 + \dfrac{(-1)^3}{3} = 1 + \dfrac{-1}{3} = \dfrac{2}{3}$

$a_7 = 1 + \dfrac{(-1)^7}{7} = 1 + \dfrac{-1}{7} = \dfrac{6}{7}$

3. $a_1 = \dfrac{1^1}{1!} = \dfrac{1}{1} = 1$

$a_2 = \dfrac{2^2}{2!} = \dfrac{2 \cdot 2}{1 \cdot 2} = 2$

$a_3 = \dfrac{3^3}{3!} = \dfrac{3 \cdot 3 \cdot 3}{1 \cdot 2 \cdot 3} = \dfrac{9}{2}$

$a_7 = \dfrac{7^7}{7!} = \dfrac{7^6}{1 \cdot 2 \cdot 3 \cdot 4 \cdot 5 \cdot 6} = \dfrac{117,649}{720}$

All the sequences of the preceding example were defined by formulas for calculating a_n in terms of n. In many applications, it is most convenient to define a sequence by specifying how to calculate a_n in terms of preceding terms of the sequence $a_1, a_2, \ldots, a_{n-1}$. Such sequences are said to be defined **recursively**.

For instance, let a_n denote the amount of money in a bank account at the end of n months, where the account earns 0.5% interest per month. Then the amount a_n is related to the amount at the end of the preceding month a_{n-1} by the formula

$$a_n = 1.005a_{n-1}$$

If the amount in the account at the end of month 1 is \$1000, then $a_1 = 1000$. The two equations, $a_n = 1.005a_{n-1}$ and $a_1 = 1000$, suffice to define the sequence of bank balances recursively.

➤ **EXAMPLE 2**
Calculating Terms
from One Another

Determine the first five terms of the previously defined bank balance sequence.

Solution

The value of a_1 gives the first term. Each successive term may be computed from its predecessor using the given formula. Here are the calculations for the first five terms:

$$a_1 = 1000$$
$$a_2 = 1.005 \cdot a_1 = 1.005 \cdot 1000 = 1005$$

$$a_3 = 1.005 \cdot a_2 = 1.005 \cdot 1005 = 1010.025$$
$$a_4 = 1.005 \cdot a_3 = 1.005 \cdot 1010.025 = 1015.0751$$
$$a_5 = 1.005 \cdot a_4 = 1.005 \cdot 1015.0751 = 1020.1505$$

The sum of the first n terms of a sequence is called the *nth partial sum* and is denoted S_n. That is,

$$S_n = a_1 + a_2 + \cdots + a_n$$

Thus, for example, if we consider the sequence $a_n = 2^{-n}$, then

$$S_1 = a_1 = 2^{-1} = \frac{1}{2}$$

$$S_2 = a_1 + a_2$$
$$= 2^{-1} + 2^{-2}$$
$$= \frac{1}{2} + \frac{1}{4} = \frac{3}{4}$$

$$S_3 = a_1 + a_2 + a_3$$
$$= 2^{-1} + 2^{-2} + 2^{-3}$$
$$= \frac{1}{2} + \frac{1}{4} + \frac{1}{8} = \frac{7}{8}$$

A sequence is a function whose domain is the set of positive integers. For each value n of the subscript, there is a value a_n of the corresponding term of the sequence. In dealing with sums of consecutive terms of a sequence, it is often convenient to use a special notation:

Definition 2
Summation Notation,
Summation Index

Suppose that A and B are integers with $A < B$. The symbol

$$\sum_{n=A}^{B} a_n$$

is another way of denoting the sum:

$$a_A + a_{A+1} + \cdots + a_B$$

The notation

$$\sum_{n=A}^{B} a_n$$

says to add up the terms a_n, where the variable n starts at A (written underneath the $\sum$) and ends with B (written above the $\sum$). The variable n is called the **summation index**.

The next example provides some practice in using summation notation.

> **EXAMPLE 3**
Evaluating Sums Given
by Summation Notation

Evaluate the following expressions:

1. $\displaystyle\sum_{n=1}^{5} n$ 2. $\displaystyle\sum_{n=2}^{5} n^2$ 3. $\displaystyle\sum_{n=0}^{3} \binom{3}{n}$

Solution

1. We write out the terms of the summation by substituting, in succession, the values $n = 1, 2, 3, 4, 5$:

$$\sum_{n=1}^{5} n = 1 + 2 + 3 + 4 + 5 = 15$$

2. The terms of the summation are obtained by evaluating n^2 in succession, for $n = 2, 3, 4, 5$:

$$\sum_{n=2}^{5} n^2 = 2^2 + 3^2 + 4^2 + 5^2 = 54$$

3. The terms of the summation are obtained by substituting, in succession, $n = 0, 1, 2, 3$:

$$\sum_{n=0}^{3} \binom{3}{n} = \binom{3}{0} + \binom{3}{1} + \binom{3}{2} + \binom{3}{3}$$
$$= 1 + 3 + 3 + 1$$
$$= 8$$

Summation notation provides a convenient way of writing many mathematical facts in a shorthand form. For instance, the definition of the nth partial sum of a sequence $a_1, a_2, \ldots, a_n, \ldots$ may be written in summation notation as

$$S_n = \sum_{j=1}^{n} a_j$$

Note that in this formula, we used j as the summation index. The right side of the previous equation says to add up all the terms a_j for j beginning with 1 and ending with n. Note that we couldn't use n to denote the summation index, since this letter is already in use, namely, as the index of the last term to sum. You may use any variable name you wish for the summation index so long as it can't be confused with a variable in use for some other purpose (such as n in the previous equation). For instance, we could have just as easily denoted the summation index k and written the last formula in the form

$$\sum_{k=1}^{n} a_k$$

Since the summation index may be replaced by any unused variable, it is often referred to as a **dummy variable**. The most common are i, j, and k.

Following are some further applications of summation notation. Here is a formula we proved by induction in the preceding section:

$$1 + 2 + 3 + \cdots + N = \frac{N(N+1)}{2}$$

In summation notation, this formula may be written

$$\sum_{n=1}^{N} n = \frac{N(N+1)}{2}$$

The binomial theorem states that for a positive integer N, we have

$$(a + b)^N = a^N + \binom{N}{1}a^{N-1}b + \binom{N}{2}a^{N-2}b^2 + \cdots + \binom{N}{N-1}ab^{N-1} + b^N$$

In terms of summation notation, this theorem may be written

$$(a + b)^N = \sum_{n=0}^{N} \binom{N}{n}a^{N-n}b^n$$

Indeed, the first term in the previous summation corresponds to $n = 0$:

$$\binom{N}{0}a^{N-0}b^0 = 1 \cdot a^N \cdot 1 = a^N$$

The second term corresponds to $n = 1$:

$$\binom{N}{1}a^{N-1}b^1 = \binom{N}{1}a^{N-1}b$$

The final term corresponds to $n = N$:

$$\binom{N}{N}a^{N-N}b^N = 1 \cdot a^0 b^N = b^N$$

As you can see, the terms in the summation notation correspond exactly to the terms in the right side of the binomial theorem. As another example of an application of the summation notation, consider the polynomial

$$a_0 + a_1 x + a_2 x^2 + \cdots + a_N x^N$$

In terms of the summation notation, this polynomial may be written

$$\sum_{n=0}^{N} a_n x^n$$

Exercises 8.3

Determine terms 1, 2, 3, and 8 of the sequences defined by:

1. $a_n = n + 2$
2. $a_n = n - 4$
3. $a_n = \dfrac{n+1}{n-2}$

4. $a_n = \dfrac{n-1}{n+1}$
5. $a_n = 2^{n-3}$
6. $a_n = 3^{n-4}$

7. $a_n = n^{-2}$
8. $a_n = n^3$

9. $a_n = (-2)\sqrt[n]{n}$
10. $a_n = (-1)^{n-1}\sqrt{2n}$

11. $a_n = 3^{n-1} - 3^n$
12. $a_n = 2^{n+1} - 2^{n-1}$

13. $a_n = \dfrac{n^2-1}{n^2-2}$
14. $a_n = \dfrac{n^3+1}{n^3-2}$

15. $a_n = -n\cos n\pi$
16. $a_n = -n\sin n\pi$

17. $a_n = \sin\left(\dfrac{-3n\pi}{2}\right)$
18. $a_n = \cos\left(\dfrac{5n\pi}{2}\right)$

Find the first five terms for the following recursively defined sequences:

19. $a_1 = 2, a_n = a_{n-1} + 5, n > 1$

20. $a_1 = -3, a_n = 2a_{n-1}, n > 1$

21. $a_1 = 9, a_n = \dfrac{1}{3}a_{n-1}, n > 1$

22. $a_1 = 1, a_n = \left(-\dfrac{1}{3}\right)^n a_{n-1}, n > 1$

23. $a_n = n + \dfrac{1}{n}$
24. $a_n = n^3 + 1$

25. $a_n = \left(-\dfrac{1}{2}\right)^n (n^{-1})$
26. $a_n = \dfrac{1}{(n+1)!}$

27. $a_n = \dfrac{1}{2n}\log 1000^n$
28. $a_n = n\log 10^{n-1}$

29. $a_1 = 3, a_{n+1} = \dfrac{1}{a_n}, n \geq 1$

30. $a_1 = -2, a_{n+1} = -a_n^{-1}, n \geq 1$

31. $a_1 = -2, a_{n+1} = \log 100^{a_n}, n \geq 1$

32. $a_n = -n\sin(-n\pi)$

Evaluate the following expressions:

33. $\sum_{n=1}^{4}(n-2)$ 34. $\sum_{n=1}^{3}(2n+1)$ 35. $\sum_{n=1}^{3}(n^2+1)$

36. $\sum_{n=1}^{5}(-2)^n$ 37. $\sum_{n=2}^{4}n^{-1}$ 38. $\sum_{n=0}^{3}(4n-1)$

39. $\sum_{n=0}^{7}\log 100^n$ 40. $\sum_{n=1}^{6}\log 1000^{n-1}$

41. $\sum_{n=3}^{10}\left(-\cos\frac{n\pi}{2}\right)$ 42. $\sum_{n=0}^{6}\cos(-n\pi)$

43. $\sum_{n=1}^{50}3$ 44. $\sum_{n=2}^{200}-1$

45. $\sum_{n=1}^{10}[1+(-1)^n]$ 46. $\sum_{n=1}^{20}(-2n)$

For each of the following sequences, find the general or nth term. Answers may vary.

47. $1, 3, 5, 7, 9, \ldots$

48. $3, 9, 27, 81, 243, \ldots$

49. $2, -4, 8, -16, 32, -64, \ldots$

50. $1, -\dfrac{1}{2}, \dfrac{1}{3}, -\dfrac{1}{4}, \dfrac{1}{5}, -\dfrac{1}{6}, \ldots$

51. $\dfrac{1}{2}, \dfrac{1}{8}, \dfrac{1}{32}, \dfrac{1}{128}, \ldots$ 52. $17, 21, 25, 29, \ldots$

Rewrite each of the following summations using sigma notation.

53. $1 + \dfrac{1}{2} + \dfrac{1}{3} + \dfrac{1}{4} + \dfrac{1}{5}$

54. $3 + 6 + 12 + 24 + 48 + 96$

55. $1 + 2 + 3 + 4 + 5 + 6 + 7 + 8 + 9 + 10$

56. $-2 + 4 - 8 + 16 - 32 + 64 - 128 + 256$

57. $-3 + 9 - 27 + 81 - 243 + 729$

58. $1 + 5 + 9 + 13 + 17$

Solve the following equations for x:

59. $\sum_{n=4}^{6}nx = 30$ 60. $\sum_{n=1}^{12}(nx-5) = 64$

61. $\sum_{n=3}^{5}(nx+3) = 45$ 62. $\sum_{n=1}^{7}(-1)^n x = -2$

63. $\sum_{n=0}^{5}n(n-x) = 25$ 64. $\sum_{n=1}^{6}(x-3n) = -3$

8.4 ARITHMETIC SEQUENCES

In this section and the next, we study two particular types of sequences that play an important role in both pure and applied problems, namely, the arithmetic and geometric sequences.

Definition 1
Arithmetic Sequence

An **arithmetic sequence** is a sequence of real numbers in which the difference between consecutive terms is constant.

➤ **EXAMPLE 1**
Proving a Sequence
Is Arithmetic

Prove that the following sequence is arithmetic:

$$6, 11, 16, \ldots, 5n + 1, \ldots$$

Solution

Let's compute the difference between two typical consecutive terms, namely, a_{n+1} and a_n. We have

$$a_{n+1} - a_n = [5(n+1) + 1] - (5n+1)$$
$$= (5n+6) - (5n+1)$$
$$= 5$$

This computation shows that the differences between consecutive terms are constant and, in fact, all equal to 5. Thus, the sequence is arithmetic.

Suppose that an arithmetic sequence has first term a_1 and that the difference between consecutive terms is d. Then the terms of the sequence are

$$a_1$$
$$a_2 = a_1 + d$$
$$a_3 = a_2 + d$$
$$= (a_1 + d) + d$$
$$= a_1 + 2d$$
$$a_4 = a_3 + d$$
$$= (a_1 + 2d) + d$$
$$= a_1 + 3d$$

These formulas generalize to the following result.

---

nTH TERM OF AN ARITHMETIC SEQUENCE

Suppose that an arithmetic sequence has first term a_1 and difference between consecutive terms d. Then the nth term of the sequence is given by the formula

$$a_n = a_1 + (n - 1)d$$

Proof   We can proceed by induction. We have shown that the formula holds for $n = 1, 2, 3, 4$. Assume now that the formula holds for n. That is, assume that

$$a_n = a_1 + (n - 1)d$$

We want to prove that

$$a_{n+1} = a_1 + [(n + 1) - 1]d = a_1 + nd$$

Since the sequence is arithmetic with difference d, we see that

$$a_{n+1} - a_n = d$$
$$a_{n+1} = a_n + d$$
$$= [a_1 + (n - 1)d] + d \quad \text{(By assumption)}$$
$$= a_1 + nd$$
$$= a_1 + [(n + 1) - 1]d$$

This last equation is the assertion of the formula for $n + 1$. This completes the inductive step and the proof of the formula.   ◆

➤ **EXAMPLE 2**
Evaluation of the nth Term of an Arithmetic Sequence

Suppose that the first term of an arithmetic sequence is 5 and the difference between consecutive terms is 7. What is the value of the 30th term of the sequence?

Solution

By the formula for the nth term of an arithmetic sequence, with $a_1 = 5$ and $d = 7$, we see that

$$a_{30} = a_1 + (30 - 1)d$$
$$= 5 + 29 \cdot 7$$
$$= 208$$

> **EXAMPLE 3**
Calculating First Term
and Difference of an Arithmetic
Sequence

Suppose that $a_1, a_2, \ldots$ is an arithmetic sequence. Suppose that $a_{15} = 40$ and $a_{20} = 50$. Determine the first term and the difference d between consecutive terms of the sequence.

Solution

According to the formula for the nth term of an arithmetic sequence and the given information, we have

$$a_{15} = 40$$
$$a_1 + (15 - 1)d = 40 \qquad (1)$$
$$a_1 + 14d = 40$$
$$a_{20} = 50$$
$$a_1 + 19d = 50 \qquad (2)$$

Subtracting equation (1) from equation (2), we have

$$(19 - 14)d = 50 - 40$$
$$5d = 10$$
$$d = 2$$

Substituting this value of d into equation (1) gives us

$$a_1 + 14d = 40$$
$$a_1 + 14 \cdot 2 = 40$$
$$a_1 = 12$$

Thus, the sequence has first term 12 and difference 2.

The following result provides a simple formula for calculating the sum of an arithmetic sequence.

SUM OF AN ARITHMETIC SEQUENCE

Let $a_1, a_2, a_3, \ldots$ be an arithmetic sequence with difference d, and let S_n be its nth partial sum. Then

$$S_n = \frac{n}{2}(a_1 + a_n)$$
$$= na_1 + \frac{n(n - 1)}{2}d$$

Proof   From the definition of the nth partial sum, we have

$$S_n = a_1 + a_2 + \cdots + a_n$$

Let's now substitute the arithmetic sequence formula for each term on the right side:

$$S_n = a_1 + (a_1 + d) + (a_1 + 2d) + \cdots + [a_1 + (n - 1)d]$$
$$= (a_1 + a_1 + \cdots + a_1) + [d + 2d + \cdots + (n - 1)d]$$
$$= na_1 + [1 + 2 + \cdots + (n - 1)]d \qquad (3)$$

In section 2 we proved the formula

$$1 + 2 + \cdots + n = \frac{n(n+1)}{2}$$

Replacing n by $n - 1$ in this formula, we see that

$$1 + 2 + \cdots + (n-1) = \frac{(n-1)(n-1+1)}{2} = \frac{n(n-1)}{2}$$

Substituting this formula into equation (3), we have

$$S_n = na_1 + \frac{n(n-1)}{2}d$$

This yields the second formula stated. To get the first formula, rewrite the last formula as follows:

$$S_n = \frac{n}{2}[2a_1 + (n-1)d]$$

$$= \frac{n}{2}[a_1 + a_1 + (n-1)d]$$

$$= \frac{n}{2}(a_1 + a_n)$$

This is the desired statement.   ◆

➤ **EXAMPLE 4**
Calculating Sum
of an Arithmetic Sequence

Compute the sum of the first 50 terms of the arithmetic sequence $1, 4, 7, 10, \ldots,$ $3n - 2, \ldots.$

Solution

The sequence has $a_1 = 1, d = 3$. Therefore, by the preceding result, we have

$$S_n = na_1 + \frac{n(n-1)}{2}d$$

$$S_{50} = 50 \cdot 1 + \frac{50 \cdot (50-1)}{2} \cdot 3 = 3725$$

EXAMPLE 5
Seats in an Auditorium

A concert auditorium has 30 rows of seats. The first row contains 50 seats. As you move to the rear of the auditorium, each row has two more seats than the previous one. How many seats are in the auditorium?

Solution

Let a_n denote the number of seats in the nth row. Then the sequence $a_1, a_2, \ldots,$ a_{30} is an arithmetic sequence with first term 50 and difference 2. The number of seats in the auditorium is the sum of the first 30 terms of the sequence, or S_{30}. By the formula for the sum of an arithmetic sequence, this quantity equals

$$S_n = na_1 + \frac{n(n-1)}{2}d$$

$$S_{30} = 30 \cdot 50 + \frac{30 \cdot 29}{2} \cdot 2 = 2370$$

That is, the auditorium has 2370 seats.

➤ **EXAMPLE 6**
Evaluating the Sum of Terms
in an Arithmetic Sequence

Evaluate the sum

$$\sum_{n=1}^{10} (5n + 3)$$

Solution

The numbers being summed are

$$5 \cdot 1 + 3, 5 \cdot 2 + 3, \ldots, 5 \cdot 10 + 3$$

an arithmetic sequence with first term 8 and last term 53. By the formula for the sum of an arithmetic sequence, the sum of these 10 terms is equal to

$$S_{10} = \frac{10}{2}(a_1 + a_{10})$$

$$= 5 \cdot (8 + 53)$$

$$= 305$$

Exercises 8.4

Prove that the following sequences are arithmetic.

1. $1, 3, 5, 7, \ldots, 2n - 1, \ldots$

2. $4, 5, 6, \ldots, n + 3, \ldots$ 3. $4, 7, 10, \ldots, 1 + 3n, \ldots$

4. $0, 2, 4, \ldots, 2n - 2, \ldots$ 5. $7, 8, 9, \ldots, n + 6, \ldots$

6. $9, 14, 19, \ldots, 4 + 5n, \ldots$

7. $1, -1, -3, \ldots, 3 - 2n, \ldots$

8. $3, -1, -5, \ldots, 7 - 4n, \ldots$

Find the indicated term for the given arithmetic sequences.

9. $a_1 = 2, d = -3$; find a_{10}.

10. $a_1 = -5, d = 7$; find a_{12}.

11. $a_1 = 0, d = 9$; find a_{30}.

12. $a_1 = 4, d = -5$; find a_{40}.

13. $a_1 = 9, d = -4$; find a_8.

14. $a_1 = -7, d = -4$; find a_{20}.

15. $a_1 = -3, d = -9$; find a_{19}.

16. $a_1 = 0, d = \frac{1}{2}$; find a_{35}.

17. $a_1 = 3, d = -\frac{1}{2}$; find a_{40}.

18. $a_1 = -\frac{1}{2}, d = 2$; find a_8.

For each of the following arithmetic sequences, find the first term, a_1, and the difference d between consecutive terms of the sequence.

19. $a_{10} = -26, a_{14} = -7$ 20. $a_{18} = 49, a_{20} = 28$

21. $a_{15} = 16, a_{19} = -12$ 22. $a_{10} = 3, a_{16} = 21$

23. $a_5 = 3, a_{50} = 30$ 24. $a_5 = \frac{1}{2}, a_{20} = \frac{7}{8}$

25. $a_{17} = -40, a_{28} = -73$ 26. $a_{10} = -11, a_{40} = -71$

27. $a_{20} = 66, a_{59} = 222$

28. $a_{16} = 60 \ 1/2, a_{41} = 160 \ 1/2$

The following exercises refer to arithmetic sequences.

29. If $d = 4$ and $a_8 = 33$, what is a_1?

30. If $a_1 = 8$ and $a_{11} = 26$, what is d?

31. If $a_7 = \frac{7}{3}$ and $d = -\frac{2}{3}$, what is a_{15}?

32. If $d = -7$ and $a_3 = 37$, what is a_{17}?

33. If $a_1 = -2, d = 7$, and $a_n = 138$, what is n?

34. If $a_3 = -7, d = -5$, and $a_n = -142$, what is n?

35. If $a_2 = 5$ and $a_4 = 1$, what is a_{10}?

36. If $a_4 = 5$ and $d = -2$, what is a_1?

37. If $a_3 = 5, d = -3$, and $a_n = -76$, what is n?

38. If $a_3 = 25, d = -14$, and $a_n = -507$, what is n?

Compute the indicated sums for the following arithmetic sequences.

39. $3, 5, 7, \ldots, 2n + 1, \ldots$; find S_{20}.

40. $2, 5, 8, \ldots, 3n - 1, \ldots$; find S_{30}.

41. $-1, -4, -7, \ldots, 2 - 3n, \ldots$; find S_{19}.

42. $7, 9, 11, \ldots, 5 + 2n, \ldots$; find S_{35}.

43. $2, 2\,1/2, 3, \ldots, \dfrac{3+n}{2}, \ldots$; find S_{44}.

44. $1\,1/2, 1/2, -1/2, \ldots, \dfrac{5-2n}{2}, \ldots$; find S_{50}.

45. $-4, -1, 2, \ldots, 30 - 7, \ldots$; find S_7.

46. $3, 9, 15, \ldots, 6n - 3, \ldots$; find S_{15}.

47. $a_1 = 10, d = -3$; find S_{30}.

48. $a_1 = 15, d = 9$; find S_{28}.

49. $a_5 = -15, d = 6$; find S_{45}.

50. $a_3 = -12, d = -2$; find S_{89}.

♦ **Applications**

51. A log pyramid has 16 logs on the bottom row, 15 on the next row, and so on, until there is just one log on the top row. How many logs are in the pyramid?

52. A student doing some typing finds that she can type 5 words per minute faster each half hour that she types. If she starts at 35 words per minute at 6:30 P.M., how fast is she typing at 10:00 P.M.?

53. A well-drilling firm charges $2.50 to drill the first foot, $2.65 for the second foot, and so on in arithmetic sequence. At this rate, what would be the cost to drill the last foot of a well 350 feet deep?

54. The developer of a housing development finds that the sale of the first house gives him a $200 profit. His profit on the second sale is $325, and he finds that each additional house sold increases his profit per house by $125. How many houses did he sell if his profit on the last house sold was $6450 ?

Solve the following arithmetic sequences for the indicated terms.

55. If $a_1 = 1$, $a_n = 408$, and $S_n = 2454$, find n and d.

56. If $a_1 = 4$ and $S_{13} = 247$, find d.

57. If $d = \dfrac{1}{2}$, $n = 81$, and $S_n = 2025$, find a_n.

58. If $n = 14$, $a_n = 53$, and $S_n = 378$, find d.

Write the following expressions in summation notation.

59. $4 + 7 + 10 + 13$

60. $3 + \dfrac{5}{2} + 2 + \dfrac{3}{2} + 1$

61. $-9 - 3 + 3 + 9 + 15 + 21 + 27$

62. $-2 - 8 - 14 - 20 - 26 - 32$

63. Show that if a and b are real numbers and r and s are integers such that $r \le s$, then $\sum_{i=r}^{s}(ai + b)$ is an arithmetic sequence having $s - r + 1$ terms; a is the common difference when $r < s$.

64. Consider the sequence for which $a_n = n^2$ starting with $n = 0$. If $b_n = a_n - a_{n-1}$, show that $b_1, b_2, b_3, \ldots$ is an arithmetic sequence.

65. Prove that an expression for the nth term of an arithmetic sequence defines a linear function, given a_1 and d.

66. Prove that for any three consecutive terms of any arithmetic sequence the middle term is the average of the other two terms.

67. Given that r, s, t, and u are the first four terms of an arithmetic sequence, then show that $r + u - s = t$.

8.5 GEOMETRIC SEQUENCES AND SERIES

Let's now proceed to the second special type of sequence that arises most commonly in applications, namely, the geometric sequence.

Definition 1
Geometric Sequence

A sequence $a_1, a_2, \ldots, a_n$ is called a **geometric sequence** provided that each term is a fixed multiple of its immediate predecessor. That is, there is a fixed real number r such that for each positive integer n, we have

$$a_{n+1} = r a_n$$

The number r is called the **ratio**.

An example of a geometric sequence is

$$\frac{1}{2}, \frac{1}{2^2}, \frac{1}{2^3}, \ldots$$

Here $r = 1/2$. Indeed, we have $a_n = 1/2^n$, so that

$$a_{n+1} = \frac{1}{2^{n+1}}$$

$$= \frac{1}{2 \cdot 2^n}$$

$$= \frac{1}{2} \cdot \frac{1}{2^n}$$

$$= \frac{1}{2} \cdot a_n$$

The next example provides another illustration of a geometric sequence as well as practice in proving that a sequence is geometric.

➤ **EXAMPLE 1**

Proving a Sequence Is Geometric

Consider the sequence given by

$$a_n = (-1)^{n-1}\left(\frac{4}{3}\right)^n$$

Prove that this is a geometric sequence.

Solution

In looking for a possible value of r, we need to compute the ratio

$$\frac{a_{n+1}}{a_n} = \frac{(-1)^{(n+1)-1}\left(\frac{4}{3}\right)^{n+1}}{(-1)^{n-1}\left(\frac{4}{3}\right)^n}$$

$$= (-1)\left(\frac{4}{3}\right)^{(n+1)-n}$$

$$= -\left(\frac{4}{3}\right)^1$$

$$= -\frac{4}{3}$$

Since this ratio is a constant (it does not depend on the value of n), the sequence is geometric with ratio $-4/3$.

Our development of geometric sequences parallels our discussion of arithmetic sequences of the preceding section, in which we derived a formula for the nth term of an arithmetic sequence in terms of the first term, the index, and the difference. Here is the analogous result for geometric sequences.

---

nTH TERM OF A GEOMETRIC SEQUENCE

Suppose that $a_1, a_2, \ldots, a_n, \ldots$ is a geometric sequence with ratio r. Then

$$a_n = a_1 r^{n-1}$$

Proof Let's proceed by induction. The initial case corresponds to $n = 1$, in which case the result asserts that

$$a_1 = a_1 r^{1-1}$$
$$= a_1 r^0$$
$$= a_1$$

So the result is true in case $n = 1$. Let's now assume that the result holds for a particular value of n. That is, we assume that

$$a_n = a_1 r^{n-1}$$

We must deduce from this the truth of the result for case $n + 1$. That is, we must prove that

$$a_{n+1} = a_1 r^{[(n+1)-1]} = a_1 r^n$$

To do so, let's first recall that the sequence is geometric, so that

$$a_{n+1} = r a_n$$

Into this last equation, substitute the assumed formula for a_n. This gives us

$$a_{n+1} = r a_n$$
$$= r \cdot a_1 r^{n-1}$$
$$= a_1 r^n$$
$$= a_1 r^{(n+1)-1}$$

This is the statement of the result for $n + 1$. So the inductive step is proved and with it the general formula. ◆

The next two examples provide applications of the preceding result.

➤ **EXAMPLE 2**
Calculating a Term
of a Geometric Sequence

Suppose that a geometric sequence has first term 3 and ratio 2/3. Determine the fifth term of the sequence.

Solution

By the preceding result, the nth term a_n of the sequence is given by the formula

$$a_n = a_1 r^{n-1}$$
$$= 3 \cdot \left(\frac{2}{3}\right)^{n-1}$$

In particular, for $n = 5$, this equation yields

$$a_5 = 3 \cdot \left(\frac{2}{3}\right)^{5-1}$$
$$= 3 \cdot \left(\frac{2}{3}\right)^4$$
$$= 3 \cdot \frac{16}{81}$$
$$= \frac{16}{27}$$

➤ **EXAMPLE 3**

Formula for the *n*th Term

Suppose that $a_1, a_2, \ldots$ is a geometric sequence. Further, suppose that $a_2 = 5/4$, $a_5 = 5/32$. Determine a formula for a_n.

Solution

Let r denote the ratio of the sequence. Then we have

$$a_2 = \frac{5}{4}$$

$$= a_1 r^{2-1}$$

$$\frac{5}{4} = a_1 r \tag{1}$$

and

$$a_5 = \frac{5}{32}$$

$$\frac{5}{32} = a_1 r^4 \tag{2}$$

Dividing equation (2) by the corresponding terms of equation (1), we have

$$\frac{5/32}{5/4} = \frac{a_1 r^4}{a_1 r}$$

$$\frac{1}{8} = r^3$$

$$r = \frac{1}{2}$$

Inserting this value of r into equation (1) gives

$$\frac{5}{4} = a_1 r$$

$$= a_1 \cdot \left(\frac{1}{2}\right)$$

$$a_1 = \frac{5}{2}$$

Inserting the values of a_1 and r into the formula for the *n*th term of a geometric sequence, we find the desired formula for a_n, namely,

$$a_n = a_1 r^{n-1}$$

$$= \frac{5}{2} \cdot \left(\frac{1}{2}\right)^{n-1}$$

➤ **EXAMPLE 4**

Compound Interest

Suppose that a bank account earns compound interest at a rate of i per period (year, month, day). Further, suppose that at the start of period 1, the account has balance P. Then the balance a_n in the account at the end of period n is given by the formula

$$a_n = P(1 + i)^n$$

1. Show that the sequence $a_0, a_1, a_2, \ldots, a_n$ is a geometric sequence with ratio $1 + i$.

2. Suppose that $400 is deposited in a bank account paying 8% interest compounded monthly. Calculate the amount in the account at the end of 5 years.

Solution

1. Note that the first index of this sequence is $n = 0$. In dealing with financial calculations, it is usually most convenient to begin sequences with the 0th term rather than the first. This allows the sequence to start its description with time 0 (the beginning of the first time period). Note that if n is any nonnegative integer, then

$$
\begin{aligned}
a_{n+1} &= P(1 + i)^{n+1} \\
&= P(1 + i)^n \cdot (1 + i) \\
&= (1 + i)a_n
\end{aligned}
$$

Thus, the sequence is geometric with ratio $1 + i$.

2. The 8% interest refers to an annual rate. Since the interest is compounded monthly, the length of a single financial period is a month, and the amount of interest for one month is $i = 8\%/12 = 2/3\% = 1/150$. Five years corresponds to 60 months, that is, 60 financial periods. So the balance in the account at the end of five years is

$$
\begin{aligned}
a_{60} &= 400\left(1 + \frac{1}{150}\right)^{60} \\
&= \$595.94
\end{aligned}
$$

(We have used a scientific calculator to calculate the numerical value of the expression.)

Let's now turn to the problem of calculating the partial sum S_n of a geometric sequence. We have the following formula:

> ### PARTIAL SUM OF A GEOMETRIC SEQUENCE
>
> Suppose that $a_1, a_2, \ldots, a_n, \ldots$ is a geometric sequence and that S_n is its nth partial sum. Then
>
> $$ S_n = a_1 \frac{1 - r^n}{1 - r} \quad (r \neq 1) \qquad (3) $$
>
> $$ = na_1 \quad (r = 1) \qquad (4) $$

Proof   We have

$$
\begin{aligned}
S_n &= a_1 + a_2 + a_3 + \cdots + a_n \\
&= a_1 + a_1 r + a_1 r^2 + \cdots + a_1 r^{n-1} \\
&= a_1(1 + r + r^2 + \cdots + r^{n-1})
\end{aligned}
$$

In case $r = 1$, each of the terms within the parentheses on the right is equal to 1, and the formula for S_n reads

$$
S_n = a_1(1 + 1 + 1 + \cdots + 1) = na_1
$$

Let's now assume that $r \neq 1$. We may now write the right side in the form

$$S_n = a_1(1 + r + r^2 + \cdots + r^{n-1})$$

$$= a_1(1 + r + r^2 + \cdots + r^{n-1})\frac{1-r}{1-r}$$

$$= a_1\frac{1-r^n}{1-r}$$

This last formula is the result claimed in equation (3) and completes the proof. ◆

➤ **EXAMPLE 5**

Sum of a Geometric Sequence

Determine the value of the following sum:

$$3 + 3\left(\frac{3}{4}\right) + 3\left(\frac{3}{4}\right)^2 + \cdots + 3\left(\frac{3}{4}\right)^9$$

Solution

The sum is the partial sum S_{10} of a geometric series with $a_1 = 3$, $r = 3/4$. Therefore, by equation (3), we have

$$S_{10} = a_1\frac{1 - r^{10}}{1-r}$$

$$= 3 \cdot \frac{1 - \left(\frac{3}{4}\right)^{10}}{1 - \frac{3}{4}}$$

$$= 12 \cdot \left[1 - \left(\frac{3}{4}\right)^{10}\right] = 11.324$$

➤ **EXAMPLE 6**

Value of a Pension Fund

Each year for 20 years, a woman deposits $5000 into a pension fund. The money earns 9% compounded annually. How much money is in the pension fund after 20 years?

Solution

There are 20 deposits to the pension fund, each earning interest for a different length of time. The last deposit earns interest for 0 years, the next-to-last deposit for 1 year, and the first deposit for 19 years. Let a_1 denote the amount of money generated by the last deposit, a_2 the amount of money generated by the next-to-last, and a_{20} the amount generated by the first deposit. The amount of money represented by a_n earns interest for $n - 1$ years at 9% compounded annually. So by our previous discussion, we have

$$a_n = 5000(1 + 0.09)^{n-1}$$

$$= 5000 \cdot 1.09^{n-1}$$

The amount in the pension fund at the end of 20 years equals the sum of the amounts attributable to each of the deposits. This equals

$$a_1 + a_2 + a_3 + \cdots + a_{20}$$

This sum is the partial sum S_{20} of a geometric series with $a_1 = 5000, r = 1.09$. By equation (3), this sum equals

$$a_1 \frac{1 - r^n}{1 - r} = 5000 \cdot \frac{1 - 1.09^{20}}{1 - 1.09}$$

$$= -\frac{5000}{0.09}(1 - 1.09^{20})$$

$$= \$255,800.60$$

That is, the pension fund balance after 20 years equals $255,800.60.

Now let us consider the infinite geometric sequence

$$1, \frac{1}{2}, \frac{1}{2^2}, \frac{1}{2^3}, \cdots$$

Here are its first few partial sums:

$$S_1 = 1$$

$$S_2 = \frac{3}{2}$$

$$S_3 = \frac{7}{4}$$

$$S_4 = \frac{15}{8}$$

$$S_5 = \frac{31}{16}$$

As more and more terms of the sequence are included, the partial sums approach ever closer to the value 2. The nth partial sum may be obtained as the sum of a geometric progression.

$$1 + \frac{1}{2} + \frac{1}{2^2} + \cdots + \frac{1}{2^{n-1}} = \frac{1 - \dfrac{1}{2^n}}{1 - \dfrac{1}{2}}$$

As n increases, $(\frac{1}{2})^n$ approaches 0, so that the expression on the right approaches

$$\frac{1}{1 - \dfrac{1}{2}} = 2$$

Similar reasoning works to obtain the sum of any infinite geometric series with ratio r for which $|r| < 1$. Namely we have:

SUM OF AN INFINITE GEOMETRIC SEQUENCE

Suppose that $|r| < 1$. The partial sums of the infinite geometric sequence $a, ar, ar^2, \ldots, ar^n$ approach the limiting value $a/(1 - r)$. This fact is expressed using the notation

$$\sum_{n=0}^{\infty} ar^n = \frac{a}{1 - r}$$

> **EXAMPLE 7**

Calculating the Sum of an Infinite Geometric Series

Calculate the sum of the infinite geometric sequence

$$3, -3 \cdot \frac{1}{5}, 3 \cdot \left(\frac{1}{5}\right)^2, -3 \cdot \left(\frac{1}{5}\right)^3, \ldots$$

Solution

The geometric sequence has $a_1 = 3$, $r = -1/5$. Therefore, by the formula for the sum of an infinite geometric sequence, the partial sums of the sequence approach a limiting value, and this value equals

$$\frac{a}{1-r} = \frac{3}{1 - \left(-\dfrac{1}{5}\right)}$$

$$= \frac{3}{\dfrac{6}{5}}$$

$$= \frac{5}{2}$$

Note that if $|r| \geq 1$, then the partial sums of the geometric series do not approach a limiting value.

Infinite sums are discussed in detail as part of a calculus course. The preceding discussion provides a taste of what you will learn later.

Exercises 8.5

Prove that the following are geometric sequences.

1. $a_n = 3\left(-\dfrac{1}{3}\right)^{n-1}$

2. $a_n = 24\left(\dfrac{1}{2}\right)^{n-1}$

3. $a_n = 5(4)^{n-1}$

4. $a_n = 2^{n-1}$

5. $a_n = 3^{-n}$

6. $a_n = \left(-\dfrac{1}{2}\right)^{n+1}$

Find the indicated term for each of the following geometric sequences.

7. $a_1 = \dfrac{1}{2}, r = \dfrac{1}{2}$; find a_{10}.

8. $a_1 = \dfrac{2}{3}, r = \dfrac{2}{3}$; find a_9.

9. $a_1 = 1, r = -\sqrt{2}$; find a_7.

10. $a_1 = 81, r = -\dfrac{1}{3}$; find a_{12}.

11. $54, 36, 24, \ldots$; find a_8.

12. $32, 16, 8, \ldots$; find a_{15}.

13. If $a_2 = 3$ and $r = 2$, find a_{13}.

14. If $a_4 = 64$ and $r = -4$, find a_{10}.

15. If $a_4 = 6$ and $a_5 = 12$, find a_{14}.

16. If $a_6 = 9$ and $a_7 = 3$, find a_2.

17. $300, -30, 3, \ldots$; find a_4.

18. $-\sqrt{5}, 5, -5\sqrt{5}, \ldots$; find a_{20}.

19. $2, \dfrac{2}{3}, \dfrac{2}{9}, \ldots$; find a_5.

20. $-162, 54, -18, 6, \ldots$; find a_8.

◆ Applications

21. On a merit salary plan, yearly raises are determined as a percent of the present salary. If a worker begins with a \$15,000 yearly salary and receives merit increases of 8.5% for nine years, what will be her yearly salary for her ninth year?

22. In place of a \$1 million salary, a baseball pitcher agreed to accept \$10 for his first win, \$20 for his second, \$40 for his third, and so on. How many games must he win before his earnings exceed the \$1 million salary? What would his salary be if he were to win one more game?

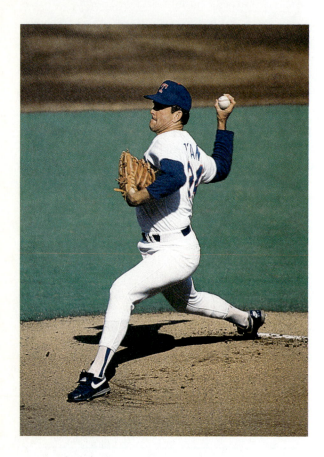

23. If a new printing press cost $50,000 and had a yearly depreciation of 25% of its value at the beginning of the year, then what is its value after five years?

24. Highway Appliance Stores have CD players that can be purchased on a daily installment plan. You pay 1 cent the first day, 3 cents the second day, 9 cents the third day, and so on for 10 days. How much will the CD player cost?

25. A girl states that one of her 10th-generation ancestors was on the Mayflower. How many ancestors does she have in the 10th generation, assuming that she has no duplication?

26. If the interest on a 10-year certificate of deposit was set at 9% compounded semiannually, what would be the value after 10 years of an investment of $8000?

27. The city of Akron, Ohio, lost 1/100 of its population each year from 1972 through 1980. If the population in 1972 was 350,000, what was the population at the end of 1980?

28. If the federal government were to put an extra $1 billion into the economy such that each business and individual saves 25% of its income and spends the rest, so that of the $1 billion 75% is respent by individuals and

businesses, then 75% of this amount is respent, and so forth, what is the total increase in spending due to the government action?

29. Which term of the geometric sequence $2, -6, 18, \ldots$ is 162?

30. If $a_7 = 256$ and $a_1 = 4$, find a_5.

31. If $r = 3$ and $a_5 = 324$, find a_1.

32. Which term of the geometric sequence $-243, 81, -27, \ldots$ is $1/9$?

33. Fill in the blanks for this geometric sequence: _____, -5, _____, _____, _____, -405.

34. Fill in the blanks for this geometric sequence: _____, _____, 1, _____, _____, 729.

In the following geometric sequences three of the five real numbers a_1, a_n, r, n, and S_n are given. Find the other two.

35. $a_1 = 5, r = -2, S_n = -25$

36. $a_1 = 5, a_n = 320, r = 2$

37. $a_1 = 4, a_n = 324, r = 3$

38. $n = 9, r = 2, S_n = 1022$

39. $n = 5, r = -3, S_n = 244$

40. $a_1 = 10, r = 3, n = 5$

41. $a_n = 384, n = 7, r = 2$

42. $a_n = 729, n = 7, r = \dfrac{3}{2}$

43. $a_1 = 162, r = -\dfrac{1}{3}, n = 8$

44. $a_1 = 4, r = -\dfrac{3}{2}, n = 6$

45. $a_n = 9, r = -\sqrt{3}, S_n = 4 - 4\sqrt{3}$

46. $a_n = -625\sqrt{5}, r = -\sqrt{5}, S_n = 780 - 781\sqrt{5}$

Find a formula for the following geometric sequences.

47. $a_2 = -4, a_5 = \dfrac{4}{27}$ 48. $a_3 = 6, a_6 = 18\sqrt{3}$

49. $a_3 = 50, a_6 = -6550$ 50. $a_2 = 3, a_5 = 54$

♦ Applications

51. To begin a new business, a man must borrow $80,000 at 12% interest compounded monthly. If the loan is to be paid off in one lump sum at the end of two years, what will be the amount of the payment?

52. To buy a used car, you borrow $1000 at 10% compounded monthly. You also elect to pay off this loan in one lump-sum payment at the end of four years. What will be the amount of your payment?

53. For retirement, you elect to deposit $1000 a year in a savings account that pays 11% compounded annually. How much will be in your account at the end of 40 years?

54. A couple with a 1-year-old decides to put $500 a year into a savings account paying 12% compounded annually for their son's college money. At the end of 18 years, what will be the amount in this account?

Solve the following infinite sequence problems, if possible.

55. Find the sum of $1 - 1/2 + 1/4 - 1/8 + 1/16 - \cdots$.

56. Find the sum of $1 - 1/5 + 1/25 - 1/125 + \cdots$.

57. Find the sum of $-5/3 - 10/9 - 20/27 - 40/81 \cdots$.

58. Find the sum of $2 + 2/3 + 2/9 + 2/27 + 2/81 + \cdots$.

59. Find the sum of $9/10 + 9/100 + 9/1000 + 9/10,000 + \cdots$.

60. Find the sum of $1/2 + 1/3 + 2/9 + 4/27 + \cdots$.

61. Find the sum of $1 + 2 + 4 + 8 + \cdots$.

62. Find the sum of $-1 + 1 - 1 + 1 - 1 + \cdots$.

63. If $s = 14$ and $a_1 = 21$, then $r = ?$

64. If $r = 2/3$ and $s = 18$, then $a_1 = ?$

65. If $r = -1/2$ and $s = 20$, then $a_1 = ?$

66. If $s = 35$ and $a_1 = 7$, then $r = ?$

67. If $a_1 = 4$ and $r = 1/2$, then $s = ?$

68. If $a_1 = 1/3$ and $s = 2/3$, then $r = ?$

69. A superball is dropped 40 feet and rebounds on each bounce 2/5 of the height from which it fell. How far will it travel before coming to rest?

70. If a ball is dropped from a height of 12 meters and rebounds 1/3 of the distance it fell on each bounce, how far will it travel before coming to rest?

71. Here is a diagram of steps that might be taken in getting some grapes from the grower to you:

Grower → Trucker → Regional market → Trucker → Wholesaler → Trucker → Retailer → You

If each person or organization in this chain makes a 25% profit, if all grapes are sold, and if the grower sells his grapes for $0.20 per pound, how much would you have to pay per pound?

72. If a dentist earns $40,000 during his first year of practice, and if each succeeding year his salary increases 10%, what would be the total of the dentist's salary over the first nine years?

73. In Exercise 71, if the grower receives $10,000 for the grapes she sold to the first trucker, then what is the sum of all the profits made once you buy these grapes?

74. In Exercise 72, how many years would the dentist have to work if the salary total is to exceed $1 million?

75. Show that there is no infinite series for which $a_1 = 2$ and $s = 1$.

✍ In Your Own Words

76. Can the sum of an infinite geometric series be less than the first term? Justify your answer.

77. Prove that every rational number a/b can be expressed as a repeating decimal.

78. Prove that every repeating decimal represents a rational number.

79. True or false: Suppose that to proceed to the door of the room you walk half the distance to the door, then stop, then move half the remaining distance to the door, then stop, and so forth. You never cover the total distance to the door and hence can never leave the room. Explain your answer.

80. Explain the differences between arithmetic and geometric sequences.

8.6 PERMUTATIONS AND COMBINATIONS

Many applied problems require you to count the number of possible occurrences of a particular type. Here are some examples:

1. A communications network consists of 500 cables from point A to point B and 300 cables from point B to point C. How many ways are there to choose a transmission path from point A to point C?

2. A state lottery requires you to pick a sequence of five digits. How many different outcomes of the lottery are possible?

3. A certain genetic combination consists of eight traits, for each of which there are four possibilities. How many genetic combinations are possible?

The field of mathematics that deals with counting problems is called **combinatorics**. In this section, we provide a brief introduction to combinatorics and consider a few of the simplest types of counting problems.

Elementary Counting Problems

One of the basic techniques for solving counting problems is to break the problem down into a sequence of independent choices, each of whose possibilities can be counted. In counting such sequences, the following basic result is used:

FUNDAMENTAL PRINCIPLE OF COUNTING

Suppose that tasks $A_1, A_2, \ldots, A_n$ are independent of one another. Further, suppose that it is possible to perform task A_1 in m_1 ways, task A_2 in m_2 ways, $\ldots$, and task A_n in m_n ways. Then the number of possible ways to perform the sequence of tasks $A_1, A_2, \ldots, A_n$, is equal to $m_1 \cdot m_2 \cdots \cdots m_n$.

We will assume the truth of the fundamental principle of counting. The next examples illustrate how it may be applied in solving counting problems.

➤ **EXAMPLE 1**
Counting License Plates

Car license plates in a certain state consist of three letters followed by 3 digits. How many different license plates are possible?

Solution

The process of selecting a license plate may be viewed as a sequence of six tasks, corresponding to the selection of the six letters/digits. Each of the first three tasks may be accomplished in 26 different ways. Each of the last three tasks may be accomplished in 10 different ways. By the fundamental principle of counting, the sequence of six tasks may then be accomplished in

$$26 \cdot 26 \cdot 26 \cdot 10 \cdot 10 \cdot 10 = 17,576,000$$

different ways. That is, there are 17,576,000 different possible license plates.

➤ **EXAMPLE 2**
Hexadecimal Numbers

In the *hexadecimal number system*, used extensively in computer work, there are 16 digits, namely, 0–9 and A–F. A typical hexadecimal number consists of a sequence of hexadecimal digits. How many four-digit hexadecimal numbers are there whose first digit is not 0 or 1 ?

Solution

The specified four-digit hexadecimal number may be viewed as a sequence of four independent tasks, corresponding to choosing each of the four letters. The first task may be chosen in 14 ways (the hexadecimal digit can't be 0 or 1). Each of the remaining tasks may be accomplished in 16 different ways. So the sequence of tasks may be accomplished in

$$14 \cdot 16 \cdot 16 \cdot 16 = 57,344$$

ways. That is, there are 57,344 such hexadecimal numbers.

Permutations

One of the most common types of counting problems involves permutations, which are defined as follows:

Definition 1
Permutation

Suppose that we are given n different objects. A selection of r of them arranged in a particular order is called a **permutation of n objects taken r at a time.**

For instance, suppose that $n = 3$ and that the objects are the three letters A, B, and C. Then there are six permutations of these objects taken two at a time, namely, AB, BA, AC, CA, BC, and CB.

The number of permutations of n objects taken r at time is denoted $_nP_r$. From the previous example, we see that $_3P_2 = 6$. We may compute $_nP_r$ from n and r using the following result:

FUNDAMENTAL THEOREM OF PERMUTATIONS

Suppose that n and r are positive integers with $r \le n$. Then
$$_nP_r = n(n-1)(n-2)\cdots(n-r+1) \quad \text{(a total of } r \text{ factors)}$$

Proof   Let us consider the task of choosing a permutation of n objects taken r at a time. This can be viewed as a sequence of r independent tasks. Select the first object in the permutation; this may be done in n ways. Next, select the second object in the permutation; since one object has already been used, this second task may be done in $n - 1$ ways. Next, select the third object in the permutation; since two objects have already been used, the second task may be done in $n - 2$ ways, and so on. The final task is selecting the rth object. The preceding tasks have used up $r - 1$ objects, so the final task may be done in $n - (r-1) = n - r + 1$ ways. By the fundamental theorem of counting, the sequence of r tasks may be accomplished in
$$n(n-1)(n-2)\cdots(n-r+1)$$

ways. This proves the desired result.   ◆

➤ **EXAMPLE 3**
Calculating Numbers of Permutations

Determine the values of the following:

1. $_5P_3$ 2. $_{10}P_4$ 3. $_nP_3$

Solution

1. We have
$$_5P_3 = 5 \cdot 4 \cdot 3 = 60$$

2. We have
$$_{10}P_4 = 10 \cdot 9 \cdot 8 \cdot 7 = 5040$$

3. The second subscript, 3, determines the number of factors. So we have
$$_nP_3 = n(n-1)(n-2)$$

➤ **EXAMPLE 4**
Assignment of Tasks

There are 30 students in a certain class. In how many ways can the instructor assign students to perform four tasks, labeled A–D?

Solution

Each assignment of students to tasks may be viewed as a permutation of 30 objects (students) taken 4 at a time. The ordering of the permutation creates the assignment to the tasks: The first student is assigned to task A, the second student to task B, and so forth. The number of ways of assigning the students is $_{30}P_4$. The numerical value of this symbol is

$$_{30}P_4 = 30 \cdot 29 \cdot 28 \cdot 27 = 657,720$$

That is, there are 657,720 ways to assign 30 students to four tasks.

➢ EXAMPLE 5
Communications Engineering

In a communications network, there are 13 cables connecting points 1 and 2, and there are five messages, labeled A–E, to be sent from point 1 to point 2. In how many ways can the cables for the transmission be selected if each cable can carry a single message?

Solution

Each assignment of messages to cables is a permutation of 13 objects (the cables) taken 5 at a time. The number of such permutations is equal to

$$_{13}P_5 = 13 \cdot 12 \cdot 11 \cdot 10 \cdot 9 = 154,440$$

So the messages may be sent in 154,440 different ways.

Combinations

Let's now consider a second common type of counting problem, namely, one involving combinations, defined as follows:

Definition 2
Combination

Suppose that we are given n distinct objects. A selection of r of them (without regard to order) is called a **combination of n objects taken r at a time**.

Note that a combination differs from a permutation in that it does not take into account the order of arrangement. For instance, we previously saw that the permutations of the three letters A, B, and C taken two at a time are

$$AB, \ BA, \ AC, \ CA, \ BC, \ CB$$

However, with combinations of these same three letters taken two at a time, order does not count. For instance, the two permutations AB and BA are regarded as the same combination. There are only three combinations in this example, namely:

$$AB, \ AC, \ BC$$

The number of combinations of n objects taken r at a time is denoted $_nC_r$. This number may be calculated using the following result:

FUNDAMENTAL THEOREM OF COMBINATIONS

The number of combinations of n objects taken r at a time is equal to

$$_nC_r = \frac{n!}{r!(n-r)!}$$

Proof Let's begin by forming all permutations of the n objects taken r at a time. The number of such permutations is $_nP_r$. Each permutation corresponds to a combination. However, permutations that are obtained by rearranging the same objects in differing order correspond to the same permutation. How many such rearrangements are there? We can choose the first rearrangement in r ways, the second in $r - 1$ ways, the third in $r - 2$ ways, and so forth. By the fundamental principle of counting, we see that each permutation may be rearranged in

$$r(r - 1)(r - 2) \cdot \cdots \cdot 2 \cdot 1 = r!$$

ways. So each combination is counted $r!$ times. That is, we have

$$_nP_r = r!_nC_r$$

$$_nC_r = \frac{_nP_r}{r!}$$

$$= \frac{n(n - 1)\cdots(n - r + 1)}{r!} \qquad \textit{(Fundamental theorem of permutations)}$$

$$= \frac{n(n - 1)\cdots(n - r + 1)(n - r)(n - r - 1)\cdot \cdots \cdot 1}{r!(n - r)(n - r - 1)\cdot \cdots \cdot 1}$$

$$= \frac{n!}{r!(n - r)!}$$

This proves the theorem.   ◆

Note that the right side of the formula in the theorem is just a binomial coefficient. That is, another way of stating the fundamental theorem of combinations is

$$_nC_r = \binom{n}{r}$$

Since $_nP_r = r!_nC_r$, we also conclude that

$$_nP_r = n!/(n - r)!$$

Let's now use the fundamental theorem of combinations to solve some applied counting problems.

➤ **EXAMPLE 6**
Calculating Number
of Combinations

Calculate the number of combinations of 10 objects taken 4 at a time.

Solution

The specified number equals $_{10}C_4$. And by the fundamental theorem of combinations, this symbol has the value

$$_{10}C_4 = \frac{10!}{4!(10 - 4)!}$$

$$= \frac{10 \cdot 9 \cdot 8 \cdot 7 \cdot 6 \cdot 5 \cdot 4 \cdot 3 \cdot 2 \cdot 1}{(4 \cdot 3 \cdot 2 \cdot 1) \cdot (6 \cdot 5 \cdot 4 \cdot 3 \cdot 2 \cdot 1)}$$

$$= \frac{10 \cdot 9 \cdot 8 \cdot 7}{4 \cdot 3 \cdot 2 \cdot 1}$$

$$= 210$$

Note that in the computation of the preceding example, we performed cancelation between the denominator and numerator in order to ease the burden of working with large numbers. Actually, we may perform such cancelation generally in working with $_nC_r$, as the following result shows:

FORMULA FOR COMBINATIONS

Let n and r be positive numbers with $r \le n$. Then

$$_nC_r = \frac{n(n-1)\cdots(n-r+1)}{r!}$$

Proof Start from the preceding formula for $_nC_r$:

$$_nC_r = \frac{n!}{r!(n-r)!}$$

$$= \frac{n(n-1)(n-2)\cdots(n-r+1)(n-r)(n-r-1)\cdots\cdots 2\cdot 1}{r!(n-r)(n-r-1)\cdots\cdots 2\cdot 1}$$

$$= \frac{n(n-1)(n-2)\cdots(n-r+1)}{r!}$$

This proves the desired formula.   ◆

➤ **EXAMPLE 7**
Communications Engineering

In a communications network, there are 13 cables connecting points 1 and 2. In how many ways can 4 cables be selected to send messages from point 1 to point 2?

Solution

Each selection of cables is a combination of 13 objects (the cables) taken 4 at a time. The number of such combinations equals

$$_{13}C_4 = \frac{13\cdot 12\cdot 11\cdot 10}{4!}$$

$$= 715$$

➤ **EXAMPLE 8**
Baseball

A baseball team has 14 players. How many different 9-player teams (disregarding order) are possible?

Solution

The number of such teams equals the number of combinations of 14 objects taken 9 at a time. This number equals

$$_{14}C_9 = \frac{14\cdot 13\cdot 12\cdot 11\cdot 10\cdot 9\cdot 8\cdot 7\cdot 6}{9\cdot 8\cdot 7\cdot 6\cdot 5\cdot 4\cdot 3\cdot 2\cdot 1}$$

$$= \frac{14\cdot 13\cdot 12\cdot 11\cdot 10}{5\cdot 4\cdot 3\cdot 2\cdot 1}$$

$$= 2002$$

That is, the team may be selected in 2002 ways.

➤ **EXAMPLE 9**
Poker

A standard deck of playing cards contains 52 cards. A poker hand contains 5 cards. How many different poker hands are possible?

Solution

A poker hand is a combination of 52 objects taken 5 at a time. The number of such objects equals

$$_{52}C_5 = \frac{52 \cdot 51 \cdot 50 \cdot 49 \cdot 48}{5!} = 2,598,960$$

That is, there are 2,598,960 possible poker hands.

Exercises 8.6

1. How many numbers consisting of one, two, or three digits, without repetitions, can you form using the digits 1, 2, 3, 4, 5, and 6?

2. How many truck license plates may be made if each plate will have three numbers followed by two letters followed by one number?

3. How many five-digit numbers begin with an even digit and end with an even digit?

4. In how many different orders can you arrange four books on a shelf?

5. How many possibilities are there for a license plate with three letters and two numbers?

6. If a girl has six different skirts and 10 different blouses, how many different skirt-blouse outfits are possible?

7. If six people run a race, in how many different orders can they all finish if there are no ties?

8. For a nine-man baseball team, how many different batting orders are possible?

9. Marv's Pizza Palace offers pepperoni, onions, sausage, mushrooms, and anchovies as toppings for its pizza. How many different pizzas can be made?

10. A club consists of 30 men and 70 women. In how many ways can a president, vice-president, secretary, and sergeant-at-arms be chosen if the president must be a woman, the sergeant-at-arms must be a man, and a person cannot hold more than one office?

Determine the following.

11. $_4P_4$ 12. $_7P_4$ 13. $_6P_2$ 14. $_9P_P$

15. $_3P_1$ 16. $_9P_6$ 17. $_nP_5$ 18. $_nP_4$

19. $_7P_3$ 20. $_6P_1$

♦ Applications

Use permutations to solve the following.

21. How many "words" can be formed from the letters of the word *fragments*, taken four at a time? (Any four letters is considered a word.)

22. A Navy signalman has six different flags. How many different signals can be sent by flying three flags at a time, one above the other, on a flagpole?

23. On leaving a train, a man finds that he has a nickel, a dime, a quarter, and a half-dollar in his pocket. In how many ways can he tip the porter?

24. In how many ways can three books be chosen from a group of seven different books and arranged in three spaces on a bookshelf?

25. How many permutations are there of five cards taken from a bridge deck of 52 cards? (Order is to be considered.)

26. How many arrangements can be formed from the letters of the word *hyperbola,* with all nine letters taken at a time?

27. A concert is to consist of three songs and two violin selections. In how many ways can the program be arranged so that the concert begins and ends with a song, and neither violin selection follows immediately after the other?

28. For Exercise 26, in how many of these words will the letters *h* and *y* occur next to each other?

Determine the following.

29. $_8C_3$ 30. $_9C_2$ 31. $_5C_4$ 32. $_6C_1$

33. $_{10}C_2$ 34. $_{12}C_8$ 35. $_nC_{n-1}$ 36. $_nC_4$

37. $_4C_0$ 38. $_nC_n$

♦ Applications

39. If five blue, three yellow, two black, and two white balls are to be arranged in a row, find the number of possible color arrangements for the balls.

40. In how many ways can a hand of 13 cards be selected from a standard bridge deck of 52 cards?

41. In how many ways can a committee of three be chosen from four married couples if all are equally eligible?

42. In how many ways can a selection of one or more books be made from five distinct algebra books and four distinct geometry books?

43. In Exercise 41, what would be the number if the committee must consist of two women and one man?

44. A contractor needs 4 carpenters, and 10 apply for the job. In how many ways can he pick the four that he needs?

45. Your math professor has five basic algebra books, four advanced algebra books, and eight geometry books. In how many ways can they be arranged on his shelf if books in the same category are kept next to each other?

46. How many lines are determined by 10 points, no 3 of which are collinear?

47. From a group of six persons, how many committees of three can be formed if two of the six people cannot be on the same committee?

48. In how many ways can you choose a committee of 5 girls and 7 boys from a group of 10 girls and 11 boys?

49. A math history club has a membership of four women and six men. A research committee of three is formed. In how many ways can this be done: (a) If there are to be two women and one man on the committee? (b) If there is to be at least one woman on the committee? (c) If all three are to be of the same sex?

50. Using three letters of the word *example* and two letters of *boot,* how many arrangements of five different letters are possible?

51. How many five-letter words, each consisting of three consonants and two vowels, can be formed from the letters of the word *equations*?

52. In how many ways can a coach choose a team of 5 from among 10 girls: (a) If 2 specified girls must be included? (b) If there are no restrictions?

53. In how many ways can 15 different objects be divided among A, B, and C if A must receive 2 objects, B must receive 3 objects, and C must receive 10 objects?

54. How many diagonals does a 20-sided polygon have? How many sides does a polygon with 35 diagonals have?

55. A student leadership of seven wishes to send a two-person committee to the dean to discuss new dorm policies. (a) How many different committees are possible? (b) If there are two women and five men student leaders, how many of the committees would include a woman? (c) If the leadership president must be on the committee, how many different committees are possible?

56. Prove $_{n+1}C_r = {_n}C_{r-1} + {_n}C_r$.

57. Prove $_nC_r = {_n}C_{n-r}$.

Solve the following for n.

58. $_nC_2 = {_{100}}C_{98}$

59. $_nC_5 = {_n}C_3$

60. $_nP_4 = 4({_n}P_3)$

61. $_nP_3 = ({_{n-1}}P_2)$

62. $_nP_4 = 8({_{n-1}}P_3)$

Prove the following.

63. $_nP_4 - {_n}P_3 = (n-4)({_n}P_3)$

64. $_5P_r = 5({_4}P_{r-1})$

✍ In Your Own Words

65. Describe the differences between permutations and combinations.

66. In what ways are permutations and combinations similar?

8.7 PROBABILITY

Laplace

The theory of probability is a branch of mathematics that deals with processes exhibiting uncertainty. Examples of such processes are throwing a pair of dice, playing a hand of blackjack, or observing the size of leaves on a tree. The theory of probability was introduced by the 18th-century mathematician Laplace, who was asked by a French king to develop a mathematical description of various games of chance. In the past two centuries, probability theory has developed into a major discipline that supports applications in biology, physics, psychology, and business, to mention only a few! In this section, we will provide a very brief taste of what probability theory is about. Our discussion will give us a chance to use the counting techniques we learned in the preceding section.

Sample Spaces and Events

Any process whose result is subject to chance is called an **experiment**. The result of an experiment is called an **outcome**. One example of an experiment consists of flipping a coin and observing the outcome, heads or tails. A second experiment consists of rolling a die and observing the number that appears (the possibilities are the digits 1–6). A third experiment is to observe the height of a three-year-old. The outcomes in this case are positive real numbers.

The third example of an experiment differs from the first two since it has an infinite number of possible outcomes (positive real numbers), whereas the first has only two possible outcomes and the second has only six possible outcomes. In this section, we will deal only with experiments that have a finite number of outcomes, such as the first two. Experiments with an infinite number of outcomes require more advanced mathematics to deal with and are beyond the scope of this book.

We will make a further simplifying assumption by dealing only with experiments in which all outcomes are equally likely. For instance, the first and second experiments just mentioned satisfy this condition: In tossing a fair coin, heads or tails is equally likely to occur; and in rolling a fair die, each of the six possible outcomes is equally likely to occur. However, in selecting a ball from an urn containing 100 white balls and 1 black ball, the possible outcomes are not equally likely.

Let's now use sets to describe experiments in mathematical terms.

| **Definition 1**
Sample Space | The sample space of an experiment is the set that consists of all possible outcomes. |

For example, the experiment of tossing a coin has the sample space

$$\{H, T\}$$

where H indicates heads and T indicates tails. The experiment of rolling a die consists of the sample space

$$\{1, 2, 3, 4, 5, 6\}$$

where the digits are the results of rolling the die.

➤ **EXAMPLE 1**
Determining a Sample Space

Suppose an experiment consists of flipping a coin three times. Describe the associated sample space.

Solution

Each outcome of the experiment consists of a sequence of three letters indicating, in order, the results of the three coin tosses. There are eight possible outcomes, as follows:

Sample Space $= \{HHH, HHT, HTH, HTT, THH, THT, TTH, TTT\}$

When performing an experiment, it is often necessary to determine whether or not a particular event occurs. For instance, in rolling a die, you might wish to know whether the number rolled is even. In this case, the event is "even number occurs." In terms of the sample space, this event is equivalent to the rolled number occurring in the subset {2, 4, 6} of the sample space. In a similar fashion, we may define an event in terms of the sample space as follows:

Definition 2 **Event**	Suppose that we are given an experiment and its associated sample space. An **event** is a subset of the sample space.

➤ **EXAMPLE 2**
Exhibiting Events

Suppose that an experiment consists of rolling two dice.

1. Describe the sample space.
2. Describe the event "the dice add up to 7."

Solution

1. Each outcome of the experiment consists of an ordered pair of numbers (n, m), where n and m are integers from 1 to 6. So the sample space consists of the following set of 36 ordered pairs:

$$\{(1, 1), \quad (1, 2), \quad (1, 3), \quad (1, 4), \quad (1, 5), \quad (1, 6)$$
$$(2, 1), \quad (2, 2), \quad (2, 3), \quad (2, 4), \quad (2, 5), \quad (2, 6)$$
$$(3, 1), \quad (3, 2), \quad (3, 3), \quad (3, 4), \quad (3, 5), \quad (3, 6)$$
$$(4, 1), \quad (4, 2), \quad (4, 3), \quad (4, 4), \quad (4, 5), \quad (4, 6)$$
$$(5, 1), \quad (5, 2), \quad (5, 3), \quad (5, 4), \quad (5, 5), \quad (5, 6)$$
$$(6, 1), \quad (6, 2), \quad (6, 3), \quad (6, 4), \quad (6, 5), \quad (6, 6)\}$$

2. The event "dice add up to 7" consists of those ordered pairs of the sample space in which the two entries sum to equal 7. Here is the subset E corresponding to this event:

$$E = \{(1, 6), (2, 5), (3, 4), (4, 3), (5, 2), (6, 1)\}$$

Assigning Probabilities to Events

Suppose that we are given a sample space S and an associated event E. We assign to the event a number, called its probability, as follows:

Definition 3 **Probability of an Event**	Suppose that $n(S)$ denotes the number of outcomes in the sample space S and that $n(E)$ denotes the number of outcomes in the event E. Then the **probability of E** is defined as the number $$p(E) = \frac{n(E)}{n(S)}$$

For instance, consider the sample space $S = \{H, T\}$ for the experiment of flipping a coin. The probability of the event $E = \{H\}$ (heads) is then

$$p(E) = \frac{n(\{H\})}{n(\{H, T\})} = \frac{1}{2}$$

➤ **EXAMPLE 3**
Dice

Suppose that you roll two dice. What is the probability that the dice add up to 7?

Solution

We have previously seen that the sample space for the experiment of throwing two dice is given by

$$\{(1, 1),\quad (1, 2),\quad (1, 3),\quad (1, 4),\quad (1, 5),\quad (1, 6)$$
$$(2, 1),\quad (2, 2),\quad (2, 3),\quad (2, 4),\quad (2, 5),\quad (2, 6)$$
$$(3, 1),\quad (3, 2),\quad (3, 3),\quad (3, 4),\quad (3, 5),\quad (3, 6)$$
$$(4, 1),\quad (4, 2),\quad (4, 3),\quad (4, 4),\quad (4, 5),\quad (4, 6)$$
$$(5, 1),\quad (5, 2),\quad (5, 3),\quad (5, 4),\quad (5, 5),\quad (5, 6)$$
$$(6, 1),\quad (6, 2),\quad (6, 3),\quad (6, 4),\quad (6, 5),\quad (6, 6)\}$$

Moreover, the event $E = $ *dice add up to 7* is equal to

$$E = \{(1, 6), (2, 5), (3, 4), (4, 3), (5, 2), (6, 1)\}$$

Then from the definition of probability, we have

$$p(E) = \frac{n(E)}{n(S)}$$

$$= \frac{6}{36} = \frac{1}{6}$$

That is, the probability that the dice add up to 7 is 1/6.

The probability of an event is a measure of likelihood that the event will occur. If the experiment is repeated over and over, then the proportion of repetitions in which the outcome will belong to an event E is approximately $p(E)$. Of course, the proportion will not be exactly $p(E)$, but it will usually approximate $p(E)$ very closely. To make these intuitive interpretations of the probability of an event more exact requires more advanced mathematics, most likely found in a course on probability and statistics. For now, however, this inexact, intuitive approach will suffice.

In calculating probabilities, counting methods such as introduced in the preceding section are often useful, as is shown in the next example.

➤ **EXAMPLE 4**
Coin Flipping

Suppose that you flip a coin six times in succession. What is the probability that exactly two heads will occur?

Solution

A typical outcome in the sample space for this experiment consists of a sequence of six letters, where each letter is either H or T. We could list these outcomes, but it would be quite tedious to do so. Actually, we don't need a list of the outcomes. Rather, we need to know how many there are, and this can be deduced from the fundamental principle of counting. There are six letters to be chosen, and each can be chosen in two ways. So the number of choices is

$$n(S) = 2 \cdot 2 \cdot 2 \cdot 2 \cdot 2 \cdot 2 = 64$$

Let's count the number of outcomes in the event $E = $ "exactly two heads occur." An event with two heads is completely determined once we specify which two of the six coin tosses resulted in heads. If we represent the coin tosses by the numbers 1, 2, 3, 4, 5, 6, then an outcome in E is determined by a combination of these six objects taken two at a time. For instance, the combination $\{2, 5\}$ corresponds to the event $THTTHT$, which has heads in positions 2 and 5. The number of outcomes in E is thus the number of combinations of six objects taken two at a time. That is,

$$n(E) = {}_6C_2 = \frac{6 \cdot 5}{1 \cdot 2} = 15$$

Thus, the probability of the event E is given by

$$p(E) = \frac{n(E)}{n(S)} = \frac{15}{64}$$

That is, the probability that exactly two heads occur is 15/64.

Exercises 8.7

Describe the associated sample space for each of the following.

1. A coin is tossed, and then a die is thrown.

2. Two letters are randomly chosen, one after another, from the word *tack*.

3. You are making a survey of two-children families. You want to record the sex of each child, in the order of their births.

4. You survey four-children families and record the sex of each child in the order of birth. How many of these points correspond to families having three boys and one girl?

5. Two dice, one green and one white, are tossed, and the numbers of dots on their upper faces are noted.

6. From five different cards, A, B, C, D, and E, three are selected.

7. Five cards are drawn from a deck of 52 playing cards.

8. In Exercise 6, how many of the sample points correspond to a selection including A?

9. In a game, three balls are rolled down an incline into slots numbered 1–9.

10. Four cards are drawn from a standard pinochle deck of 48 cards.

11. Two dice are viewed from an angle so that three sides are visible.

12. Four coins are flipped.

13. Two letters from the name *William* are selected.

14. Three balls are drawn one at a time from a bag of eight green and seven red balls.

15. Of 20 cartons of milk in a grocery store cooler, 5 are spoiled. You choose 3 at random.

16. Two faulty batteries are placed on a shelf with a dozen reliable ones. Two are chosen by a customer.

17. You have a penny, a nickel, a dime, and a quarter in your pocket. You take out two coins, one after the other.

18. In a group of three men and two women, a two-person committee is chosen.

19. Three dice are tossed, and the number on the top surface of each die is written down.

20. Three letters are chosen one at a time from the word *jumping*.

◆ Applications

21. For a chronic disease, there are five standard treatments: 1, 2, 3, 4, and 5. A research doctor is going to study three of the treatments. If the three are chosen at random, what is the probability that treatment 1 will be chosen for study?

22. If two cards are drawn at random from a 52-card standard deck, what is the probability of drawing one king?

23. Two balls are drawn at random from a bag containing 6 red and 10 green balls. What is the probability of drawing 2 red balls?

24. In the game in which three balls are rolled down an incline into slots numbered 1–9, what is the probability that two of the balls will fall into the same slot?

25. From a standard deck of 52 cards, what is the probability of a 5-card hand having all cards from the same suit?

26. If four coins are flipped, find the probability of not obtaining two heads and two tails.

27. In tossing two dice, what is the probability of their sum totaling 5 ?

28. Four draftees, Ed, Larry, Marvin, and William, are to be assigned at random to either the infantry or the artillery. What is the probability that Ed and Marvin will be assigned to the same service?

29. A poker hand of 5 cards is dealt from a 52-card pack of standard cards. What is the probability of getting a full house (2 of one denomination and 3 of another)?

30. In a poker hand of 7 cards taken at random from a standard 52-card deck, what is the probability of getting 4 of a kind?

31. You and your best friend are placed in a group of 10 persons. From this group, a committee of 3 will be picked for an all-expense-paid trip to Disney World. What is the probability that both you and your best friend will be chosen?

32. You and your best friend join a softball team of 12 persons. In a game, 10 of the group are picked at random to play. What is the probability that you will be picked to play but your best friend will not be picked?

33. From Exercise 31, what is the probability that your best friend will be chosen but you will not be chosen?

34. From Exercise 32, what is the probability that neither you nor your friend will be picked?

35. A slot machine in Reno contains three reels with each reel containing 20 symbols. If the first reel contains five apples, the middle reel has four apples, and the third reel has only two apples, what is the probability of getting three apples in a row in order to win $5000 from this slot machine?

36. To complete your college physical education requirements you must take two classes from the following list: bowling, volleyball, tennis, basketball, softball, swim-

ming, and soccer. If you decide that you will choose your two at random by drawing classes from a hat one at a time, what is the probability that you will take tennis and swimming?

37. If a single die is tossed and then a single card is drawn from a standard 52-card deck, what is the probability of a 5 on the die and the 2 of spades on the card?

38. In a carnival game, 20 colored balls are placed in a bag. Of this 20, 6 are white, 4 are blue, 3 are red, 3 are green, 3 are yellow and 1 is black. In order to win, 3 balls are drawn at random, and of these, exactly 2 must be the same color. Also, if the black ball is drawn, you automatically lose. What is the probability of your winning?

39. Mrs. Robinson invites 10 relatives to a family party: her mother, two uncles, three brothers, and four cousins. If the chances of any one of her guests arriving first are equally likely, what is the probability that the first guest will be an uncle?

40. At a credit union dinner, a door prize of $300 is to be given to the person whose ticket stub is drawn from a bowl. If 35 tickets numbered 1–35 were sold, what is the probability that the winning ticket bears a number that is less than 20 ?

41. Of the 500,000 state income tax returns that one branch of the state tax bureau receives, 50,000 will be analyzed thoroughly. What is the probability that a return will not be thoroughly analyzed?

42. Ten cards, the 2 through the 6 of clubs and the 2 through the 6 of diamonds, are shuffled thoroughly and then taken one by one from the top of the deck and placed on a table. What is the probability that each card is next to a card bearing the same numeral?

8.8 CHAPTER REVIEW

Important Concepts, Properties, and Formulas—Chapter 8

n factorial	Let n be a positive integer. Then **n factorial,** denoted $n!$, is the product of all positive integers from 1 to n. That is, $$n! = n(n-1)(n-2)\cdots 2\cdot 1$$	p. 394
Binomial coefficient	$$\binom{n}{k} = \frac{n!}{k!(n-k)!}$$	p. 395
Binomial theorem	Let n be a nonnegative integer. Then $$(a+b)^n = \binom{n}{0}a^n + \binom{n}{1}a^{n-1}b$$ $$+ \binom{n}{2}a^{n-2}b^2 + \cdots + \binom{n}{n}b^n$$	p. 396

Principle of mathematical induction	Suppose that for each positive integer k there is given a statement P_k. Furthermore, suppose that the following two conditions are fulfilled: *a.* P_1 is true. *b.* By assuming that P_k is true, we can deduce the truth of P_{k+1}. (Here k is any positive integer.) Then P_k is true for every positive integer k.	p. 401
Summation notation	$$\sum_{n=A}^{B} a_n$$ is another way of writing $$a_A + a_{A+1} + \cdots + a_B$$	p. 408
Arithmetic sequences	$$a_n = a_1 + d(n-1)$$ $$s_n = \frac{n}{2}(a_1 + a_n)$$ $$s_n = \frac{n}{2}[2a_1 + d(n-1)]$$	pp. 411, 413
Geometric series	$$a_n = a_1 r^{n-1}$$ $$s_n = \begin{cases} a_1 \dfrac{1-r^n}{1-r}, & r \neq 1 \\ na_1, & r = 1 \end{cases}$$	pp. 416, 417, 420
Infinite geometric series	$$a + ar + ar^2 + \cdots$$ $$S = \frac{a}{1-r} \text{ if } |r| < 1$$	p. 422
Number of permutations of n things taken k at a time	$$_nP_k = n(n-1)(n-2)\cdots(n-k+1)$$ $$= \frac{n!}{(n-k)!}$$	p. 427
Number of combinations of n things taken k at a time	$$_nC_k = \binom{n}{k}$$	p. 428

Cumulative Review Exercises—Chapter 8

Expand.

1. $(a + t)^4$

2. $(3ab^2 - 4a^2b)^3$

3. $(3ab^2 - 4a^2b)^7$

4. $\left(\sqrt{a} - \dfrac{1}{\sqrt{a}}\right)^6$

5. Find the middle term of
$$\left(t - \frac{1}{t}\right)^{10}$$

6. Find the fourth term of
$$\left(\frac{4a}{5} - \frac{5}{2a}\right)^{10}$$

Use mathematical induction to prove that the following hold for all positive integers n.

7. $2^3 + 4^3 + 6^3 + \cdots + (2n)^3 = 2n^2(n+1)^2$

8. $1 + 2 + 3 + \cdots + n = \dfrac{n^2 + n}{2}$

9. $1 + 4 + 7 + \cdots + (3n - 2) = \dfrac{n}{2}(3n - 1)$

10. $\dfrac{1}{2} + \dfrac{1}{2^2} + \dfrac{1}{2^3} + \cdots + \dfrac{1}{2^n} = 1 - \dfrac{1}{2^n}$

11. $x^n - y^n$ is divisible by $x - y$.

12. $x^{2n+1} + y^{2n+1}$ is divisible by $x + y$.

For the given $f(n)$: (a) Complete the table. (b) On the basis of the results in the table, what would you believe to be the value of $f(9)$? (c) Compute $f(9)$. (d) Make a conjecture about the value of $f(n)$.

13. $f(n) = 2 + 5 + 8 + \cdots + 3n - 1$

n	1	2	3	4	5	6
$f(n)$						

14. $f(n) = 1 + 3 + 9 + \cdots + 3^{n-1}$

n	1	2	3	4	5	6
$f(n)$						

15. Prove that if a is any real number greater than 1, then $a^n > 1$ for every positive integer n.

16. Prove that 5 is a factor of $n^4 - 1$ for all natural numbers $n \geq 2$ not divisible by 5.

Determine the first three terms and the 10th term of the sequences defined by the following:

17. $a_n = n - 2$

18. $a_n = 2n + 3$

19. $a_n = \dfrac{4}{n + 1}$

20. $a_n = n^2$

Determine the first five terms for the following defined sequences.

21. $a_n = n \log 10^{n+1}$

22. $a_n = \log 100^{n-1}$

23. $a_1 = 2, a_n = \dfrac{3}{a_{n-1}}, n > 1$

24. $a_1 = -2, a_n = \dfrac{1}{2}a_{n-1}, n > 1$

Evaluate the following expressions.

25. $\displaystyle\sum_{n=4}^{10}(n - 5)$

26. $\displaystyle\sum_{n=1}^{6}(2n - 1)$

27. $\displaystyle\sum_{n=2}^{6}\left(\dfrac{1}{n - 1}\right)^2$

28. $\displaystyle\sum_{n=4}^{10}\left[-\cos\left(\dfrac{n\pi}{2}\right)\right]$

Prove the following sequences are arithmetic.

29. $2, 3, 4, \ldots, 1 + n, \ldots$

30. $2, -2, -6, -10, \ldots, 6 - 4n, \ldots$

31. $2, 5, 8, 11, \ldots, 3n - 1, \ldots$

32. $1, \dfrac{3}{2}, 2, \dfrac{5}{2}, \ldots, \dfrac{n + 1}{2}, \ldots$

Find the indicated term for the given arithmetic sequences.

33. $a_1 = 2, d = 3$; find a_6.

34. $a_1 = -1, d = 5$; find a_9.

35. $a_1 = 4, d = -2$; find a_{10}.

36. $a_1 = 5, d = -2$; find a_7.

For each of the following arithmetic sequences, find the first term, a_1, and the difference d between consecutive terms of the sequence.

37. $a_9 = 6, a_{12} = 21$

38. $a_3 = 1, a_8 = -14$

39. $a_{10} = -32, a_{40} = -122$

40. $a_{17} = 96, a_{28} = 141$

For the following arithmetic sequences, answer the given question.

41. If $a_1 = 2, d = 2$, and $a_n = 32$, what is n?

42. If $a_3 = 9, d = -2$, and $a_n = -27$, what is n?

43. If $a_6 = 12$ and $d = -2$, what is a_1?

44. If $a_{12} = -20$ and $d = 4$, what is a_3?

Compute the indicated sums for the following arithmetic sequences.

45. $-2, -1, 0, \ldots, n - 3, \ldots$; find s_{20}.

46. $-4, -2, 0, \ldots, 2n - 6, \ldots$; find s_{30}.

47. $2, 0, -2, \ldots, 4 - 2n, \ldots$; find s_{35}.

48. $5, 7, 9, \ldots, 3 + 2n, \ldots$; find s_{25}.

♦ **Applications**

49. In building a tower, a child uses 15 cubes for the first row, and 2 fewer cubes for each row thereafter. How many blocks would there be in the eighth row?

50. How many meters would a skydiver fall during the ninth second if he falls 10 meters during the first second, 20 meters during the second second, 30 meters during the third second, and so on?

51. The population of St. John, Michigan, is 42,000. If the zoning commission permits an increase of 250 people each year, what will be the maximum population in six years?

52. How much material will be needed for the rungs of a ladder of 31 rungs if the rungs taper from 18 cm to 28 cm and the lengths form an arithmetic sequence?

Prove that the following are geometric sequences.

53. $a_n = 4\left(\dfrac{1}{4}\right)^{n-1}$

54. $a_n = -24\left(\dfrac{1}{2}\right)^{n-1}$

55. $a_n = -\dfrac{1}{2}(2^{n+1})$

56. $a_n = (-4)^{n+1}$

Find the indicated term for each of the following geometric sequences.

57. $a_1 = 2, r = 4$; find a_{10}.

58. $a_1 = \dfrac{4}{5}, r = -\dfrac{1}{5}$; find a_8.

59. If $a_6 = -9$ and $a_7 = 3$; find a_3.

60. If $a_{10} = 12$ and $a_{11} = 18$; find a_7.

61. $4, 8, 16, \ldots$; find a_n.

62. $-3, 2, -\dfrac{4}{3}, \ldots$; find a_n.

Solve the following for geometric sequences.

63. Which term of the geometric sequence $2, -6, 18, \ldots$, is -54?

64. Which term of the geometric sequence $81, -27, 9, \ldots$, is $-1/27$?

◆ **Applications**

65. If the interest of a five-year certificate of deposit was set at 8% compounded semiannually, what would be the value after five years if you invested $2000?

66. To buy that new car of your dreams, you borrow $4000 at 10% compounded monthly. You also elect to pay off your loan in one lump sum payment at the end of three years. What will be the amount of your payment?

Given three of the values: a_1, a_n, r, n, and S_n of a geometric sequence, find the missing two.

67. $a_n = 324, n = 5, r = 3$

68. $a_n = -128, a_1 = -2, n = 6$

Find a formula for the following geometric sequence.

69. $a_2 = 10, a_5 = -1250$

70. $a_1 = 4, a_5 = 324 \quad (r > 0)$

Solve the following problems for infinite sequences (if possible).

71. Find the sum of $10 - 5 + 5/2 - 5/4 + \cdots$.

72. Find the sum of $12 + 18 + 27 + \cdots$.

73. Find the sum of $24 + 16 + 32/3 + 64/9 + \cdots$.

74. Find the sum of $16 - 12 + 9 - 27/4 + \cdots$.

◆ **Applications**

75. A ball is dropped 20 feet and rebounds on each bounce 2/5 of the height from which it fell. How far will it travel before coming to rest?

76. A ball is tossed from the ground to a height of 60 feet. If it will rebound on each bounce to a height that is 3/10 of the height from which it fell, then how far will it travel before coming to rest?

Solve the following.

77. In how many different orders can you arrange five books on a shelf?

78. How many numbers consisting of two or three digits, without repetitions, can you form from the digits 1, 2, 3, 4, 5 ?

79. How many possibilities are there for a license plate with two letters followed by four numbers?

80. If four people run a race, how many different orders can they all finish in if there are no ties?

Determine the following:

81. $_3P_2$ 82. $_{10}P_3$ 83. $_9P_9$ 84. $_6P_1$

Use permutations to solve the following.

85. How many "words" of four letters can be formed from the letters of the word *empty*? (Any four letters is considered a word.)

86. How many permutations are there of 13 cards taken from a bridge deck of 52 cards?

87. How many 3-digit numbers can be formed from the 10 digits $0, 1, \ldots, 9$?

88. In arranging five pictures on a wall such that a certain one must be second, how many arrangements are possible?

Determine the following.

89. $_{10}C_4$ 90. $_7C_6$ 91. $_9C_6$ 92. $_{10}C_2$

93. You are one of 12 people in your class. How many 5-member committees can be formed that include you?

94. In a group of nine Republicans and eight Democrats, how many eight-member committees can be formed with six Republicans and two Democrats?

95. In a group of 30 members, how many different committees of 4 people can be formed?

96. How many five-letter "words," each having three consonants and two vowels, can be formed from the letters of the word *equations?*

Describe the associated sample space for each of the following.

97. Five coins are flipped.

98. Six cards are drawn from a deck of 52 different cards.

99. Three balls are drawn, one at a time, from a bag of seven blue and six red balls.

100. Two dice are thrown.

Solve the following.

101. In the Indian game of Tong, two players simultaneously show their right hands to each other, exhibiting either one, two, or three extended fingers. If each player is equally likely to extend one, two, or three fingers, then what is the probability that the total number of fingers extended is odd?

102. The numbers from 1 through 15 are printed on 15 balls, one number per ball. If one of these balls is drawn at random, what is the probability that the number of it is divisible by 5?

103. A committee of two is to be selected from Bill, Ted, Carol, Alice, and Dave. What is the probability that Dave will not be on the committee?

104. If two cards are drawn one at a time from a standard bridge deck, what is the probability that the first card is an ace and the second card is a jack?

TOPICS IN ANALYTIC GEOMETRY

A cooler tower at an electrical power generation plant, such as the one shown here, uses a hyperbolic design to efficiently dissipate excess heat.

I n this final chapter, let's return to the subject of analytic geometry and discuss topics that are often useful in other courses, especially calculus and physics.

9.1 CONIC SECTIONS, PARABOLAS

A **conic section** is a curve obtained as the intersection of a cone and a plane. There are four different types of conic sections: circles, ellipses, parabolas, and hyperbolas. Exactly which type of curve is generated depends on the relative position of the cone and the plane. Each of the four types of conic sections is shown in Figure 1.

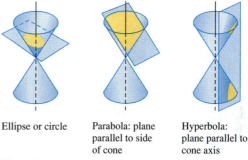

Ellipse or circle   Parabola: plane parallel to side of cone   Hyperbola: plane parallel to cone axis

Figure 1

The conic sections were first studied by the ancient Greek geometers, who were interested in them for their geometric properties. However, beginning in the 17th century, conic sections were studied in terms of their equations and were the testing ground for the new ideas of analytic geometry and calculus. Although the conic sections were originally studied from the point of view of pure mathematics, they were found to have many applications in physics, optics, and astronomy.

Rather than define the conic sections in terms of the intersection of a plane and a cone, we will give definitions of each of the conics as a **locus**, that is, as the set of points in the plane satisfying a specific geometric property. From the locus definitions, we will derive the equations and properties of each of the types of conics.

Definition and Standard Form for Parabolas

Let us start our discussion of the conics with a locus definition of a parabola.

Definition 1
Parabola, Directrix, Focus

A **parabola** is the set of all points whose distance from a fixed line D equals the distance[1] from a fixed point F not on the line. The line D is called the **directrix** and the point F is called the **focus**.

In Figure 2, we have drawn the focus and directrix of a parabola. The parabola consists of all points P for which the distance $\overline{PA}$ equals the distance $\overline{PF}$.

The line through the focus perpendicular to the directrix is shown as a dotted line. This line is called the **axis** of the parabola. (See Figure 3.) If a point is on

[1]The distance from a point P to a line L is the length of the perpendicular from P to L.

the parabola, then its reflection in the axis is also on the parabola. That is, the parabola is symmetric about its axis. The point on the axis halfway in between the directrix and the focus is clearly on the parabola. This point is called the **vertex**. The directed distance from the vertex to the focus is called the **focal length** and is usually denoted p. If the focus is either to the right of or above the vertex, then the value of p is positive; if the focus is either to the left of or below the vertex, then the value of p is negative.

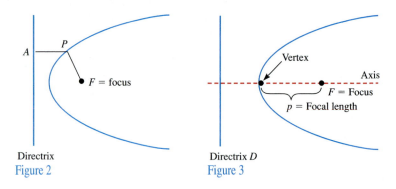

Directrix   Directrix D
Figure 2   Figure 3

Parabolas for which the directrix is either horizontal or vertical have especially simple equations, as the following result shows:

STANDARD FORMS OF THE EQUATION OF A PARABOLA

Suppose that a parabola has vertex at (h, k) and that the focal length is p.

1. If the directrix is vertical, the parabola has equation
$$(y - k)^2 = 4p(x - h)$$
(See Figure 4.)

2. If the directrix is horizontal, the parabola has equation
$$(x - h)^2 = 4p(y - k)$$
(See Figure 5.)

   In general, the graph of a parabola is bowl-shaped. The focus is within the bowl. The directrix is outside the bowl and perpendicular to the axis of the parabola.

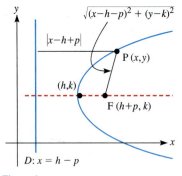

Figure 4

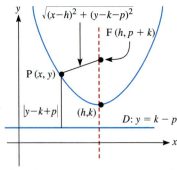

Figure 5

Proof   Let's prove statement 1. (The proof of statement 2 is similar.) Let $P(x, y)$ be a point on the parabola. Since the directrix is vertical, the distance from P to the directrix is $d_1 = |(x - h) + p|$. The focus is the point $(h + p, k)$, so that the distance from P to the focus is

$$d_2 = \sqrt{((x - h) - p)^2 + (y - k)^2}$$

(See Figure 4.) Since P is on the parabola, we must have $d_1 = d_2$:

$$\sqrt{((x - h) - p)^2 + (y - k)^2} = |(x - h) + p|$$
$$(x - h)^2 - 2p(x - h) + p^2 + (y - k)^2 = |x - h + p|^2 = ((x - h) + p)^2$$
$$(x - h)^2 - 2p(x - h) + p^2 + (y - k)^2 = (x - h)^2 + 2p(x - h) + p^2$$
$$(y - k)^2 = 4p(x - h)$$

This gives us the standard form of the equation of a parabola in case the directrix is vertical.   ♦

Parabolas and Their Equations

> **EXAMPLE 1**
Determining Equation from Focus and Directrix

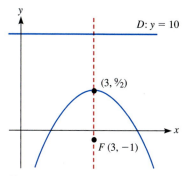

Figure 6

Determine the equation of a parabola with focus at $(3, -1)$ and directrix $y = 10$.

Solution

The directrix is horizontal, so the axis of the parabola is vertical. The focus is on the axis so that the axis has equation $x = 3$. (See Figure 6.) The vertex is midway between the focus and directrix and is therefore $(3, 9/2)$. The directed distance from the focus to the vertex is given by

$$p = -1 - \frac{9}{2} = -\frac{11}{2}$$

Since the directrix is horizontal, standard form 2 applies. The equation of the parabola is

$$(x - 3)^2 = 4\left(-\frac{11}{2}\right)\left(y - \frac{9}{2}\right)$$
$$(x - 3)^2 = -22\left(y - \frac{9}{2}\right)$$

A parabola whose directrix is parallel to one of the coordinate axes has an equation of either of the following forms:

$$y = ax^2 + bx + c \tag{1}$$
$$x = ay^2 + by + c \tag{2}$$

(To get this form of the equations, just multiply out the standard form and rearrange terms.) Conversely, if we are given an equation in one of these forms, we may complete the square and write the equation in one of the standard forms. By reading the proof of statement 1 in reverse, we see that the graph of the equation is a parabola. The next two examples illustrate how to sketch the parabola from an equation of the form (1) or (2).

➤ **EXAMPLE 2**
Determine Focus and Directrix
from Equation

Determine the focus and the directrix of the parabola with equation

$$y = 3x^2 - 6x$$

Sketch the parabola.

Solution

Complete the square on the right to obtain:

$$y = 3x^2 - 6x$$
$$y = 3(x^2 - 2x + 1) - 3$$
$$y + 3 = 3(x - 1)^2$$
$$(x - 1)^2 = \frac{1}{3}(y + 3)$$
$$(x - 1)^2 = 4 \cdot \frac{1}{12}(y + 3)$$

From the last form of the equation, we see that the graph is a parabola with vertex $(h, k) = (1, -3)$ and $p = \frac{1}{12}$. The last equation is in the second standard form, so the directrix is horizontal and the axis is vertical. The x-coordinate of the focus is the same as that of the vertex, namely, 1. The y-coordinate of the focus is obtained by adding p to the y-coordinate of the vertex: $-3 + p = -3 + \frac{1}{12} = -\frac{35}{12}$. So the focus is at $(1, -\frac{35}{12})$. The directrix is

$$y = -3 - p = -3 - \frac{1}{12} = -\frac{37}{12}$$

The graph of the equation is sketched in Figure 7.

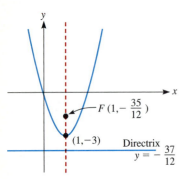

Figure 7

➤ **EXAMPLE 3**
Reducing to Standard Form

Determine the focus and the directrix of the parabola with equation

$$y^2 + 2y = -x - 5$$

Sketch the parabola.

Solution

In this example, y appears in a term of the second degree, so we complete the square on the left:

$$y^2 + 2y + 1 = (-x - 5) + 1$$
$$(y - (-1))^2 = 4 \cdot \left(-\frac{1}{4}\right) \cdot (x - (-4))$$

From the equation, we read off the vertex: $(-4, -1)$. Also from the equation, we see that $p = -\frac{1}{4}$. Since the y-term is the one of second degree, the directrix is vertical and the axis horizontal. Since the value of p is negative, the focus is to the left of the vertex, at

$$\left(-4 - \frac{1}{4}, -1\right) = \left(-\frac{17}{4}, -1\right)$$

The directrix lies a directed distance p to the right of the vertex and is therefore the line $x = -4 - p = -4 + \frac{1}{4} = -\frac{15}{4}$. Figure 8 shows the graph of the equation.

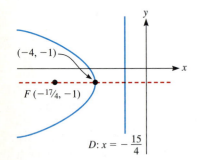

Figure 8

The Reflection Property of the Parabola

When a light or sound wave hits a surface, it is reflected in such a way that the angle made by the incoming wave equals the angle made by the reflected wave. (See Figure 9.) This is a physical law that was first observed in the 17th century. The parabola has a geometric property that, when combined with the law of reflection of light and sound waves, allows for many interesting and useful applications. Figure 10 shows a parabola with vertical axis. A series of light (or sound) waves moves parallel to the axis and is reflected off the surface of the parabola, so it can be proved that the reflected waves all pass through the focus. This result is known as the **reflection property** of the parabola.

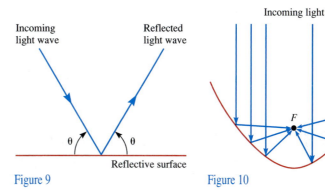

Figure 9

Figure 10

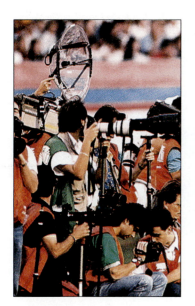

The reflection property may be used to construct a telescope. The observed light waves travel the length of the telescope tube and are reflected off a parabolic mirror. The reflected light rays all pass through the focus of the parabola, where the eyepiece of the telescope is located. In this case, the reflection property of the parabola states that the observed image is displayed at the eyepiece. (See Figure 10.)

Another application using a similar design principle is the parabolic microphone, often used in sporting events. This device consists of a parabolic dish to reflect sound to a microphone located at the focus. By concentrating diverse sound waves at the focus, the microphone can pick up and amplify sounds quite a distance away.

To prove the reflection property of the parabola requires calculus.

Exercises 9.1

Find the vertex, focus, and directrix of the following parabolas.

1. $y = 9x^2$
2. $x = 4y^2$
3. $y^2 = -2x$
4. $x^2 = -12y$
5. $y + 1 = 2(x - 1)^2$
6. $3y - 4 = 6(x + 2)^2$
7. $y^2 - (y + 1)^2 = x^2$
8. $y = 4x^2 - 4x + 1$
9. $(3x + 1)^2 = 4y + 12$
10. $(2y - 1)^2 = 3(x + 5) - 1$

11. $y^2 + 3y - 6x = 8$
12. $y^2 + 3y - 6x = 0$
13. $25x^2 + 10x - y + 5 = 0$
14. $2x^2 + 6x + 5y - 20 = 0$

Determine the equation of the parabola with the following properties.

15. vertex: $(0, 0)$
    directrix: $x = 2$
16. vertex: $(0, 0)$
    directrix: $x = -3$

17. vertex: $(2, 1)$
    directrix: $y = -1$

18. vertex: $(1, 1)$
    directrix: $y = 2$

19. vertex: $(0, 0)$
    focus: $(1, 0)$

20. vertex: $(0, 0)$
    focus: $(-2, 0)$

21. vertex: $(3, -2)$
    focus: $(3, 4)$

22. vertex: $(4, 2)$
    focus: $(4, 0)$

23. focus: $(1, 2)$
    directrix: $y = -2$

24. focus: $(1, 2)$
    directrix: $y = 5$

25. focus: $(-1, -4)$
    directrix: $x = 3$

26. focus: $(1, 0)$
    directrix: $x = 0$

27. Axis vertical and passes through the three points $(0, 0)$, $(-2, 4)$, $(3, 6)$.

28. Axis vertical, vertex at the origin, and passes through the point $(3, 10)$.

29. Axis horizontal and passes through the three points $(0, 0)$, $(2, 5)$, $(3, 12)$.

30. Axis horizontal, vertex at $(2, 1)$, and passes through the origin.

◆ Applications

31. The main span of a suspension bridge is 4000 feet long. Suppose that the towers are 500 feet high and the roadway is 300 feet above the surface of the water. The cables form a parabola. Assuming a coordinate system with horizontal axis along the roadway and vertical axis along the left tower, determine the equation of the parabola.

32. The main span of a suspension bridge is 5200 feet long. Suppose that the towers hold the ends of the cable 300 feet above the roadway. The cables form a parabola. Assuming a coordinate system with horizontal axis along the roadway and vertical axis along the left tower, determine the equation of the parabola.

33. The mirror of a telescope has diameter 6 inches and depth .01 inches at the center. Determine the distance from the center of the mirror to the focus.

34. The wall and ceiling in back of the orchestra of a symphony hall has a cross-section that is a parabola. Suppose that the conductor stands at the focus, which is 50 feet from the wall. The height of the wall above the conductor is 60 feet. Determine the equation of the parabola. Explain what the reflection property of the parabola has to do with the design of the wall.

35. The headlight of a car has a cross-section that is a parabola. The light source is located at the focus. Suppose that the light source is 2 inches from the back of the headlight and that the radius of the headlight is 6 inches. Determine the equation of the parabola. Explain what the reflection property of the parabola has to do with the operation of the headlight.

In calculus, it is proved that the slope m of the tangent line to the parabola

$$y - k = 4p(x - h)^2$$

at the point where $x = a$ is given by the formula

$$m = 8p(a - h)$$

Determine the equation of:

36. The tangent line to $y = x^2$ at $(2, 4)$.

37. The tangent line to $y = 2(x - 3)^2$ at the point where $x = 1$.

38. The line perpendicular to the tangent line to $y = 2x^2 + 3$ at the point where $x = -2$ and passing through the point of tangency.

39. The line parallel to the tangent line to $y = x^2$ at $x = 3$ and passing through the origin.

40. Use the formula for the slope of the tangent line to prove the reflection property of the parabola.

Match the equations in exercises 41–46 with their graphs.

41. $y = 9x^2$

42. $x = 4y^2$

43. $y^2 = -2x$

44. $x^2 = -12y$

45. $y + 1 = 2(x - 1)^2$

46. $(2y - 1)^2 = 3(x + 5) - 1$

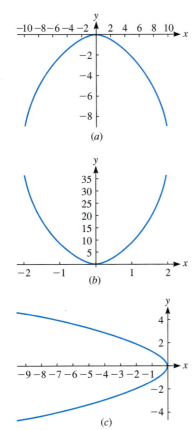

(a)

(b)

(c)

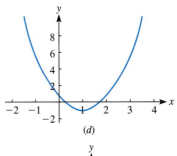

(d)

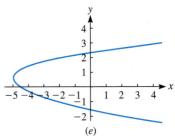

(e)

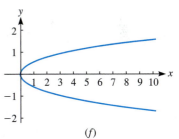

(f)

➡ **Technology**

47. **(Graphing calculator)** On one set of axes draw parabolas with vertex at the origin and focus at $(p, 0)$ for $p = 0.5, 0.75, 1, 2, 3$. Describe what happens to the graph as the focal length p increases.

Use a graphing utility to graph the parabolas in exercises 48–51. (For some of them, you will need to graph two separate equations, obtained by solving for y in terms of x.)

48. $3y^2 = 4x + 1$

49. $-\dfrac{y^2}{2} = \dfrac{4x}{7}$

50. $(y - 3)^2 = 2(x + 5) + 1$ 51. $(3y + 2)^2 = x - 3$

▌9.2 ELLIPSES

Let us now consider two additional conic sections: circles and ellipses. As we shall see, circles are special cases of ellipses, so let's start the section by concentrating on ellipses.

Definition and Standard Equation of an Ellipse

We may define an ellipse in geometric terms as follows.

> ▌**Definition 1**
> **Ellipse, Foci**
>
> An **ellipse** is the set of all points P, the sum of whose distances from two fixed points is a constant. The two points are called the **foci** (the plural of focus).

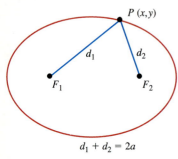

$d_1 + d_2 = 2a$

Figure 1

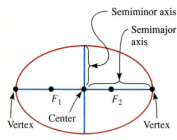

Figure 2

Refer to Figure 1. The two foci are F_1 and F_2. A typical point on the ellipse is $P(x, y)$. The distance of P to F_1 is denoted d_1, and the distance of P to F_2 is denoted d_2. The definition of an ellipse says that there is a constant $2a$, independent of the point P such that

$$d_1 + d_2 = 2a \qquad (1)$$

(The reason for writing $2a$ rather than a will become apparent shortly.)

Draw a line through the foci. (See Figure 2.) It intersects the ellipse in two points, called **vertices**. The line segment connecting the vertices is called the **major axis**. Half the length of the major axis is called the **semimajor axis** of the ellipse. The midpoint of the major axis is called the **center** of the ellipse. It is clear that the center is midway between the two foci. Draw a line segment perpendicular to the major axis through the center, with both endpoints on the ellipse. This line segment is called the **minor axis**. Half the length of the minor axis is called the **semiminor axis** of the ellipse.

Assume that the center of the ellipse is at (h, k) and that the distance from the center to each focus is c. The number c is called the **focal length** of the ellipse. Let us assume that the major axis is horizontal. Then the foci are at $(h \pm c, k)$. Suppose that the right vertex is at (l, m). It is clear that $m = k$. Moreover, the distance of the right vertex from F_1 is $(l - h) + c$, and the distance from F_2 is $(l - h) - c$. So by equation (1), we have

$$(l - h + c) + (l - h - c) = 2a$$
$$2(l - h) = 2a$$
$$l = h + a$$

That is, the coordinates of the right vertex are $(h + a, k)$. Similarly, the coordinates of the left vertex are $(h - a, k)$. Moreover, from these coordinates, we see that $2a$ is the length of the major axis of the ellipse.

Let's now write equation (1) in terms of the coordinates of the foci and the point $P(x, y)$:

$$d(F_1, P) + d(F_2, P) = 2a$$

$$\sqrt{(x - [h - c])^2 + (y - k)^2} = 2a - \sqrt{(x - [h + c])^2 + (y - k)^2}$$

$$(x - [h - c])^2 + (y - k)^2 = 4a^2 + (x - [h + c])^2 + (y - k)^2$$
$$- 4a\sqrt{(x - [h + c])^2 + (y - k)^2}$$

$$4a\sqrt{(x - [h + c])^2 + (y - k)^2} = 4a^2 - 4c(x - h)$$

$$a\sqrt{(x - [h + c])^2 + (y - k)^2} = a^2 - c(x - h)$$
$$a^2(x - [h + c])^2 + a^2(y - k)^2 = (a^2 - c(x - h))^2$$
$$a^2(x - h)^2 + a^2c^2 + a^2(y - k)^2 = a^4 + c^2(x - h)^2$$
$$(a^2 - c^2)(x - h)^2 + a^2(y - k)^2 = a^2(a^2 - c^2)$$
$$\frac{(x - h)^2}{a^2} + \frac{(y - k)^2}{a^2 - c^2} = 1 \qquad (2)$$

From Figure 3, we see that if b denotes half the length of the minor axis, then by equation (1), we have

$$2\sqrt{b^2 + c^2} = 2a$$
$$a^2 = b^2 + c^2$$

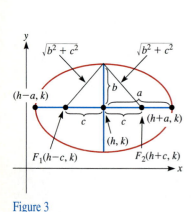

Figure 3

Therefore, we may rewrite equation (2) in the form

$$\frac{(x-h)^2}{a^2} + \frac{(y-k)^2}{b^2} = 1 \quad \text{(major axis horizontal)} \tag{3}$$

This is the standard form of the equation of an ellipse whose major axis is horizontal.

If we assume that the major axis is vertical, we arrive at the equation

$$\frac{(x-h)^2}{b^2} + \frac{(y-k)^2}{a^2} = 1 \quad \text{(major axis vertical)} \tag{4}$$

In either case, the foci are a distance c from the center along the major axis, where

$$a^2 = c^2 + b^2 \tag{5}$$

A circle is an ellipse in which the semimajor and semiminor axes are equal (to the radius). In this case, the above standard equations of an ellipse reduce to the familiar equation of a circle. Indeed, if an ellipse has $a = b = r$ and center (h, k), then the equation of the ellipse is

$$\frac{(x-h)^2}{r^2} + \frac{(y-k)^2}{r^2} = 1$$
$$(x-h)^2 + (y-k)^2 = r^2$$

the standard form of a circle with center at (h, k) and radius r.

Illustrations

➤ **EXAMPLE 1**
Determine Equation from Foci and Minor Axis

Determine the equation of the ellipse with foci at $(3, 5)$ and $(9, 5)$ and minor axis of length 10.

Solution

Since the foci have the same y-coordinate, the major axis is horizontal. The center of the ellipse is midway between the two foci and is therefore $(6, 5)$, and the distance from the center to the foci is given by $c = 3$. The semiminor axis b is given by

$$2b = 10$$
$$b = 5$$
$$a^2 = b^2 + c^2$$
$$a^2 = 5^2 + 3^2$$
$$= 25 + 9$$
$$= 34$$
$$a = \sqrt{34}$$

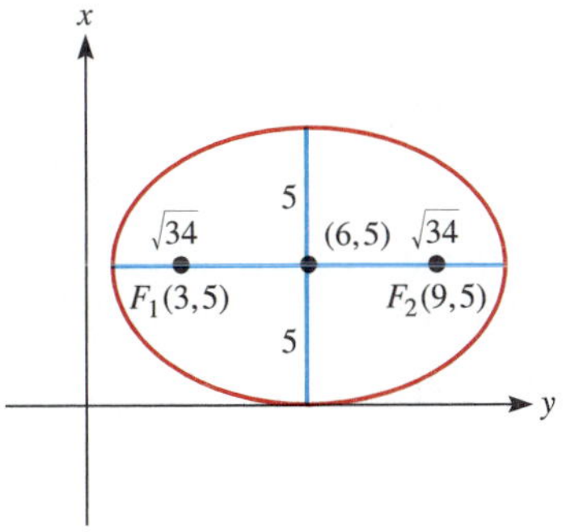

Figure 4

Therefore, the equation of the ellipse is

$$\frac{(x - 6)^2}{\left(\sqrt{34}\right)^2} + \frac{(y - 5)^2}{5^2} = 1$$

$$\frac{(x - 6)^2}{34} + \frac{(y - 5)^2}{25} = 1$$

The graph is sketched in Figure 4.

➤ EXAMPLE 2
Determine Parameters from Equation

Sketch the graph of the ellipse with equation

$$4x^2 + 25y^2 = 100$$

Determine the foci, vertices, semimajor axis, and semiminor axis.

Solution

Since both x and y appear to the second power and the coefficients of both have the same sign, the graph of the equation is an ellipse. In the standard form of the equation of an ellipse, the constant term is 1. To reduce the equation to this form, we divide both sides by 100:

$$\frac{4x^2}{100} + \frac{25y^2}{100} = 1$$

$$\frac{x^2}{25} + \frac{y^2}{4} = 1$$

$$\frac{x^2}{5^2} + \frac{y^2}{2^2} = 1$$

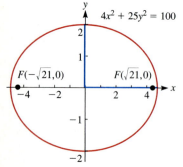

From the equation, we see that the ellipse is centered at the origin with semimajor axis along the x-axis with length 5 and semiminor axis along the y-axis with length 2. The vertices are $(\pm 5, 0)$. The foci are at a distance c from the center, where

$$c^2 + b^2 = a^2$$
$$c^2 = 5^2 - 2^2$$
$$= 21$$
$$c = \sqrt{21}$$

Figure 5

Therefore, the foci are $(\pm \sqrt{21}, 0)$. The graph is sketched in Figure 5.

➤ EXAMPLE 3
Reducing to Standard Form

Sketch the graph of the ellipse with equation

$$5x^2 + 10x + 4y^2 + 8y - 5 = 0$$

Determine the foci, vertices, semimajor axis, and semiminor axis.

Solution

Note that the equation involves both x and y raised to the second power and that the coefficients of the square terms have the same sign. This means that the graph of the equation is an ellipse. To transform the equation into standard form, we complete the square in both x and y:

$$5x^2 + 10x + 4y^2 + 8y - 5 = 0$$
$$5(x^2 + 2x) + 4(y^2 + 2y) = 5$$

$$5(x^2 + 2x + 1) + 4(y^2 + 2y + 1) = 5 + 5 + 4$$

$$5(x + 1)^2 + 4(y + 1)^2 = 14$$

$$\frac{5(x + 1)^2}{14} + \frac{4(y + 1)^2}{14} = 1$$

$$\frac{(x + 1)^2}{\frac{14}{5}} + \frac{(y + 1)^2}{\frac{7}{2}} = 1$$

$$\frac{(x + 1)^2}{\left(\sqrt{\frac{14}{5}}\right)^2} + \frac{(y + 1)^2}{\left(\sqrt{\frac{7}{2}}\right)^2} = 1$$

This last equation has a graph that is an ellipse with center $(-1, -1)$. Since $7/2 = 3.5$ is greater than $14/5 = 2.8$, we see that the semimajor axis lies in the vertical direction and the semiminor axis in the horizontal direction. The lengths of the axes are

$$\text{semimajor axis} = \sqrt{\frac{7}{2}} = \frac{\sqrt{14}}{2}, \text{ vertical}$$

$$\text{semiminor axis} = \sqrt{\frac{14}{5}} = \frac{\sqrt{70}}{5}, \text{ horizontal}$$

The foci are at a distance c from the center, where

$$c^2 = a^2 - b^2$$

$$= \left(\sqrt{\frac{7}{2}}\right)^2 - \left(\sqrt{\frac{14}{5}}\right)^2$$

$$= \frac{7}{2} - \frac{14}{5}$$

$$= \frac{7}{10}$$

$$c = \sqrt{\frac{7}{10}} = \frac{\sqrt{70}}{10}$$

Therefore, the foci are $(-1, -1 \pm \sqrt{70}/10)$. The graph is sketched in Figure 6.

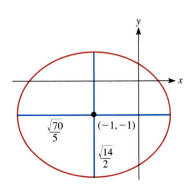

Figure 6

Eccentricity of an Ellipse

The eccentricity e measures the *ovalness* of an ellipse. The eccentricity always lies between 0 and 1. The closer the eccentricity is to 0, the closer the ellipse is to a circle. The closer the eccentricity is to 1, the more elongated the ellipse.

Definition 2 **Eccentricity**	Suppose that an ellipse has semimajor axis a and focal length c. Then the **eccentricity** e of the ellipse is defined as $$e = \frac{c}{a}$$

Note that since the focal length is less than the semimajor axis, the eccentricity as defined by the formula lies between 0 and 1.

➤ **EXAMPLE 4**
Calculating Eccentricity

Determine the eccentricity of the ellipse of Example 3.

Solution

In Example 3, we showed that $a = \sqrt{14}/2$ and $c = \sqrt{70}/10$. Therefore, we have

$$e = \frac{c}{a} = \frac{\dfrac{\sqrt{70}}{10}}{\dfrac{\sqrt{14}}{2}} = \frac{\sqrt{5}}{5}$$

➤ **EXAMPLE 5**
Eccentricity of a Circle

Show that an ellipse of eccentricity 0 is a circle.

Solution

If $e = 0$, then $c = 0$, so that

$$0 = a^2 - b^2$$
$$a^2 = b^2$$
$$a = b \qquad \textit{(since a and b are positive)}$$

That is, the semimajor and semiminor axes are equal and the ellipse is a circle.

One of the major accomplishments of physics is Kepler's Laws, which state that the planets travel in elliptical orbits around the sun, with the sun located at a focus. Using Newton's Law of Gravitation, it is possible to deduce the precise equations of the orbits in terms of the masses of the planets, the mass of the sun, and observed distances and velocities of the planets at particular times.

➤ **EXAMPLE 6**
Earth's Orbit

The eccentricity of the earth's orbit about the sun is approximately 0.0167. The closest distance between the earth and sun is approximately 93 million miles. What is the furthest distance between the earth and the sun? (See Figure 7.)

Solution

Let a be the semimajor axis of the orbit. Assume that the center of the orbit is $(0, 0)$ and the sun is at the focus $(c, 0)$. Then

$$a = c + 93$$

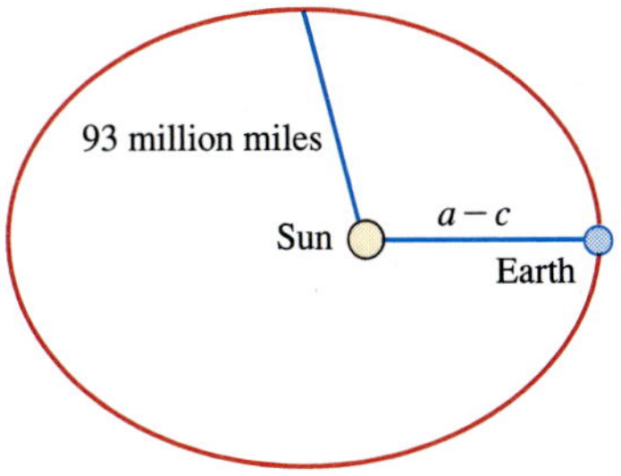

Figure 7

$$e = \frac{c}{a} = \frac{c}{c + 93} = 0.0167$$

$$c = 0.0167c + 1.5531$$

$$c = 1.5795$$

The furthest distance of the earth from the sun is

$$a + c = 2c + 93$$
$$= 96.159 \text{ million miles}$$

Exercises 9.2

Determine the equation of the ellipse having the following properties:

1. vertices: $(0, \pm 3)$
   foci: $(0, \pm 2)$

2. vertices: $(5, 1), (-5, 1)$
   foci: $(3, 1), (-3, 1)$

3. vertices: $(-1, 4), (-1, 10)$
   foci: $(-1, 5), (-1, 9)$

4. vertices: $(-3, 2), (5, 2)$
   foci: $(-1, 2), (3, 2)$

5. vertices: $(0, \pm 7)$
   semiminor axis: 3

6. vertices: $(\pm 5, 0)$
   semiminor axis: 2

7. vertices: $(-1, 2), (3, 2)$
   semiminor axis: 1

8. vertices: $(5, -3), (5, 7)$
   semiminor axis: 3

9. vertices: $(5, -3), (5, 7)$
   eccentricity: $\frac{2}{5}$

10. vertices: $(\pm 3, 0)$
    eccentricity: $\frac{3}{4}$

11. vertices: $(0, \pm 1)$
    foci: $\left(0, \pm \frac{1}{3}\right)$

12. vertices: $(\pm 5, 0)$
    foci: $(\pm 1, 0)$

13. vertices: $(5, 1), (15, 1)$
    foci: $(7, 1), (13, 1)$

14. vertices: $(3, 7), (3, -5)$
    foci: $(3, 5), (3, -3)$

15. vertices: $(0, \pm 4)$
    passing through: $(2, 0)$

16. vertices: $(\pm 3, 0)$
    passing through: $(0, 2)$

17. vertices: $(\pm 3, 0)$
    passing through: $(2, 1)$

18. vertices: $(0, \pm 5)$
    passing through: $(-2, 3)$

Graph each of the following ellipses. Determine the semimajor and semiminor axes, the foci, and the eccentricity.

19. $\frac{x^2}{9} + \frac{y^2}{4} = 1$

20. $\frac{x^2}{4} + \frac{y^2}{9} = 1$

21. $2x^2 + 4y^2 = 16$

22. $4x^2 + 2y^2 = 16$

23. $\frac{(x + 2)^2}{4} + \frac{(y - 2)^2}{16} = 1$

24. $\frac{(x + 2)^2}{16} + \frac{(y - 2)^2}{4} = 1$

25. $3(x - 1)^2 + 6(y - 2)^2 = 108$

26. $6(x - 1)^2 + 3(y - 2)^2 = 108$

27. $x^2 - 4x + 5y^2 + 30y + 24 = 0$

28. $4x^2 + 8x + 6y^2 + 24y + 4 = 0$

29. $36x^2 + 4y^2 - 144 = 0$ 30. $x^2 + 9y^2 - 9 = 0$

31. $9x^2 + y^2 - 54x - 4y + 49 = 0$

32. $4x^2 + 25y^2 + 8x + 150y + 129 = 0$

33. $\frac{(x + 1)^2}{16} + \frac{(y + 1)^2}{25} = 1$

34. $16(x + 1)^2 + 25(y + 1)^2 = 400$

35. $x^2 + y^2 + 6y = 0$ 36. $3x^2 + 3y^2 + 6x = 1$

37. The **latus recta** of an ellipse are the line segments perpendicular to the major axis, drawn through the foci, and having endpoints on the ellipse. Determine the endpoints of the latus recta of the ellipse

$$\frac{x^2}{4} + \frac{y^2}{9} = 1$$

38. Show that the length of the latus recta are $2b^2/a$.

39. Suppose that an ellipse has semimajor axis a, semiminor axis b, and eccentricity e. Prove that $b^2 = a^2(1 - e^2)$.

40. A communications satellite is launched into an earth orbit with low point 22,000 miles and high point 24,000 miles. Determine the eccentricity of the orbit. Determine the equation of the orbit.

41. Suppose that the semimajor and semiminor axis of an ellipse are both multiplied by the same factor k. Show that the eccentricity is unchanged.

Match the equation in exercises 42–47 with their graphs.

42. $x^2 + \frac{y^2}{4} = 1$

43. $\frac{x^2}{4} + y^2 = 1$

44. $\frac{x^2}{12.25} + \frac{y^2}{6.25} = 1$

45. $\frac{x^2}{6.25} + \frac{y^2}{12.25} = 1$

46. $\frac{(x + 2)^2}{4} + \frac{(y + 3)^2}{9} = 1$

47. $\frac{(x - 2)^2}{9} + \frac{(y + 2)^2}{9} = 1$

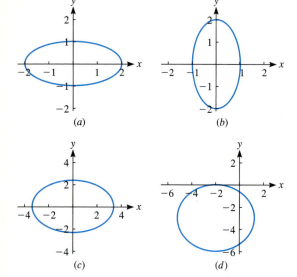

(a)

(b)

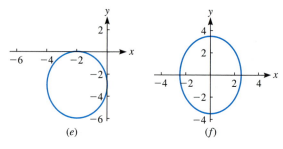

(e)

(f)

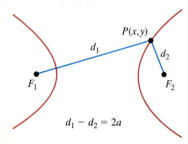

(c)

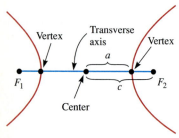

(d)

Use a graphing utility to graph the ellipses in exercises 48–51.

48. $\dfrac{x^2}{2.25} + \dfrac{y^2}{6.25} = 1$ 49. $3x^2 + 5y^2 = 13$

50. $\dfrac{(x-2)^2}{9} + \dfrac{(y+2)^2}{20.25} = 1$

51. $7(x-2)^2 + 6(y-2)^2 = 17$

9.3 HYPERBOLAS

Let's now turn our attention to the final conic section, the hyperbola. We can give a geometric definition of the hyperbola in terms of the distances to two foci, as follows:

Definition 1 Hyperbola, Foci	A **hyperbola** is the set of all points P, the difference of whose distances from two fixed points is a constant. The two points are called the **foci**.

Refer to Figure 1. The two foci are F_1 and F_2. A typical point on the hyperbola is $P(x, y)$. The distance of P to F_1 is denoted d_1, and the distance of P to F_2 is denoted d_2. The definition of a hyperbola says that there is a constant $2a$, independent of the point P such that

$$d_1 - d_2 = 2a \qquad (1)$$

Draw a line connecting the foci. (See Figure 2.) It intersects the hyperbola in two points, called the **vertices**. The line connecting the vertices is called the **transverse axis** of the hyperbola. The midpoint of the transverse axis is called the **center** of the hyperbola. The distance from the center to either of the vertices is a. (Proof similar to the case of an ellipse.) Let c be the distance of the foci from the center. Then c is called the **focal length**. Define the number b by the equation

$$b^2 = c^2 - a^2 \qquad (2)$$

Then, by reasoning as in the case of the ellipse, we can deduce:

Figure 1

Figure 2

STANDARD FORMS FOR THE EQUATION OF A HYPERBOLA		
$\dfrac{(x-h)^2}{a^2} - \dfrac{(y-k)^2}{b^2} = 1$	*(transverse axis horizontal)*	(3)
$\dfrac{(y-k)^2}{a^2} - \dfrac{(x-h)^2}{b^2} = 1$	*(transverse axis vertical)*	(4)

We will leave the derivation of these equations as an exercise.

> **EXAMPLE 1**
Determine Equation from Foci and Transverse Axis

Determine the equation of a hyperbola with foci at $(-3, 1)$ and $(5, 1)$, and transverse axis of length 4.

Solution

The transverse axis is horizontal, since the two foci lie on a horizontal line. The center is halfway between the foci, which is the point $(h, k) = (1, 1)$. Since the length of the transverse axis is 4, the vertices are $(3, 1)$ and $(-1, 1)$ and $a = 2$. The distance c of the foci from the center is 4. Thus, the value of b is given by

$$b^2 = c^2 - a^2$$
$$= 4^2 - 2^2$$
$$= 12$$
$$b = \sqrt{12} = 2\sqrt{3}$$

Thus, the equation of the hyperbola is

$$\frac{(x - 1)^2}{2^2} - \frac{(y - 1)^2}{(2\sqrt{3})^2} = 1$$
$$\frac{(x - 1)^2}{4} - \frac{(y - 1)^2}{12} = 1$$

Graphing Hyperbolas

Let's now examine the graph of a hyperbola. Let's first consider the standard form corresponding to a horizontal transverse axis:

$$\frac{(x - h)^2}{a^2} - \frac{(y - k)^2}{b^2} = 1$$

Solve the equation for $y - k$:

$$\frac{(y - k)^2}{b^2} = \frac{(x - h)^2}{a^2} - 1$$
$$\frac{y - k}{b} = \pm\sqrt{\left(\frac{x - h}{a}\right)^2 - 1}$$
$$y - k = \pm\frac{b}{a}\sqrt{(x - h)^2 - a^2}$$

The square root on the right yields a real number only if $|x - h| \geq a$. Therefore, the x-coordinates of the points on the hyperbola must satisfy the inequality

$$|x - h| \geq a$$

which is equivalent to

$$x \leq h - a \quad \text{OR} \quad x \geq h + a$$

That is to say, a point on the hyperbola is either to the right of the right vertex or to the left of the left vertex. As the value of x increases in absolute value, so does the value of

$$\frac{b}{a}\sqrt{(x - h)^2 - a^2}$$

Moreover, the $\pm$ indicates that for each value of x there are two points on the graph, symmetrically placed, above and below the line $y = k$. The graph is as

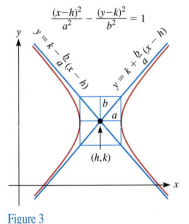

Figure 3

shown in Figure 3. We may write the equation of the hyperbola in the form

$$y - k = \pm\frac{b}{a}\sqrt{(x-h)^2\left(1 - \frac{a^2}{(x-h)^2}\right)}$$

$$= \pm\frac{b}{a}(x-h)\sqrt{1 - \frac{a^2}{(x-h)^2}}$$

As x approaches either infinity or minus infinity, the expression under the square root approaches 1, so that the value of y approaches

$$k \pm \frac{b}{a}(x-h)$$

This means that as x approaches either infinity or minus infinity, the graph approaches the lines with equations

$$y = k + \frac{b}{a}(x-h)$$

$$y = k - \frac{b}{a}(x-h)$$

These lines are called **asymptotes** of the hyperbola. The lines are drawn in Figure 3. Note that if we draw a rectangle centered at (h, k) of width $2a$ and height $2b$, then the asymptotes are the diagonals of the rectangle. (See Figure 3.) The vertical line through the center extended b units above and below is called the **conjugate axis** of the hyperbola.

Repeating this reasoning for a hyperbola with vertical transverse axis, we see that the graph is as shown in Figure 4. In this case, the asymptotes are the equations

$$y = k + \frac{a}{b}(x-h)$$

$$y = k - \frac{a}{b}(x-h)$$

The conjugate axis in this case is horizontal.

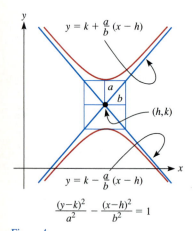

$$\frac{(y-k)^2}{a^2} - \frac{(x-h)^2}{b^2} = 1$$

Figure 4

➤ **EXAMPLE 2**

Graphing Hyperbolas

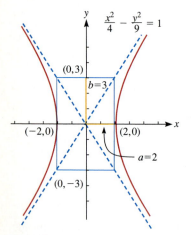

Figure 5

Sketch the graphs of the following equations:

1. $\dfrac{x^2}{4} - \dfrac{y^2}{9} = 1$ 2. $(x-3)^2 - \dfrac{(y-1)^2}{4} = 1$ 3. $y^2 - \dfrac{x^2}{4} = 1$

Solution

1. We first write the equation in one of the two standard forms:

$$\frac{x^2}{2^2} - \frac{y^2}{3^2} = 1$$

We see that $a = 2$ and $b = 3$. Moreover, since the positive sign is on the x-term, the axis is horizontal. The center is $(0, 0)$. The vertices are $(\pm a, 0) = (\pm 2, 0)$. Draw the transverse axis of length $2a = 4$ and the conjugate axis of length $2b = 6$, centered at $(0, 0)$. Around the axes, draw the associated rectangle and its diagonals. Next, draw the hyperbola, symmetric with respect to the transverse and conjugate axes and asymptotic to the diagonal lines. (See Figure 5.)

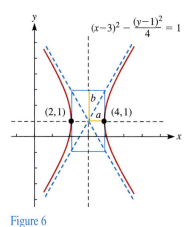

Figure 6

2. Write the equation in the standard form

$$\frac{(x-3)^2}{1^2} - \frac{(y-1)^2}{2^2} = 1$$

From this form, we determine that $a = 1$ and $b = 2$. Since the positive sign is on the x-term, the transverse axis of the hyperbola is horizontal. Moreover, the center of the hyperbola is the point $(3, 1)$. We plot the center and then draw the transverse axis of length $2a = 2$ and the conjugate axis of length $2b = 4$. We use the axes to draw a rectangle. The diagonals of the rectangle are the asymptotes. Next, we draw the hyperbola, symmetric with respect to both axes and asymptotic to the diagonal lines. (See Figure 6.)

3. We write the equation in the standard form

$$\frac{y^2}{1^2} - \frac{x^2}{2^2} = 1$$

In this case, we have $a = 1$ and $b = 2$. Since $h = 0$ and $k = 0$, the center of the hyperbola is the point $(0, 0)$. Moreover, since the positive sign is on the y-term, the transverse axis is vertical. The transverse axis has length $2a = 2$. The conjugate axis is of length $2b = 4$ and is horizontal. To sketch the graph, we plot the center and draw the axes. We use the axes to draw a rectangle. The diagonals of the rectangle are the asymptotes. Next, draw the hyperbola, symmetric with respect to both axes and asymptotic to the diagonal lines. (See Figure 7.)

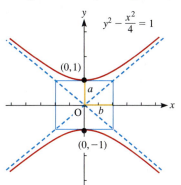

Figure 7

➤ **EXAMPLE 3**
Reducing to Standard Form

Sketch the graphs of the following equations:

1. $x^2 - 2y^2 = 1$ 2. $x^2 - 2x - y^2 + 6y - 7 = 0$

Solution

1. We write the equation in the standard form

$$x^2 - \frac{y^2}{\dfrac{1}{2}} = 1$$

$$\frac{x^2}{1^2} - \frac{y^2}{\left(\dfrac{1}{\sqrt{2}}\right)^2} = 1$$

From the last equation, we see that the hyperbola has a horizontal transverse axis of length $2a = 2$, and a vertical conjugate axis of length $2b = 2/\sqrt{2} = \sqrt{2}$. To sketch the graph, we first draw in the rectangle determined by the vertices $(\pm 1, 0)$ and the points $(0, \pm \sqrt{2}/2)$. Next, we draw the diagonals of the rectangle. These are the asymptotes of the hyperbola. Finally, we draw the hyperbola. (See Figure 8.)

2. We begin by putting the equation into standard form. To do this, we complete the square with respect to both x and y:

$$(x^2 - 2x) - (y^2 - 6y) = 7$$
$$(x^2 - 2x + 1) - (y^2 - 6y + 9) = 7 + 1 - 9$$
$$(x - 1)^2 - (y - 3)^2 = -1$$
$$(y - 3)^2 - (x - 1)^2 = 1$$
$$\frac{(y - 3)^2}{1^2} - \frac{(x - 1)^2}{1^2} = 1$$

From this equation, we see that the hyperbola is centered at $(1, 3)$. Since the variable y appears with a positive sign, the transverse axis of the hyperbola is vertical. Moreover, the lengths of the two axes are both 2. This provides us with the data to sketch the graph. (See Figure 9.)

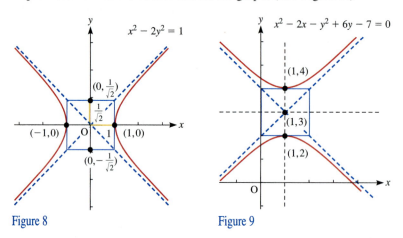

Figure 8   Figure 9

The Eccentricity of a Hyperbola

Paralleling the definition in the case of an ellipse, we define the eccentricity of a hyperbola as the focal length divided by the length of the semitransverse axis (half the length of the transverse axis), that is,

$$e = \frac{c}{a} \qquad (5)$$

Since the foci are located within the interior of the *wings* of the hyperbola, we see that $c > a$, so that $e > 1$. The eccentricity measures the *spread* of the wings of a hyperbola. The larger the eccentricity, the wider the spread of the wings.

We have defined the eccentricity of a hyperbola and an ellipse. Using analogous reasoning, we may define the eccentricity of a parabola. Namely, the eccentricity is the ratio of the focal length to the semiaxis. In the case of a parabola, the semiaxis is the line segment from the vertex to the focus. The eccentricity of a parabola is therefore 1.

We can thus classify the conics by their respective eccentricities as follows:

Circle: $e = 0$
Ellipse: $0 < e < 1$
Parabola: $e = 1$
Hyperbola: $e > 1$

It can be proved, using calculus and elementary physics, that the orbit of a comet or an asteroid is always a conic section. The eccentricity may be interpreted as the total energy of the comet. Roughly speaking, the faster the comet travels, the greater the eccentricity. Thus, for example, a slow-moving comet (relatively speaking) will have eccentricity less than 1 and travel in an ellipse. With precisely the correct speed, the eccentricity becomes 1 and the orbit changes to a parabola. If the comet is moving faster still, the eccentricity is greater than 1 and the orbit is a hyperbola.

General Equations the Second Degree

In the first three sections of this chapter, we have studied the various conic sections. By completing the square, any equation of the form

$$Ax^2 + By^2 + Cx + Dy + E = 0, \qquad A, B \text{ not both } 0.$$

may be reduced to one of the standard forms for equations of conics. Here is how to tell which conic will result in terms of the coefficients of the equation:

GRAPHS OF SECOND DEGREE EQUATIONS

Parabola: Either A or B zero.
Ellipse: A and B nonzero and of the same sign, $A \neq B$.
Circle: $A = B$.
Hyperbola: A and B nonzero and of opposite signs.

➤ EXAMPLE 4
Equations of Second Degree

Determine which type of conic is the graph of each equation.

1. $5x^2 + 3y^2 - 7x + 3y = 1$
2. $-5x^2 + 3y^2 + 12x - 4 = 0$
3. $10x^2 + 10y^2 + 3x - 2y - 100 = 0$
4. $x + 7y^2 + 3y + 5 = 0$

Solution

1. In this case, we have $A = 5$ and $B = 3$. Since both coefficients are nonzero and of the same sign, the graph of the equation is an ellipse.
2. In this case, we have $A = -5$ and $B = 3$. Since the coefficients are nonzero and of opposite sign, the graph of the equation is a hyperbola.
3. In this case, $A = B = 10$. Since the coefficients are nonzero and equal, the graph of the equation is a circle.
4. In this case, $A = 0$ and $B = 7$. Since one of the coefficients is 0, the graph of the equation is a parabola.

Exercises 9.3

Determine the center, vertices, foci, and asymptotes of the following hyperbolas. Sketch the graph with the aid of this information.

1. $x^2 - y^2 = 1$

2. $y^2 - x^2 = 1$

3. $\dfrac{x^2}{4} - y^2 = 1$

4. $\dfrac{y^2}{9} - \dfrac{x^2}{4} = 1$

5. $\dfrac{x^2}{100} - \dfrac{y^2}{36} = 1$

6. $y^2 - \dfrac{x^2}{3} = 1$

7. $(x - 1)^2 - (y - 2)^2 = 1$

8. $4(x + 3)^2 - 9y^2 = 36$

9. $\dfrac{(x - 4)^2}{9} - \dfrac{(y - 1)^2}{25} = 1$

10. $\dfrac{x^2}{\frac{1}{3}} - \dfrac{y^2}{\frac{1}{2}} = 1$

11. $4x^2 - 9y^2 = 1$

12. $2x^2 - 3y^2 = 1$

13. $5(x + 6)^2 - 3(y - 4)^2 = 15$

14. $10y^2 - 3(x + 2)^2 = 30$

15. $x^2 - 6x - 4y^2 + 12y = 30$

16. $2y^2 - 3x^2 + 12x - y = 0$

17. $(x + 2)^2 - 3(y - 4)^2 = 1$

18. $16x^2 - 48x - 2y^2 - 10 = 0$

19. $9y^2 + 18y - 4x^2 + 24x - 28 = 0$

20. $100x^2 + 200x - y^2 = 200$

Determine the equation of the hyperbola having the following properties:

21. center: $(0, 0)$
    transverse axis: 4, horizontal
    conjugate axis: 2

22. center: $(0, 0)$
    transverse axis: 10, horizontal
    conjugate axis: 20

23. center: $(1, 1)$
    transverse axis: 2, vertical
    conjugate axis: 3

24. center: $(-1, -2)$
    transverse axis: 3, vertical
    conjugate axis: 1

25. vertices: $(\pm 3, 0)$
    foci: $(\pm 5, 0)$

26. vertices: $(\pm 8, 0)$
    foci: $(\pm 12, 0)$

27. vertices: $(-5, 1), (3, 1)$
    foci: $(-7, 1), (5, 1)$

28. vertices: $(-2, -2), (-2, 4)$
    foci: $(-2, -5), (-2, 7)$

29. vertices: $(\pm 2, 0)$
    asymptotes: $y = \pm 2x$

30. vertices: $(0, \pm 1)$
    asymptotes: $y = \pm 2x$

31. vertices: $(0, \pm 2)$
    passing through $(5, 3)$

32. vertices: $(1, -2), (1, 5)$
    passing through $(-1, 8)$

Without completing the square, determine which conic is the graph of the equation.

33. $x^2 + 3y^2 + 12x - 14y - 30 = 0$

34. $x + 12y^2 - 13y + 128 = 0$

35. $y^2 = 3x^2 + 10$

36. $100x^2 - 300x + 100y^2 - 200x + 50 = 0$

37. $-30x = y^2 - 3y + 100$

38. $y^2 + 10y = 100 - 12x^2$

39. Suppose that rescuers are searching a wooded area for a trapped camper. Two teams hear the camper's shouts. Suppose that the two rescue teams are 5 miles apart and hear the shouts of the camper 3 seconds apart. Assuming a coordinate system that has origin at the rescue team that hears the call first, and positive x-axis extending from the first rescue team to the second, write an equation describing the possible location of the camper. (The equation is a hyperbola.) Assume that sound travels at 1100 feet per second.

40. (Continuation of the preceding exercise.) Suppose that a third rescue team, located at $(0, 3$ miles$)$, hears the camper 1 second after the first rescue team. Locate the camper.

41. Suppose that the transverse axis and conjugate axis are both multiplied by the same factor. Show that the eccentricity is unchanged.

42. Suppose that the transverse axis and the conjugate axis are both multiplied by the same factor. Show that the asymptotes remain unchanged.

43. Prove that a hyperbola never crosses its asymptotes.

✎ In Your Own Words

44. Describe how the shape of an ellipse changes as its eccentricity increases.

45. Describe how the shape of an ellipse changes as the foci get closer to the vertices.

Match the equations in exercises 46–51 with their graphs.

46. $x^2 - \dfrac{y^2}{4} = 1$

47. $\dfrac{x^2}{1} - \dfrac{y^2}{1} = 1$

48. $\dfrac{x^2}{12.25} - \dfrac{y^2}{6.25} = 1$

49. $\dfrac{y^2}{6.25} - \dfrac{x^2}{12.25} = 1$

50. $\dfrac{(x-2)^2}{9} - \dfrac{(y-2)^2}{16} = 1$

51. $\dfrac{(y-2)^2}{9} - \dfrac{(x-2)^2}{9} = 1$

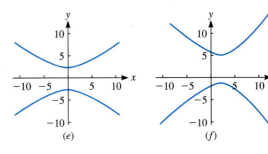
(e)   (f)

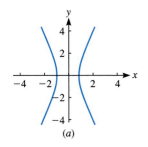
(a)   (b)

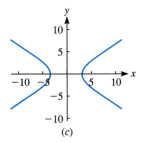
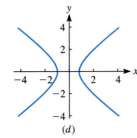
(c)   (d)

➡ **Technology**

Use a graphing utility to graph the hyperbolas in exercises 52–55. (You will need to graph them as two separate equations, obtained by solving for y in terms of x.)

52. $\dfrac{x^2}{2.25} - \dfrac{y^2}{6.25} = 1$

53. $3y^2 - 5x^2 = 13$

54. $\dfrac{(x-2)^2}{9} - \dfrac{(y+2)^2}{20.25} = 1$

55. $7(y-2)^2 - 6(x-2)^2 = 17$

▌9.4 PARAMETRIC EQUATIONS

In the early chapters of the book, we introduced graphs of equations of the form $f(x, y) = 0$, where $f(x, y)$ is an expression in x and y. We have graphed many equations and arrived at an interesting collection of graphs, not the least of which are the conic sections considered in the preceding sections.

Rather than specifying a curve by an equation relating x and y, it is often expedient to relate both x and y to a third variable, which we will call t.

Definition 1 **Parametric Equations** **for a Curve**	A set of **parametric equations** in the **parameter** t is a pair of equations $$x = f(t), \quad y = g(t) \tag{1}$$ where t ranges over an interval (finite or infinite). Each value of t corresponds to a point $(x, y) = (f(t), g(t))$. As t ranges over the interval, the set of points (x, y) traces out a curve in the plane.

In many examples, the parameter has a physical interpretation. For instance, t may represent time, so that the point $(f(t), g(t))$ may be interpreted as the position of a point at time t.

A set of parametric equations for a curve gives much more information than just the set of points describing the curve. As t runs through a specified set of values, the parametric equations specify the speed with which the curve is traced out, the direction in which it is traced, and even if certain portions of the curve are traced over themselves more than once.

➤ **EXAMPLE 1**
Motion of a Particle

The position of a particle at time t is given by the parametric equations

$$x = 2t, y = 3t, \quad 0 \le t \le 5$$

Describe the motion of the particle.

Solution

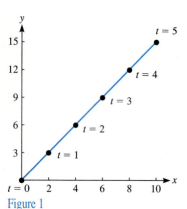

Figure 1

At time $t = 0$ the particle starts at the point $(x, y) = (2 \cdot 0, 3 \cdot 0) = (0, 0)$. The following table lists the location of the point at various times t:

t	x	y
0	0	0
1	2	3
2	4	6
3	6	9
4	8	12
5	10	15

We plot the various points in Figure 1. We see that the graph is a line. As time runs from 0 to 5, the point moves along the line from $(0, 0)$ to $(10, 15)$.

➤ **EXAMPLE 2**
Graphing Parametric Equations

Graph the curve described by the parametric equations

$$x = t, y = \frac{3}{2}t, \quad 0 \le t \le 10$$

Solution

The curve is the same curve shown in Figure 1. However, the present curve is traced out at half the speed. That is, for any given time t, the location of the present curve is the same as the location on the previous curve at time $t/2$. This example shows that there is no single set of parametric equations that describes a particular curve.

➤ **EXAMPLE 3**
Translating Parametric Equations into Rectangular Coordinates

Write the following set of parametric equations in rectangular coordinates:

$$x = \sqrt{t}, y = 3t + 1, \quad t \ge 0$$

Solution

We may solve the first equation for t in terms of x:

$$x = \sqrt{t}$$
$$t = x^2$$

Now substitute the expression for t into the equation for y:

$$y = 3t + 1 = 3x^2 + 1$$

From the first equation, the values of x are all nonnegative. In fact, each nonnegative real number is a value of x for some value of t. Therefore, when we graph the nonparametric equation $y = 3x^2 + 1$, we must subject x to the restriction $x \geq 0$, which is inherited from the parametric equations.

➤ **EXAMPLE 4**

Parametric Form of a Circle

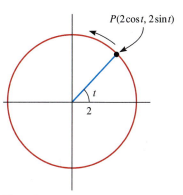

Figure 2

Graph the curve described by the parametric equations

$$x = 2\cos t, \, y = 2\sin t, \quad 0 \leq t \leq 2\pi$$

Solution

To solve this problem, let's pull a rabbit out of a hat. Draw a circle of radius 2 about the origin. From the definitions of the trigonometric functions sine and cosine, the coordinates of a point P on the circle determined by an angle t in standard position are $(\cos t, \sin t)$. So if we interpret the parameter t as the angle shown in Figure 2, we see that the parametric equations describe the point P. As t ranges over the interval $[0, 2\pi]$, the point described by the parametric equations traces out the circle shown, in the counterclockwise direction. This example illustrates how the parameter may often be interpreted in geometric terms, as an angle, length, or area.

In some instances, it is possible to solve the parametric equation for t in terms of either x or y and then eliminate the parameter t, thereby obtaining an equation for the curve in nonparametric form.

➤ **EXAMPLE 5**

Translating Parametric Equations into Rectangular Coordinates

Write the following equations in rectangular coordinates:

$$x = e^t, \qquad y = t^2$$

Solution

Note that the value of x is positive. Moreover, solving the first equation for t in terms of x, we have

$$t = \ln x$$

Substituting this value into the equation for y, we have

$$y = t^2 = (\ln x)^2, \quad x > 0$$

Using a Graphing Utility to Graph Parametric Equations

Optional

Most graphing utilities allow you to graph parametric equations. For instance, the TI-81 can plot parametric equations by entering the **MODE** menu and selecting `Param` instead of `Function`. Or in Mathematica, you would use the `ParametricPlot` command. This is very useful since the graphs of some equations, such as those for many conic sections are not easily produced with just one equation on most graphing utilities.

On the TI-81, after setting the **MODE**, return to the **Y=** screen. The screen will now look like the one shown in Figure 3. At this screen you can enter up to three parametrically defined equations. For example, to graph

$$x = 3\cos t, \, y = 3\sin t, \quad 0 \leq t \leq 2\pi$$

enter $3\cos t$ for X_{1T} and $3\sin t$ for Y_{1T}. Then graph the equation. You may need to adjust your viewing box using the `Square` command in the **ZOOM** screen to get a graph that looks like Figure 4.

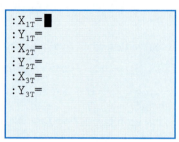

Figure 3

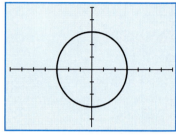

Figure 4

Exercises 9.4

Sketch the graphs of the following parametric equations.

1. $x = 3t, y = 5t - 1, \quad 0 \le t \le 1$

2. $x = 2t - 1, y = t + 1, \quad -1 \le t \le 1$

3. $x = 2t, y = e^t, \quad t \ge 0$

4. $x = 2t, y = \ln t, \quad 0 < t < 5$

5. $x = 2\cos\theta, y = 2\sin\theta, \quad 0 \le \theta \le 2\pi$

6. $x = 3 + 2\sin\theta, y = 1 + 2\cos\theta, \quad 0 \le \theta \le 2\pi$

7. $x = \sin\theta, y = \cos\theta, \quad 0 \le \theta \le 2\pi$

8. $x = \sin 3t, y = \cos 3t, \quad 0 \le t \le \pi$

9. $x = \cos\theta, y = \sin\theta, \quad -\pi \le \theta \le \pi$

10. $x = 3\sin\theta, y = -3\cos\theta, \quad 0 \le \theta \le 6\pi$

11. $x = 2t, y = 3t, \quad t = 1$ to $t = 0$

12. $x = 2 + \cos t, y = -1 + 3\sin t, \quad 0 \le t \le 2\pi$

13. $x = \dfrac{1}{t}, y = t, \quad t > 0$

14. $x = \dfrac{1}{t^2 + 1}, y = \dfrac{t}{t^2 + 1}, \quad$ all real t

15. $x = \sqrt{t}, y = \dfrac{1}{t}, \quad t > 0$

16. $x = \cos t, y = \sin t, \quad 0 \le t \le \dfrac{\pi}{2}$

17–32. For each of the parametric equations in exercises 1–16, write the equation in rectangular coordinates. Be careful to state the correct domain in each case.

Show that the conic sections may be written in the following parametric form.

33. Circle: $(x - h)^2 + (y - k)^2 = r^2$:
$$x = h + r\cos\theta, y = k + r\sin\theta, \quad 0 \le \theta \le 2\pi$$

34. Ellipse: $\dfrac{(x - h)^2}{a^2} + \dfrac{(y - k)^2}{b^2} = 1$:
$$x = h + a\cos\theta, y = k + b\sin\theta, \quad 0 \le \theta \le 2\pi$$

35. Hyperbola: $\dfrac{(x - h)^2}{a^2} - \dfrac{(y - k)^2}{b^2} = 1$:
$$x = h + a\sec\theta, y = k + b\tan\theta,$$
$$\begin{cases} -\dfrac{\pi}{2} < \theta < \dfrac{\pi}{2} & \text{(left branch)} \\ \dfrac{\pi}{2} < \theta < \dfrac{3\pi}{2} & \text{(right branch)} \end{cases}$$

36. Derive a set of parametric equations for a circle of radius 2 and center $(3, -1)$.

37. Derive a set of parametric equations for the unit circle.

38. Derive a set of parametric equations for an ellipse with equation $x^2/4 + y^2/25 = 1$, traced out twice in the clockwise direction, starting at the point for which $x = 0$.

39. Derive a set of parametric equations for an ellipse with center at the origin, horizontal semimajor axis 2 and semiminor axis 1.

40. Derive a set of parametric equations for an ellipse with center at $(5, -2)$, vertical semimajor axis 3 and semiminor axis 2.

41. Derive a set of parametric equations for an ellipse with foci at $(5, 1), (-3, 1)$ and semimajor axis 10.

42. Derive a set of parametric equations for the hyperbola $x^2 - y^2 = 1$.

43. Derive a set of parametric equations for the right branch of the hyperbola $(x - 1)^2/3 - (y - 1)^2/6 = 1$.

44. Prove that the line segment connecting (x_1, y_1) to (x_2, y_2) may be written parametrically in the form
$$x = x_1 + t(x_2 - x_1), y = y_1 + t(y_2 - y_1), \quad 0 \le t \le 1$$

45. Write parametric equations of a line segment going from $(2, 3)$ to $(-1, 4)$.

46. Describe the graph of the parametric equations

$$x = 3 + 4t, y = -1 + 2t, \quad 0 \le t \le 1$$

47. Describe the graph of the parametric equations obtained by replacing t by $1 - t$ in the preceding exercise.

Suppose that $x = f(t), y = g(t), 0 \le t \le 1$ is a set of parametric equations for a curve C. Describe the graph of the following equations.

48. $x = f(2t), y = g(2t), \quad 0 \le t \le \dfrac{1}{2}$

49. $x = f(1 - t), y = g(1 - t), \quad 0 \le t \le 1$

50. $x = g(t), y = f(t), \quad 0 \le t \le 1$

51. $x = f(t) + 1, y = g(t) + 2, \quad 0 \le t \le 1$

⮕ **Technology**

Use the parametric plotting features of your graphing utility and the formulas from exercises 33–35 to graph the conic sections in exercises 52–59.

52. $\dfrac{x^2}{16} + \dfrac{y^2}{36} = 1$

53. $(x - 2)^2 + (y + 3)^2 = 11$

54. $\dfrac{(x - 2)^2}{49} + \dfrac{(y + 2)^2}{20} = 1$

55. $7(x - 2)^2 + 6(y - 2)^2 = 17$

56. $\dfrac{x^2}{2} - \dfrac{y^2}{9} = 1$

57. $7(y + 1)^2 - 5(x + 2)^2 = 13$

58. $\dfrac{(x - 2)^2}{9} - \dfrac{(y + 2)^2}{20} = 1$

59. $(y - 2)^2 + (x - 2)^2 = 17$

▌ 9.5 CHAPTER REVIEW

Important Concepts, Properties, and Formulas—Chapter 9

Parabolas	A parabola is the set of all points whose distance from the focus equals the distance from the directrix. Standard form of equations of parabolas: $\quad (x - h)^2 = 4p(y - k) \quad$ axis vertical $\quad (y - k)^2 = 4p(x - h) \quad$ axis horizontal where $p = $ directed focal length.	p. 443
Ellipses	An ellipse is the set of all points the sum of whose distance from the two foci is a given constant. Standard form equation of an ellipse: $\dfrac{(x - h)^2}{a^2} + \dfrac{(y - k)^2}{b^2} = 1,$ If $a > b$ then major axis horizontal If $a < b$ then major axis vertical If c is the focal length then $\qquad a^2 = b^2 + c^2$	p. 450
Hyperbolas	A hyperbola is the set of all points the difference of whose distances from the two foci is a given constant. Standard form equations of hyperbolas: $\dfrac{(y - k)^2}{a^2} - \dfrac{(x - h)^2}{b^2} = 1, \quad$ vertical transverse axis $\dfrac{(x - h)^2}{a^2} - \dfrac{(y - k)^2}{b^2} = 1, \quad$ horizontal transverse axis	pp. 455, 457

| Hyperbolas (continued) | If c is the focal length then $$c^2 = a^2 + b^2$$ Asymptotes are the lines: $$y - k = \pm\frac{b}{a}(x - h) \quad \text{(horizontal transverse axis)}$$ $$y - k = \pm\frac{a}{b}(x - h) \quad \text{(vertical transverse axis)}$$ | |
| Parametric equation of a curve | $$x = f(t), y = g(t), a \le t \le b$$ | p. 462 |

Cumulative Review Exercises—Chapter 9

Sketch the graphs of the following conics. For each, determine the foci, directrices, and eccentricity. For ellipses, determine the semimajor and semiminor axes. For hyperbolas, determine the transverse and conjugate axes and asymptotes.

1. $y^2 = 3x + 1$
2. $x^2 = 2y - 3$
3. $2x^2 - y^2 = 4$
4. $y^2 - 3x^2 = 10$
5. $x^2 + (y - 3)^2 = 10$
6. $2(x - 1)^2 + 5(y - 3)^2 = 25$
7. $\dfrac{x^2}{25} - y^2 = 6$
8. $\dfrac{(x - 1)^2}{3^3} + \dfrac{(y + 2)}{4^3} = 1$
9. $-x^2 + 3x + y = 1$
10. $x^2 = y^2 + 3$
11. $y^2 = -5x - 8$
12. $5x^2 - 3x + 2y^2 + 6y + 3 = 0$
13. $x^2 - 2y^2 + 3y = 0$
14. $(x + 1)^2 + 2(y - 3)^2 + 3x = 0$

Graph the following parametric equations.

15. $x = -t + 1, y = 2t, \quad 0 \le t \le 4$
16. $x = t^2, y = -t, \quad 0 \le t \le 1$
17. $x = -1 + 2\sin t, y = 4 + 2\cos t, \quad 0 \le t \le \pi$
18. $x = \sin t, y = t, \quad 0 \le t \le \pi$

Determine the rectangular form of the following parametric equations.

19. $x = 3z - 1, y = 2z + 3, \quad 0 \le z \le 5$
20. $x = t^2, y = \dfrac{1}{t - 1}, \quad t > 1$
21. $x = t, y = \sqrt{t - 1}, \quad t \ge 1$
22. $x = -1 + \cos t, y = 5 + \cos t, \quad 0 \le t \le 2\pi$
23. $x = 4\sin t, y = 3\cos t, \quad 0 \le t \le 2\pi$
24. $x = 5\ln(t - 1), y = 3t + 1, \quad 1 < t \le 5$

APPENDIXES

Appendix A Table of Exponential Functions

x	e^x	e^{-x}	x	e^x	e^{-x}	x	e^x	e^{-x}
1	2.71828	0.36788	3.4	29.96410	0.03337	6.8	897.84729	0.00111
0.1	1.10517	0.90484	3.5	33.11545	0.03020	6.9	992.27472	0.00101
0.2	1.22140	0.81873	3.6	36.59823	0.02732	7	1096.63316	0.00091
0.3	1.34986	0.74082	3.7	40.44730	0.02472	7.1	1211.96707	0.00083
0.4	1.49182	0.67032	3.8	44.70118	0.02237	7.2	1339.43076	0.00075
0.5	1.64872	0.60653	3.9	49.40245	0.02024	7.3	1480.29993	0.00068
0.6	1.82212	0.54881	4	54.59815	0.01832	7.4	1635.98443	0.00061
0.7	2.01375	0.49659	4.1	60.34029	0.01657	7.5	1808.04241	0.00055
0.8	2.22554	0.44933	4.2	66.68633	0.01500	7.6	1998.19590	0.00050
0.9	2.45960	0.40657	4.3	73.69979	0.01357	7.7	2208.34799	0.00045
1	2.71828	0.36788	4.4	81.45087	0.01228	7.8	2440.60198	0.00041
1.1	3.00417	0.33287	4.5	90.01713	0.01111	7.9	2697.28233	0.00037
1.2	3.32012	0.30119	4.6	99.48432	0.01005	8	2980.95799	0.00034
1.3	3.66930	0.27253	4.7	109.94717	0.00910	8.1	3294.46808	0.00030
1.4	4.05520	0.24660	4.8	121.51042	0.00823	8.2	3640.95031	0.00027
1.5	4.48169	0.22313	4.9	134.28978	0.00745	8.3	4023.87239	0.00025
1.6	4.95303	0.21090	5	148.41316	0.00674	8.4	4447.06675	0.00022
1.7	5.47395	0.18268	5.1	164.02191	0.00610	8.5	4914.76884	0.00020
1.8	6.04965	0.16530	5.2	181.27224	0.00552	8.6	5431.65959	0.00018
1.9	6.68589	0.14957	5.3	200.33681	0.00499	8.7	6002.91222	0.00017
2	7.38906	0.13534	5.4	221.40642	0.00452	8.8	6634.24401	0.00015
2.1	8.16617	0.12246	5.5	244.69193	0.00409	8.9	7331.97354	0.00014
2.2	9.02501	0.11080	5.6	270.42641	0.00370	9	8103.08393	0.00012
2.3	9.97418	0.10026	5.7	298.86740	0.00335	9.1	8955.29270	0.00011
2.4	11.02318	0.09072	5.8	330.29956	0.00303	9.2	9897.12906	0.00010
2.5	12.18249	0.08208	5.9	365.03747	0.00274	9.3	10938.01921	0.00009
2.6	13.46374	0.07427	6	403.42879	0.00248	9.4	12088.38073	0.00008
2.7	14.87973	0.06721	6.1	445.85777	0.00224	9.5	13359.72683	0.00007
2.8	16.44465	0.06081	6.2	492.74904	0.00203	9.6	14764.78157	0.00007
2.9	18.17415	0.05502	6.3	544.57191	0.00184	9.7	16317.60720	0.00006
3	20.08554	0.04979	6.4	601.84504	0.00166	9.8	18033.74493	0.00006
3.1	22.19795	0.04505	6.5	665.14163	0.00150	9.9	19930.37044	0.00005
3.2	24.53253	0.04076	6.6	735.09519	0.00136	10	22026.46579	0.00005
3.3	27.11264	0.03688	6.7	812.40583	0.00123			

Appendix B   Table of Natural Logarithms

x	$\ln(x)$	x	$\ln(x)$	x	$\ln(x)$
0.1	−2.30259	3.5	1.25276	6.8	1.91692
0.2	−1.60944	3.6	1.28093	6.9	1.93152
0.3	−1.20397	3.7	1.30833	7	1.94591
0.4	−0.91629	3.8	1.33500	7.1	1.96009
0.5	−0.69315	3.9	1.36098	7.2	1.97408
0.6	−0.51083	4	1.38629	7.3	1.98787
0.7	−0.35667	4.1	1.41099	7.4	2.00148
0.8	−0.22314	4.2	1.43508	7.5	2.01490
0.9	−0.10536	4.3	1.45862	7.6	2.02815
1	0.00000	4.4	1.48160	7.7	2.04122
1.1	0.09531	4.5	1.50408	7.8	2.05412
1.2	0.18232	4.6	1.52606	7.9	2.06686
1.3	0.26236	4.7	1.54756	8	2.07944
1.4	0.33647	4.8	1.56862	8.1	2.09186
1.5	0.40547	4.9	1.58924	8.2	2.10413
1.6	0.47000	5	1.60944	8.3	2.11626
1.7	0.53063	5.1	1.62924	8.4	2.12823
1.8	0.58779	5.2	1.64866	8.5	2.14007
1.9	0.64185	5.3	1.66771	8.6	2.15176
2	0.69315	5.4	1.68640	8.7	2.16332
2.1	0.74194	5.5	1.70475	8.8	2.17475
2.2	0.78846	5.6	1.72277	8.9	2.18605
2.3	0.83291	5.7	1.74047	9	2.19722
2.4	0.87547	5.8	1.75786	9.1	2.20827
2.5	0.91629	5.9	1.77495	9.2	2.21920
2.6	0.95551	6	1.79176	9.3	2.23001
2.7	0.99325	6.1	1.80829	9.4	2.24071
2.8	1.02962	6.2	1.82455	9.5	2.25129
2.9	1.06471	6.3	1.84055	9.6	2.26176
3	1.09861	6.4	1.85630	9.7	2.27213
3.1	1.13140	6.5	1.87180	9.8	2.28238
3.2	1.16315	6.6	1.88707	9.9	2.29253
3.3	1.19392	6.7	1.90211	10	2.30259
3.4	1.22378				

Appendix C   Table of Common Logarithms

x	$\log_{10} x$	x	$\log_{10} x$	x	$\log_{10} x$
0.1	−1.00000	3.5	0.54407	6.8	0.83251
0.2	−0.69897	3.6	0.55630	6.9	0.83885
0.3	−0.52288	3.7	0.56820	7	0.84510
0.4	−0.39794	3.8	0.57978	7.1	0.85126
0.5	−0.30103	3.9	0.59106	7.2	0.85733
0.6	−0.22185	4	0.60206	7.3	0.86332
0.7	−0.15490	4.1	0.61278	7.4	0.86923
0.8	−0.09691	4.2	0.62325	7.5	0.87506
0.9	−0.04576	4.3	0.63347	7.6	0.88081
1	0.00000	4.4	0.64345	7.7	0.88649
1.1	0.04139	4.5	0.65321	7.8	0.89209
1.2	0.07918	4.6	0.66276	7.9	0.89763
1.3	0.11394	4.7	0.67210	8	0.90309
1.4	0.14613	4.8	0.68124	8.1	0.90849
1.5	0.17609	4.9	0.69020	8.2	0.91381
1.6	0.20412	5	0.69897	8.3	0.91908
1.7	0.23045	5.1	0.70757	8.4	0.92428
1.8	0.25527	5.2	0.71600	8.5	0.92942
1.9	0.27875	5.3	0.72428	8.6	0.93450
2	0.30103	5.4	0.73239	8.7	0.93952
2.1	0.32222	5.5	0.74036	8.8	0.94448
2.2	0.34242	5.6	0.74819	8.9	0.94939
2.3	0.36173	5.7	0.75587	9	0.95424
2.4	0.38021	5.8	0.76343	9.1	0.95904
2.5	0.39794	5.9	0.77085	9.2	0.96379
2.6	0.41497	6	0.77815	9.3	0.96848
2.7	0.43136	6.1	0.78533	9.4	0.97313
2.8	0.44716	6.2	0.79239	9.5	0.97772
2.9	0.46240	6.3	0.79934	9.6	0.98227
3	0.47712	6.4	0.80618	9.7	0.98677
3.1	0.49136	6.5	0.81291	9.8	0.99123
3.2	0.50515	6.6	0.81954	9.9	0.99564
3.3	0.51851	6.7	0.82607	10	1.00000
3.4	0.53148				

ANSWERS TO ODD-NUMBERED EXERCISES

CHAPTER 1

Section 1.1, page 13

1. $24, 0, \sqrt[5]{32}, \sqrt{16}, -234$ 3. $2\pi, \sqrt{5}$

5. $0.\overline{285714}$ 7. $1.91\overline{6}$ 9. $\dfrac{7}{8}$

11. $\dfrac{89}{78}$ 13. 6 15. $\dfrac{27}{2}$

17. $\dfrac{1}{45}, -45$ 19. $-\dfrac{1}{3.2}, 3.2$ 21. $-\dfrac{3}{2}, \dfrac{2}{3}$

23. $62.5, -0.016$ 25. Rational.   27. Rational.

29. Rational.

31. Commutative property of addition.

33. Distributive property.

35. Multiplicative identity.

37. Additive inverse.

39. Associative property of addition.

41. Additive inverse.

43. Multiplicative inverse.

45. Additive identity.

47. $<$ 49. $>$ 51. $=$ 53. $>$

55. $<$ 57. $>$ 59. 11 61. 15

63. 21 65. 6 67. $\dfrac{1}{4}$ 69. 11

71. 10 73. 0 75. $\sqrt{10}$ 77. $\sqrt{5} - 2$

79. $>$ 81. $<$ 83. $ac - bc$

85. $-ac - bc$ 87. $\dfrac{pa - pb}{cq}$ 89. $\dfrac{a^2 - 9}{3a}$

91. 3 93. $\dfrac{23}{20}$ 95. 27.5

97. $3 \cdot 3 \cdot 11$ 99. $2 \cdot 71$ 101. $5 \cdot 5 \cdot 7$

103. $2 \cdot 2 \cdot 3 \cdot 3 \cdot 3 \cdot 5 \cdot 5$ 105. $\{x : x < 0\}$

107. $\{x : x \geq -3\}$ 115. $\{x : x \geq 0\}$

117. *a.* ± 3

    b. 0

    c. No solutions.

119. $\dfrac{1}{2 + 3} \neq \dfrac{1}{2} + \dfrac{1}{3}$ 121. $\dfrac{3927}{1250}$

123. $-x - 3$ 125. $-x^2 + x - 4$

Section 1.2, page 25

1. 1 3. $\dfrac{1}{a^3}$ 5. q^4 7. $-12x^{11}$

9. $\dfrac{36}{y^2}$ 11. $9a^2b^8$ 13. $6^4x^4y^4$ 15. $\dfrac{1}{t^5}$

17. $\dfrac{a^5}{b^4}$ 19. $\dfrac{b^8}{9x^4}$ 21. $\dfrac{8b^3}{27a^3}$ 23. $\dfrac{yz^8}{2x^3}$

25. $-\dfrac{25}{8}$ 27. $\dfrac{1}{3}$ 29. 81 31. $-\dfrac{1}{256}$

33. -5 35. 32 37. 9 39. 27

41. 16 43. $y^{2/3}$ 45. $\dfrac{1}{y^{17/6}}$ 47. $\dfrac{1}{a^{16}}$

49. $x^{1/10}$ 51. $\dfrac{1}{a^6}$ 53. $5\sqrt{3}$ 55. $2\sqrt[3]{2}$

57. $7x$ 59. $4\sqrt{2}(t + 1)$ 61. $3\sqrt{2}x$

63. $2xy^2$ 65. $\dfrac{7\sqrt{2a}}{b^3}$ 67. $3\sqrt[3]{2}xy^{4/3}$

69. 3 71. $x^2\sqrt{x}$ 73. $5ab^{5/3}$

75. $2\sqrt{2}(x + 1)^{3/2}$ 77. $\dfrac{3ac^2}{b}$

79. 2×10^9 81. 1.6×10^{-18}

83. 3.56×10^{12} 85. $\$12,000,000,000$

87. $.00000000000000000000000383$

89. 0.00000000000000000578

91. 1.29316×10^{19} miles

93. 2.387×10^5 miles   95. 66.779 ft.

97. 11.355 sec.   99. 5041 m/sec.

101. 5.7557% 103. 329.67 m/sec.

105. For example, $a = 5, b = 5$.

107. 1 109. $-5^5x^{5a}y^{5b}$ 111. 26

113. $\dfrac{929}{24}$ 115. 401π

117. $2.4045175 \times 10^{-6} \times \pi$

Section 1.3, page 34

1. Yes; leading coefficient $= 34$; degree $= 5$.

3. No.   5. No.   7. No.

9. $17x^3 - 7x^2 + 3x - 1$ 11. $33x^3 - 41x^2 + 11x - 8$

13. $-\dfrac{1}{4}x^4 + \dfrac{1}{4}x^3 - \dfrac{5}{3}x^2 - \dfrac{3}{4}x + \dfrac{6}{5}$

15. $\dfrac{2}{5}x^4 - \dfrac{13}{6}x^3 - \dfrac{3}{4}x^2 - \dfrac{1}{8}x + \dfrac{1}{3}$

17. $4x^2 + 2x^3y - 5x^2y^2 - 4xy^3 + y^4$

19. $4p^4 - 9p^3q^2 + 2p^2q^3 - 3q^4$

21. $0.6x^4 + 2.7xy + 1.8x^2 + 0.5y^2$

23. $1.1x^2 - 0.8xy + 1.9y^2 + 1.2$

25. $8x\sqrt[3]{x} - 6x\sqrt{x} + 9x\sqrt[4]{x}$

27. $-3x\sqrt{y} - 5y\sqrt{x} + 6\sqrt{xy}$

29. $-x^3 + x^2 + 5x - 1$ 31. $8x^2 + 26x + 15$

33. $81a^2 - 4$ 35. $a^3 - 8$

37. $6a^3b^4 + 4a^4b^3 - 15a^2b^3 - 10a^3b^2$

39. $25x^2 - 30xy + 9y^2$

41. $4a^2b^2 + 12ab^2c + 9b^2c^2$

43. $16x^4 - 56x^3y + 49x^2y^2$

45. $a^4 - 2a^2b^2 + b^4$ 47. $a^2 + 2ab + b^2 - y^6$

49. $2x^2y^2 - 6x^2y\sqrt{2} + 9x^2$ 51. $3a^2 - 2b^2$

53. $\dfrac{16}{25}x^2 - \dfrac{9}{49}y^2$ 55. $\dfrac{4}{9}t^4 - \dfrac{4}{5}t^2y^3 + \dfrac{9}{25}y^6$

57. $\dfrac{1}{8}x^2 - x^3y - \dfrac{3}{8}x^2y^2 + 4xy^3 - \dfrac{1}{2}y^4$

59. $-2x^4 - 9x^3 + 2x^2 + 24x - 9$

61. $A^4 + 4A^3B + 6A^2B^2 + 4AB^3 + B^4$

63. $x^4 + 32x^3 + 384x^2 + 2048x + 4096$

65. $1296x^8 + 864x^6 + 216x^4 + 24x^2 + 1$

67. $3x(1 - xy)$ 69. $(x - 7)(x + 3)$

71. $2(2x - 3)(5x - 2)$ 73. $2ab(7a - 6b)$

75. $(y - w)(x - z)$ 77. $(x - 2)(x + 2)(2x + 1)$

79. $(23 - 18x)(23 + 18x)$ 81. $4x(2a - 3x + a^2 - 5)$

83. $(x - 3)(x^2 + 3x + 9)$ 85. $2a(2x + 3)^2$

87. $(2x + 5)(3x + 7)$

89. $(0.5x - 0.7y)(0.5x + 0.7y)$

91. $(-2t + 3)(t - 4)$ 93. $2(a - 2)(a^2 + 2a + 4)$

95. $(x\sqrt{0.3} - y\sqrt{0.2})(x\sqrt{0.3} + y\sqrt{0.2})$

97. $(3x^2 - 2y)(9x^4 + 6x^2y + 4y^2)$

99. $(x - 2)(x + 2)(x - 3)(x + 3)$

101. $(a + 4)(a^2 + 2a + 4)$ 103. $(x + y - 5)(x + y + 5)$

105. $3(x - \sqrt{2}y)(x + \sqrt{2}y)(x^4 + 2x^2y^2 + 4y^4)$

107. $\dfrac{1}{72}(4x - 3)(3x + 2)$

109. a. $65p + 20c$ 111. a. $65\pi p + 100\pi w$
b. $74p + 24c$ b. $240\pi p + 700\pi w$
c. $139p + 44c$ c. $305\pi p + 800\pi w$

113. $x^2 + 10x + 24$ 115. $4x^4$

117. $9x^{2a} - 12x^ay^c + 4y^{2c}$ 119. $20t^{2n} + 3t^n - 35$

121. x^{2m^2+2mn} 123. $(a^n - b^n)(a^n + b^n)$

125. $(6t^n - 5)(6t^n - 5)$

127. $\left(\dfrac{x^4}{10} + 1\right)\left(\dfrac{x^8}{100} - \dfrac{x^4}{10} + 1\right)$

129. $9x^2 + 30x + 25 \neq 9x^2 + 25$

131. $9x^2 - 30xy + 25y^2 \neq 9x^2 - 25y^2$

Section 1.4, page 47

1. $\dfrac{x + 1}{x^2 + x + 1}$ 3. $\dfrac{(x - 1)(x + 1)}{(2x + 1)(x + 3)}$

5. a^2 7. $\dfrac{1}{r + s}$ 9. $\dfrac{x - 1}{x^2 + x + 1}$

11. $\dfrac{-5x + 2}{(x^2 - 2x + 4)^2}$ 13. 1

15. $\dfrac{t - 1}{t + 1}$ 17. $\dfrac{(x + 2)(7x^2 + 3)}{6(x^2 + 2)(x^2 - 3)}$

19. $\dfrac{6x - 10}{2x + 1}$ 21. $\dfrac{x - 7 - 3y}{x - 3}$ 23. $\dfrac{3t^2 - 4t + 7}{t^2 + 3t - 5}$

25. $\dfrac{-2x^2 + 4x - 6}{x^2 - 3x - 8}$ 27. $\dfrac{t - 3}{(t + 3)(t + 1)}$

29. $\dfrac{8x - 29}{(x + 5)(x - 3)}$ 31. $\dfrac{x^3 - 3x - 4}{(x + 1)(x - 1)(x^3 + 1)}$

33. $\dfrac{(n + 3)(n + 1)(n + 2)}{6}$ 35. $\dfrac{a^2 + b^2}{a + b}$

37. $\dfrac{y(x + z) + xz}{(x + z)}$ 39. $\dfrac{-1}{x^2 + xh}$

41. $-\dfrac{3x^2 + 3xh + h^2}{x^3(x + h)^3}$ 43. 1

45. $2a + h$ 47. $2x + 3 + h$

49. $\dfrac{1}{(x + 1)(x + h + 1)}$ 51. .5545

53. a. 2.2222×10^{-5}
b. $L = 10.0533$ m

55. $\dfrac{\sqrt{3xy}}{2y}$ 57. $4(\sqrt{6} - \sqrt{5})$ 59. $\dfrac{\sqrt{a} + \sqrt{b}}{a - b}$

61. $\dfrac{(\sqrt{a} + \sqrt{2})(\sqrt{a} + \sqrt{b})}{a - b}$ 63. $\dfrac{1}{\sqrt{a + h} + \sqrt{a}}$

65. $\dfrac{a - 2}{a - 2\sqrt{2a} + 2}$ 67. $\dfrac{1}{\sqrt{n}(\sqrt{n + 1} + \sqrt{n})}$

69. $\dfrac{1}{1 + \sqrt{x}}$ 71. $-1 + \dfrac{1}{\sqrt{x + h} + \sqrt{x}}$

73. $2\sqrt{x^2 - 4}$ 75. $\dfrac{(x - 2)^2n}{(n + 1)^2}$

77. $60(x^3 + 15x)^{18}[59x^4 + 600x^2 + 1425]$

79. $\dfrac{\sqrt{x^2 + 1}}{-(x^2 + 1)}$ 81. $\dfrac{5x^2 - 60}{3(x^2 - 4)^{4/3}}$

83. A^2−2*A*B+B^2   85. (2/3)*x

87. A^2/B−3*D+(A/B)^3

89. $3a^4 - 5d$ 91. $a - \dfrac{2}{b}$

93. $4x^3 - 3x^2 + 2x - 17$

95. $-\dfrac{4}{3}, -\dfrac{4}{3}, \dfrac{10}{3}, \dfrac{10}{3}, -\dfrac{16}{3}, -\dfrac{16}{3}$

Section 1.5, page 51

1. $\sqrt{x+y}$

3. $x - 0.37x$, or $.63x$

5. 700π ft^2

7. All real numbers.

9. $-54.7, 4^{5/2}, 17, -4, -\frac{1}{7}, 6^0, -3^2, 9^{3/2}, 67.2\%,$
   $0.892, 2.\overline{345}$

11. Associative for multiplication.

13. Additive inverse.

15. Distributive property.

17. $>$

19. $=$

21. $6x^3 - x^2 - 10x - 3$

23. $\dfrac{6 - t^2}{3t}$

25. $\dfrac{a-3}{2a-1}$

27. $x^{2/3}(2 + x + 2x^{5/6})$

29. $\dfrac{x-9}{(x+9)(x+7)}$

31. $(x-9)(x-24)x$

33. 0

35. $\dfrac{3}{x-1}$

37. $\dfrac{-32b^7}{a^2}$

39. $\dfrac{2}{5}xy$

41. $\dfrac{9x^{18}z^{22}}{16y^8}$

43. $z^{4/3}$

45. $(a^2 + b^2)^{3/2}$

47. $\dfrac{b^3}{a^2}$

49. 8

51. $\dfrac{45}{8}$

53. $\dfrac{1}{ab(a+b)}$

55. $\dfrac{1}{a^2b^2 + ab + 1}$

57. $\dfrac{3a}{4}$

59. $(x+y)^6\sqrt{x+y}$

61. $16x^4y^6\sqrt{\dfrac{\sqrt[5]{5^4x^4y^3}}{y}}$

63. $\dfrac{2x \cdot n^5}{(n+1)^5}$

65. $2^3 \cdot 3 \cdot 19$

67. $(2x+1)(2x-3)2$

69. $(5a+1)(25a^2 - 5a + 1)$

71. $(a-b)^3$

73. $(x^a + 3)(x^{2a} - 3x^a + 9)$

75. $2x(2x+1)(x+7)$

77. $-1(x+11)(x-5)$

79. $(a+b)(x+y)^2$

81. $4(3t - 2m)(3t + 2m)$

83. $\left(\dfrac{a}{4} + \dfrac{b}{2} - \dfrac{ab}{3}\right)\left(\dfrac{a}{4} + \dfrac{b}{2} + \dfrac{ab}{3}\right)$

85. $(x - \sqrt[3]{7})(x^2 + \sqrt[3]{7}x + \sqrt[3]{49})$

87. $\dfrac{1 - 2\sqrt{a} + a}{1 - a}$

89. $\dfrac{\sqrt[3]{450}}{5}$

91. $\dfrac{1}{5\sqrt[4]{t^3}}$

93. $-\dfrac{n}{\sqrt{n^2 - n} + n}$

95. 2.467×10^3

97. 0.0648

99. a. 1.674×10^{-24} g; b. 4.482×10^{-23} g
    c. 3.442×10^{-22} g

101. $\sqrt{x^2 - \left(\dfrac{x}{2}\right)^2} = \sqrt{3}\left(\dfrac{x}{2}\right)$

103. $a\sqrt{2}$

105. $(X - Y)/(A + B)$

107. $45 \cdot 63^2 + 2a - b^2$

109. $3xy + 4\sqrt{xy} + 6x\sqrt{y} + 2y\sqrt{x}$

111. $x^{2\sqrt{3}} + 2x^{\sqrt{2}+\sqrt{3}} + x^{2\sqrt{2}}$

113. $-\dfrac{-13\sqrt{7} - 8\sqrt{42} - 6\sqrt{3} + 22\sqrt{2}}{215}$

115. 8

CHAPTER 2

Section 2.1, page 63

1. $x = 13$

3. $y = 4.35$

5. $-\dfrac{83}{40}$

7. $t = 2$

9. $x = 35$

11. $x = -25$

13. 5000

15. 650

17. $x = -8$

19. $x = 8$

21. $x = -7$

23. $x = 5$

25. $x = 1$

27. -2

29. $-\dfrac{600}{23}$

31. $x = -\dfrac{29}{3}$

33. $y = \dfrac{14}{17}$

35. $x = -12, 2$

37. $x = 11$

39. $x = 6$

41. No solution.

43. $y = 6$

45. $x = -\dfrac{1}{2}, -\dfrac{1}{4}$

47. $t = 1, -\dfrac{1}{2}$

49. No solutions.

51. No solutions.

53. No solutions.

55. $t = \dfrac{1}{7}$

57. $t = 1$

59. Yes.

61. No.

63. No.

65. Yes.

67. No.

69. Yes.

Section 2.2, page 71

1. 273.17

3. 1.7

5. 13%

7. $\$92$

9. $\$1500$

11. 414 m^2, $l = 23$ m, $w = 18$ m

13. 36 cm^2

15. $s = 16, m = 19$

17. $90, 91, 92$

19. $128, 130, 132$

21. 26 ft., 52 ft., 19 ft.

23. 40 minutes

25. 0.039 hr.

27. 30 minutes

29. $v = \dfrac{F}{pq}, p = \dfrac{F}{qv}$

31. $y = -\dfrac{2x+5}{2}, x = -\dfrac{2y+5}{2}$

33. $x = \dfrac{144}{8 + 3\pi}$

35. $r = \sqrt{\dfrac{A}{\pi}}$

37. $b = \sqrt[3]{\dfrac{p}{a}}$

39. $d = \dfrac{k-p}{hq + (1/2)v^2}, q = \dfrac{k - p - (1/2)dv^2}{hd}$

41. $a_1 = \dfrac{s(1-r)}{1 - r^n}$

43. $m_2 = \dfrac{m_1v_1 - m_1u_1}{u_2 - v_2}, v_1 = \dfrac{m_1u_1 + m_2u_2 - m_2v_2}{m_1}$

45. $\lambda = \pm\dfrac{1}{2}$

47. a. $\$8745.28$

49. $\$32,754.34$

b. $I = n\left(\left(\dfrac{A}{P}\right)^{1/nt} - 1\right)$

c. 8%

51. a. $\$1909.20$; b. $\$1946.38$; c. $\$1955.23$;
    d. $\$1958.69$; e. $\$1959.58$

53. $60,000/7$ gallons of mixture A, $80,000/7$ gallons of
    mixture B

55. 6500 at 6%, 5500 at 9%

57. 7800

59. $V = 9700 - t\left(\dfrac{9700 - 1300}{6}\right)$
    a. $V = 9700 - 1400t$; b. Yes, it is linear.
    c. $\$9700, \$8300, \$6900, \$5500, \$2700, \$1300$

61. *a.* 68°F; *b.* 100°C; *c.* −40°

63. $x = \pm\dfrac{2\sqrt{3}}{3}r$ 65. $h = 2\sqrt[3]{6}$

Section 2.3, page 86

1. $\pm\sqrt{3}$ 3. No solutions. 5. No solutions.

7. 3 or 1 9. $\dfrac{2}{3}$ or $-\dfrac{4}{3}$ 11. $-\dfrac{5}{2} \pm \dfrac{\sqrt{65}}{2}$

13. No solutions. 15. $-\dfrac{2}{3} \pm \dfrac{\sqrt{7}}{3}$ 17. $\dfrac{10}{3}$ or −1

19. 0 and $-\dfrac{4}{3}$ 21. No solutions. 23. $\dfrac{-1 \pm \sqrt{33}}{4}$

25. 0 or $\dfrac{5}{9}$ 27. $\dfrac{3}{2}$ or $\dfrac{-7}{3}$ 29. No solutions.

31. No solutions. 33. $\dfrac{43 \pm 5\sqrt{85}}{6}$

35. $\dfrac{-\sqrt{2} \pm \sqrt{2 + 4\sqrt{3}}}{2}$ 37. No solutions.

39. $\dfrac{-2}{7}$ and $\dfrac{4}{3}$ 41. $\dfrac{17}{2} \pm \dfrac{\sqrt{293}}{2}$

43. $\dfrac{5 \pm \sqrt{73}}{6}$ 45. $\dfrac{-3 \pm \sqrt{329}}{40}$

47. $\dfrac{-15 \pm 5\sqrt{67}}{29}$ 49. 100

51. No solutions. 53. 9 55. 11

57. $r = \dfrac{1}{2}\sqrt{\dfrac{340}{3\pi}} \approx 3$ m, $h = \sqrt{\dfrac{340}{3\pi}} \approx 6$ m

59. 8 and 15 ft.

61. *a.* 13 seconds; *b.* $\dfrac{1}{2}, \dfrac{11}{2}$ seconds 63. 130

65. *a.* 808.8 m; *b.* 100°C; *c.* 98.9867°C; *d.* 97.976°C
67. 5% 69. 8%

71. *a.* $h = \dfrac{\sqrt{3}a}{2}$; *b.* $A = \dfrac{ha}{2}$; *c.* $P = 3a$; *d.* 30 cm

73. *a.* 20; *b.* 12; *c.* No. 75. $n = \dfrac{1 \pm \sqrt{1 + 8G}}{2}$

77. $t = \dfrac{-V_0 \pm \sqrt{V_0^2 + 64(h - H)}}{-32}$

79. $r = \dfrac{-h \pm \sqrt{h^2 + 2S/\pi}}{2}$ 81. $0, -3, \dfrac{4}{5}$

83. 1 85. −3, −1, 1 87. 0, 1

89. −1, 1 91. 4 93. $\dfrac{-5 \pm \sqrt{221}}{14}$

Section 2.4, page 96

1. $\dfrac{35}{6}$ 3. 0.7 5. $\dfrac{16}{3}$

7. No solutions. 9. $\dfrac{17}{4}$ 11. No solutions.

13. No solutions. 15. $4 + 2\sqrt{7}$ 17. $\dfrac{1}{16}$

19. 19 21. 5 23. No solutions.
25. 8, −2 27. 9 29. 9, 16 31. ±2

33. 4, −1; 2, 1 35. 256, 225 37. $\dfrac{5}{6}, \dfrac{4}{3}$

39. $\dfrac{1}{4}, 9$ 41. −4, 2, −1 43. $0, \dfrac{6}{5}$

45. 1 47. $\dfrac{6}{5}$ hr. 49. $\dfrac{2}{19}$ hr. 51. 36 hr.

53. 42 mph 55. 20 mph or $\dfrac{500}{7}$ mph

57. *a.* $s = \dfrac{\sqrt{\pi d}}{2}$; *b.* 179.02 mm; *c.* 451.35 mm

59. Yes, if the hole is 2 m below the top.
61. $3 + \sqrt{5}$ cm 63. 133.034 ft.
65. −1 67. 1

69. $x = a - b, -a$ provided $a \neq -\dfrac{b}{2}$

71. No solution. 73. ±6 75. $0, \pm\sqrt[4]{\dfrac{3}{5}}$

Section 2.5, page 107

1. $c > 0$ 3. $a \geq 0$ 5. $-3 \leq m$ 7. $a \leq b$
9. $[-2, 3)$ 11. $[-4, 5]$ 13. $[-2, 3)$ 15. $[-2, \infty)$
17. $\{x : -1 \leq x \leq 3\}$ 19. $\{x : x > -2\}$

21. $\{x : x \leq 2\}$ 23. $[-1, \infty)$ 25. $\left(-\infty, -\dfrac{3}{4}\right)$

27. $(-\infty, 4]$ 29. $\left[-\dfrac{4}{5}, \infty\right)$ 31. $[4, \infty)$

33. $(-\infty, -2.6)$ 35. $[-4, 5]$ 37. $\left[1, \dfrac{13}{3}\right]$

39. $(-5, -2]$ 41. $[25, \infty)$ 43. $\left[-8, \dfrac{14}{3}\right)$

45. $\left(\dfrac{14}{3}, \infty\right)$ 47. $\left(-\infty, -\dfrac{3}{2}\right)$

49. $(-\infty, -5) \cup (3, \infty)$ 51. $(-\infty, 3] \cup [4, \infty)$

53. $(-5, 5)$ 55. $(-\infty, 0) \cup \left(\dfrac{2}{3}, \infty\right)$

57. $(-3, -1) \cup (2, \infty)$ 59. $(-\infty, -5] \cup [0, 2]$
61. $(-\infty, -2) \cup (-1, 1)$ 63. $(-\infty, -1) \cup (3, \infty)$

65. $\left[\dfrac{3}{2}, 4\right)$ 67. $(-\infty, -4) \cup (-1, 2)$

69. $(-\infty, 0) \cup (0, \infty)$ 71. $0 < h \leq 33\dfrac{1}{3}$ in.

73. $52 < S < 92$ 75. *a.* $x \geq 167$; *b.* $x \geq 167$

77. $8 \leq n \leq 17, n$ an integer 79. $p \leq \dfrac{1}{2}$

81. $(-\infty, 0) \cup (0, \infty)$

83. $(-\infty, -1) \cup \left(\dfrac{1}{5}, 2\right) \cup (2, \infty)$

85. $(0, 2\sqrt{2}) \cup (2\sqrt{2}, \infty)$

87. $(-\infty, -4) \cup (-1, 2)$ 89. $(-3, -1) \cup (2, \infty)$

91. $(-5, 2]$ 93. $\left[-8, \dfrac{14}{3}\right)$

Section 2.6, page 112

1. $\{-3, 3\}$ 3. $\{-5, -1\}$ 5. $\{-4, 4\}$ 7. $\{9, -6\}$
9. $[-23, 23]$ 11. $(-\infty, -14) \cup (14, \infty)$
13. $(-\infty, -2) \cup (6, \infty)$ 15. $(-1, 5)$

17. $\left[-6, \dfrac{22}{3}\right]$ 19. $\varnothing$ 21. $(-\infty, \infty)$

23. $\left(-\infty, \dfrac{2}{3}\right) \cup \left(\dfrac{2}{3}, \infty\right)$ 25. $\left(-\dfrac{23}{12}, \dfrac{1}{4}\right)$

27. $\varnothing$ 29. $[-3, -2] \cup [6, 7]$ 31. $\varnothing$

33. $[-1, 1]$ 35. $\varnothing$ 37. $\left[-\dfrac{1}{2}, \dfrac{7}{2}\right]$ 39. $[-4, 4]$

41. $\left(-\infty, -\dfrac{1}{2}\right)$

43. $(-\infty, -1) \cup (1, 4) \cup (6, \infty)$

45. $[-2, 2]$ 47. $x - \delta < x < x + \delta$

51. $|x - 5| \le r$ 53. $|x - 2| \ge 6$

55. $[\$2{,}775{,}000, \$3{,}225{,}000]$ 57. $(6, 14)$

Section 2.7, page 120

1. $\sqrt{3}i$ 3. $9i$ 5. $7i\sqrt{2}$ 7. $-7i$
9. $4 - 2i\sqrt{15}$ 11. $2i(1 + \sqrt{3})$ 13. $-i$
15. 1 17. -1 19. i
21. $8 + 0i = 8$ 23. $1 - 26i$ 25. $0 + 0i = 0$
27. 0 29. 1 31. $5 - 8i$

33. $2 - \dfrac{\sqrt{6}}{2}i$ 35. $\dfrac{4}{5}$ 37. $5 + 6i$

39. 6 41. $2 + 4i$ 43. $-7 + 10i$
45. $14 - 2i$ 47. 29 49. $6 + 8i$

51. $21 - 20i$ 53. $\dfrac{1}{10} + \dfrac{7}{10}i$ 55. $\dfrac{1}{2} - \dfrac{\sqrt{3}}{2}i$

57. $-3 - 5i$ 59. $-\dfrac{1}{2} + \dfrac{1}{2}i$ 61. $\dfrac{2}{25} + \dfrac{11}{25}i$

63. $\dfrac{1}{5} + \dfrac{2}{5}i$ 65. $\dfrac{1}{34} + \dfrac{2}{17}i$ 67. $\dfrac{3}{5} + \dfrac{1}{5}i$

69. $\dfrac{-1}{2}i$ 71. $\pm\sqrt{5}i$ 73. $\pm 8i$

75. $x^2 + 4$ 77. $x^2 - \dfrac{2}{3}x + \dfrac{1}{3}$

79. $(x - 2i)(x + 2i)$ 81. $(x - \sqrt{3}i)(x + \sqrt{3}i)$

83. $(a - bi)(a + bi)$ 85. $\left(-\dfrac{9}{5}, 8\right)$

87. $\left(3, \dfrac{3}{2}\right), \left(-3, \dfrac{3}{2}\right)$ 99. $4\overline{\alpha}^2 - 25$

Section 2.8, page 122

1. $(-\infty, \infty)$ 3. $-\dfrac{16}{17}$

5. $\left\{\dfrac{-5 \pm \sqrt{41}}{8}\right\}$ 7. $\dfrac{1 \pm \sqrt{4001}}{20}$ 9. $\left\{-\dfrac{1}{2}, -1\right\}$

11. $4, \dfrac{2}{3}$ 13. No solutions.   15. $\{16, 25, 0\}$

17. -10 19. No solutions.   21. $\pm\sqrt{5}, \pm\sqrt{2}$

23. $\left\{\dfrac{4}{3}, -12\right\}$ 25. $\{\pm 3, -4\}$ 27. $(-5, \infty)$

29. $[-9, 9)$ 31. $(-1, 6)$

33. $(-\infty, -7) \cup (5, \infty)$ 35. $\left(\dfrac{1}{3}, 7\right)$

37. $\pm\dfrac{5}{2}$ 39. $\sqrt{\dfrac{Q}{m}} = n$ 41. $L = \dfrac{T}{4v^2 - 1}$

43. $n = \dfrac{-1 \pm \sqrt{1 + 8S}}{2}$ 45. $x = \dfrac{y \pm \sqrt{y^2 - 4}}{2}$

47. 200 km/hr   49. $\dfrac{5000}{17}$ ft.^{3} 51. 9%

53. $a.$ 5.5125 m; $b.$ 0.2485 m; $c.$ $\dfrac{T^2}{4\pi^2}(9.8) = L$

55. 2.374% 57. $-38.2° < F < 674.6°$

59. $\$65{,}000$ at 6%, $\$55{,}000$ at 9%

61. $a.$ 10 seconds
    $b.$ The ball will never reach a height of 1216 ft.

63. $-i$ 65. $2 + \dfrac{\sqrt{10}}{4}i$ 67. $-12 + 14i$

69. $-\dfrac{2}{41} + \dfrac{23}{41}i$ 71. -2 73. $\pm 5i$

75. $(x - 3i)(x + 3i)$ 77. $\left(-\dfrac{7}{2}, -\dfrac{5}{2}\right)$

79. $x^2 + 9$

81. $a.$ $p \le \dfrac{1}{4}$; $b.$ $p > \dfrac{1}{4}$ 83. $6, -3 \pm 3\sqrt{3}i$

85. $1 \pm \sqrt{2}$ 87. $(-2, 0) \cup (2, \infty)$

CHAPTER 3

Section 3.1, page 130

1.

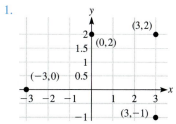

3. $a.$ Quadrant I; $b.$ Quadrant IV; $c.$ y-axis; $d.$ x-axis

5. $\sqrt{61}$ 7. $2\sqrt{2}$ 9. 6

11. $\sqrt{9a^2 + 4}$ 13. $\sqrt{a^2 + b^2}$ 15. $\sqrt{a + b}$

17. 1 19. $\sqrt{a^2 + b^2}$ 21. $\left(-1, -\dfrac{1}{2}\right)$

23. $(1, -4)$ 25. $(1, 5)$ 27. $\left(\dfrac{a}{2}, 4\right)$

29. $\left(\dfrac{a}{2}, \dfrac{b}{2}\right)$ 31. $\left(\dfrac{\sqrt{a}}{2}, \dfrac{\sqrt{b}}{2}\right)$ 33. $\left(0, -\dfrac{5}{3}\right)$ 17.

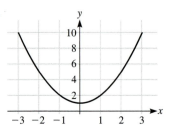

35. $(14.5, 0)$ 37. $-2 \pm \sqrt{33}$ 39. 21

41. $\left(0, -\dfrac{61}{20}\right)$ 43. No.   45. Yes.

47. $x - y = 2$

49. Radius $= \dfrac{5\sqrt{2}}{2}$; center $= (7, 10)$.

55. The points $(a, 0)$ greater than $\sqrt{19}$ from $(0, 3)$ are $(a, 0)$, such that $|a| > \sqrt{10}$.

19.

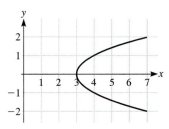

Section 3.2, page 148

1. Yes: $-2(2) - 16(-1) = 12$.
3. No: $2(1) - 3(-4)^2 \neq 46$.
5. Yes: $-(-2) = \sqrt{4}$.
7. Yes: $(-2 - 3)^2 + 2(-2 - 3) = 15$.

21.

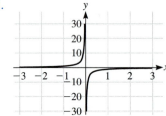

9.

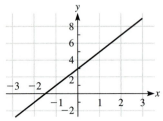

23.

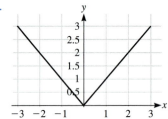

11.

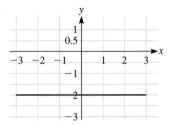

25.

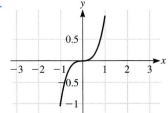

13.

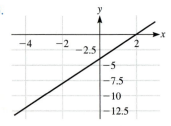

27.

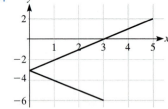

15.

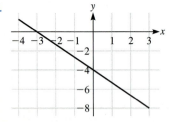

29.

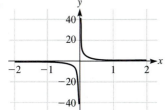

31.

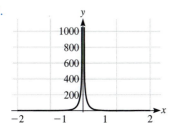

33.

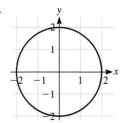

35. *a.*

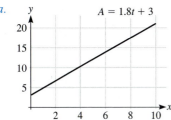

$A = 1.8t + 3$

b. The y-intercept corresponds to $t = 0$. This is the area covered by the organism at time 0. There is no t-intercept.

37.

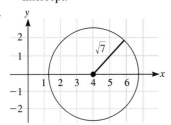

39.

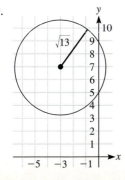

41.

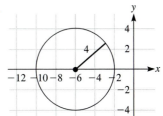

43.

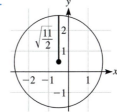

45. $(x + 6)^2 + (y - 1)^2 = 4$
47. $x^2 + y^2 = 39$
49. $(x + 2)^2 + (y - 3)^2 = 4$
51. $(x - 11)^2 + (y + 11)^2 = 121$
53. $(x + 1)^2 + (y - 6)^2 = 20$
55. $(x + 2)^2 + (y + 3)^2 = 9$
57. $(x - 7)^2 + (y - 10)^2 = \dfrac{25}{2}$

59. $(-5, 6), r = 9$ 61. $\left(\dfrac{3}{2}, -2\right), r = \dfrac{15}{2}$

63. Yes.   65. No.   67. Yes.   69. No.
71. The solution set is the set of all points on and enclosed by the circle whose center is at $(-3, -5)$ and whose radius equals 6.

Section 3.3, page 161

1. 0.3   3. -6 5. $\dfrac{1}{2}$ 7. $\dfrac{2}{3}$

9. *a.* y decreases by 36; *b.* y increases by 600.
11. *a.* The slope is the amount that the area increases each year.
    b. Area A is increased by 18 square miles.
13. *a.* The slope is the price of one set.
    b. Revenue increases $818,000.
15. *a.* If the total length increases by 1 mm, the tail length increases by 0.143 mm.
    b. The longer tailed snake is 559.44 mm longer than the other snake.

17. $-\dfrac{1}{2}$ 19. -3

21. Slope undefined.   23. Slope undefined.
25. The slope represents the value the machine loses per year.
27. $S = 24.000 + 0.38x$, S = salary, x = sales

29. $y = -\dfrac{1}{2}x$ 31. $y = -3$ 33. $y = -4x$

35. $y + 6 = -\dfrac{2}{3}(x + 2)$ 37. $y = -6$

39. Slope $= -\dfrac{3}{2}$; y-int: $(0, 3)$.

41. Slope $= \dfrac{10}{3}$; y-int: $\left(0, -\dfrac{1}{2}\right)$.

43. Slope undefined; no y-int.

45. Neither.   47. Neither.   49. Perpendicular.

51. Parallel.   53. $y = \dfrac{1}{3}x + 6$ 55. $x = 5$

57. $y = 5$ 59. $y = 1 - 2x$

61. $y = 2$ 63. $y = -\dfrac{9}{2}x - \dfrac{17}{2}$

65. Parallel line: $y = 9x - 56$; perpendicular line:
$y = -\dfrac{1}{9}x - \dfrac{4}{3}$.

67. *a.* x-intercept $= 4$, y-intercept $= -2$
    b. x-intercept $= 3$, y-intercept $= -4$

69. $-\dfrac{2}{3}(x + 2) = y + 2$ 71. $\dfrac{1}{2}$

73. *a.* $N = .9Y + 2.2$ (1983 corresponds to $Y = 0$.)
    b. 13.9 million

75. *a.* $P - 1.29 = -0.004(N - 60)$
    b. \$1.13. The predicted price is \$0.06 less than the listed price, \$1.19.

Section 3.4, page 171

1. -5 3. $-8a - 5$ 5. 309

7. $3(a - h)^2 + (a - h) - 1$ 9. 5   11. 3.56

13. 0 is not in the domain of g.

15. 11   17. -1

19. $-2h^3 + 9h^2 - 12h + 4$ 21. 27

23. $(1 + a + h)^3$ 25. 1

27. $4a^2 + 8ah + 4h^2 + 4a + 4h + 1$

29. 12   31. 12

33. Domain $=$ all real numbers; range $=$ all real numbers.

35. Domain $=$ all real numbers; range $= \{12\}$.

37. Domain $=$ all real numbers except 3; range $=$ all real numbers except 0.

39. Domain $= \left\{x : x \le \dfrac{3}{2}\right\}$; range $=$ set of all nonnegative real numbers.

41. Domain, range $=$ all real numbers.

43. Domain $=$ all real numbers; range $=$ set of all nonnegative real numbers.

45. Domain $= \{x : -2 \le x \le 2\}$; range $= \{x : 0 \le x \le 2\}$.

47. Domain $= \{x : x \le 2\} \cup \{x : x \ge 3\}$; range $= \{y : y \ge 0\}$.

49. 2, 4   51. $2a - 5 + h, 4a - 10$

53. $\dfrac{2}{\sqrt{2a + 2h - 3} + \sqrt{2a - 3}}$,
$\dfrac{4}{\sqrt{2a + 2h - 3} + \sqrt{2a - 2h - 3}}$

55. $0, 0$ 57. $4a^3 + 6a^2h + 4ah^2 + h^3, 8a^3 + 8ah^2$

59. Domain $=$ all nonzero real numbers.

61. Domain $= \{x : x \ne -1, -2\}$.

63. Domain $=$ all real numbers.

65. Domain $=$ all real numbers.

67. Domain $= \{x : x \ne 1\}$.

69. Domain $= \left\{x : x \ne \dfrac{5}{4}\right\}$.

71. Domain $=$ all real numbers.

73. $D(t) = 5t\sqrt{290}$ 75. $S(x) = x^2 + \dfrac{1200}{2}$

77. $A(x) = 14x - x^2\left(\dfrac{1}{2} + \dfrac{\pi}{8}\right)$

79. $C(x) = 4000(3 - \sqrt{x^2 - 1}) + 6000x$

81. $S(t) = -16t^2 - 9.5t + 1800$

Section 3.5, page 185

1. Yes.   3. No.

5. $0, 2, -2, 5$ 7. $2, 4, 0, -2$

9. Domain $= [-2, 5]$; range $= [-2, 5]$.

11. Domain $= [-2, 5.5]$; range $= [-5.5, 4]$.

13. Domain $= [-1, 1]$; range $= [-4, 12]$.

15. Domain $= [-1, 1]$; range $= [0, 1]$.

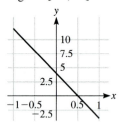

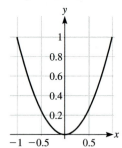

17. Domain $=$ all reals; range $= [-3, \infty)$.

19. Domain $= [-2, 2]$; range $= [-9, 7]$.

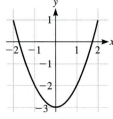

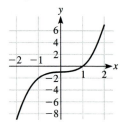

21. Domain $= [-2, 2]$; range $= [0, 2]$.

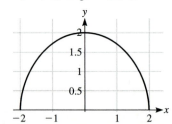

23. Domain = all positive reals; range = all positive reals.

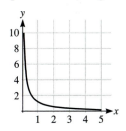

25.

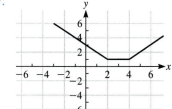

27.

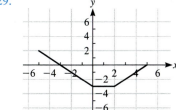

29.

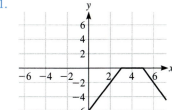

31.

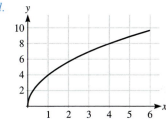

33.

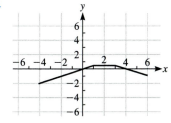

35. *a.*

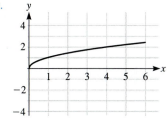

b.

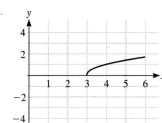

c.

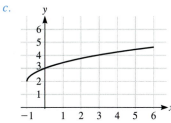

d.

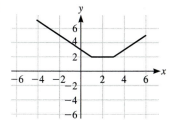

e.

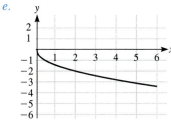

f.

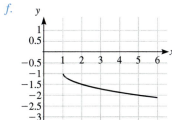

37. Both.   39. Neither.   41. Neither.   43. *x*-axis
45. Yes.   47. Yes.   49. No.   51. Yes.

53. Neither. **55.** Neither. **57.** Even. **59.** Neither.
61. Odd. **63.** Even.

65.

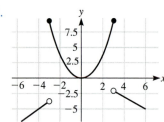

67.

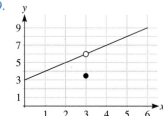

69.

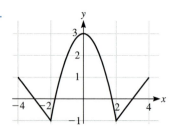

71.

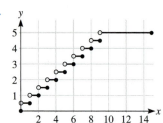

73.

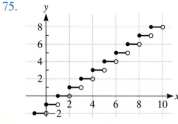

75.

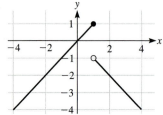

77.

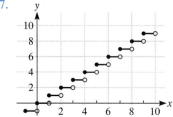

79. *a.* $0, 6, 2, 2$
  b. Domain = all reals;
     range = $\{0, 1, 2, 3, 4, 5, 6, 7, 8, 9\}$.
  c. The graph resembles an ascending staircase.
81. Symmetric with respect to the x-axis when $b = 0$. Symmetric with respect to the y-axis when $a = 0$.

Section 3.6, page 194

1. D_f = set of all real numbers, D_{ff} = set of all real numbers

  D_g = set of all real numbers, $D_{t/g}$ = set of all real numbers except -5

  D_{f+g} = set of all real numbers, $D_{g/f}$ = set of all real numbers except 4

  D_{f-g} = set of all real numbers, $D_{f \circ g}$ = set of all real numbers

  D_{fg} = set of all real numbers, $D_{g \circ f}$ = set of all real numbers

3. D_f = set of all real numbers, D_{ff} = set of all all real numbers

  D_g = set of all real numbers, $D_{f/g}$ = set of all real numbers except 0

  D_{f+g} = set of all real numbers, $D_{g/f}$ = set of all real numbers except 1 and $-\dfrac{3}{2}$

  D_{f-g} = set of all real numbers, $D_{f \circ g}$ = set of all real numbers

  D_{fg} = set of all real numbers, $D_{g \circ f}$ = set of all real numbers

5. -1 **7.** $x^2 - 2x - 4$
9. 72 **11.** undefined
13. $2x^3 + 3x^2 - 2x - 3$ **15.** $\dfrac{2x + 3}{x^2 - 1}$
17. $4x^2 + 12x + 8$ **19.** $2(2x + 3) + 3 = 4x + 9$
21. x, x **23.** x, x **25.** x, x **27.** $|x|, x$
29. x, x **31.** $3, -5$
33. $f(x) = x^5$ **35.** $f(x) = \dfrac{1}{x^4}$
  $g(x) = 4x^3 - 1$ $g(x) = x + 5$
37. $f(x) = \dfrac{x + 1}{x - 1}$ **39.** $f(x) = \left(\dfrac{1 + x}{1 - x}\right)^4$
  $g(x) = x^3$ $g(x) = x^3$
41. $f(x) = \sqrt{x}$, $g(x) = \dfrac{x + 2}{x - 2}$
43. $f(x) = x^{63}$, $g(x) = x^5 + x^4 + x^3 - x^2 + x - 2$

45.

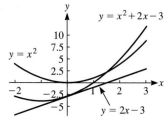

47.

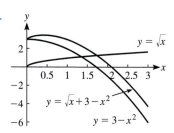

49. *a.* $a(t) = 125t$; *b.* $p(a) = \sqrt{a^2 + 40,000}$
    c. $p(a(t)) = \sqrt{15,625t^2 + 40,000}$
    The distance from the tower after t hours.

51. *a.* $b(t) = 90 - 3t$; *b.* $h(b) = \sqrt{b^2 + 81,000}$
    c. $(h \circ b)(t) = \sqrt{9t^2 - 540t + 16,200}$

53. $P(x) = -0.5x^2 + 40x - 3$

55. $m = \dfrac{1}{3}$, $b = \dfrac{2}{3}$

63.

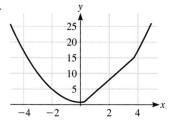

65.

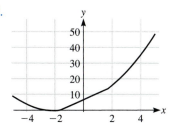

67.

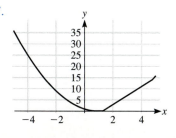

69.

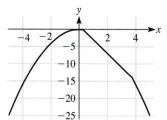

71.

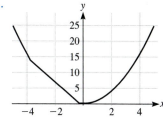

73.

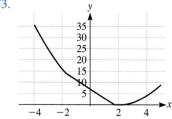

Section 3.7, page 203

1. No.

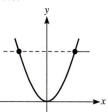

3. No.

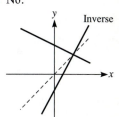

5. No.

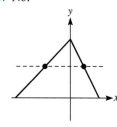

7. $f^{-1}(x) = \dfrac{5}{8} - \dfrac{x}{8}$.

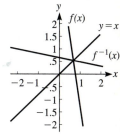

9. No inverse.

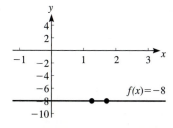

11. No inverse.

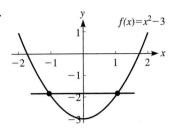

13. $f^{-1}(x) = \sqrt{4 - x^2}$,
    $0 \le x \le 2$.

15. No inverse.

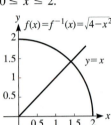

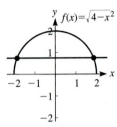

17. Yes.   19. Yes.   21. Yes.
23. No.   25. Yes.   27. No.

29. $f^{-1}(x) = \dfrac{1}{2}(x + 3)$ 31. $f^{-1}(x) = \dfrac{3}{2}\left(x - \dfrac{3}{5}\right)$

33. $f^{-1}(x) = \sqrt{x}, x \ge 0$ 35. $f^{-1}(x) = x^2 - 2, x \ge 0$

37. $f^{-1}(x) = \sqrt[5]{x}$ 39. $f^{-1}(x) = \dfrac{2}{x}$

41. $\dfrac{x + 3}{2 - 3x}$ 43. $f^{-1}(x) = -\sqrt{25 - x^2}, 0 \le x \le 5$

45. $x^3 - 8$

47. $(f \circ f^{-1})(739) = 739, (f^{-1} \circ f)(5.00023) = 5.00023$

51. $\dfrac{x - b}{m}$ 53. No.

Section 3.8, page 208

1. *a.* $y = \dfrac{1}{70}x$; *b.* $y = 5$ 3. *a.* $y = 20x^2$
                                                  b. $y = 273.8$

5. *a.* $y = \dfrac{.98}{x^2}$ 7. *a.* $y = \dfrac{xz}{w}(3.25)$
   b. $y = 98$ *b.* $y = 62.4$

9. *a.* $y = \dfrac{xz}{w^2}(22.62857143)$; *b.* $y = 15,713.28$

11. *a.* $y = \dfrac{z^3}{x^2}\left(\dfrac{2}{9}\right)$; *b.* $y = \dfrac{1}{6}$

13. $A = \pi r^2$
    Constant of variation is π.
    Varies directly as r^2.

15. $d = 65t$
    Constant of variation is 65.
    d varies directly as *t*.

17. 28.16 g   19. 22°C   21. 235.6942 ft.

23. *a.* $s = \sqrt{30d}, d = 53.3333$ ft., $s = 65.03845$ mph
    b. $s = \sqrt{12d}, d = 252.08333$ ft., $s = 49.95998$ mph

25. 360°   27. 66.3158 mph
29. 17.320508 in.   31. -3.59375×10^{-13} dyn

33. *a.* y varies directly as $\sqrt{x}$; *b.* $\dfrac{1}{\sqrt{k}}$

Section 3.9, page 213

1. $\sqrt{34}$ 3. $\dfrac{3}{5}$

5. $\left(x - \dfrac{1}{2}\right)^2 + \left(y + \dfrac{11}{2}\right)^2 = \dfrac{17}{2}$

7. $y + 6 = -\dfrac{3}{2}(x - 5)$

9. Neither.   11. $\left(\dfrac{53}{12}, 0\right)$ 13. $\left(\dfrac{5}{2}\sqrt{2}, -\dfrac{\sqrt{5}}{2}\right)$

15.

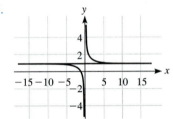

17.

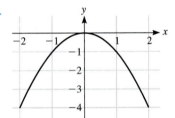

19.

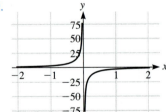

21.

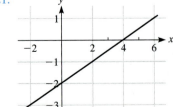

23.

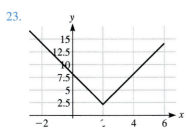

25.

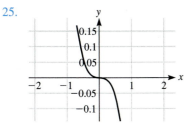

27.

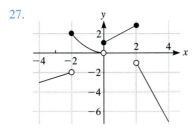

29.

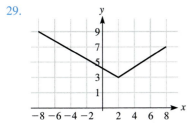

31.

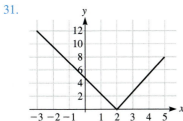

33. Symmetric with respect to origin.
35. Symmetric with respect to x-axis.
37. Symmetric with respect to origin.
39.   41.

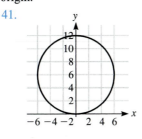

43. Yes.   45. No.   47. -6 49. -4

51. $(a - 1)^2 + 3(a - 1) - 4$

53. $2a + 3 + h$ 55. 1

57. Domain = all reals; range = all reals.

59. Domain = $\{x : x \le 6\}$; range = $[0, \infty]$.

61. Domain = $(-\infty, \infty)$; range = $\{56.7\}$.

63. Domain = $\{x : |x| \le 8\}$; range = $[0, 8]$.

65. Domain = $(-\infty, 0) \cup (0, \infty)$

67. Domain = $[-1, \infty)$.   69. Domain = $(-\infty, \infty)$.

71. Neither.   73. Even.   75. Odd.   77. Even.

79. Even.   81. All real numbers.

83. *a.* $\{x : x \ne \pm 1\}$ 85. *a.* All reals.

     b. $\dfrac{2x + 3}{x^2 - 1}$ *b.* $4x^2 + 12x + 8$

87. $(f \circ g)(x) = \dfrac{9}{x^4} - \dfrac{6}{x^2} + 1$

     $(g \circ f)(x) = \dfrac{3}{(x^2 - 2x + 1)^2}$

89. $(f \circ g)(x) = x^{33}$; $(g \circ f)(x) = x^{33}$

91. $(f \circ g)(x) = \dfrac{(10/x^2) + 7}{(15/x^2) - 4} = \dfrac{10 + 7x^2}{15 - 4x^2}$

     $(g \circ f)(x) = \dfrac{5(3x - 4)^2}{(2x + 7)^2}$

93. $f(x) = \sqrt[5]{x}, g(x) = \dfrac{x^3 + 1}{x^3 - 1}$

95. $f(x) = |x|, g(x) = \dfrac{x^2 - 5}{x^2 + 5}$

97. $3x + 4y = -23$ 99. $y - 3 = 3(x + 2)$

101. $L = \dfrac{R}{d^2}$

     The loudness decreases by a factor of $\dfrac{1}{9}^{\text{th}}$ if d is tripled.

103. $G = N(700 - N)k$ or; $G = (700N - N^2)k$

105. *a.* salary
$$= \begin{cases} 25{,}000 & (S \le 400{,}000) \\ 25{,}000 + .24(S - 400{,}000) & (S > 400{,}000) \end{cases}$$
$$S = \text{amount of sales in dollars}$$

     b.

| 200,000 |
| 175,000 |
| 150,000 |
| 125,000 |
| 100,000 |
| 75,000 |
| 50,000 |
| 25,000 |
| 0 |

400,000

107. V would have to increase by a factor of 1.041976.

109. $.55x^2 + \dfrac{246.4}{x}$ 111. $\dfrac{\pi h^3}{9}$

113. $f(x) = \begin{cases} 3 & (x = -3) \\ 1 & (-3 < x \le 0) \\ \sqrt{x} & (x > 0) \end{cases}$

115. $f \circ g(x) = \begin{cases} (2x-7)^2 & (-3 \leq x \leq 0) \\ 1-x^3 & (0 < x \leq 2) \\ \dfrac{x}{3-x} & (2 < x) \end{cases}$

$g \circ f(x) = \begin{cases} 2x^2-7 & (-3 \leq x \leq 0) \\ (1-x)^3 & (0 < x \leq 2) \\ 3x-5 & (2 < x) \end{cases}$

117. Yes.   119. Yes.

121. $\dfrac{2x+4}{2x+3}$ 123. $f^{-1}(x) = \sqrt[3]{x+1}$

CHAPTER 4

Section 4.1, page 225

1. (b) 3. (d) 5. (c)

7. $y = \dfrac{5}{2}x^2$ 9. $y = -7x^2$

11. $y = -\dfrac{1}{3}(x-2)^2 + 4$

13. *a.* $(0, 0)$; *b.* Downward.

c.
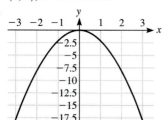

15. *a.* $(2, 0)$; *b.* Upward.

c.
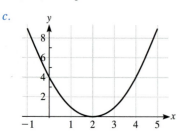

17. *a.* $(3, 0)$; *b.* Upward.

c.
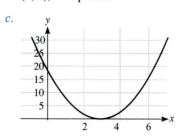

19. *a.* $(-3, -2)$; *b.* Downward;

c.
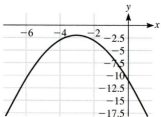

21. *a.* $f(x) = (x+3)^2 - 4$; *b.* $(-3, -4)$;
c. Upward;   min $= -4$;

e.
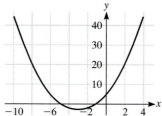

23. *a.* $f(x) = -\dfrac{1}{4}(x+6)^2 + 8$; *b.* $(-6, 8)$;

c. Downward;   max $= 8$;

e.
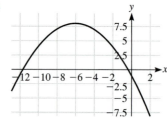

25. max $= 0$ 27. min $= 3$

29. max $= \dfrac{121}{8}$ 31. min $= -\dfrac{169}{12}$

33. min ≈ -435.99 35. max ≈ 5.84

37. 15 and 15   39. $b = h = 83$ cm

41. *a.* $A(x) = -2x^2 + 150x$; *b.* 37.5 yd.; *c.* 2812.5 yd.^{2}

43. Dimensions are 18.5 ft. $\times$ 18.5 ft.; area $= 342.25$ ft.^{2}

45. 1250

47. *a.* $R(x) = 1200 - 10x - x^2$; *b.* 0; *c.* 1200

49. *a.* $A(x) = \dfrac{4+\pi}{16\pi}x^2 - \dfrac{7}{2}x + 49$

b. $\dfrac{28\pi}{4+\pi} \approx 12.32$ in.; *c.* 27.44 in^2

51. *a.* He should wait 5 weeks.; *b.* \$6000

53. $-\dfrac{2}{7}$ 55. -4.5

57. maximum value $= \dfrac{(t^2-4)(t-2)}{4} - 16$

59. $\left(\dfrac{3}{2p}, \dfrac{4p^2-q}{4p} \right)$

61.

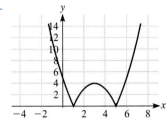

Section 4.2, page 233

1.

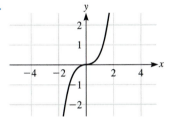

3.

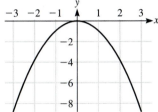

5.

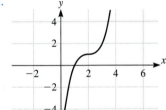

7.

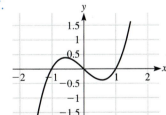

9.

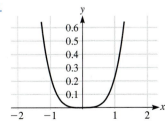

11.

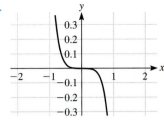

13.

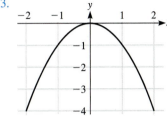

15.

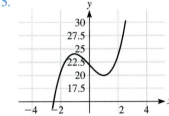

17.

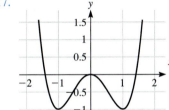

19.

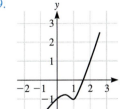

21.

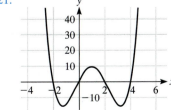

23.

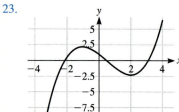

25.

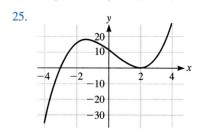

27. *a.* $V(x) = 4x^3 - 32x^2 + 64x$

b.

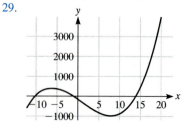

c. $x \approx 1.5$ gives the approximate maximum; max ≈ 38

29.

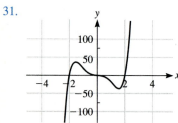

31.

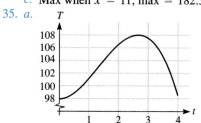

33. *a.* $P(x) = -\dfrac{1}{3}x^3 + 5x^2 + 11x - 100$

b.

c. Max when $x = 11$; max $= 182.\overline{3}$.

35. *a.*

b. max ≈ 108; variable at max ≈ 2.65

37. (*b*)   **39.** (*f*)   **41.** (*e*)

Section 4.3, page 242

1. $q(x) = x^4 + x^2 + 2x + 4, \quad r(x) = 3$
3. $q(x) = 2x^2 - 3x - 7, \quad r(x) = 20$
5. $q(x) = x^3 + \dfrac{1}{3}x^2 + \dfrac{2}{9}x + \dfrac{22}{7}, \quad r(x) = -\dfrac{118}{27}$
7. $q(x) = x^3 - 2x^2 + 3x - 4, \quad r(x) = 6x - 13$
9. $q(x) = x^2 - 5x + 25, \quad r(x) = 0$
11. $q(x) = x^2 - 2, \quad r(x) = 6$
13. $q(x) = x, r(x) = 2x$
15. $q(x) = x^5 + yx^4 + y^2x^3 + y^3x^2 + y^4x + y^5, \quad r(x) = 0$
17. $-\dfrac{5}{2}(x + 1)(x - 2)$ 19. $\dfrac{6}{5}(x + i)(x - i)$
21. $4(x + 1)(x)(x - 1)$
23. $(-2)x(x + 2)(x + 1)(x - i)(x + i)$
25. $\dfrac{2}{25}(x - 1 + \sqrt{2})(x - 1 - \sqrt{2})(x - 4 + 3i)(x - 4 - 3i)$
27. $(x - 2 + i)$ and $(x - 2 - i)$ are its factors.
29. *a.* -31; *b.* -31; *c.* -1; *d.* -1
31. *a.* -2; *b.* -2
33. $q(x) = 4x^4 - 8x^3 + 18x^2 - 36x + 71, r(x) = -137$
35. $q(x) = 2x^2 + 5x - 3, \quad r(x) = 0$
37. $q(x) = x^2 - 5x + 25, \quad r(x) = 0$
39. $q(x) = 3x^3 - 2x^2 - \dfrac{61}{3}x - \dfrac{50}{9}, \quad r(x) = \dfrac{224}{27}$
41. $q(x) = x^5 + x^4y + x^3y^2 + x^2y^3 + xy^4 + y^5, \quad r(x) = 0$
43. $0, 30, 402$ 45. $4, 0, 34$
47. $41, -351, 2353, \dfrac{43}{27}$ 49. Yes; no.
51. Yes; no.   53. Yes; yes.
55. $\{-2, -3, 1, 5\}$ 57. $\{-4, -2, 1, 3\}$
59. $\{-4 < x < -2\} \cup \{1 < x < 3\}$ 61. 1
63. $q(x) = x - 1, r(x) = 2$
65. $q(x) = 4(x+1)^2(x-5)^3 + 3(x-5)^4(x+1), \quad r(x) = 0$
67. *b.* $q(x) = \dfrac{5}{2}x^2 - \dfrac{7}{4}x + \dfrac{9}{8}, r(x) = -\dfrac{83}{8}$

Section 4.4, page 252

1. $6x^2\left[x - \dfrac{5 + i\sqrt{95}}{12}\right]\left[x - \dfrac{-5 - i\sqrt{95}}{12}\right]$
3. $(x - i)(x + i)(x + 1 + \sqrt{2}i)(x + 1 - \sqrt{2}i)$
5. $(x - 1)\left(x - \dfrac{-1 + i\sqrt{3}}{2}\right)\left(x - \dfrac{-1 - i\sqrt{3}}{2}\right)$

7. $(x + 3)(x + \sqrt{5}i)(x - \sqrt{5}i)$

9. $\left(x - \dfrac{-1 + i\sqrt{11}}{2}\right)^2 \left(x - \dfrac{-1 - i\sqrt{11}}{2}\right)^2$

11. 0, multiplicity $= 1$; 2, multiplicity $= 2$; $\dfrac{5}{2}$, multiplicity $= 3$

13. 2, multiplicity $= 2$; -2, multiplicity $= 5$

15. 2, multiplicity $= 3$; -1, multiplicity $= 1$

17. 0, multiplicity $= 1$; 3, multiplicity $= 2$
    i, multiplicity $= 1$; $-i$, multiplicity $= 1$

19. 0, multiplicity $= 5$; $\dfrac{-1 + \sqrt{5}}{2}$, multiplicity $= 3$;
    $\dfrac{-1 - \sqrt{5}}{2}$, multiplicity $= 3$

21. 3 23. 2

25. $\pm 2, 1 \pm \sqrt{3}i, -1 \pm \sqrt{3}i$ 27. $1, \dfrac{-1 \pm \sqrt{3}i}{2}$

29. $-2, 1 \pm \sqrt{3}i$ 31. $\pm \sqrt{2}, \pm \sqrt{2}i$

33. $0, \pm i$ 35. $\pm 2i, \dfrac{1 \pm \sqrt{5}}{2}$

37. $\pm \sqrt{3}, \pm i$ 39. $2, -3$

41. At most 2 positive; at most 2 negative.

43. At most 2 positive; at most 2 negative; at least 6 nonreal.

45. At most 2 positive; at most 2 negative.

47. At most 3 positive; at most 2 negative.

49. At most 2 positive; at most 2 negative, at least 6 nonreal.

51. At most 2 positive; at most 2 negative; 0 a zero of multiplicity 2.

53. $-2 + i$ and -1 55. $0, -1$, and $2 \pm \sqrt{2}$

57. 3 59. $1 - 2i, 1$, and 2

Section 4.5, page 259

1. $2, 3, \dfrac{1}{2}$ 3. None.   5. $\dfrac{4}{5}$ 7. None.

9. Lower bound $= -1$; upper bound $= 2$.

11. Lower bound $= -4$; upper bound $= 1$.

13. Lower bound $= -2$; upper bound $= 1$.

15. Lower bound $= -3$; upper bound $= 3$.

17. $1, \dfrac{1}{3}, \dfrac{2}{3}$ 19. $\pm 5, \dfrac{1}{4}$ 21. $\pm \dfrac{5}{2}$ 23. None.

25. $-2, -\dfrac{1}{3}, \dfrac{1}{2}$ 27. $f(-1) = -1; f(0) = 1$

29. $f(2) = -1; f(3) = 7$ 31. $f(0) = -2; f(1) = 9$

33. $f(-1) = -3; f(-2) = 9; f(1) = -3; f(2) = 9$

35. $-.6823$ 37. 2.2134

39. 0.2346 41. 1.3802 and -0.8191

43. $-0.6985, 1.3645, 1.8186$ 45. $a.$ 25; $b.$ 13

47. $x \approx 7.862$ 49. $t \approx .9507$ 51. $2, -1$

53. $2, \dfrac{-1 + \sqrt{5}}{2}$ 55. $1, 2, -1 \pm \sqrt{2}$

57. $\dfrac{1}{2}, -\dfrac{2}{3}, \pm i$ 59. $4, 2, -\dfrac{1}{2}, -1, \dfrac{3}{2}$

61. $\pm\sqrt{\dfrac{1 \pm \sqrt{53}i}{3}}$ 63. $3, 2, -1$

65. $\pm\sqrt{\dfrac{-3 \pm \sqrt{89}}{2}}$ 67. $0, \pm 2, \pm \dfrac{1}{2}$

Section 4.6, page 271

1. $a.$ $x = 0, y = 0$ 3. $a.$ $x = 0, y = 0$
   $b.$ $b.$

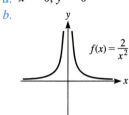
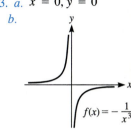

5. $a.$ $x = -1, y = 0$
   $b.$

7. $a.$ $x = -2, y = 0$
   $b.$

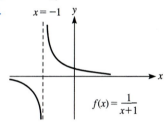

9. $a.$ $x = -3, y = 0$
   $b.$

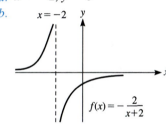

11. $a.$ $y = -2, x = -1$
    $b.$

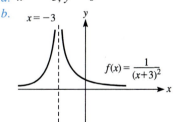

13. *a.* $x = -\dfrac{10}{3}$, $y = -\dfrac{2}{3}$

   b.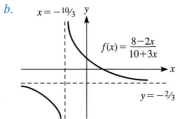

23. *a.* $x = 2$, $x = -1$, $y = 1$

   b.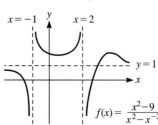

15. *a.* $y = 0$

   b.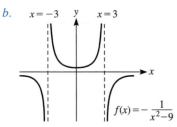

25. *a.* $x = -4$, $y = x - 6$

   b.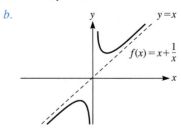

17. *a.* $x = \pm 3$, $y = 0$

   b.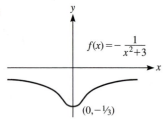

27. *a.* $x = 0$, $y = x$

   b.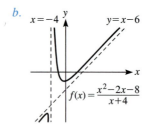

19. *a.* $x = \dfrac{1}{2}$, $x = -3$, $y = 0$

   b.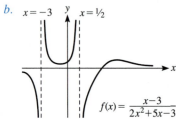

29. *a.* $x = \pm\dfrac{\sqrt{3}}{3}$, $y = \dfrac{1}{3}$

   b.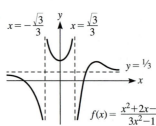

21. *a.* $x = 3$, $y = x + 3$

   b.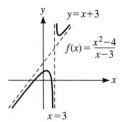

31. *a.* $x = 0$

   b.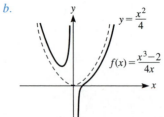

33. *a.* None.

b.

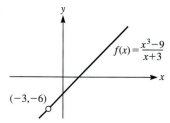

$$f(x) = \frac{x^3 - 9}{x + 3}$$

$(-3, -6)$

35. *a.* None.

b.

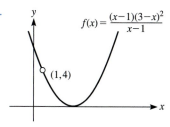

$$f(x) = \frac{(x-1)(3-x)^2}{x-1}$$

$(1, 4)$

37. *a.* None.

b.

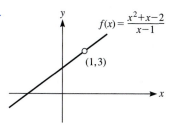

$$f(x) = \frac{x^2 + x - 2}{x - 1}$$

$(1, 3)$

39. *a.* $x = \pm 1$, $y = 4$

b.

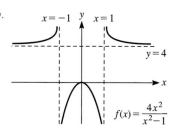

$x = -1$ $x = 1$
$y = 4$

$$f(x) = \frac{4x^2}{x^2 - 1}$$

41. *a.* $y = x + 2$, $x = \pm 5$

b.

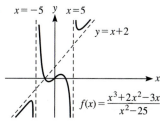

$x = -5$ $x = 5$
$y = x + 2$

$$f(x) = \frac{x^3 + 2x^2 - 3x}{x^2 - 25}$$

43. *a.*

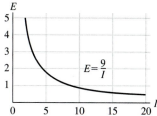

$E = \dfrac{9}{I}$

b.

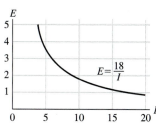

$E = \dfrac{18}{I}$

c.

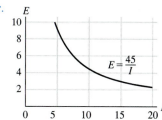

$E = \dfrac{45}{I}$

45.

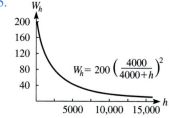

$$W_h = 200 \left(\frac{4000}{4000 + h} \right)^2$$

47. *a.* $W_h = 200 \left(\dfrac{4000}{4000 + h} \right)^2$

b.

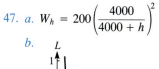

$$L = \frac{0.1}{d^2}$$

c. $4000(\sqrt{2} - 1)$ miles

49.

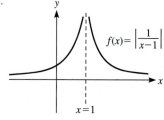

$$f(x) = \left| \frac{1}{x - 1} \right|$$

$x = 1$

51.

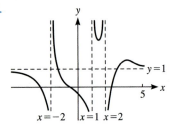

53.

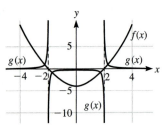

55.

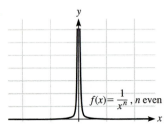

57.

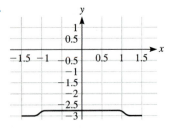

59.

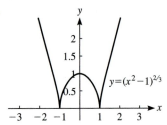

61. (*d*)   63. (*f*)   65. (*e*)

67.

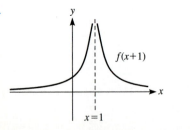

69.

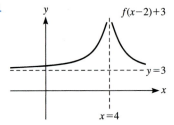

71.

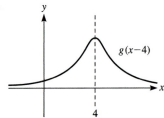

Section 4.7, page 275

1. *a.* $f(x) = \left(x - \dfrac{3}{2}\right)^2 - \dfrac{9}{4}$; *b.* $\left(\dfrac{3}{2}, -\dfrac{9}{4}\right)$;

   c. Upward; *d.* min $= -\dfrac{9}{4}$

   e.

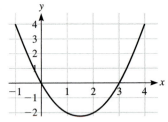

3. *a.* $f(x) = 2(x + 3)^2 - 16$; *b.* $(-3, -16)$

   c. Upward; *d.* min $= -16$

   e.

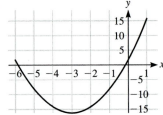

5. *a.* $f(x) = -\dfrac{1}{3}\left(x - \dfrac{3}{5}\right)^2 - \dfrac{17}{25}$; *b.* $\left(\dfrac{3}{5}, -\dfrac{17}{25}\right)$

   c. Downward; *d.* max $= -\dfrac{17}{25}$

   e.

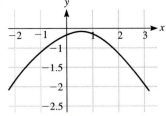

7. $\min = -\dfrac{1198}{3}$

9. $\max = \dfrac{11}{4}\sqrt{2}$

11. $0.35

13.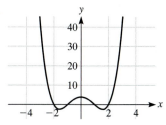

15.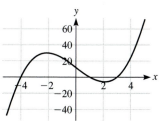

17.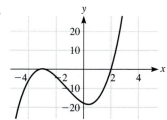

19.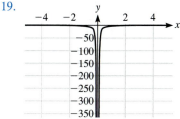

21.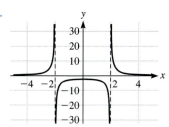

23.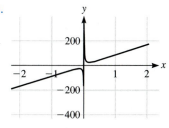

25.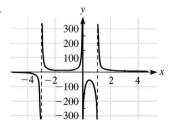

27.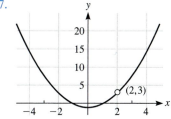

29.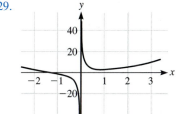

31. $q(x) = x^4 + 4x^3 + 4x^2 + 2x + 8, \quad r(x) = 5$

33. $f(x) = 10x^2 + \dfrac{5}{3}x - \dfrac{10}{3}$

35. $f(x) = \dfrac{10}{9}x^2 - \dfrac{10}{9}x + \dfrac{25}{9}$

37. $f(x) = \dfrac{3}{20}x^5 - \dfrac{3}{20}x^4 - \dfrac{3}{20}x^3 - \dfrac{21}{20}x^2 - 3x - \dfrac{9}{5}$

39. $x^3 - 3x^2 - 2x + 1$ remainder 0

41. $x^3 + (-2 - i)x^2 + (2i - 1)x + i$ remainder 0

43. $-4, \ 41, \ 2 - 2i$

45. Yes; no; no.   47. $-1, 2, 5$

49. 0 of multiplicity 2, 1 of multiplicity 4   51. 5

53. 1 positive real zero, 1 negative real zero, 2 nonreal zeros.

55. At most 3 positive, at most 2 negative real zeros.

57. $\pm\sqrt{\dfrac{3}{2}}, \ -1 + i\sqrt{2}$ 59. $2 \pm 2i\sqrt{3}$

61. $\pm 1, \ \pm 2, \ \pm 3, \ \pm 6, \ \pm\dfrac{1}{2}, \ \pm\dfrac{3}{2}, \ \pm\dfrac{1}{4}, \ \pm\dfrac{3}{4}$

63. $2, -2$ 65. None.   67. None.

69. $1, \ \dfrac{-1 \pm i\sqrt{3}}{2}$ 71. $1 \pm \sqrt{2}, 3$

73. $0, 1 \pm 2\sqrt{2}$ 75. $-1, \ \dfrac{1}{2}, \ 1 \pm i$

77. $f(-1) = 2 > 0, f(0) = -4 < 0$

79. -0.7320 81. $2, \ \dfrac{21 - \sqrt{177}}{4}$

83.

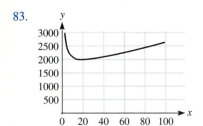

85. x intercepts are $2, 4, -\dfrac{2}{3}$.

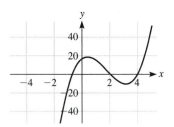

87. No x intercepts.

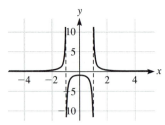

89. x intercept is 0.

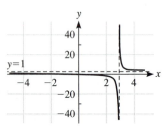

91. $c = 226.12305$

CHAPTER 5

Section 5.1, page 286

1. *a.* 64; *b.* 73.5; *c.* 77.71; *d.* 77.816;
   e. 77.8702; *f.* 77.87995; *g.* 77.880163; *h.* 77.88;
3. *a.* $f(1) = 3; f(2) = 9; f(3) = 27; f(0) = 1$
   $$f(-1) = \frac{1}{3}; f(-2) = \frac{1}{9}; f(-3) = \frac{1}{27};$$
   $$f\left(\frac{1}{2}\right) = \sqrt{3}$$

b.

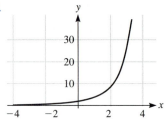

5.

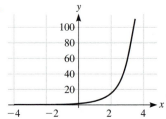

7.

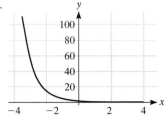

9.

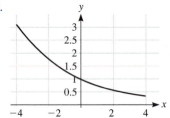

11.

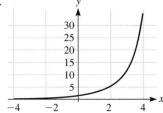

13.

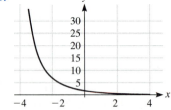

15.

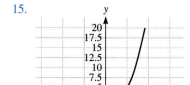

17.

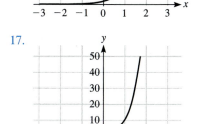

19.

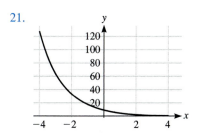

21.

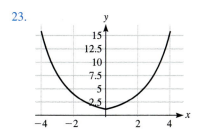

23.

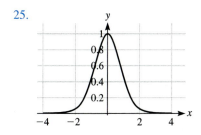

25.

27.

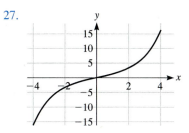

29.

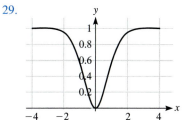

31. a. $10,000(1.08)^t$
   b. 1 yr. $\rightarrow$ \$10,800   3 yr. $\rightarrow$ \$12,597.12
      2 yr. $\rightarrow$ \$11,664   10 yr. $\rightarrow$ \$21,589.25
   c.

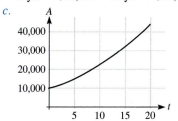

33. a. $A_0(.99997)^t$
   b. 1 yr. $\rightarrow$ 399.988 kg   200 yr. $\rightarrow$ 397.607 kg
      2 yr. $\rightarrow$ 399.976 kg
35. a. $16,370,000(1.0305955)^t$
   b. 23 yr. $\rightarrow$ 32,740,000   10 yr. $\rightarrow$ 22,127,445
      46 yr. $\rightarrow$ 65,480,000   15 yr. $\rightarrow$ 25,726,019
37. a. 2 cm $\rightarrow$ 56.25%   10 cm $\rightarrow$ 5.63%
      3 cm $\rightarrow$ 42.19% x cm $\rightarrow$ $(.75)^x \cdot 100\%$
   b. $(.75)^x \cdot (100\%)$
   c.

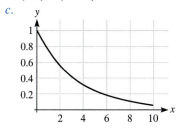

39. (i) 41. (a) 43. (c) 45. (e)
47. (f)

Section 5.2, page 294

1. e^3 3. $e^3 + 2e^7$ 5. 0.9999092
7. $e^2 - 2e + 1$ 9. e^{4x} 11. $e^{2x} - 2 + e^{-2x}$
13. $e^{3x} - 3e^{2x} + 3e^x - 1$ 15. $4e^{2x-2} + 5e^{x-2}$

17. $e^{2x} - 9$ 19. e^x 21. 1.2761926

23. *a.* 1 sec. → 29.003 mph   10 sec. → 138.346 mph
     2 sec. → 52.749 mph   20 sec. → 157.069 mph
     5 sec. → 101.139 mph

b.

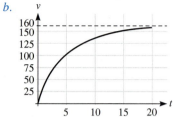

25. $P(t) = .5(1 - e^{-0.08t})$

     a. 1 → .0384   10 → .2753   50 → .4908
       2 → .0739   20 → .399

     b.

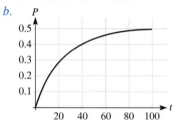

27. $P(t) = 1 - e^{-0.3t}$

     a. 1 → 25.918%   5 → 77.687%   12 → 97.268%
       2 → 45.119%   10 → 95.021%

     b.

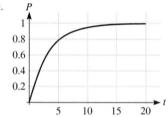

29. $Y(n) = 30(1 - e^{-0.04n})$

     a. 1 → 1.176   10 → 9.89
       2 → 2.3065   20 → 16.52
       5 → 5.438   30 → 20.964
       6 → 6.4   40 → 23.943

     b.

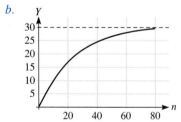

31. $N(t) = \dfrac{540}{1 + 4.4e^{-3t}}$

     a. 10 → 540   30 → 540
       20 → 540   100 → 540

b.

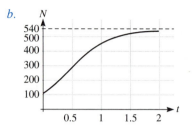

33. *a.* CE; *b.* ≈ 4.722×10^{-5} coulombs

35. $\dfrac{4}{(e^x + e^{-x})^2}$ 37. $\dfrac{(e^x + e^{-x-h})(e^h - 1)}{2h}$

39. $e^{-3x} \cdot \dfrac{e^{-3h} - 1}{h}$

Section 5.3, page 304

1. $2^3 = 8$ 3. $3^5 = 243$ 5. $2^{-3} = \dfrac{1}{8}$ 7. $x^{-12} = 3$

9. $\log_4 x = 2$ 11. $\log_{10} 0.001 = -3$

13. $\log_4 2 = \dfrac{1}{2}$ 15. $\log_x 0.01 = -3$

17. 1   19. 1   21. 0   23. 0

25. 4   27. −3   29. 0.8241   31. 3

33. $2x - 1$ 35. x^4

37. *a.*

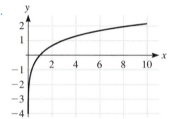

b.

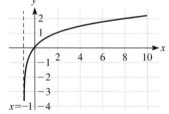

c.

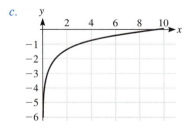

39.

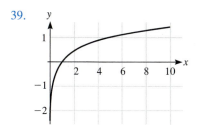

41.

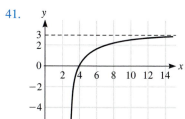

43.

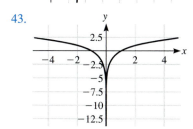

45.

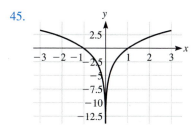

47.

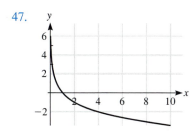

49. $-\log_{10} 3$ 51. $\log_a 3$ 53. 1

55. $\log_a(x + 3)$ 57. 1   59. $\log_a(x - y)$

61. $\log_b x + \dfrac{1}{2} \log_b y - 2 \log_b z$

63. $2 \log_2 x + 2 \log_2 y + 4 \log_2 z$

65. $4 \log_3 x + \log_3 y - 2 \log_3 z$

67. $2 \log_{10} x + \dfrac{2}{3} \log_{10} y - 4 \log_{10} z$

69. $\dfrac{3}{2} \log_{10} x + \dfrac{1}{4} \log_{10} y + \dfrac{5}{4} \log_{10} z$

71. $\log_a(m^3 - n^3) = \log_a(m - n) + \log_a(m^2 + mn + n^2)$

73. $\dfrac{1}{3}\left[\log_a(x - 1) + \log_a(x - 5)\right]$

75. $2 \log_b(x^3 + 8) = 2 \log_b(x + 2) + 2 \log_b(x^2 - 2x + 4)$

77.

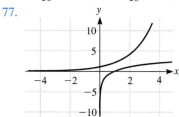

79. .7124   81. $-.2084$ 83. .4354   85. -1

87. 2.0104   89. .41355   91. .32664

93. $-.0360667$ 95. 17   97. 18

99. $-\dfrac{33}{\log_2 0.4} = 24.96354$ days

101. $\dfrac{9}{\log_2(21.5/14.5)} \approx 15.8$ yr.

103. $x = 5$ 105. $x = 0$ 107. $x = 4$

109. $x = 2197$ 111. $x = 6$ 113. $x = \sqrt[3]{5}$

115. $x = \pm\sqrt{5 + \log_5 14}$ 117. $-\dfrac{1}{3}$

119. $x = \log_b 5$ 121. $x = \log_5 7$ 123. -2.5

125. $x = \dfrac{7}{4}$ 127. $\dfrac{-1 \pm \sqrt{7}}{2}$ 129. $x = \pm 3$

131. False.   133. False.   135. False.   137. True.

139. $x \geq 1$ 141. $-3 < x < -2$

143. $x < 0$ 145. $x < 2$

147. $3 \log_a(x + y)$ 151. $\log_3 x$

Section 5.4, page 311

1. 1   3. 3   5. -1 7. -3

9. 1.92153   11. .538573734   13. $-.33913$

15. 6.828144   17. 49.9615   19. -3.0654

21. -22.9164 23. 130 decibels

25. 10^{-15} watts per m^2

29. $I = I_0 \cdot 10^{D/10}$

31. a. 1; b. 2; c. 3; d. 4; e. 5; f. 8

33. The magnitude is 2 greater on the Richter scale.

37. $I = I_0 10^R$

39. a. 1.585×10^{-5} mol/ℓ; b. 6.3096×10^{-5} mol/ℓ

41. $H^+ < 10^{-7}$ 43. 0   45. -7

47. 1   49. $7k$ 51. $\dfrac{1}{625e^8}$

53. 3   55. 128

57. $\ln(x^2)^{x^{-1}} = \ln x^{2/x}$ 59. $\left[\ln(n^{3n})\right]^{-1}$

61. $\dfrac{x}{y^2}$ 63. $\ln 4$ 65. $\ln(18)^{32x \, 2^x}$

67. $\ln \sqrt[3]{\dfrac{3 + \sqrt{10}}{2}}$ 69. 2.523 days

71. a. $1995 \to 260{,}249{,}015$ $2001 \to 274{,}688{,}828$
   b. 77.0163534 yr.

73. *a.* 0.0150964837; *b.* $P(t) = 23,669,000e^{0.015t}$
    c. 32,011,520
75. \$1582.02   77. 28.2%
79. *a.* 9.477121255   81. *a.* \$80,000
    b. 10.477121255   *b.* \$1465.25
83. ≈ 48.008444 85. ≈ 2.010382
87. ≈ -1.5058883 89. $\dfrac{\ln 4578}{\ln 10}$
91. $\dfrac{\ln x^2}{\ln 10}$ 93. $\dfrac{\log 0.987}{\log e}$ 95. $\dfrac{1}{2}\dfrac{\log y}{\log e}$
97. e^{2e^x} 99. 0   101. No solutions.
103. 3.33228   105. (*c*)   107. (*a*)

Section 5.5, page 318

1. $\pm\sqrt{\dfrac{\ln 4}{\ln 7}}$ 3. $\dfrac{2}{5-\ln 4}$ 5. $\dfrac{1}{3e^6-1}$

7. $\ln 100$ 9. $\ln 100$ 11. $\dfrac{\ln 1.845}{10}$

13. $\dfrac{1}{10}\ln\dfrac{5}{3}$ 15. $\dfrac{-3\pm\sqrt{13}}{2}$ 17. $\dfrac{2}{3}$

19. 1   21. 2   23. $\sqrt{41}$

25. $\dfrac{96}{31}$ 27. $\dfrac{1}{8}, 4$ 29. $\dfrac{1+\sqrt{5}}{2}$

31. $10^{1/4}, 10^{-2/3}$ 33. 5   35. 8

37. $\ln 2$ 39. No solutions.   41. $\ln(8+\sqrt{65})$

43. $\dfrac{1}{2}\ln 3$ 45. $9, \dfrac{1}{9}$ 47. 16

49. $1000, \dfrac{1}{1000}$ 51. $3^{1/(\log_2 3-1)}$ 53. $\dfrac{1}{k}\ln\dfrac{P}{P_0}$

55. $-\dfrac{1}{k}\ln\dfrac{P-N}{AN}$ 57. $\ln(x\pm\sqrt{x^2-1})$

59. $\ln\sqrt{\dfrac{1+x}{1-x}}$ 61. $-\dfrac{L}{R}\ln\left(1-\dfrac{RI}{E}\right)$

69. $3e^{-2x/e}$ 71. $e^{-2/3}e^{-4x/3}$ 73. $-4e^{-2x/3}$

Section 5.6, page 320

1. $e^{0.6931} = 2$ 3. $(0.5)^2 = 0.25$
5. $p = e^k$ 7. $\log_2 2 = 1$ 9. $\ln 7.3891 = 2$
11. *a.* $N(t) = 20\cdot 2^{t/23}$
    b. 23 years $\to$ 40 million   1995 $\to$ 27,034,143
       46 years $\to$ 80 million   2000 $\to$ 31,430,689
    c. ≈ 7.4 yr.
13.

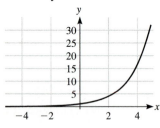

15.

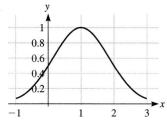

17.

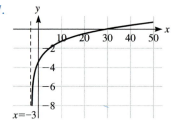

19.

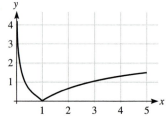

21. 2   23. $\dfrac{\sqrt{2}}{2}$ 25. $\dfrac{1}{4}$ 27. 2

29. 2   31. $-\dfrac{5}{6}$ 33. $-\dfrac{1}{2}\ln 0.01$

35. 5   37. $4, \dfrac{1}{8}$ 39. $10^4, 10^{-4}$

41. $16, \dfrac{1}{16}$ 43. e^{2x} 45. $e^{2x}-e^{2y}$

47. *a.* $t = 0 \to 0, 1 \to 2.083317, 2 \to 3.9496,$
       $5 \to 8.461004, 10 \to 13.343$
    b.

    c. 6.3 yr.
49. $5x - 2$ 51. 1   53. 3   55. 4
57. $\log_a\dfrac{x^3+x^2y}{y^3}$ 59. $\log_a\dfrac{\sqrt[3]{xv^3}}{\sqrt[6]{w}}$
61. $\log_3 x + \log_3(y+z)$ 63. 2
65. b 67. $\dfrac{1}{3}$ 69. $\dfrac{1}{2}$ 71. 18
73. ≈ 40.3817 days   75. 11.086186

77. *a.* $1 \rightarrow 1.4816$ mg $4 \rightarrow 0.60239$ mg
$2 \rightarrow 1.0976$ mg $6 \rightarrow 0.330598$ mg
b. ≈ 2.3105 hr

79. 11.75 yr.

81. *a.* 0.012; *b.* $P(t) = 2,286,000e^{0.012t}$; *c.* 3,276,591

83. $\frac{1}{2} \ln\left(\frac{y+1}{y-1}\right)$ 85. $\sqrt[n]{\frac{y}{a}}$

87. $\frac{\ln(A/P)}{n \ln(1 + i/n)}$ 89. 5.005×10^{-4} 91. $\frac{1}{h} \ln\left(1 + \frac{h}{x}\right)$

CHAPTER 6

Section 6.1, page 331

1. $x = -1, y = \frac{1}{3}$ 3. $x = 5, y = -4$

5. $x = -8, y = 14$

7. $(x, y) = \left(\frac{-6 + \sqrt{71}}{5}, \frac{3 + 2\sqrt{71}}{5}\right),$
$\left(\frac{-6 - \sqrt{71}}{5}, \frac{3 - 2\sqrt{71}}{5}\right)$

9. $(x, y) = \left(\frac{-12 + 2\sqrt{21}}{3}, \frac{-3 + 2\sqrt{21}}{3}\right),$
$\left(\frac{-12 - 2\sqrt{21}}{3}, \frac{-3 - 2\sqrt{21}}{3}\right)$

11. No real solutions.

13. $(x, y) = \left(\frac{-4 + \sqrt{26}}{5}, \frac{-2 + 3\sqrt{26}}{10}\right),$
$\left(\frac{-4 - \sqrt{26}}{5}, \frac{-2 - 3\sqrt{26}}{10}\right)$

15. $(x, y) = (\pm\sqrt{11}, \pm 2\sqrt{3})$ (all four sign combinations)

17. $(x, y) = (-2, -4), (2, 4), (4, 2)$

19. 700 balcony tickets, 900 main floor tickets

21. Aluminum = 13, antimony = 51.

23. \$4000 at 12%, 1000 at 6%

25. 4, 6   27. 0

29. 614 or 423   31. 20 cm $\times$ 34 cm

33. 5 and 1 or -5 and -1 35. \$120

37. Fixed cost = \$100; cost per guest = \$1.25.

Section 6.2, page 335

1. $x = 1, y = 2$ 3. $x = 7, y = -6$

5. No solutions.   7. $x = -\frac{14}{3}, y = \frac{9}{4}$

9. $x = \frac{26}{11}, y = -\frac{28}{11}$ 11. $x = \frac{5}{3}, y = \frac{2}{5}$

13. Length = 24 m; width = 18 m.

15. 170 dimes, 80 quarters

17. Old mower = 6 hours; new mower = 4 hours.

19. 2 mph   21. $x = a + b, y = a - b$

Section 6.3, page 345

1. $x = 1, y = 2, z = 1$

3. $x = 5, y = -1, z = -2$ 5. $x = -4, y = 2, z = 3$

7. $\begin{bmatrix} -1 & 2 & 3 \\ 4 & 1 & 0 \\ 0 & -1 & 5 \end{bmatrix}$ 9. $\begin{bmatrix} 4 & 1 & 0 \\ -(1/2) & 1 & 3/2 \\ 0 & -1 & 5 \end{bmatrix}$

11. $\begin{bmatrix} 4 & 1 & 0 \\ 9 & 0 & -3 \\ 0 & -1 & 5 \end{bmatrix}$ 13. $\begin{bmatrix} 4 & 1 & 0 \\ -1 & 2 & 3 \\ -6 & 2 & 11 \end{bmatrix}$

15. $\begin{bmatrix} 1 & 2 & 0 & 3 \\ 0 & 1 & -(1/3) & 4 \\ 0 & 0 & 1 & -(33/7) \end{bmatrix}$

17. $\begin{bmatrix} 1 & 1 & 2 \\ 0 & 0 & 1 \end{bmatrix}$ 19. $\begin{bmatrix} 1 & 3 & 2 & 1 \\ 0 & 1 & 1 & 0 \\ 0 & 0 & 1 & 1 \end{bmatrix}$

21. $\begin{bmatrix} 1 & 5 & 6 \\ 0 & 1 & 1 \end{bmatrix}$ 23. $\begin{bmatrix} 1 & 2 & -3 & -15 \\ 0 & 1 & 2 & 11 \\ 0 & 0 & 1 & 5 \end{bmatrix}$

25. $\begin{bmatrix} 1 & 2 & 3 & 0 \\ 0 & 1 & 1 & 1 \\ 0 & 0 & 1 & -(5/3) \end{bmatrix}$

27. $x = -\frac{17}{3}, y = \frac{28}{3}, z = -\frac{13}{3}$

29. $x = 2, y = -1, z = 0$

31. $x = 2 - \frac{7z}{5}, y = -1 + \frac{3z}{5}, z = $ any real number

33. $x = -1, y = 2, z = -2$

35. $x = 1, y = -2, z = 3, w = -4$

37. $x = -\frac{1}{3}, y = \frac{8}{3}, z = -\frac{5}{3}$

39. $x = 2y + 4, y = $ any number

41. $x = \frac{2}{11}z + 1, y = \frac{7}{11}z + 2, z = $ any number

43. $x = 1, y = 0, z = -1, w = 2$ 45. $x = 1, y = 2$

47. \$12,000 at 8%, \$15,000 at 6.9%, \$23,000 at 12%

49. 80 Sonys, 5 Sanyos

Section 6.4, page 352

1.
3.
5.

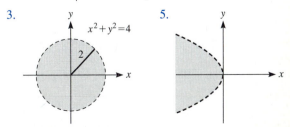

7.

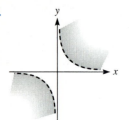

9.

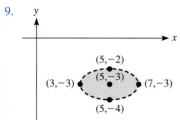

11.

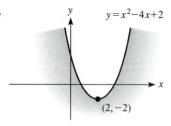

13.

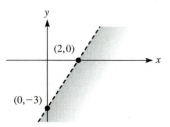

15.

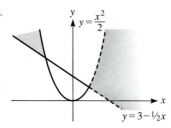

17.

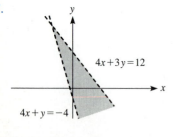

19.

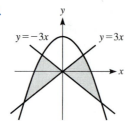

21.

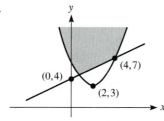

23.

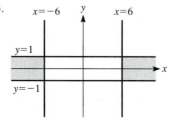

25.

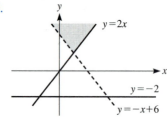

27.

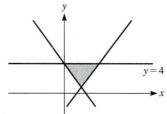

29.

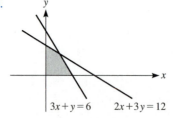

31.

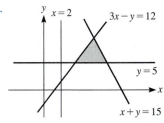

33. *a.* x = recreational skis, y = racing skis

$x \le 60, y \le 45, 3x + 4y \le 240, x \ge 0, y \ge 0$

b.

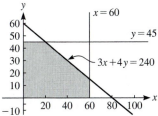

35. *a.* $x \ge 0, y \ge 0, x + y \le 16, 3x + 6y \le 60$

b.

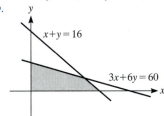

37. *a.* $2w + t \ge 60, t \ge 0, w \ge 0$

b.

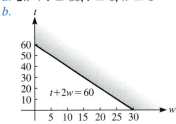

39. *a.* $0 \le L \le 74, 0 \le w \le 50$

b.

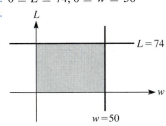

Section 6.5, page 355

1. $(x, y) = (4, 8)$ maximum value = 52

$(x, y) = (1, 1)$ minimum value = 8

3. $(x, y) = (0, 10)$ minimum value = 30

5. $(x, y) = \left(\dfrac{5}{2}, 6\right)$ maximum value = 97

$(x, y) = \left(\dfrac{37}{7}, \dfrac{16}{7}\right)$ minimum value = $\dfrac{562}{7}$

7. $(x, y) = (7, 3)$ maximum value = 5

$(x, y) = (4, 6)$ minimum value = -10

9. $(x, y) = (5, 5)$ maximum value = 30

$(x, y) = (1, 2)$ minimum value = 7

11. $(x, y) = (8, 0)$ maximum value = 64

$(x, y) = (0, 0)$ minimum value = 0

13. five 19-inch and four 13-inch televisions give a profit of $460

15. 464

17. Deluxe = 300; top = 200;
maximum profit = $1300/week

19. 134.5 acres in corn, 177.5 acres in beans, maximum profit = $50,248.50

21. 818 ft.^{2}

23. 100 units of lumber, 300 units of plywood, maximum profit = $11,000

Section 6.6, page 358

1. $x = \dfrac{71}{17}, y = -\dfrac{26}{17}$ **3.** $(x, y) = (4, -3), (-3, 4)$

5. $(x, y, z) = (1, 2, 3); \left(\dfrac{1 + \sqrt{5}}{2}, 1 + \sqrt{5}, \dfrac{9 - 3\sqrt{5}}{2}\right);$

$\left(\dfrac{1 - \sqrt{5}}{2}, 1 - \sqrt{5}, \dfrac{9 + 3\sqrt{5}}{2}\right)$

7. 20 dimes and 10 quarters **9.** 3 and 11

11. $x = 7, y = 3, z = -1$ **13.** $x = 1, y = 1$

15. $x = -6 + \dfrac{3y}{2}, y = $ any real number

17. $14,000 at 12.5% and $46,000 at 14%

19.

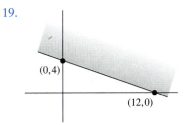

21.

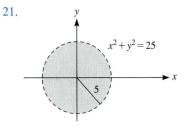

23.

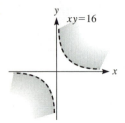

25.

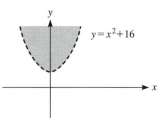

27.

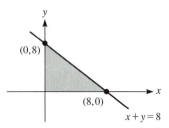

29.

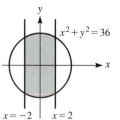

31.

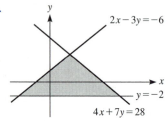

33.

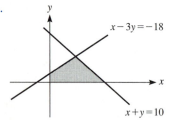

35. $(3, 1)$, minimum $= 9$; $(8, 1)$, maximum $= 19$

37. Minimum value 9/8 at $x = 0$, $y = 3/8$, maximum value 107/9 at $x = 25/9$, $y = 19/9$.

39. $(2, 6)$, maximum $= 26$; $(3, 1)$, minimum $= 7$

41. 17 plain drawers, 13 fancy drawers

CHAPTER 7

Section 7.1, page 365

1. $\dfrac{2}{x-1} + \dfrac{4}{x+1}$

3. $\dfrac{1}{x} + \dfrac{2}{x+1} - \dfrac{3}{x+2}$

5. $\dfrac{4/3}{x-2} + \dfrac{13/3}{2x-1}$

7. $-\dfrac{1}{x-4} + \dfrac{2x-9}{x^2+1}$

9. $\dfrac{4}{x-3} + \dfrac{1}{3x+2}$

11. $1 + 3x + \dfrac{2}{x-1} + \dfrac{2}{2+x}$

13. $3x + 1 + \dfrac{4}{x-2} + \dfrac{2}{x-1}$

15. $\dfrac{4}{x-3} + \dfrac{1}{x+2}$

17. $\dfrac{2}{x} - \dfrac{4}{x-1} + \dfrac{7}{(x-1)^2}$

19. $\dfrac{3}{2x-1} - \dfrac{2}{x+2} + \dfrac{10}{(x+2)^2}$

21. $1 + 3x$

23. $\dfrac{x+1}{x^2+2} - \dfrac{2x}{(x^2+2)^2}$

25. $2x + 3 + \dfrac{5}{3x+1} + \dfrac{1}{x-1}$

27. $\dfrac{3}{x+2} - \dfrac{2}{(x+2)^2} + \dfrac{1}{(x+2)^3}$

29. $\dfrac{5}{x+1} - \dfrac{5}{x+2} - \dfrac{5}{(x+2)^2}$

31. $\dfrac{1}{4a^2(x+a)} + \dfrac{1}{4a^2(x-a)} - \dfrac{x}{2a^2(a^2+x^2)}$

33. $\dfrac{6}{x+1} - \dfrac{5}{x-1}$

35. $2 - \dfrac{1}{x} + \dfrac{3}{x+1}$

37. $x = -\dfrac{1}{9}, y = \dfrac{1}{9}, z = \dfrac{14}{9}$

Section 7.2, page 371

1. $\begin{bmatrix} 1 & 3 \\ 1 & 1 \end{bmatrix}$

3. $\begin{bmatrix} 2 & 6 \\ 4 & 12 \end{bmatrix}$

5. $\begin{bmatrix} -5 & 1 \\ 1 & -2 \end{bmatrix}$

7. $\begin{bmatrix} 7 & 0 \\ 2 & 9 \end{bmatrix}$

9. $\begin{bmatrix} 0 & 3 \\ 0 & -1 \end{bmatrix}$

11. $\begin{bmatrix} -3 & 3 \\ 0 & 0 \end{bmatrix}$

13. $\begin{bmatrix} 5 & -15 \\ 0 & 10 \end{bmatrix}$

15. $\begin{bmatrix} -2 & 6 \\ -1 & -3 \end{bmatrix}$

17. $\begin{bmatrix} 4 & 2 \\ -1 & 0 \end{bmatrix}$

19. $\begin{bmatrix} -14 & -4 \\ -1 & -2 \end{bmatrix}$

21. $\begin{bmatrix} 13 \\ 25 \end{bmatrix}$

23. $\begin{bmatrix} -10 & -7 & -9 \\ -2 & -19 & -14 \end{bmatrix}$

25. Undefined.

27. $\begin{bmatrix} -1 & -3 & 14 \\ 6 & 9 & 16 \end{bmatrix}$

29. Undefined.

31. $\begin{bmatrix} -3 & 2 & 18 \end{bmatrix}$

33. Undefined.

35. 3×5

37. Undefined.

39. 1×3

41. $x = 3, y = 1, z = 4$

43. $x = 18, y = 3, z = 3, w = 3$

Section 7.3, page 380

1. -7 3. -5 5. -14 7. -13
9. -5 11. -14 13. 8 15. -6
17. 0 19. 720 21. -410 23. 9072
25. -153 27. $x = 3, y = 2$
29. $x = -6, y = 1$ 31. $x = 1, y = 2$
33. Zero determinant, Cramer's rule not applicable.
35. $x = -\dfrac{1}{3}, y = \dfrac{8}{3}, z = -\dfrac{5}{3}$
37. $x = 2, y = -1, z = 0$
39. Nonsquare matrix. Cramer's rule not applicable.
41. $x = 0, y = 2, z = -2, w = 1$
43. $\begin{vmatrix} x & y & 1 \\ 3 & 4 & 1 \\ 4 & 8 & 1 \end{vmatrix} = 0$ 45. $\begin{vmatrix} x & y & 1 \\ 0 & 0 & 1 \\ 5 & 2 & 1 \end{vmatrix} = 0$
47. No.   49. No.   51. 1 53. $\dfrac{37}{2}$
55. 12 57. $a = \pm 4, -1$
59. $-10x - 2y - 4z$

Section 7.4, page 386

1. $\begin{bmatrix} 7/16 & 5/32 \\ -(1/16) & -(3/32) \end{bmatrix}$ 3. $\begin{bmatrix} -1 & -2 \\ -(1/2) & -(3/2) \end{bmatrix}$
5. Inverse doesn't exist.
7. $\begin{bmatrix} 7/15 & -(1/5) & 1/15 \\ -(2/3) & 0 & 1/3 \\ -(3/5) & 2/5 & 1/5 \end{bmatrix}$
9. $\begin{bmatrix} 15 & 4 & -5 \\ -12 & -3 & 4 \\ -4 & -1 & 1 \end{bmatrix}$
11. $\begin{bmatrix} 3/8 & -(1/4) & 1/8 \\ -(1/8) & 3/4 & -(3/8) \\ -(1/4) & 1/2 & 1/4 \end{bmatrix}$
13. $\begin{bmatrix} 1/2 & 1/2 & -(1/4) & 1/2 \\ -1 & 4 & -(1/2) & -2 \\ -(1/2) & 5/2 & -(1/4) & -(3/2) \\ 1/2 & -(1/2) & 1/4 & 1/2 \end{bmatrix}$
15. Yes.   17. Yes.   19. Yes.
21. $\begin{bmatrix} x \\ y \end{bmatrix} = \begin{bmatrix} 4 \\ 1 \end{bmatrix}$ 23. $\begin{bmatrix} x \\ y \end{bmatrix} = \begin{bmatrix} 61/19 \\ 6/19 \end{bmatrix}$
25. $\begin{bmatrix} x \\ y \end{bmatrix} = \begin{bmatrix} 1/2 \\ -1 \end{bmatrix}$ 27. $\begin{bmatrix} x \\ y \end{bmatrix} = \begin{bmatrix} -2 \\ 5 \end{bmatrix}$
29. $\begin{bmatrix} x \\ y \\ z \end{bmatrix} = \begin{bmatrix} -(9/4) \\ 15/4 \\ 10 \end{bmatrix}$ 31. $\begin{bmatrix} x \\ y \\ z \end{bmatrix} = \begin{bmatrix} 0 \\ 1 \\ -1 \end{bmatrix}$
33. $A^{-1} = \begin{bmatrix} a^{-1} & 0 \\ 0 & b^{-1} \end{bmatrix}$ provided $ab \neq 0$

Section 7.5, page 389

1. $-\dfrac{4}{a} + \dfrac{1}{a+1} - \dfrac{3}{(a+1)^2}$
3. $-\dfrac{4}{(3x-1)} + \dfrac{5}{2x-1}$ 5. $\begin{bmatrix} 23 & 13 & 26 \\ 4 & 15 & -1 \\ -5 & -5 & -6 \end{bmatrix}$

7. $\begin{bmatrix} 25 & 20 \\ 1 & 3 \end{bmatrix}$ 9. $\begin{bmatrix} 18 & 13 & 19 \\ 1 & -7 & 4 \end{bmatrix}$
11. $\begin{bmatrix} 6 \\ -9 \end{bmatrix}$ 13. $\begin{bmatrix} 0 & 0 & 0 \\ 12 & 3 & 0 \end{bmatrix}$
15. Not defined.   17. 1×1
19. $x = 3, y = 3, w = 7, z = 2$
21. 5 23. -13 25. -5 27. -85
29. -33 31. 157
33. 9 35. $\begin{bmatrix} 2 & -5 \\ -1 & 3 \end{bmatrix}$
37. No inverse exists.   39. $\begin{bmatrix} 1 & -2 & 3 & 8 \\ 0 & 1 & -3 & 1 \\ 0 & 0 & 1 & -2 \\ 0 & 0 & 0 & -1 \end{bmatrix}$
41. Not inverses.   43. $x = 5, y = -3$
45. $x = -1, y = 5, z = -3$

CHAPTER 8

Section 8.1, page 399

1. 1 3. $40,320$ 5. 15 7. 21
9. 6 11. 126 13. $35x^4y^3$
15. $3^{10} \cdot 3003t^5$ 17. $5376x^3t^9$
19. $504t^{5/2}, -504\sqrt{2}t^2$ 21. b^{200}
23. $p^5 - 5p^4q + 10p^3q^2 - 10p^2q^3 + 5pq^4 - q^5$
25. $a^{10} + 15a^8b + 90a^6b^2 + 270a^4b^3 + 405a^2b^4 + 243b^5$
27. $\dfrac{t^8}{256}$
29. $256 + 512t + 448t^2 + 224t^3 + 70t^4 + 14t^5$
$+ \dfrac{7}{4}t^6 + \dfrac{1}{8}t^7 + \dfrac{1}{256}t^8$
31. $\dfrac{81a^8}{16b^4} - \dfrac{9a^5}{b} + 6a^2b^2 - \dfrac{16b^5}{9a} + \dfrac{16b^8}{81a^4}$
33. $a^{15} - 5a^9 + 10a^3 - 10a^{-3} + 5a^{-9} - a^{-15}$
35. $\dbinom{n}{0} + \dbinom{n}{1} + \dbinom{n}{2} + \cdots + \dbinom{n}{n}$
37. $56 - 32\sqrt{3}$
39. $a^4 - 4a^3\sqrt{3} + 18a^2 - 12\sqrt{3}a + 9$
41. $1024a^{10} - 5120a^9b + 11,520a^8b^2 - 15,360a^7b^3$
43. $x^{-4} - 24x^{-3/2} + 252x$
45. $\dfrac{-5940x^3}{y^9} + \dfrac{594x^2}{y^{10}} - \dfrac{36x}{y^{11}} + \dfrac{1}{y^{12}}$
47. $\dbinom{12}{8}\left(\dfrac{3a^{-2}}{2}\right)^4 \cdot \left(\dfrac{a}{3}\right)^8 = \dfrac{55}{144}$
49. 1.18428 by binomial theorem, 1.18430 by calculator
51. $\dbinom{50}{10}(.02)^{10}(.98)^{40}$
53. a. $a^7 + 7a^6b + 21a^5b^2 + 35a^4b^3 + 35a^3b^4 + 21a^2b^5$
$+ 7ab^6 + b^7$
b. $a^8 + 8a^7b + 28a^6b^2 + 56a^5b^3 + 70a^4b^4 + 56a^3b^5$
$+ 28a^2b^6 + 8ab^7 + b^8$
c. $a^9 + 9a^8b + 36a^8b^2 + 84a^6b^3 + 126a^5b^4$
$+ 126a^4b^5 + 84a^3b^6 + 36a^2b^7 + 9ab^8 + b^9$

d. $a^{10} + 10a^9b + 45a^8b^2 + 120a^7b^3 + 210a^6b^4$
   $+ 252a^5b^5 + 210a^4b^6 + 120a^3b^7 + 45a^2b^8 + 10ab^9$
   $+ b^{10}$

55. $5x^4 + 10x^3h + 10x^2h^2 + 5xh^3 + h^4$

57. $[a + (b + c)]^n = a^n + na^{n-1}(b + c)$

   $+ \binom{n}{2}a^{n-2}(b + c)^2 + \cdots + (b + c)^n$

59. $1 - x + x^2 + \cdots - x^7$

61. $1 + \frac{1}{2}x - \frac{1}{8}x^2 + \frac{1}{16}x^3 - \frac{5}{128}x^4 + \frac{7}{256}x^5 - \frac{21}{1024}x^6$
   $+ \frac{33}{2048}x^7$

63. $\frac{2(n + 2)^2}{n^2(n + 1)}$

Section 8.3, page 410

1. $3, 4, 5, 10$
3. -2, undefined, $4, \frac{3}{2}$

5. $\frac{1}{4}, \frac{1}{2}, 1, 32$
7. $1, \frac{1}{4}, \frac{1}{9}, \frac{1}{64}$

9. $-2, -2\sqrt{2}, -2\sqrt[3]{3}, -2\sqrt[8]{8}$

11. $-2, -6, -18, -4374$

13. $0, \frac{3}{2}, \frac{8}{7}, \frac{63}{62}$
15. $1, -2, 3, -8$

17. $1, 0, -1, 0$
19. $2, 7, 12, 17, 22$

21. $9, 3, 1, \frac{1}{3}, \frac{1}{9}$
23. $2, \frac{5}{2}, \frac{10}{3}, \frac{17}{4}, \frac{26}{5}$

25. $-\frac{1}{2}, \frac{1}{8}, -\frac{1}{24}, \frac{1}{64}, -\frac{1}{160}$
27. $\frac{3}{2}, \frac{3}{2}, \frac{3}{2}, \frac{3}{2}, \frac{3}{2}$

29. $3, \frac{1}{3}, 3, \frac{1}{3}, 3$
31. $-2, -4, -8, -16, -32$

33. 2
35. 17
37. $\frac{13}{12}$
39. 56

41. 0
43. 150
45. 10

47. $a_n = 2n - 1$
49. $a_n = (-1)^{1+n}2^n$

51. $a_n = 2^{-(2n-1)}$
53. $\sum_{n=1}^{5}\frac{1}{n}$
55. $\sum_{n=1}^{10}n$

57. $\sum_{n=1}^{6}(-3)^n$
59. $x = 2$
61. $x = 3$
63. $x = 2$

Section 8.4, page 415

1. Difference is 2.
3. Difference is 3.

5. Difference is 1.
7. Difference is -2.

9. -25
11. 261
13. -19
15. -165

17. -16.5
19. $a_1 = -\frac{275}{4}, d = \frac{19}{4}$

21. $a_1 = 114, d = -7$
23. $a_1 = \frac{3}{5}, d = \frac{3}{5}$

25. $a_1 = 8, d = -3$
27. $a_1 = -10, d = 4$

29. 5
31. -3
33. 21
35. -11

37. 30
39. 440
41. -532
43. 561

45. 35
47. -1005
49. 4185
51. 136

53. $\$54.85$
55. $n = 6$
57. $a_{81} = 45$

59. $\sum_{d=0}^{3}(4 + 3d)$
61. $\sum_{d=0}^{6}(-9 + 6d)$

Section 8.5, page 423

1. $\frac{a_{n+1}}{a_n} = -\frac{1}{3}$
3. $\frac{a_{n+1}}{a_n} = 4$
5. $\frac{a_{n+1}}{a_n} = \frac{1}{3}$

7. $\frac{1}{1024}$
9. 8
11. $\frac{256}{81}$

13. 6144
15. 6144
17. $-\frac{3}{10}$
19. $\frac{2}{81}$

21. $\$28,809.07$
23. $\$11,865.23$
25. 1024

27. $322,960$
29. 5
31. 4

33. $\frac{5}{3}, 15, -45, 135$ or $-\frac{5}{3}, -15, -45, -135$

35. $n = 4, a_n = -40$
37. $n = 5, S_n = 484$

39. $a_n = 324, a_1 = 4$
41. $a_1 = 6, S_n = 762$

43. $a_n = -\frac{2}{27}, S_n = \frac{3280}{27}$
45. $a_1 = 3, n = 3$

47. $a_n = (12) \cdot \left(-\frac{1}{3}\right)^{n-1}$

49. $r = -\sqrt[3]{131}, a_n = -\frac{50(\sqrt[3]{-131})^n}{131}$

51. $\$101,578.77$
53. $\$581,826.07$
55. $\frac{2}{3}$

57. -5
59. 1
61. Sum does not exist.

63. $-\frac{1}{2}$
65. 30
67. 8

69. $93\frac{1}{3}$ ft.
71. $\$0.95$
73. $\$30,146.97$

Section 8.6, page 431

1. 156
3. $20,000$
5. $17,576,000$

7. 720
9. 32
11. 24

13. 30
15. 3

17. $n(n - 1)(n - 2)(n - 3)(n - 4)$

19. 210
21. 3024
23. 15

25. $311,875,200$
27. 12
29. 56

31. 5
33. 45
35. n
37. 1

39. $\frac{12!}{5!3!2!2!} = 166,320$
41. 56

43. 24
45. $696,729,600$
47. 16

49. a. 36; b. 100; c. 24
51. 4800

53. $30,030$
55. a. 21; b. 11; c. 6

59. 8
61. No solutions.

Section 8.7, page 436

1. $\{(H, 1), (H, 2), (H, 3), (H, 4), (H, 5), (H, 6), (T, 1), (T, 2), (T, 3), (T, 4), (T, 5), (T, 6)\}$

3. $\{MM, MF, FF, FM\}$

5. $\{2, 3, 4, 5, 6, 7, 8, 9, 10, 11, 12\}$

7. $_{52}C_5 = 2,598,960$ hands
9. 729

11. Ordered pairs of the numbers: {653, 654, 623, 624, 153, 154, 123, 124}; 64 elements in the sample space.

13. {WI, WL, WA, WM, IW, IL, II, IA, IM, LW, LI, LL, LA, LM, AW, AI, AL, AM, MW, MI, ML, MA}

15. {GGG, BGG, BBG, BBB}

17. {PN, PD, PQ, NP, ND, NQ, DN, DP, DQ, QP, QN, QD}

19. {3, 4, . . . , 18}

21. $\dfrac{3}{5}$ 23. $\dfrac{1}{8}$ 25. $\dfrac{33}{16,660}$ 27. $\dfrac{1}{9}$

29. $\dfrac{4}{465}$ 31. $\dfrac{1}{15}$ 33. $\dfrac{7}{30}$ 35. $\dfrac{1}{200}$

37. $\dfrac{1}{312}$ 39. $\dfrac{1}{5}$ 41. $\dfrac{9}{10}$

Section 8.8, page 438

1. $a^4 + 4a^3t + 6a^2t^2 + 4at^3 + t^4$

3. $2187a^7b^{14} - 20{,}412a^8b^{13} + 81{,}648a^9b^{12}$
$- 181{,}440a^{10}b^{11} + 241{,}920a^{11}b^{10} - 193{,}536a^{12}b^9$
$+ 86{,}016a^{13}b^8 - 16{,}384a^{14}b^7$

5. -252

13. $a.$

n	1	2	3	4	5	6
$f(n)$	2	7	15	26	40	57

$b.$ 126; $c.$ 126;

$d.$ $f(n) = 2n + \dfrac{3}{2}n(n-1)$

17. $-1, 0, 1, 8$ 19. $2, \dfrac{4}{3}, 1, \dfrac{4}{11}$

21. $a_n = n(n+1)$, $a_1 = 2$, $a_2 = 6$, $a_3 = 12$, $a_4 = 20$, $a_5 = 30$

23. $a_1 = 2$, $a_2 = \dfrac{3}{2}$, $a_3 = 2$, $a_4 = \dfrac{3}{2}$, $a_5 = 2$

25. 14   27. $\dfrac{5269}{3600}$

29. Common difference = 1.

31. Common difference = 3.

33. 17   35. -14

37. $d = 5$, $a_1 = -34$ 39. $d = -3$, $a_1 = -5$

41. 16   43. 22   45. 150

47. -1120 49. 1   51. 43,500

53. Common ratio = $\dfrac{1}{4}$.   55. Common ratio = 2.

57. 2^{19} 59. 243   61. 2^{n+1}

63. The fourth term.   65. $2960.49

67. $a_1 = 4$, $s_5 = 484$ 69. $a_n = (-2) \cdot (-5)^{n-1}$

71. $\dfrac{20}{3}$ 73. 72   75. $46\dfrac{2}{3}$ ft.   77. 120

79. 6,760,000   81. 6

83. $9! = 362{,}880$ 85. 120

87. 900 (0 cannot be first digit)

89. 210   91. 84   93. 330   95. 27,405

97. {HHHHH, HHHHT, HHHTH, . . . , TTTTT}, 32 sample points

99. {BBB, BBR, BRB, RBB, BRR, RBR, RRB, RRR}

101. $\dfrac{4}{9}$ 103. $\dfrac{3}{5}$

CHAPTER 9

Section 9.1, page 447

1. vertex: (0, 0)
   focus: (0, 1/36)
   directrix: $y = -(1/36)$

3. vertex: (0, 0)
   focus: $(-1/2, 0)$
   directrix: $x = 1/2$

5. vertex: $(1, -1)$
   focus: $(1, -7/8)$
   directrix: $y = -9/8$

7. vertex: $(0, -1/2)$
   focus: $(0, -1)$
   directrix: $y = 0$

9. vertex: $(-1/3, -3)$
   focus: $(-(1/3), -(26/9))$
   directrix: $y = -(28/9)$

11. vertex: $(-41/24, -3/2)$
    focus: $(-(5/24), -3/2)$
    directrix: $x = -77/24$

13. vertex: $(-(1/5), 4)$
    focus: $(-(1/5), (401/100))$
    directrix: $y = (399/100)$

15. $y^2 = -8x$ 17. $(x-2)^2 = 8(y-1)$

19. $y^2 = 4x$ 21. $(x-3)^2 = 24(y+2)$

23. $(x-1)^2 = 8y$ 25. $(y+4)^2 = -8(x-1)$

27. $\left(x - \dfrac{1}{4}\right)^2 = \dfrac{5}{4}\left(y + \dfrac{1}{20}\right)$

29. $\left(y - \dfrac{71}{6}\right)^2 = -\dfrac{140}{3}\left(x - \dfrac{5041}{1680}\right)$

31. $(x - 2600)^2 = 8000y$

33. 225 inches   35. $y^2 = 8x$

37. $y - 8 = -8(x-1)$ 39. $y = 6x$

41. (b) 43. (c) 45. (d)

Section 9.2, page 455

1. $\dfrac{x^2}{5} + \dfrac{y^2}{9} = 1$

3. $\dfrac{(x+1)^2}{5} + \dfrac{(y-7)^2}{9} = 1$

5. $\dfrac{x^2}{9} + \dfrac{y^2}{49} = 1$

7. $\dfrac{(x-1)^2}{4} + (y-2)^2 = 1$

9. $\dfrac{(x-5)^2}{21} + \dfrac{(y-2)^2}{25} = 1$

11. $\dfrac{9x^2}{8} + y^2 = 1$

13. $\dfrac{(x-10)^2}{25} + \dfrac{(y-1)^2}{16} = 1$

15. $\dfrac{x^2}{4} + \dfrac{y^2}{16} = 1$ 17. $\dfrac{x^2}{9} + \dfrac{5y^2}{9} = 1$

19. $a = 3, b = 2, c = \sqrt{5}, e = \dfrac{\sqrt{5}}{3}$; foci: $(\pm\sqrt{5}, 0)$

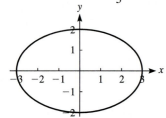

21. $a = 2\sqrt{2}, b = 2, c = 2, e = \dfrac{\sqrt{2}}{2}$; foci: $(\pm 2, 0)$

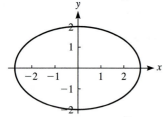

23. $a = 4, b = 2, c = 2\sqrt{3}, e = \dfrac{\sqrt{3}}{2}$; foci: $(-2, 2 \pm 2\sqrt{3})$

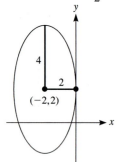

25. $a = 6, b = 3\sqrt{2}, c = 3\sqrt{2}, e = \dfrac{\sqrt{2}}{2}$;
foci: $(1 \pm 3\sqrt{2}, 2)$

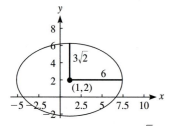

27. $a = 5, b = \sqrt{5}, c = 2\sqrt{5}, e = \dfrac{2\sqrt{5}}{5}$;
foci: $(2 \pm 2\sqrt{5}, -3)$

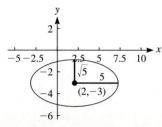

29. $a = 6, b = 2, c = 4\sqrt{2}, e = \dfrac{2\sqrt{2}}{3}$;
foci: $(0, \pm 4\sqrt{2})$

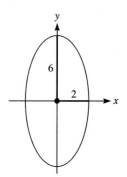

31. $a = 6, b = 2, c = 4\sqrt{2}, e = \dfrac{2\sqrt{2}}{3}$;
foci: $(3, 2 \pm 4\sqrt{2})$

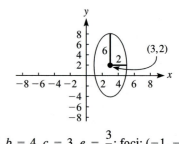

33. $a = 5, b = 4, c = 3, e = \dfrac{3}{5}$; foci: $(-1, -1 \pm 3)$

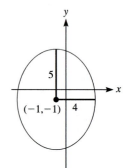

35. $a = b = 3, c = 0, e = 0$; foci coincide: $(0, -3)$

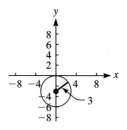

37. Endpoints of latus recta are $\left(\dfrac{4}{3}, \sqrt{5}\right), \left(\dfrac{4}{3}, -\sqrt{5}\right),$
$\left(\dfrac{-4}{3}, \sqrt{5}\right),$ and $\left(\dfrac{-4}{3}, -\sqrt{5}\right).$

43. (a)   45. (f)   47. (e)

Section 9.3, page 462

1. center: $(0, 0)$, $e = \sqrt{2}$; vertices: $(\pm 1, 0)$;
   foci: $(\pm \sqrt{2}, 0)$; asymptotes: $y = \pm x$

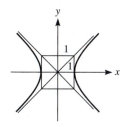

3. center: $(0, 0)$, $e = \sqrt{5}/2$; vertices: $(\pm 2, 0)$;
   foci: $(\pm \sqrt{5}, 0)$; asymptotes: $y = \pm(x/2)$

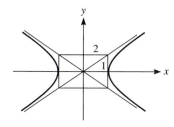

5. center: $(0, 0)$, $e = \sqrt{34}/5$; vertices: $(\pm 10, 0)$;
   foci: $(\pm 2\sqrt{34}, 0)$; asymptotes: $y = \pm 3x/5$

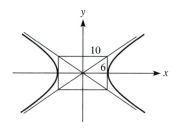

7. center: $(1, 2)$, $e = \sqrt{2}$; vertices: $(1 \pm 1, 2)$;
   foci: $(1 \pm \sqrt{2}, 2)$; asymptotes: $y - 2 = \pm(x - 1)$

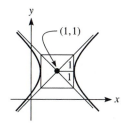

9. center: $(4, 1)$, $e = \sqrt{34}/3$; vertices: $(1, 1)$, $(7, 1)$;

foci: $(4 \pm \sqrt{34}, 1)$;
asymptotes: $y - 1 = \pm\left[\dfrac{5}{3}(x - 4)\right]$

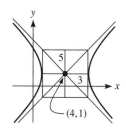

11. center: $(0, 0)$, $e = \sqrt{13}/3$;
    vertices: $(1/2, 0)$, $(-1/2, 0)$
    foci: $(\pm \sqrt{13}/6, 0)$; asymptotes: $y = \pm 2x/3$

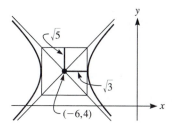

13. center: $(-6, 4)$, $e = \sqrt{24}/3$; vertices: $(-6 \pm \sqrt{3}, 4)$;
    foci: $(-6 \pm \sqrt{8}, 4)$; asymptotes: $y - 4 = \pm \sqrt{5/3}(x + 6)$

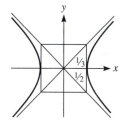

15. center: $(3, 3/2)$, $e = \sqrt{5}/2$; vertices: $(3 \pm \sqrt{30}, 3/2)$;
    foci: $(3 \pm 5\sqrt{6}/2, 3/2)$;
    asymptotes: $y - 3/2 = \pm(1/2)(x - 3)$

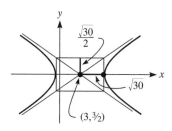

17. center: $(-2, 4)$; vertices: $(-2 \pm 1, 4)$;
    foci: $(-2 \pm 2\sqrt{3}/3, 4)$;

asymptotes: $y - 4 = \pm\dfrac{1}{\sqrt{3}}(x + 2)$

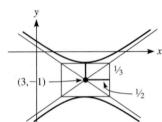

19. center: $(3, -1)$; vertices: $(3, -1 \pm 1/3)$;
foci: $(3, -1 \pm \sqrt{13}/6)$;
asymptotes: $y + 1 = \pm(2/3)(x - 3)$

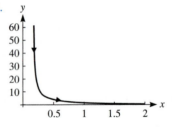

21. $\dfrac{x^2}{4} - \dfrac{y^2}{1^2} = 1$ 23. $\dfrac{(y - 1)^2}{1^2} - \dfrac{(x - 1)^2}{(3/2)^2} = 1$

25. $\dfrac{x^2}{9} - \dfrac{y^2}{16} = 1$

27. $\dfrac{(x + 1)^2}{16} - \dfrac{(y - 1)^2}{20} = 1$

29. $\dfrac{x^2}{4} - \dfrac{y^2}{16} = 1$ 31. $\dfrac{y^2}{4} - \dfrac{x^2}{20} = 1$

33. Ellipse. 35. Hyperbola. 37. Parabola.

39. $\dfrac{256\left(x - \dfrac{5}{2}\right)^2}{25} - \dfrac{256y^2}{1575} = 1$

47. (d) 49. (e) 51. (f)

Section 9.4, page 466

1.
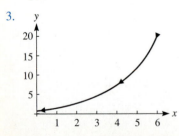

3.

5.
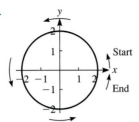

7.
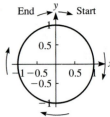

9.
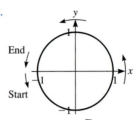

11.
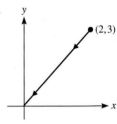

13.
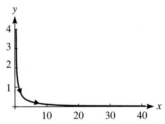

15.

17. $y = \dfrac{5}{3}x - 1, \quad 0 \le x \le 3$

19. $x = e^{x/2}, \quad x \ge 0$

21. $x^2 + y^2 = 4$, Domain: $-2 \le x \le 2$

23. $x^2 + y^2 = 1$, Domain: $-1 \le x \le 1$

25. $x^2 + y^2 = 1$, Domain: $-1 \le x \le 1$

27. $y = \dfrac{3x}{2}, \quad 0 \le x \le 2$

29. $y = \dfrac{1}{x}, \quad x > 0$ 31. $y = \dfrac{1}{x^2}, \quad x > 0$

37. $x = \cos t, y = \sin t, \quad 0 \le t \le 2\pi$

39. $x = 2\cos t, y = \sin t, \quad 0 \le t \le 2\pi$

41. $x = 1 + 10\cos t, y = 1 + \sqrt{84}\sin t,$
$\quad 0 \le t \le 2\pi$

43. $x = 1 + \sqrt{3}\sec t, y = 1 + \sqrt{6}\tan t, \dfrac{\pi}{2} < t < \dfrac{3\pi}{2}$

45. $x = 2 - 3t, y = 3 + t, \quad 0 \le t \le 1$
47. Line segment from $(7, 1)$ to $(3, -1)$.
49. Same graph traced in the opposite direction.
51. Graph is translated horizontally 1 unit to the right and vertically 2 units upward.

Section 9.5, page 468

1.
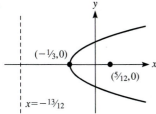

3.
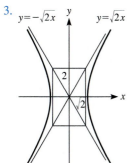

5.
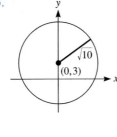

7.
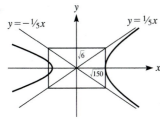

9.
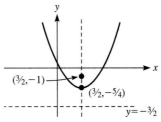

11.
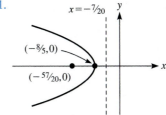

13.
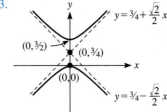

15.
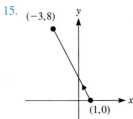

17.
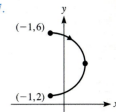

19. $y = \left(\dfrac{2}{3}\right)x + \dfrac{11}{3}, \quad -1 \le x \le 14$

21. $y = \sqrt{x - 1}, \quad x \ge 1$ 23. $\dfrac{x^2}{16} + \dfrac{y^2}{9} = 1$

INDEX

Z

OF THE ELEMENTS

NONMETALS

		IIIA	IVA	VA	VIA	VIIA	0
						1.0079 **H** 1	4.00260 **He** 2
		10.81 **B** 5	12.011 **C** 6	14.0067 **N** 7	15.9994 **O** 8	18.9984 **F** 9	20.179 **Ne** 10
IB	IIB	26.9815 **Al** 13	28.086 **Si** 14	30.9738 **P** 15	32.06 **S** 16	35.453 **Cl** 17	39.948 **Ar** 18
63.546 **Cu** 29	65.38 **Zn** 30	69.72 **Ga** 31	72.59 **Ge** 32	74.9216 **As** 33	78.96 **Se** 34	79.904 **Br** 35	83.80 **Kr** 36
107.868 **Ag** 47	112.40 **Cd** 48	114.82 **In** 49	118.69 **Sn** 50	121.75 **Sb** 51	127.60 **Te** 52	126.9046 **I** 53	131.30 **Xe** 54
196.9665 **Au** 79	200.59 **Hg** 80	204.37 **Tl** 81	207.2 **Pb** 82	208.9804 **Bi** 83	(210) **Po** 84	(210) **At** 85	(222) **Rn** 86

157.25 **Gd** 64	158.9254 **Tb** 65	162.50 **Dy** 66	164.9304 **Ho** 67	167.26 **Er** 68	168.9342 **Tm** 69	173.04 **Yb** 70	174.97 **Lu** 71
(245) **Cm** 96	(245) **Bk** 97	(248) **Cf** 98	(253) **Es** 99	(254) **Fm** 100	(256) **Md** 101	(253) **No** 102	(257) **Lr** 103

Techniques and
Experiments for
Organic Chemistry

"I love crystals; the beauty of their form—and their formation; liquids, dormant, distilling, sloshing!; the fumes; the odors,—good and bad; the rainbow of colors; the gleaming vessels, of every size, shape and purpose. Much as I might think about chemistry, it would not exist for me without these physical, visual, tangible, sensuous things. . . ."

R.B. Woodward

About the cover: Crystals of 2,4-Dinitro-4'-methyldiphenylamine, shown on the cover of this book, can be made according to the procedures described in Experiment 51.